A history of the late medieval siege, 1200–1500

Sieges were the predominant form of warfare across the medieval world and siege methods and technology developed alongside improvements in defence. This book goes back to the original sources to present a comprehensive view of the whole subject, tracing links across continents and analysing the relationship with changes in the design of town and castle defences, and linking contemporary historical accounts with archaeological studies. It considers the most important questions raised by siege warfare: who designed, built and operated siege equipment? How did medieval commanders gain their knowledge? What were the roles of theoretical texts and the developing science of siege warfare? How did nomadic peoples acquire siege skills? Were castles and town walls built purely of a military purpose, or did they play a symbolic role also?

The volume begins with the Mongol conquests in Asia and Europe and the thirteenth-century apogee of pre-gunpowder siege warfare, before examining the slow impact of guns and the cumulatively massive changes in attack and defence of the fifteenth century.

A history of the late medieval siege 1200–1500

Peter Purton

THE BOYDELL PRESS

First published 2010
The Boydell Press, Woodbridge

ISBN 978 1 84383 449 6

The Boydell Press is an imprint of Boydell & Brewer Ltd
PO Box 9, Woodbridge, Suffolk IP12 3DF, UK
and of Boydell & Brewer Inc.
668 Mt Hope Avenue, Rochester, N.Y. 14620, USA
website: www.boydellandbrewer.com

A CIP record for this book is available
from the British Library

Designed and typeset in Adobe Jenson with Duc de Berry display by
The Stingray Office, Chorlton-cum-Hardy, Manchester

Contents

Maps

Plates

Except for those marked *, all photographs are © the author. Extended captions and photo credits accompany the plates themselves.

Preface

THE INTRODUCTION of new siege artillery, the trebuchet, around the turn of the twelfth and thirteenth centuries made a large impact on the practice of medieval siege warfare for the next two centuries.[1]

By the thirteenth century, the world was much changed from early medieval times. The small, fragile and little-governed successor states to the western Roman empire had been transformed into large, powerful monarchies. From these states had come most of the crusaders who by bloody force of arms (and through division among the Muslim powers) had carved out Christian states in the Middle East and had everywhere turned back the tide of Muslim conquests that had previously created an entirely new world from North Africa to Delhi and Samarkand. Now, another new era was about to dawn. Politically, the Mongol conquests would once more alter the face of Asia from China to Mesopotamia and beyond. In Europe, old dynasties would fall, and the days when kings of Germany had been big players would end. Eastern Rome (Byzantium) would begin its slide from insignificance to extinction. The Plantagenets would extend their realm within Britain, then in the fourteenth century turn on their French neighbours to initiate the horrors of the Hundred Years' War. In Eastern Europe, a new power emerged with the unique theocracy of the Teutonic Knights. In Asia, Mongol conquest of China led to even further efforts to extend their empire to the south and east before their ejection by the Ming in the next century. Meanwhile, for the people, the fourteenth century well deserved its traditional description as a time of misery with repeated plagues and famines leading to population collapse and shrinking economies. None of these catastrophes diverted the warrior class from pursuit of its favourite occupation: war. Towns and cities, which had grown in number, size and importance during the previous century, survived and in some places (Germany, Italy, Flanders) became quasi-independent political forces in their own right. The direction of urban development was quite different in Muslim lands (where they had long been centres of government) and in China, but cities remained the target of many attacks, and their defence a key concern of rulers.

The Mongol conquests provided the geographical links between the Christian and Muslim worlds on the one hand and the rest of Asia on the other. Those links were sometimes personal (as with Marco Polo), but they also served to confirm the ubiquity of siege techniques despite political, social and cultural differences between these worlds. Just as in early medieval times,[2] one sign of this had been

[1] See my *History of the early medieval siege*, pp. 326, 348, 382–7.

[2] See ibid., 32–6.

the transmission from China to the West of the technology of the mangonel (traction-propelled stone thrower), so, in the thirteenth century, first the trebuchet would move in the reverse direction and then, with enormous if slow-developing significance, gunpowder technology would make the trip westwards, and be transformed when it arrived. For these reasons, this book takes account of the evidence of siege warfare in China, India and Indo-China, although because only those sources translated into Western languages have been accessible, I have been able to give them only relatively limited coverage.

The same cannot be said of Europe, where, in contrast to early medieval times when surviving records are few, the problem now becomes one of selection from an enormous quantity of material: not just chronicles, but government records, correspondence, biographies and much else besides offer insights, both deliberately and by chance, into the conduct of sieges and the development of fortifications. Wherever possible, I have tried to study material from opposite sides of a conflict to minimise the effects of bias. Structures, too, have survived in greater number, and although it remains necessary to distinguish between work from different periods (something not always achieved, or at least not agreed among students), it is usually easier than before to determine what was there during particular sieges. Likewise, it is easier to draw conclusions about the evolution of defensive architecture than before. Deciding on the *function* of particular structures, however, is another matter, addressed in its place when necessary.

The sieges described are those where primary sources exist, and where more modern archaeological or architectural descriptions have been used to round out the picture a reference is given in the footnote. Generally, secondary accounts have been called upon where they are the only way to reach original material. The purpose of this book is to redraw the history of the medieval siege through consideration of contemporaneous material and studies of the fortifications themselves.

This approach means that as well as deciding to use only dates in the form of the Christian calendar for reasons of simplicity, compromises have also been necessary in presenting names of places and people. Complete consistency in either would be neither possible nor desirable: should it be Byzantium, Constantinople or Istanbul, or all three at the relevant times? Is it Owen Glendower (as he was called in older accounts) or Owain Glyn Dŵr (as he is called in modern histories)?

Where indicating measures, I have used imperial and metric, converting one to the other as necessary, but weights are a different problem, given the many variations between units given the same name. A similar problem affects money. It is safest for the reader not to attempt comparisons between currencies of different places, rather to consider as equivalent only those within the same country: and even that could vary considerably.

This book ends with the beginning of a new era, known broadly as "early modern". It presents a picture of siege warfare that closely mirrors this evolution: by 1500, the attack and defence of strong places had been transformed by comparison with 1200. It is a rich and fascinating history, illustrated here through what actually happened.

Acknowledgements

I AM GRATEFUL to many people for their help in this study. Richard Eales, John Kenyon, Denys Pringle and the late Andrew Saunders were kind enough to read draft chapters and offer helpful criticism, or pointers to more recent studies. Most of their suggestions have been taken on, and I alone am responsible for remaining errors of fact, or outdated theories. It would be impossible to keep up to date with every development in studies across three continents covering a thousand years of history, but they have saved me from many mistakes. Tomás Durdik and Olaf Wagener helped with information on central European defences, and Rikke Olsen, Sarah Speight and Tomas Torbus provided material in advance of publication. Others have assisted with information on many different aspects: I would mention here just Brian Ayers, Pam Braddock, Lawrence Butler, Steve Freeman, P Humphries, and Rob Morgan. Boris Krivchenko organised the translation of an important article in Russian for me. British Library staff have coped with my demands over more than a decade, while the Institute of Archaeology (University College London) and the Warburg Institute have allowed me access to their collections of journals. Other research was done at King's College London, the School of Oriental and African Studies, the Bodleian Library and the School of Slavonic and East European Studies. Richard Eales, Pamela Marshall and Sarah Speight led study tours enabling me to visit many medieval fortifications discussed here, while friends and, especially, my extraordinarily willing mother, chauffeured me to stone fragments and half-buried mounds aplenty with patience and self-sacrifice, even if they could not always understand why I wished to go there. To all of them I express my sincere gratitude.

Abbreviations

The following abbreviations are used for frequently cited titles in both the footnotes and, where appropriate, the bibliography.

AC	*Archaeologia cambrensis*
AM	*Archéologie médiévale*
AQ	*Ausgewählte Quellen zur deutsche Geschichte des Mittelalters*
BAR	*British archaeological reports*
BM	*Bulletin monumental*
ByS	*Brenhinned y Saesson*
ByT(H)	*Brut y Tywysogyon* (Hergest version)
ByT(P)	*Brut y Tywysogyon* (Peniarth version)
CFHB	*Corpus fontium historiae byzantinae*
CG	*Château Gaillard*
CSHB	*Corpus scriptorum historiae byzantinae*
EHR	*English historical review*
MGH	*Monumenta Germaniae historica* (followed by title of section)
MGH SrG	*Monumenta Germaniae historica: scriptores rerum germanicarum*
MGH SS	*Monumenta Germaniae historica: scriptores*
NF	neue Folge
NS, ns	new series
PG	*Patrologia cursus completus: series graeca*, ed. J P Migne
PL	*Patrologia cursus completus: series latina*, ed. J P Migne
RA	*Revue archéologique*
RCHM	*Royal commission on (ancient and) historical monuments*
RH	*Revue historique*
RHC	*Recueil des historiens des Croisades*
RHC A	*Recueil des historiens des Croisades: historiens arméniens*
RHC Oc	*Recueil des historiens des Croisades: historiens occidentaux*
RHC Or	*Recueil des historiens des Croisades: historiens orientaux*
RIS	*Rerum italicarum scriptores*, ed. L Muratori (and successors)
RS	Rolls series
sa	*sub anno*
SSCISAM	*Settimane di studi del Centro italiano di studi sull'alto Medioevo*
VCH	*Victoria county history*

Maps

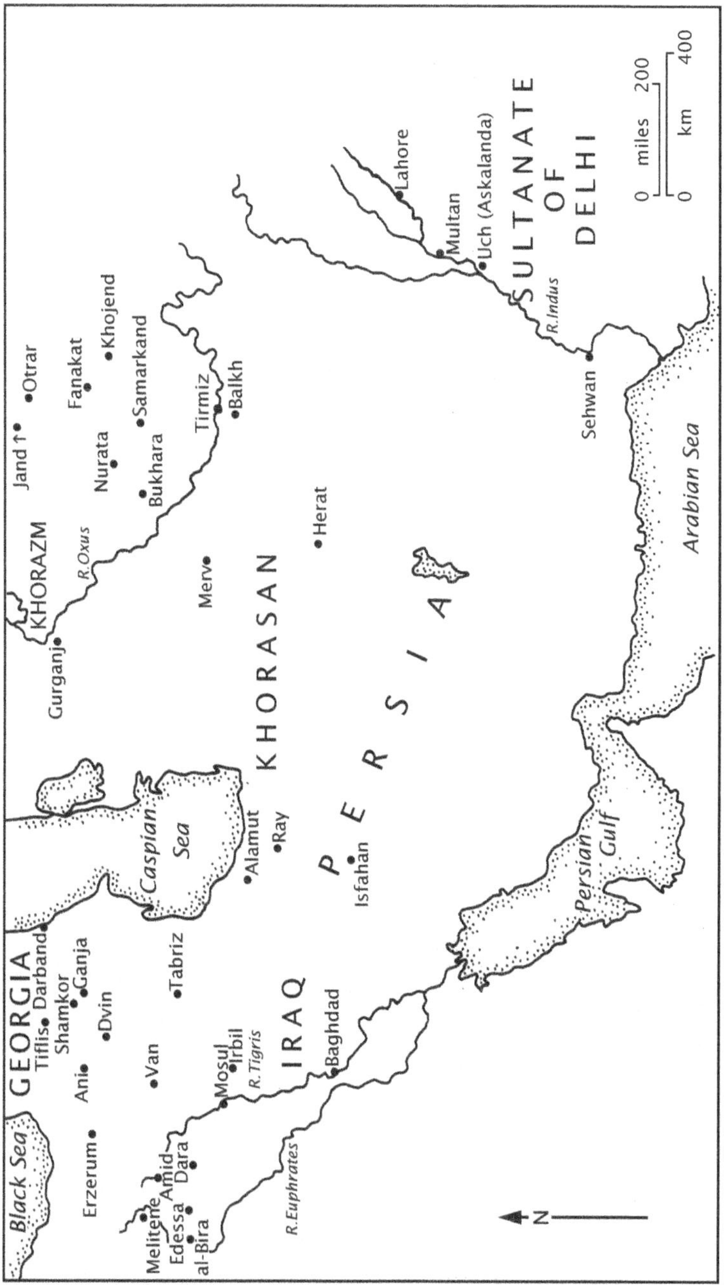

Map 1: Central Asia

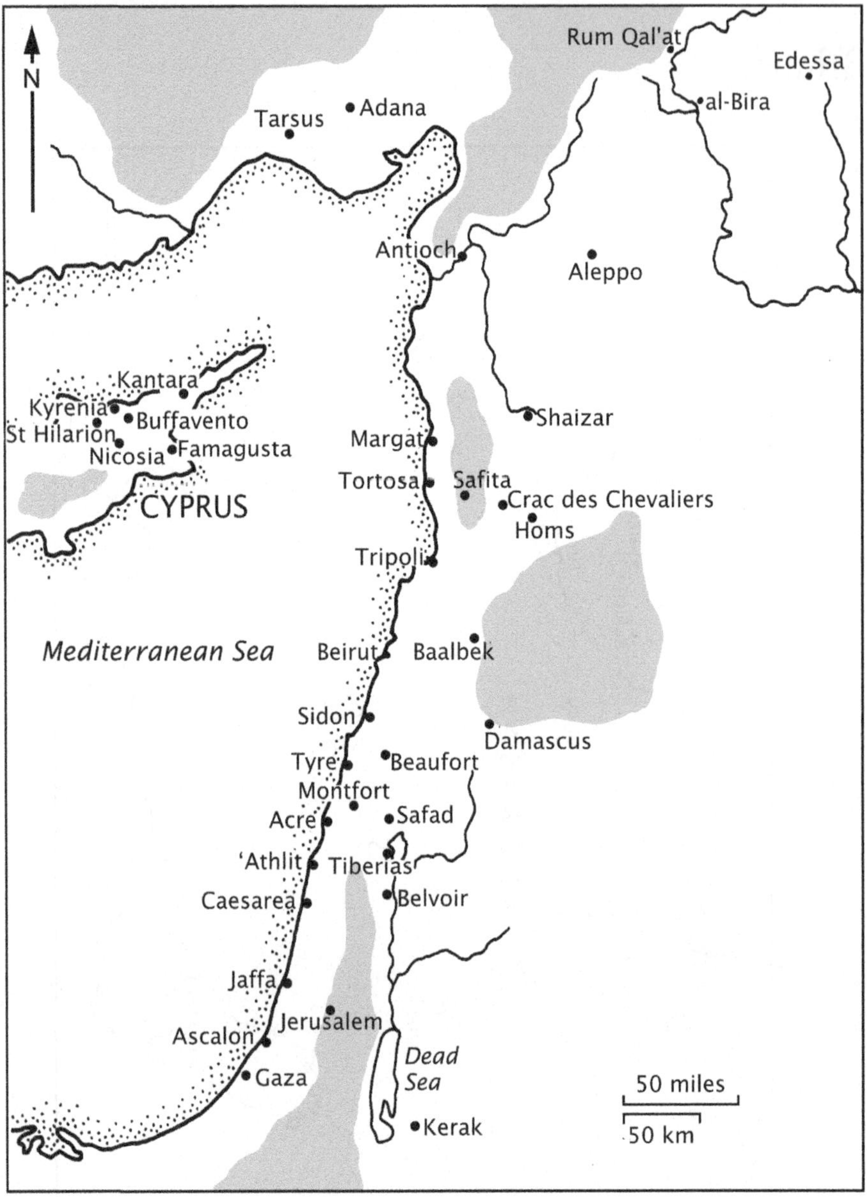

Map 2: Middle East

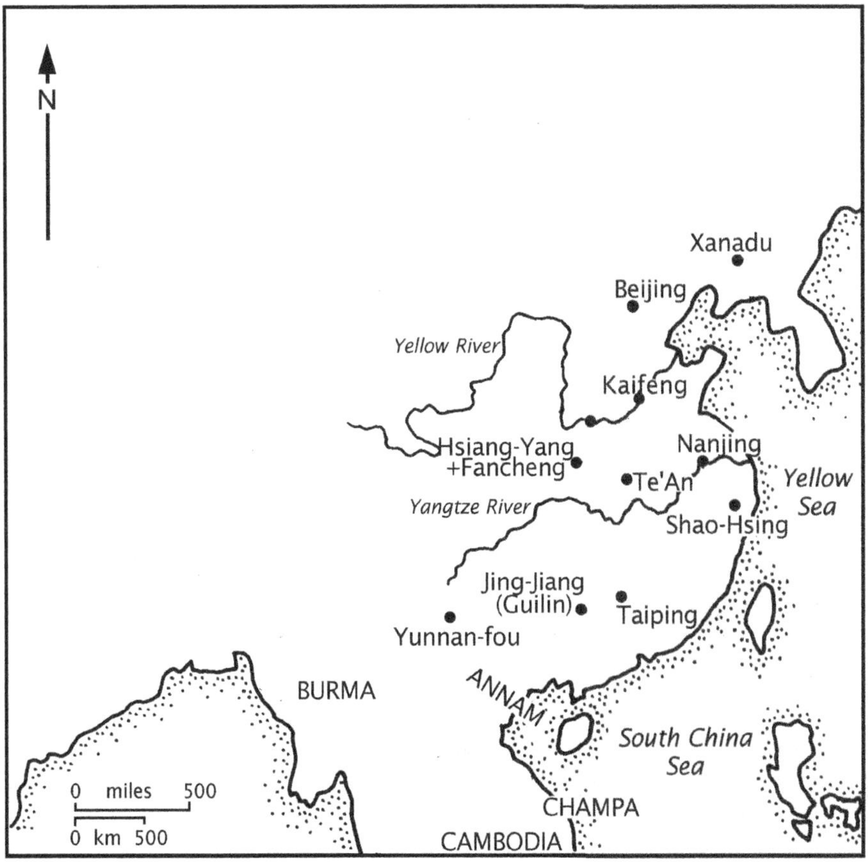
N
Xanadu
Beijing
Yellow River
Kaifeng
Hsiang-Yang
+Fancheng
Nanjing
Te'An
Yellow
Sea
Yangtze River
Shao-Hsing
Jing-Jiang
(Guilin)
Taiping
Yunnan-fou
BURMA
ANNAM
South China
Sea
CHAMPA
CAMBODIA
0 miles 500
0 km 500

Map 3: China

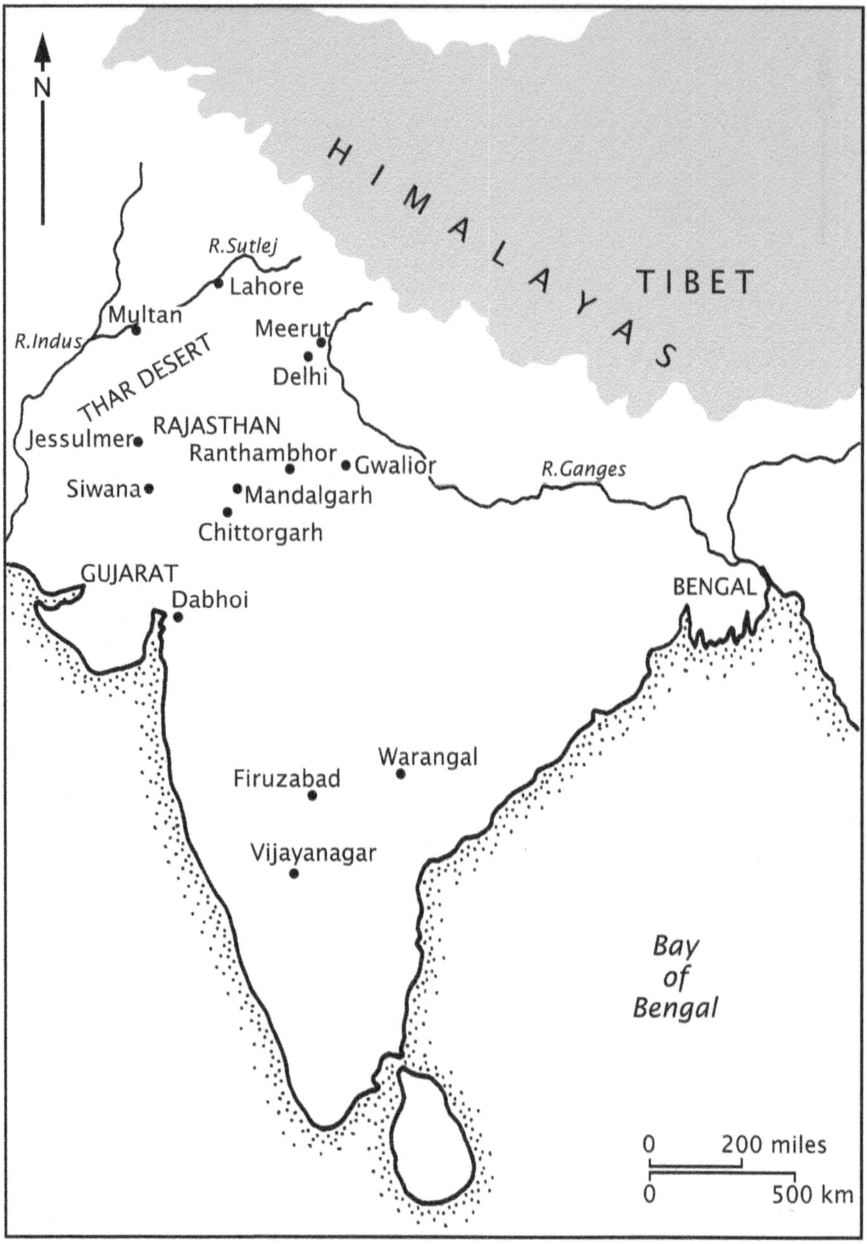
N
HIMALAYAS
TIBET
R.Sutlej
Lahore
Multan
R.Indus
Meerut
Delhi
THAR DESERT
RAJASTHAN
Jessulmer
Ranthambhor
Gwalior
R.Ganges
Siwana
Mandalgarh
Chittorgarh
GUJARAT
Dabhoi
BENGAL
Warangal
Firuzabad
Vijayanagar
Bay of Bengal
0 200 miles
0 500 km

Map 4: India

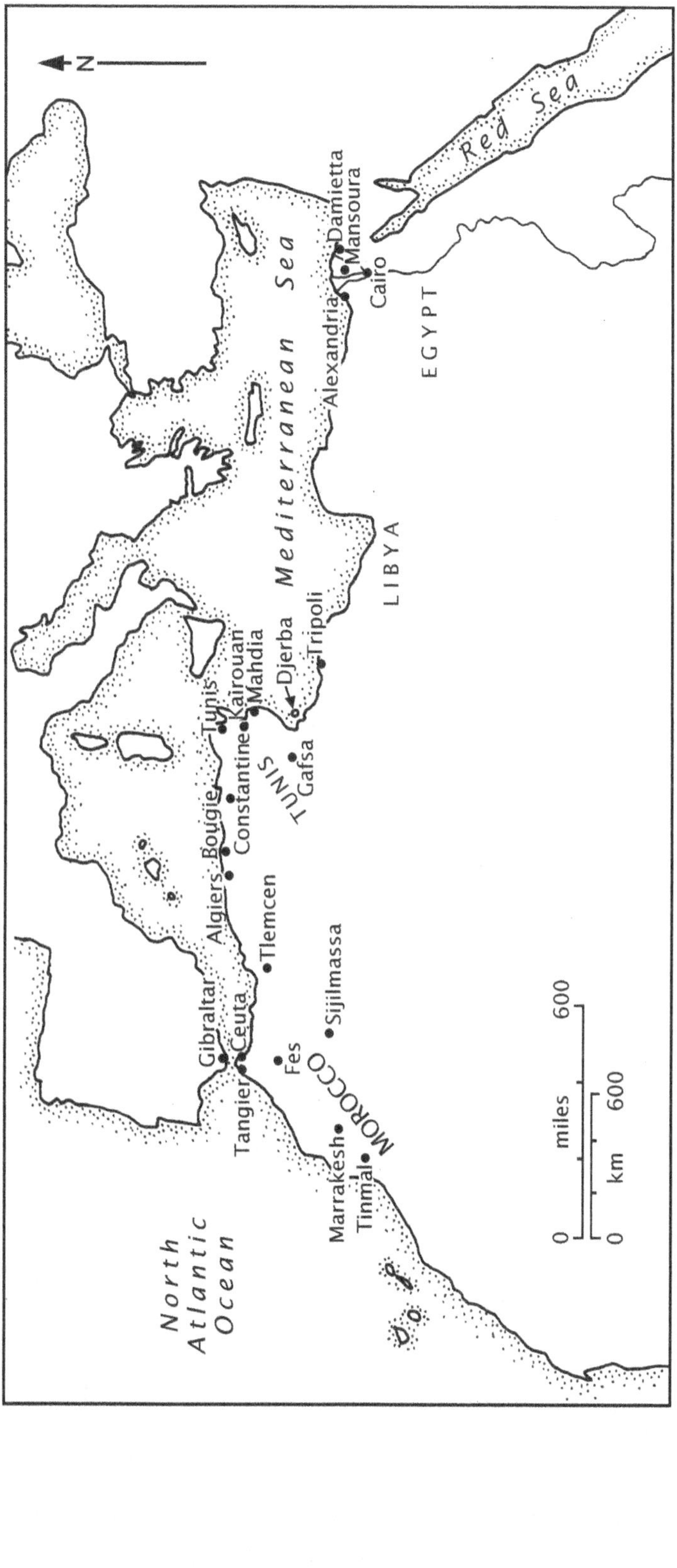

Map 5: North Africa

SAVOY
HUNGARY
N
Udine
Cividale
Aquileia
Bergamo
Brescia
Ivrea
Milan
Verona
Venice
Padua
Chioggia
Turin
Saluzzo
Genoa
Parma
Bologna
Ravenna
Zadar
Pisa
Florence
Split
Siena
Ancona
Adriatic Sea
Foligno
CORSICA
Viterbo
Rome
Manfredonia
Foggia
Barletta
Gaeta
Capua
Aversa
Melfi
Naples
Brindisi
Matera
SARDINIA
Tyrrhenian Sea
Otranto
Cagliari
Messina
Trapani
SICILY
Mediterranean Sea
Catania
Syracuse
Tunis
100 miles
100 km

Map 6: Italy

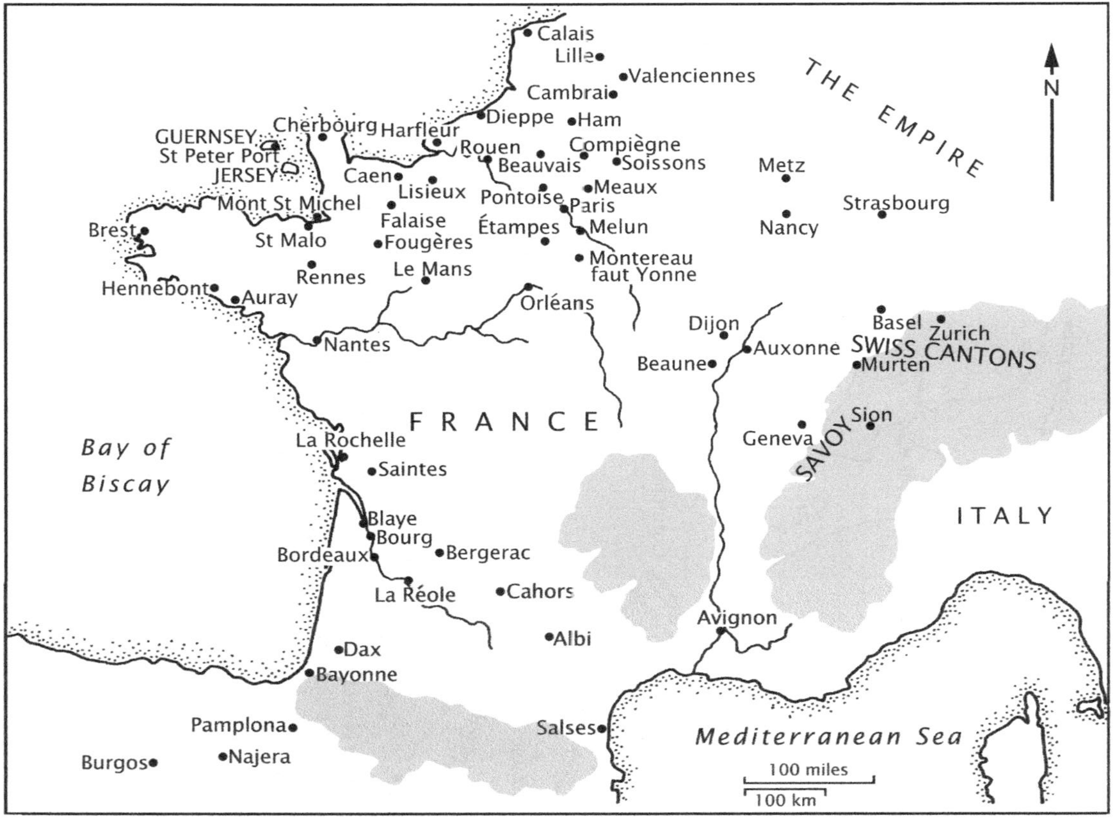
N
THE EMPIRE
Calais
Lille
Valenciennes
Cambrai
Dieppe
Ham
Cherbourg
Harfleur
GUERNSEY
St Peter Port
JERSEY
Rouen
Compiègne
Beauvais
Soissons
Metz
Caen
Lisieux
Meaux
Pontoise
Paris
Mont St Michel
Strasbourg
Brest
St Malo
Falaise
Étampes
Melun
Nancy
Fougères
Montereau faut Yonne
Rennes
Le Mans
Hennebont
Auray
Orléans
Dijon
Basel
Zurich
Nantes
Auxonne
SWISS CANTONS
Beaune
Murten
F R A N C E
Sion
Geneva
SAVOY
La Rochelle
Bay of Biscay
Saintes
ITALY
Blaye
Bourg
Bordeaux
Bergerac
La Réole
Cahors
Avignon
Albi
Dax
Bayonne
Pamplona
Salses
Mediterranean Sea
Burgos
Najera
100 miles
100 km

Map 7: France

Map 8: Low Countries

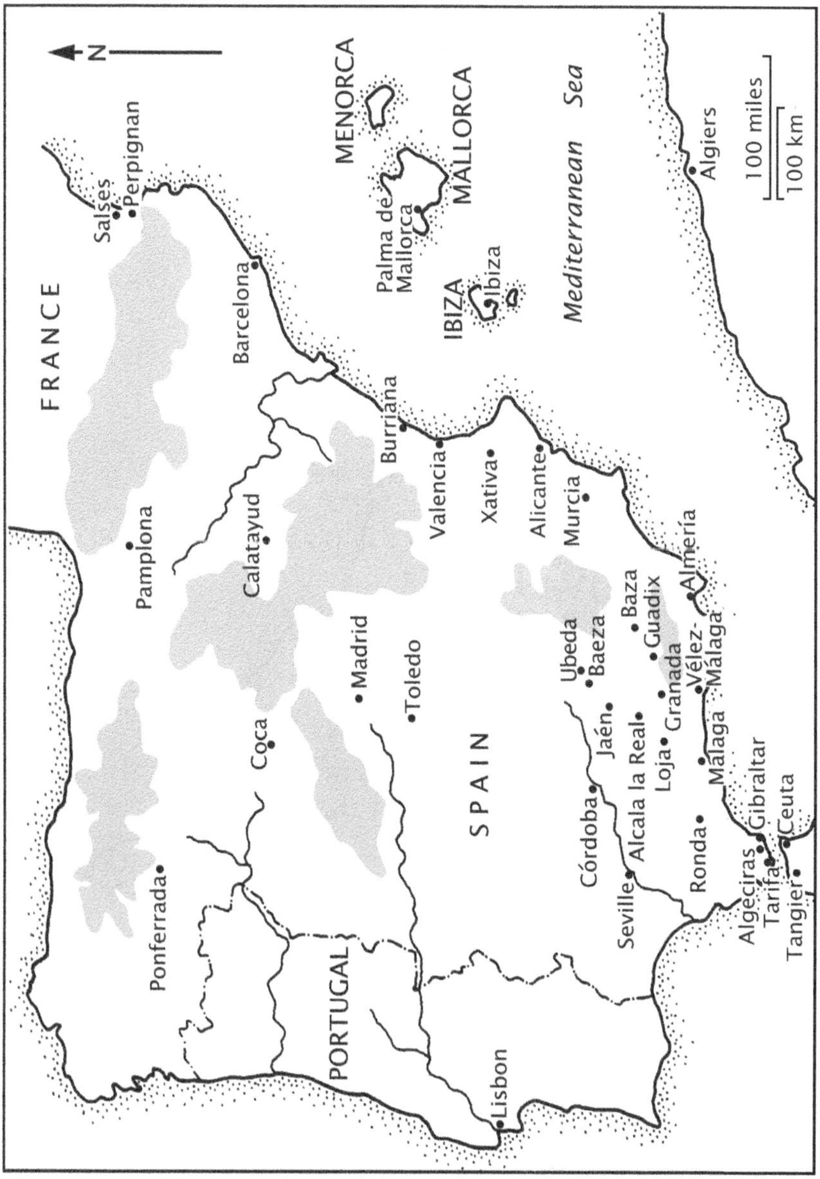
N
FRANCE
Salses
Perpignan
MENORCA
MALLORCA
Palma de Mallorca
IBIZA
Ibiza
Mediterranean Sea
Algiers
100 miles
100 km
Barcelona
Pamplona
Calatayud
Burriana
Valencia
Xativa
Alicante
Murcia
Almería
Baza
Guadix
Madrid
Toledo
Ubeda
Baeza
Vélez-Málaga
Granada
Jaén
Alcala la Real
Loja
Málaga
Gibraltar
Ceuta
Coca
SPAIN
Córdoba
Ronda
Seville
Algeciras
Tarifa
Tangier
Ponferrada
PORTUGAL
Lisbon

Map 9: Iberia

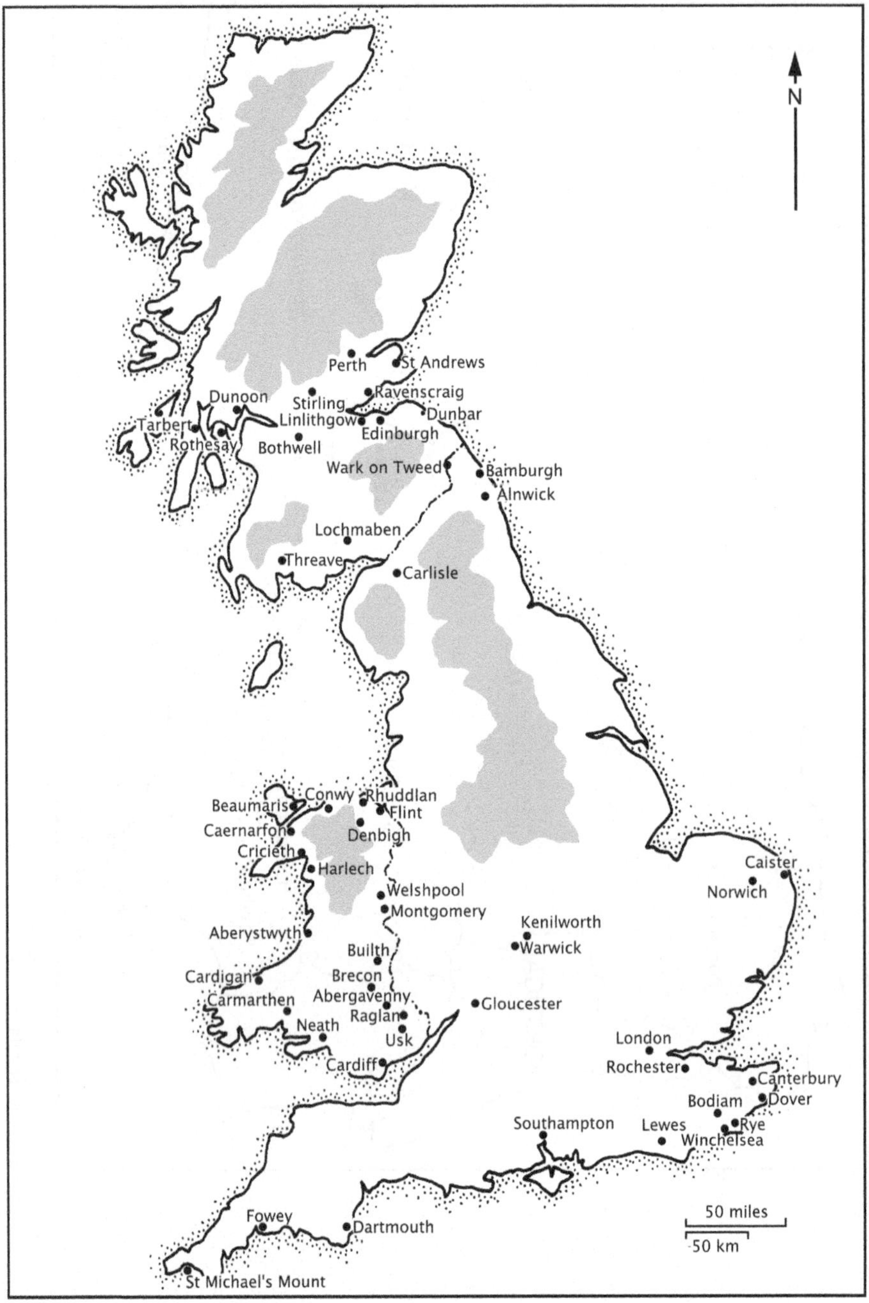

Map 10: England, Wales, Scotland

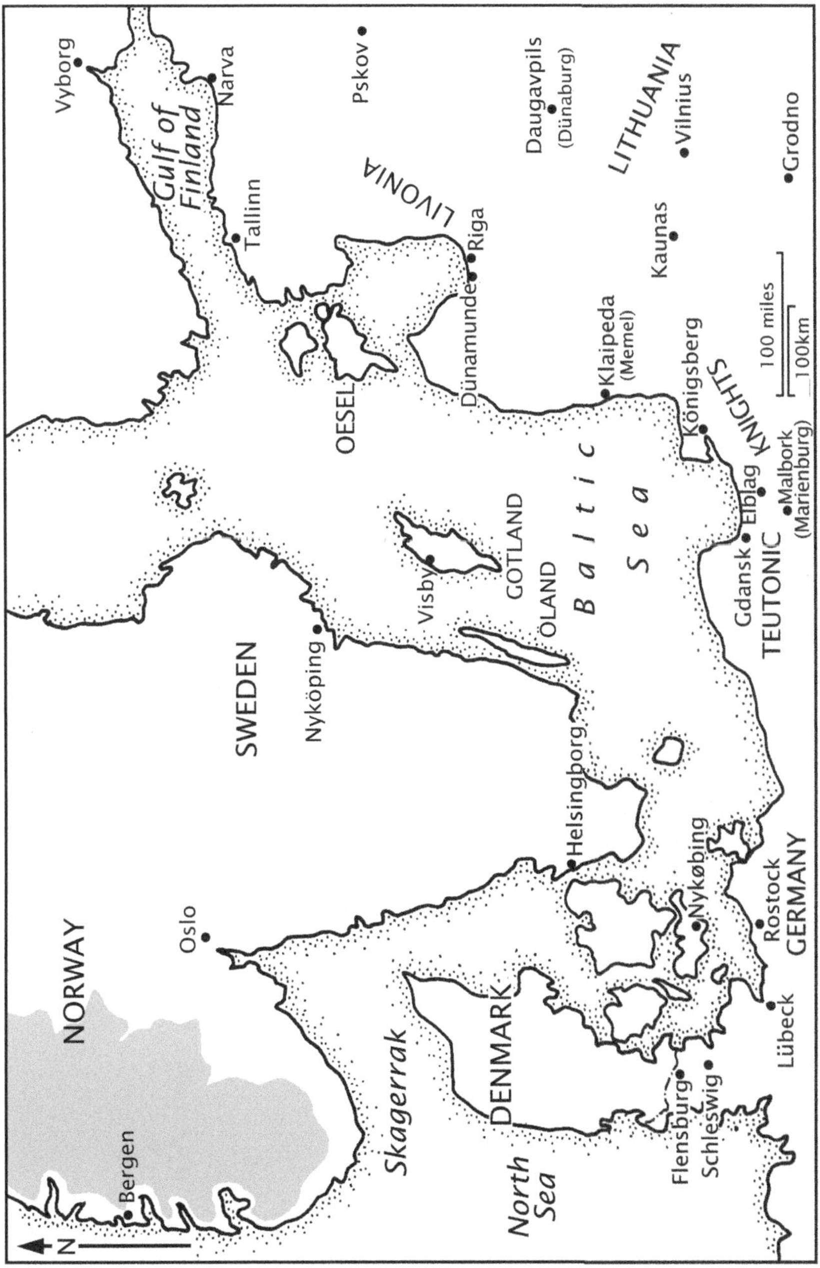

Map 11: The Baltic

Map 12: Central Europe

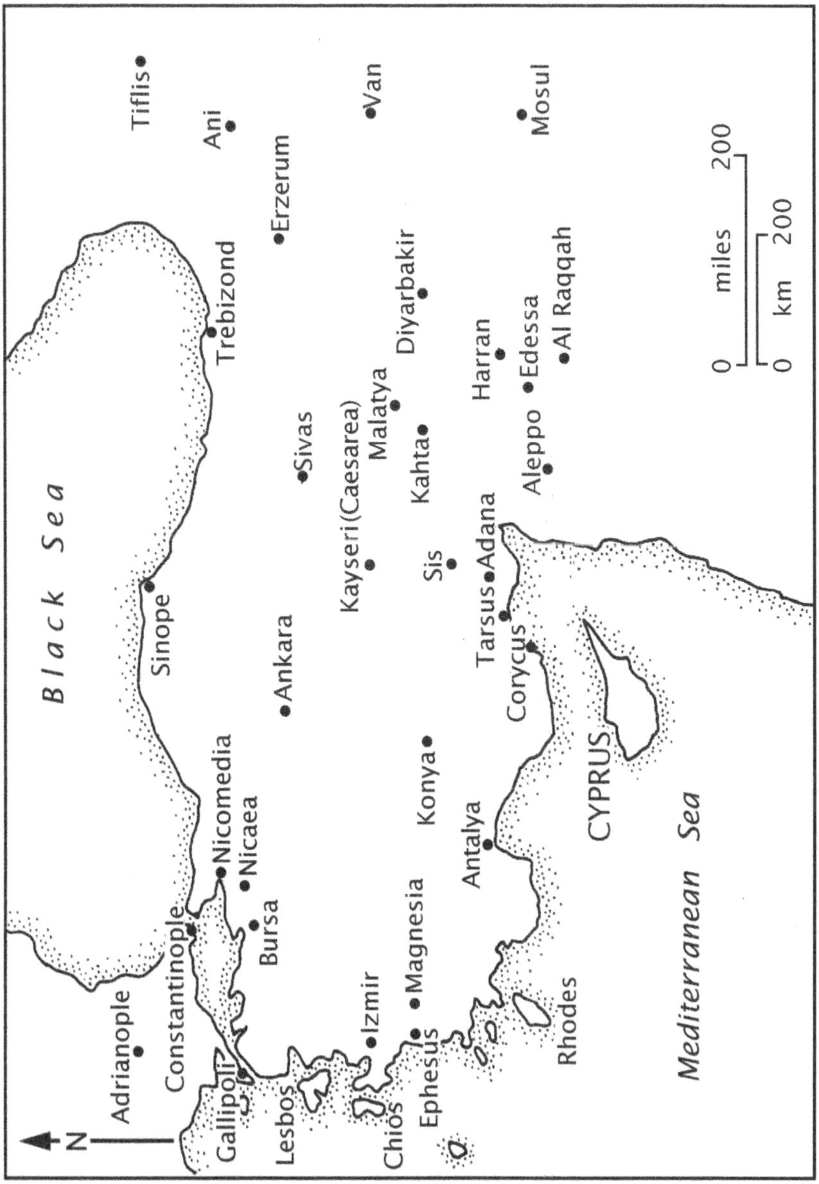

Map 13: Asia Minor

Map 14: The Balkans

Map 15: Russia

1 The age of the Mongol conquests

The Mongol empire

At the end of the twelfth century, an extraordinary leader was gradually uniting the tribes of Mongolia into a single entity, displaying complete ruthlessness and remarkable skill, including military prowess of the highest order. The man who would be known as Genghis Khan first created a Mongol federation,[1] of which he was acclaimed the Khan in 1206, then turned his attentions to his neighbours. At the outset, the Mongols had few apparent advantages over their more settled enemies. Their nomadic and pastoralist existence meant that Mongol people were physically very tough and capable of great feats of endurance, and it also meant that every adult male could be mobilised as a warrior, thus giving a Mongol chief access to a vastly greater proportion of fighting men than either an urban or peasant society, off-setting the much lower density of population that could be sustained by the steppes.[2] They were also very mobile, and capable of travelling immense distances. However, in compensation, Mongol armies should have been helpless in front of stone walls defended by crossbows and catapults, there not being much call for siege warfare in the steppes. It is testimony to the genius of Genghis Khan that not only were his armies invincible in open battle, but very quickly he learnt how to conquer walled towns.

In rapid succession, the political structures of central Asia, then the Muslim states in Transoxiana and Khurasan to their south, and in between, the Jurchen state of northern China, fell under Genghis' sway. The great Khan's sons then continued the expansion, the heartlands of Islam being overrun as far as Syria and Egypt, the Russian principalities annihilated and turned into Mongol tributaries,

[1] There are several different ways of transcribing the name of Genghis Khan and the Mongols themselves were often called Tartars or Tatars further west, although the Tartars were in fact one of the tribal groupings defeated by Genghis. I use these spellings for the sake of recognisability.

[2] For discussion of the size of the population, and the number of soldiers available to the Mongols, see J Mason Smith, 'Demographic considerations in Mongol siege warfare', *Archivium Ottomanicum*, 13, 1998. This scholar put the size of the Mongol federation in 1250 at around 4–5 million, compared to the 100 million in the Chinese empire. From this number, however, the Khans controlled armed forces amounting to up to 850,000 men. Back in 1200, at the beginning of Genghis' career, and before the subjection of all of the nomadic tribes to his rule, he could draw on armed forces of around 135,000 from a population of 7–800,000.

and central Europe itself given a taste of Mongol military power. Genghis' grandson Kubilai completed the conquest of China, establishing a new dynasty, and in the second half of the thirteenth century an empire the size of which the world has never seen before or since had been brought into being. The achievement was remarkable, but it was brought about over the corpses of an untold number of people: anyone brave enough to stand up to the Mongols' demand for their unconditional surrender became (by Genghis' definition) a rebel, and the massacres carried out by the victors were utterly merciless. Centuries-old centres of civilisation were left abandoned and burnt, their inhabitants turned into whitening heaps of bones: except for the artisans who could be of use to their barbaric conquerors, craftsmen who included the builders and operators of siege equipment. Often, where others were spared from the initial slaughter, they were forced to act as human shields for the Mongols in the assault of the next target.

The Mongols' neighbours appeared to have colluded in their own destruction, fighting among themselves, even forming alliances with the Mongols against an intervening enemy, or making what proved to be the fatal mistakes of executing the Khan's ambassadors. But the steppe nomads had always been there, and raids across the frontiers were normal. There was no reason to expect that Genghis' federation would stay together longer than any of its predecessors, or that the Khan would display such effective military genius.

Chinese siege warfare at the time of Genghis Khan

Since the collapse of the Tang dynasty in 907, no one power had emerged to re-establish a united domain in China. In the south, the Sung dynasty had established an empire. Further north, the Manchurian people known as the Jurchen had taken control, overthrowing the Liao dynasty of the Khitan people in 1125, establishing their own dynasty of the Kin. Further west, the Tangut people created the kingdom of Hsi-Hsia (from 1038). Although to the Sung, both these states were barbarian, in fact they both very quickly adopted the culture of the Chinese empire, making use of one eternal element of Chinese civilisation, its vast civil service (or bureaucracy, depending on how one views it), able to adapt without difficulty to serving new political masters while also preserving the methods and traditions of imperial government. The Hsi-Hsia continued to rely on armoured cavalry as their mainstay, since they lived in prime horse-rearing country, although they also had fortified towns. The Jurchen of Kin used the Chinese population for infantry and siege engineers. They also built substantial stone fortifications. These comprised both long walls, designed to thwart the nomads to the north, and cities defended with strong ramparts, towers and ditches. They learnt siege skills too, and captured the Sung's eastern capital of K'aifeng in 1126/7. All of these kingdoms were vast by comparison with the states of Europe, with populations supporting armies of hundreds of thousands, although the cost of maintaining such forces was as draining as it had been for the Roman empire.

The Sung regarded themselves as the heirs of the empire and most closely maintained its traditions. The military writings of previous eras were preserved and reproduced as the *Seven military classics*. Siege artillery reached its greatest development under the Sung, with a variety of manually-operated lever stone-throwers built according to standard designs and deployed in vast numbers for differentiated functions, although as in Europe and the Muslim lands, only the largest (those propelled by an arm of seven or more bamboo poles bound together) could have an effect on the fortification itself. Crossbows also continued to be central to Sung armies. Naphtha and other forms of incendiaries were introduced in the tenth century, presumably from contact with the Arabs, and in the text of 1044, the *Wu Ching Tsung Yao*, there is the first mention of gunpowder. This was used in the form of fire-lances (of very short range) and of bombs propelled from the artillery. These became considerably more powerful in the early thirteenth century as the proportion of saltpetre in the mixture was raised.[1]

A detailed account of the capture of K'aifeng demonstrated that despite being "barbarians", the Kin were no less adept at siege warfare than their Sung rivals. K'aifeng was a large city, consisting of a series of roughly rectangular stone walls surrounding in turn an inner "palace-town" housing the emperor's large household and government, an inner (old) and an outer (new) town, all three surrounded by wet ditches fed by several rivers. The armed forces on both sides were very large, the defenders being able to call upon more than 100,000 men, in separate commands in each part of the town. Few seem however to have been professional soldiers.

The city was not initially in a defensible condition. Woods and houses around the perimeter were not removed in time to allow the emperor to establish a clear field for a battery of 100 catapults, as a result of which the Kin were able to set up their own siege camp in December 1126 and quickly completed lines of encirclement. Massive assaults followed, under cover of bombardments of stones and arrows, against the gates, the defenders resisting with success, throwing up earthworks to shield the gates and protecting the towers with nets to dampen the effect of the striking stones. The population was put to work breaking up stones for ammunition, and the chancellor was placed in charge of organising wagons to transport the stone balls for the 7-poled catapults. Frequent mention is made of the awful winter weather, including the need for sorties to break the ice forming on the moats.[2] The Kin erected siege towers 30 metres high from which they could see, and shoot into, the town. An attempt at escalade was defeated. A further assault lasted two whole weeks, and at the end stone balls were piled metres deep at the foot of the wall. The Kin may have lost 10,000 men in this unsuccessful attack, including generals killed by catapult stones, but they kept up the pressure. Under

[1] On the above see Needham and Yates, *Science and Civilisation in China*, references given above, and for a clear summary, C Peers, M Perry, *Imperial Chinese Armies 590–1260 AD*, 1996, 20–42.

[2] S Werner, *Die Belagerung von K'ai Feng im Winter 1126/7*, trans. of Hsu Meng-Hsin, *Son-Ch'ao Pei Ming Hui-pien* (chapters 64–9), 1992, 72–80, 97–103, 105–15. The report was compiled in 1196.

cover of no fewer than 200 catapults, some of the seven-pole weapons hurling stones of 50 pounds (21 kg), they managed to fill the outer ditch and attack the wall with siege towers and mantlets. The defence responded with iron rams, trying to topple the siege towers, and with their own batteries of catapults, the heaviest requiring teams of 250 men and hurling 100 pound (42 kg) shots up to 77 metres. Finally, on 9 January, the Kin managed to scale the walls, and Sung resistance crumbled, the inner defences being useless in the chaos and panic.[1] References to mutinies among defending troops suggest that despite their large numbers, there had been other significant problems for the Sung in this siege.

There was another major war between the Sung and the Kin in 1206, and from this period there have also been translated direct accounts by participants in the sieges of Hsiang-yang and Te-an, two cities in the north of the Sung territory, although this time the outcome was different. Both sieges began in December 1206. The Sung defences in both places were in need of repair and undermanned, the system of watchtowers in decay and the number of professional solders very limited. In both towns, a vigorous defender was appointed (or at least they emerge as such from the accounts, written from close to home!), who set about raising armed forces including militias provided by the tea merchants, and ordered the ditches cleared (or flooded), magazines established, and catapults readied.[2]

At Hsiang-yang, the Kin first tried to divert the river but were prevented. The gates of the town were reinforced with earth banks to prevent them being burnt, and houses outside the walls were demolished. On 1 January 1207, there was an assault with wheeled ladders and siege towers supported by catapults. The defenders resisted with fire arrows and stones and defeated the attack, following up with a sortie that burnt many machines. The besiegers deployed 10 large catapults (7 or 9 poled) to hurl 40 pound (17 kg) stones at a corner tower. The defenders protected the masonry with horse hides, and during the night sortied, setting off "thunder weapons" (*p'i-li p'ao*) causing the Kin to flee and allowing the Sung soldiers to kill the catapult operators and destroy the machines.[3] A similar pattern was seen in the coming weeks, until later in February, the Kin began the construction of an earthen ramp. A sortie successfully destroyed 150 metres of it, but they rebuilt it and in ten days it was nearly as high as the town wall. The defenders concentrated on attacks designed to destroy it by firing the timbers used to provide it with an internal framework, and succeeded. The thwarted enemy tried a new assault on 14 March, and the defenders used the darkness to dig a substantial new ditch outside, 450 metres long, five wide and eight deep. The besiegers had great numbers of stone throwers operating day and night, and several thousand men were permanently employed supplying them with stones made of basalt. The commander of the town, Chao Wan, was careful not to provide the enemy

[1] Werner, 135–44.

[2] C Hana, *Bericht über die Verteidigung der Stadt Te-an während der periode K'ai-Hsi 1205–8 (K'ai Hsi Te-an Shou Ch'eng Lu,* by Wang Chih Yuan), 1970, 126–31; H Franke, ed., Hsiang-yang Shou-Ch'eng Lu, nu Chao Wan-nien, *Studien und Texte zur Kriegsgeschichte der Südlichen Sungzeit,* 1987, 158–66.

[3] Franke, 169–84.

with free ammunition, and ordered that the defending catapults use balls made of dried clay. This must also have been a gigantic logistics operation, as the number of catapults, large and small, was increased from 16 to 98, including many with 9 or 10 poles. The siege was abandoned after 90 days at the end of March.[1]

Events at Te-an followed a similar path. Here, the enemy tried initial surprise attacks on the gates, and was driven off by arrows and *p'ao* catapults. On 26 December, they tried to fill the ditch with bamboo, timber and straw, and the defending general, Wang, had it burnt by fire arrows. On 9 January, the enemy attacked a gate with rams and siege towers and were driven off by crossbows. Fire arrows and shots from Whirlwind catapults (*Feng p'ao*) were the main elements of the defence. The Kin then set up a battery of five 7 or 9 poled catapults to bombard a tower, which the defenders protected with nets and sandbags, while the catapultist Chang Pin was ordered to deploy a 7-poled heavy weapon, and 10 Whirlwinds, against the battery, which were devastatingly effective.[2] Attacks continued on several fronts and were resisted, the defenders using gunpowder weapons to create panics. Frequent sorties took place, usually of the professional solders in the garrison, well provided with crossbows, and these must have been well organised since the Kin had large numbers of cavalry who were not recorded as intervening. Another catapultist, Ch'en Yua, used a big 9-poled machine to target the enemy's best operators, hitting them and causing them to abandon their weapon. At another point, the Kin tried to mine under the town's outer wall, but were driven back. On 31 January 1207, Wang granted a contract for the supply of 20,000 hempen arrows, presumably for use as fire arrows for the crossbows. In February, messengers brought the news that a relief force was on its way, and (following the repulse of yet another assault with siege towers), the Kin tried to get in on 22 March by disguising troops to appear like the relief column: a ruse that did not deceive the wise commander. Finally, the siege was abandoned after 108 days.[3]

A number of points stand out. The quantity of siege equipment deployed by these large states was immense, and the numbers of people used to build, maintain and operate them ran into thousands on both sides. The short range of the catapults however left them extremely vulnerable to the arrows and bolts of the defenders, and also to being destroyed by sorties. We need to be suspicious about some of the detail: these were accounts written by the victors. The close adherence in both cases to the recommendations of the Chinese military manuals is also striking.[4] Generals trained in a particular method would be likely to follow it. Particularly noticeable were the skilled deployment of artillery, crossbows and incendiaries, and the regular resort to an active defence in the form of destructive sorties. Both the Sung and the Kin, then, knew their military classics, and could call on great arsenals. Both towns held out, too, in a way with which a European

[1] Franke, 185–209.

[2] Hana, 136–54. The account is a day by day, blow by blow account by Wang's son.

[3] Hana, 155–212.

[4] R Sawyer, trans., *The seven military classics of ancient China*, 1993. The six ancient tracts, including that of Sun Tzu, had been collected together in 1078 and a Tang dynasty text added.

leader might have been familiar, the attackers being prevented (through a skilled and aggressive defence) from reaching or breaching the wall. How would they manage against the Mongols who did not (yet) have such arsenals?

The first Mongol invasions

In the same year as these two unsuccessful sieges, Genghis Khan completed his consolidation of the Mongol tribes into one large political unit. There had been little cause to resort to the attack or defence of fortifications in the process. References to possible captures by Genghis' forces of hostile fortifications contain little explicit information, and the "fortresses" referred to are probably best understood as a palisade. This is certainly what one would expect from a nomadic tribe in the steppe. Nor is it clear what the ramparts were made of, timber not necessarily being available in sufficient quantity to construct a wooden defence, although time was surely not sufficient to create mortared stonework. In 1202 Genghis himself is recorded as constructing a wall, implying the use of stone, while in his battles with the Naiman people in 1204 and 1205, his armies had to take some kinds of fortresses. In the latter year, another of the tribes, the Merkid, rebelled and "fortressed a fortress" on top of a hill, which Genghis sent a contingent to destroy.[1] In 1205 again, he invaded Hsi-Hsia, and it is recorded that he destroyed two of their fortresses. But when in 1206 he faced the large walled cities of the Kin with their professional garrisons and supplies of artillery, this represented a new challenge altogether. But the Khan's early attacks were raids. In subsequent years most of his energy went into suppressing the Hsi-Hsia kingdom. Here he met with a setback in 1209, as the defenders broke the dams of the Huang-ho river to force him to abandon his attack on their capital at Chung hsing-fu. But in 1210–13 his campaigns against the Kin involved the capture of many towns. His son To-lui was first to mount the ramparts of Te-hsing fu, suggesting a direct assault. By 1215, it was recorded that he had taken no fewer than 862 towns of the Kin, although some such as the capital Chung-tu had resisted successfully.[2] No detail is given, and the success rate might signify that weaker defences were simply overrun by weight of numbers or else that the defenders capitulated at once: the sheer number of towns, which almost defies belief, suggests this must have been the case. However, as the territory and peoples of the Kin fell under Mongol rule, the astute Khan no doubt began very quickly to recruit Chinese/Kin engineers and specialists to serve his army, because there is clear evidence from other sources that he used sophisticated siege equipment including artillery very shortly after.

[1] *Secret History of the Mongols*, ed. Cleaves, 1982, 132–3; Cingis Khan, *Die Geschichte seines Lebens nach den chinesischen Reichsannalen*, ed. Krause, 1922, 19, 27; Cheng-wou ts'in tcheng lou, *Histoire des campagnes de Genghis Khan*, ed. Pelliot, Hambis, 1951, 296.

[2] *Secret History*, 183–7; Cingis Khan, *Die Geschichte*, 30–35

Transoxiana and beyond

The first recorded victims of this additional weaponry in Genghis' armoury were the Muslims of Transoxiana. Sultan Maḥmūd, the Kwarazmshah in distant Khurasan, was a sufficiently powerful prince to treat the Mongol demand for his submission with contempt. No doubt he considered the large forces he could mobilise adequate to defend his domain, and the walls of his fortresses and cities strong enough to resist the horsemen from the steppes. In 1218, Genghis arrived. The contemporary sources are few, but there is the biography of Mahmoud's son Djelal ed-Din written by one of his men, and the later history of 'Ala ad-Din 'Ata Malik Juvaini, writing from inside the Mongol court.

Having first taken and destroyed some small towns that had rebelled, and wiping out the inhabitants, Genghis took on Maḥmūd's town of Otrar with a force of 60,000 men. The town had outworks, wall and citadel and had been prepared for defence. Genghis sent contingents off to attack other towns including Khojend and Fanakat while he marched on towards Bukhara and his sons Ögedei and Chaghatai remained to take Otrar. This place held on for five months until the Mongols broke into the town and the garrison in the citadel died to the last. The defences were levelled, and artisans were carried off. Meanwhile, Suqnaq (present day Sunak-Kurgan in Kazakhstan) was annihilated to the last person when it resisted, other towns capitulated, and the sultan's commander at Jand (near modern Kzyl Orda) retreated across the river Syr Darya. The Mongols arrived there on 21 April 1219 and set up rams, catapults and ladders while filling the moat. The defenceless inhabitants watched from the ramparts. Their one attempt at resistance confirmed that there were no fighting men left in the place: when they tried to loose a mangonel, its stone fell back and smashed the iron ring of the machine itself. Capitulation followed. At Fanakat, 5000 Mongols defeated the garrison and massacred the survivors, except for the artisans. At Khojend, a garrison of 1000 was ensconced in a citadel on an island in the river, where it could be reached "neither by bowshot nor by mangonel". So the Mongols forced thousands of people to collect stones and use them to fill in the river. The resourceful governor built a fleet of boats protected with clay kneaded with vinegar from which the battle was carried to the Mongols, and with which he made his escape down river. The Mongol army tried several times to block him, using mangonels and ballistas, without success. Eventually taking to the desert, he escaped.[1] At one town, the Mongols' battery of 20 mangonels was specifically described as having been transported there by hand by the local country people.[2]

Genghis and To-lui arrived before Zarnuq, which capitulated. The khan pressed some of the young men into his service for the next battle. The town of Nur (Nurata, in Uzbekistan) was similarly prudent, although the castle of princess Zarqa of Yamama was captured by a contingent that approached carrying

[1] 'Ata Malik Juvaini, trans. Boyle, 1997, 81–94; *Secret History*, 257; Minhāj ud-Din, *Tabakat-i-Nasiri*, trans. Raverty, 1881, 969.

[2] Mohammed en-Nesawi, *Histoire du Sultan Djelal ed-Din Mankobirti, prince du Kharezam*, trans. Ottoudas, 1895, 85–8.

cut-down trees in front as a disguise, and she was killed. In March 1220, Genghis arrived before Bukhara. 20,000 of the sultan's troops marched out to do battle and were wiped out, whereupon the town opened its gates. The garrison in the citadel held out, and Genghis had the town burnt to the ground. Then he attacked with mangonels and arrows, using the Bukharans to fill the ditch. The defenders replied with ballistas and pots of naphtha but in vain and 30,000 died in the massacre. The walls were then flattened. Samarkand was the next target, held by a large army of Turkish and Iranian (Tajik) solders. In the time available, the defenders strengthened and raised the walls of the citadel and erected outworks. The defences seem to have been neglected by the Sultan, and his governors had even removed the city's gates in 1212 as the result of a revolt. The city walls stretched some 40 kms. The citadel, to the north of the town, was on a plateau and had become increasingly difficult to supply with water over the centuries. The battle raged for two months in 1220 with both sides using mangonels and naphtha. The townsfolk capitulated but the citadel was stormed and the garrison annihilated – alternatively, the urban militia was overwhelmed outside the walls, and the sultan's Turkish garrison then capitulated.[1] The other oasis city, Merv, was annihilated early in 1221, even though it surrendered. It may be that the massive circuit of walls and the powerful citadel reinforced by the Seljuk rulers in the twelfth century had been allowed to decay, something which mud brick defences did rapidly if not maintained.[2]

Sultan Maḥmūd was now in no doubt as to the immense power of the enemy he had enraged. He ordered the towns of Khurasan strengthened and provided with ditches and appointed commanders to each. One Mongol army attacked the capital town of Kharazm itself, Gurganj (Urganch, modern Kuhna-Urganch, south of the Aral sea), then a large and cultured city, which, having been deserted by Maḥmūd, elected its own king. Chaghatai and Ögedei led the siege, first collecting timber with which they constructed mangonels, siege towers and protected shelter. They had a problem with missiles. There being no suitable stone around, they manufactured them from mulberry wood, which after being soaked in water became "as hard as stone". Then they organised their non-Mongol levies to collect anything with which to fill the ditch, and also used them to breach the outworks. Ferocious street fighting followed and the entire town and most of its inhabitants suffered destruction. Meanwhile, Genghis headed south to the Oxus and attacked Tirmuz, strongly built on the river, whose garrison was "reduced to helplessness by the stones of the catapults" (Minhāj) and which was stormed in

[1] The different versions are explored, alongside a report of an archaeological investigation of the medieval city, by F Grenet, M Isamiddinov, 'La prise de Samarcande par Gengis-Khan', *L'Histoire*, 149, 11/1991, 8–12, with a plan of the city in 1220 at 11.

[2] Juvaini, 97–109, 115–22, 153–68; Cingis Khan, *Die Geschichte*, 35–6; E Haenisch, trans., 'Die letzten Feldzüge Cinggis Han und sein Tod', *Asia Major*, IX, 1933, 527–8 (a composite Chinese/Mongolian account); Minhaj ud-Din, 975–80; Muhammad ibn Ja'far Narshakhi, *History of Bukhara*, trans. Frye, 1954, 24–5, 34–5. The walls of Bukhara had fallen into disrepair during the twelfth century but had been restored from 1207 when Sultan Muhammad had taken the town. Herrman, *Monuments of Merv*, 1999, 14–18; the same, report in *Iran*, XXXV, 1997, 19–22.

11 days, everyone being slaughtered. The city of Balkh (now in Afghanistan) surrendered, but because there was still an enemy in the field, Genghis had every last citizen massacred and the town flattened. Ṭālaqān also fell after being besieged with numerous mangonels, with which a bastion was battered down, and a breach created after 15 days. It suffered the same fate as the others, as did numerous other towns. Sistan (Herat) fell to To-lui after 11 months, with the usual results, and at the sieges of other towns in this province it was noted that Genghis had deployed a corps of 10,000 *mandjaniki* to operate his siege train (according to Minhaj).

The Mongols were victorious whether fighting in plain or in mountain, towns and fortresses however strong falling to their relentless assault. By the end of 1222, this great kingdom had been reduced to a human desert.[1] The Mongols had not been totally successful, the occasional fortress holding out long enough to cause them to abandon their siege, but such events were few and far between. In 1227, Genghis was dead, but if anyone thought the worst was over, his successor Ögedei had other ideas.

Mongol siege skills

Genghis' siege train at the outset must have been composed of experts from his now large reserve of Chinese subjects. There is no hint in the accounts of the initial onslaught of anything other than professional operation. This is confirmed by information in the voluminous records of the Yuan dynasty established by Kubilai Khan. A man called Ambughai was made imperial agent of catapult operators (*Darughachi*) in 1214, and charged with training 500 selected men, presumably (although this cannot be proven) Chinese. Xue Taraqui was appointed Marshal of Catapult Operators, and on his death in 1233, he was followed by Zhang Batur. It seems that skills were passed on from father to son. But once he had begun his latest conquests, he used large numbers of local levies whose participation was hardly voluntary. Some were explicitly used for labouring, carrying out the very dangerous task of filling in the ditches and moats. Others were described as leading the attack on the ramparts themselves, Genghis' "cannon-fodder". Although it is not stated, it would also be reasonable to assume that among the tens of thousands of craftsmen and artisans uprooted from their homes were engineers and carpenters. The figure of 10,000 given for the catapult corps might appear excessive, but if it included all those whose task it was to convey the equipment and to prepare and transport the ammunition, it becomes a little more probable.[2]

[1] Juvaini, 123–33; Minhāj ud-Din, 1000–1054; Mohammed en-Nesawi, 153–6 (Gurganj, where it is named Kwarezm after its province); *Secret History*, 200 (Herat).

[2] Chi-ch'ing Hsiao, *The military establishment of the Yuan dynasty*, 1978, 134, citing the Yuan Shih compiled after the rule of the Yuan dynasty (1260–1368). For the evidence for the establishment of this under Genghis, see the information from the Yuan Shih cited in T Allsen, The circulation of military technology in the Mongolian Empire, in N di Cosmo, ed., *Warfare in Inner Asian History*, 2002, 265 ff., 277–81; and using Russian studies conveying the same information (albeit with different versions of the individual names), W Swietoslawski, *Arms and Armour of the Nomads of the Great Steppe in the times of the Mongol Expansion (12th–14th centuries)*, 1999, 67–69.

But it is not just a question of finding foreign experts to fill a gap in his military repertoire: the Mongols displayed the same overall command of siege warfare as they did of other forms. While it may not be true that Mongol commanders necessarily outnumbered their opponents (at least in China), they nonetheless knew well how to deploy their skilled, disciplined and mobile armies to maximum effect. Their repertoire included considerable technical and engineering skills, including understanding and using rivers: how to cross them, regardless of width, how to block them, and even how to divert their course away from a besieged town, as will be shown.

Despite the guaranteed horror of the consequences of a failed resistance, and the very few places where resistance was successful (even for a short time), large numbers of towns and forts did try to hold out. The tactic of guaranteed terror did not always secure the rapid surrender it was intended to achieve. In numbers of cases, too, immediate surrender might secure the lives of the inhabitants, but would not prevent the ruthless sack and plunder of the place, and the dispersal or enslavement of the people. While numbers of places were left genuinely untouched, this was not a random choice: Genghis also had to reward his army with the riches that came from plunder. In such circumstances, next to the loyalty of garrisons to their lords, seen several times in the kingdom of the Kwarazmshah, there might also have been little incentive not to fight.

Conquests continue

The kingdom of the Kin was still not entirely destroyed, but the Sung decided to complete the task, removing the last buffer between themselves and the new empire. In 1233 the target was the last capital of Kin, Ts'ai-Chou (Hunan province), and the story was told later by Wang E, a scribe of Kubilai Khan. The city was attacked by the Sung from the south, and by a Mongol army from the north. The Kin emperor enlisted 10,000 of the best of the provincial soldiers he had summoned, and appointed officers to various key tasks including the repair of arms and armour, and the force responsible for checking for spies. Another was responsible for a market set up where physicians would dispense medicine and health foods to soldiers and civilians. The late harvest of rice was collected or destroyed and all grain in the city requisitioned as the Sung army neared in October. All brewing was also prohibited.[1] The position of the emperor was clearly desperate. Repeated pleas for aid from the countryside went unanswered, and as the vast hosts of besiegers appeared, there were desertions from the garrison. On 23 November there was archery and crossbow practice inside the inner city, at respectively 120 and 200 paces range, illustrating the significant capabilities and usefulness of the Chinese crossbow. In December, the Mongols defeated the Kin forces assembling in the open country, and in January 1234 the besiegers broke the banks of the rivers on which the city stood, weakening the defences on the

[1] Hok-Lam Chan, *The fall of the Jurchen Chin. Wang E's memoir on Ts'ai Chou under the Mongol siege (1233/34)*, 1993, 77, 89–96.

western and southern sides. An assault was then launched with powerful artillery support, the corner tower on the south being almost knocked down. The defence was active enough to prepare palisades and ditches to protect the outer walls, but the enemy used the self same palisades as shelter so that during the next assault more than 100 paces of the rampart fell into Mongol hands. The emperor abdicated on 9 February and the city fell, along with the Kin empire.[1] Perhaps, like the Kwarazmshah, the Sung believed themselves powerful enough to stand against the Mongol empire.

Khan Ögedei's armies meanwhile were also spreading the Mongol world empire to the south and west, the forces available now being large enough to tackle several rival states simultaneously and with confidence. The next expansion began.

Russia, Eastern Europe and the Golden Horde

Russia

The Russian princely family, while still acknowledging some degree of deference to the Grand Prince in Kiev, had spread through the generations and each new son demanded a separate apanage from which to support them in their own princely dignity. There were constant family feuds and warfare. The rise of the Mongol empire in the steppes had no impact on these princelings' attitude to each other. There had always been raids by nomads from the east. The impact of the arrival of a Mongol army from the east led by Batu was however a shock in a different league. Yet there had been a warning. In 1236, Batu had crushed the kingdom of the Volga Bulgars and destroyed its "great city" (Kazan). It was the winter of 1237 when the Mongol host emerged from the forests and invaded Riazan. The prince went down to defeat. A siege of Riazan followed immediately, starting on 6 December. The Mongols surrounded the town with an encircling fortification and pressed their attack. Despite a brave resistance, the town fell to assault on 27 December, it was burnt and the people massacred. Koloma and Moscow suffered similar fates, and on 3 February the army arrived before Vladimir. In the few weeks they had had, the townspeople had strengthened their fortifications. The Mongols sent a force to burn Suzdal, then began serious preparations at Vladimir. For the first time in Russia, they were described as building scaling ladders and "wall battering" weapons (*porok*) which may have been either battering rams or artillery, but was probably the latter. Having secured themselves by throwing up a stockade around the whole town (which must have been a considerable undertaking), an attack from all sides began on 8 February. Stones fell on the walls and inside the town "like rain". The wall and gates were broken down and the city fell.[2]

The Mongols then divided into separate columns and ravaged the whole

[1] Chan, 99–113.

[2] *Nikonian Chronicle*, II, 307–12; Hypatian Chronicle, section translated in M Laran, J Saussay, *La Russie ancienne*,1975, 90.

territory, taking every other town on the Volga and in the land of Rostov and Suzdal. On 3 March, an army assembled by the princes was defeated at the river Sit and the Grand Prince himself was captured. Torchok faced a siege of two weeks and a non-stop bombardment, under the effect of which it fell, and everyone was killed (5 March). Although this town belonged to Novgorod, the Mongols did not attempt that city. This was apparently because of the onset of flooding caused by spring rains. Batu instead went on to assault Kozel'sk, where he stormed the earthen walls having broken down the wall. The townspeople resisted heroically with what they had, and in a sortie destroyed some of the Mongol catapults and killed (allegedly) 4,000 Mongols. It did them no good and they were all killed after seven weeks. There was then a pause in the Mongol onslaught, during which time Prince Iaroslav rebuilt Suzdal and Prince Alexander (Nevsky) of Novgorod threw up a new fortress on the river Shalon.[1] Clearly, these were constructions of earth and timber.

In 1240, Batu marched on Kiev, where there were extensive defences surrounding the inner and outer cities and the magnificent cathedral. The city resisted his demand for surrender, but it seems they abandoned the outer, newer town, and withdrew into the higher, inner city. The Mongols carried out their normal siege technique of surrounding the whole place, something which on its own is evidence of the very large numbers they must have deployed. The batteries of catapults were then erected, concentrating on the Polish Gate, because of the existence of natural cover there. In a day and night bombardment, the artillery apparently breached the walls and after a ferocious battle in the breach, numbers told and the Mongols broke in. Defenders threw up new defences around the church, but this collapsed under the weight of people, and the whole town fell to the usual sack and massacre, either on 19 November or 6 December. Batu then came to Kolakozhno in Volhynia. Here, twelve catapults were deployed but the walls were too strong. The defences were then tricked into surrender and all massacred. The town of Kremenets then proved, unusually, to be too strong, so the Mongols pressed onto Vladimir (in Volhynia) itself, which met the same fate as most of the others, followed by Galich (Galicia) and Cherven, both of which fell to attack by catapult. The Russian compiler had grown weary of the endless tale of catastrophe by now, noting "but why should we count them all? One after another, he (Batu) occupied a numberless multitude of Russian cities".[2]

Despite a lack of detailed contemporary evidence, a clear picture emerges. In a short campaign, every Russian principality south of Novgorod was overthrown. Numerous towns were captured, many by siege. The efforts of the princes to stem the invasion were futile and the brave defence of many of the towns equally so.

[1] *Nikonian Chronicle*, II, 315–18; *Chronicle of Novgorod*, 82–4; Perfecky, *The Galician-Volhynian Chronicle*, 46–7; Hypatian Chronicle, in Laran, Saussay, *La Russie ancienne*, 92–3.

[2] *Nikonian Chronicle*, II, 319–21; Perfecky, 48–9; Thomae, Historia Pontificarum Salonitarum et Spalatinorum, *MGH SS*, XXIX, 1892, 585 (for Suzdal); B Grekov, A Iaboubovski, trans. Thuret, *La Horde d'Or*, 1939, 202–6 (citing the Laurentian and Pskov chronicles for Kiev).

The Mongols already had the siege skills acquired in the east, and would have had access to plentiful supplies of timber to build machines on site, their speed of movement precluding the use of a mobile siege train. This means they must have had a force of catapult builders as well as operators, unless the machines were dismantled and carried in carts. The success of their attacks, and the rapidity with which they were able to prepare assaults after arrival, indicates a high degree of skill. But it also suggests the relative weakness of the defences they were facing. Most of the towns were small, and their defences rarely comprised more than a bank (sometimes only a low one at that) and a timber palisade. Kiev was different, but insufficient detail survives of its capture to analyse the event properly. Various stratagems seem to have been employed, and some use of incendiaries to burn wooden towers. The shortage of stone would have been a problem for the artillery, leading to a reliance on clay shot. The Mongols were nothing if not versatile in military matters.[1]

The destruction of Hungary

In 1241, the kingdom of Hungary stretched from west of the Danube as far as Galicia and Russia in the east, and including Transylvania to the south-east and Croatia to the south-west. Its founding dynasty, the Arpadian, still ruled and Béla IV had been king for six years. There had been internal peace for some time. There were perhaps only 10 stone castles in the whole country by 1241, and most of these were in the west. The country was divided as in the eleventh century into counties, each administered from a royal centre with some kind of fortification: a visiting Arab had said there were 78 of these, so the proportion that had been modernised was very low.[2] The frontier passes across forest and mountain to the east still relied on barriers erected with trees to block the routes. These *indagines* had been sufficient to thwart attacks in previous centuries. This time, the Mongols (or Tartars as the Europeans called them) brushed past them as it they were not there. One column of Batu's horde swept north-west through Poland, devastating the province of Sandomir and utterly crushing an army of Poles, Teutonic Knights and other Christian rulers, at the battle of Legnica in April.[3] 40,000 men led by Batu himself descended upon the Hungarian plain, where it was joined by the victors from Legnica. Once again, there had been warning signs: the conquest

[1] B Spuler, *Die Goldene Horde in Russland 1223–1502*, 1943, 382–4. Spuler suggests that sieges presented difficulties to the Golden Horde. The evidence of the conquest itself does not perhaps bear out this generalisation.

[2] Fugedi, *Castles and Society in medieval Hungary*, 43–7; I Hrbek, 'Ein Araber Bericht über Ungarn (Abu Hamid al-andalusi al Garnati)', *Acta Orientalia Academiae Scientarum Hungaricae*, v, 1955, 208 (late twelfth century). See also L Gero, *Castles in Hungary*, trans. Gulyas, Gaster, 1969, 5–16. The total of stone castles does not include the numerous stone constructions, including rectangular courtyard with Bergfried castles, built by the Templars and Hospitallers in Croatia, part of the Hungarian monarchy since 1102: Z Tomčič, 'Fortifications of the orders of Knights in medieval Croatia', *Castrum Bene*, 5, 1996, 201–17.

[3] Annales Sandivogii, *MGH SS*, XXIX, 428; Annales Capitoli Posnaniensis, *MGH SS*, XXIX, 440–1.

of the Ruthenians (western Russians) had been noted by the chroniclers, and the Cuman people of the east had also been driven westwards by the Mongol advance, although the Hungarians had fallen out with them violently. The undefended towns of the east of the country were abandoned in a hurry, the town of Vác where the church was "fortified as if it were a castle *(quasi castrum)*" (Roger) was captured and destroyed, and the frontier forces defeated. Bela's forces were routed by the Mongols, superior in number, skills and leadership, in April 1241, in a battle on the river Sajo in which they erected seven mangonels to bombard the Hungarian army in its field fortifications, a rare example of stone-throwing catapults in use in a battle, and another illustration of Mongol versatility. King Bela escaped and fled, first to Austria, where the duke, in the spirit of Christian comradeship, extracted three counties from him as his price, and then to Dalmatia. The Mongols meanwhile spread out across the kingdom "like locusts" (Thomas of Spalato) and destroyed all in their path.

Vardin resisted, the bishop ordering the erection of additional wooden defences and the repair of the castle and town walls, which had fallen away on one side. The Mongols deployed seven stone-throwers against the damaged area and brought it crashing down, followed by the inevitable storm and massacre. The Cistercian monastery of Egres was defended in vain, once again leading the Mongols to deploy machines: the defenders surrendered but were massacred anyway. Esztergom was attacked in January 1242. This was provided with substantial walls, towers and ditches, although the chronicle refers to the towers as being wooden. Above the town stood a powerful royal palace, originally dating back to King Stephen, with strong defences and a twelfth century donjon, (although it has all been built over several times since). The Mongols erected a barricade as high as the town's walls themselves, and from behind this shelter a battery of 30 mangonels was deployed, while the ditches were filled with bundles (*fasciculas*). First the machines bombarded the walls, destroying the wooden defences, possibly signifying that these were wooden hoards erected to cover the foot of the walls. Meanwhile, other machines joined in the effort to fill the ditch by propelling sacks filled with earth. In despair, the citizens abandoned the town and retired to the citadel, burning the houses behind them. From here, a large garrison containing many crossbowmen held off the attack, and the Mongols decided to withdraw.

Elsewhere, measures were being taken to defend the remaining corners of the kingdom from the onslaught. Thomas of Spalato noted that his city prepared engines and placed them in appropriate sites, but the Mongol detachment that approached was interested in finding King Bela, so did not stay in the area: however, rumours were rife in the town, including that the invaders had "numerous and enormous machines". Behind the river Danube, King Bela reported in his appeals for help to the pope, he had ordered strong garrisons into a long list of fortresses, and the order of the Hospitallers was helping him.[1] On this occasion, the Mon-

[1] Thomae, Historia, 585–95; Rogerii, Miserabile Carmen super destructione regni Hungariae per Tartaros facta, *MGH SS*, XXIX, 554–62; also in *Scriptores Rerum Hungaricum*, ed. Szentpetery, II, 545–85; and see Knauz, *Monumenta Ecclesia Strigensis*, I, 342–3; letters of King Bela in Gockenjan, Sweeney, *Der Mongolsturm*, 1985, 293–4, 306–10. On

gols did not stay: on the death of Khan Ögedei, Batu returned to Mongolia. In their short stay, the devastation wreaked upon Hungary had been appalling: the minimum modern estimate is that between 20% and 50% of the entire population was wiped out, either directly or in the famine that followed immediately. Among the other consequences was the decision of King Bela to encourage a great programme of castle building, including by the nobility, so that at least 63 new stone castles were built between 1242 and 1270 (of which 30 were private),[1] on sites chosen for being on high ground with narrow approaches. The devolution of authority to the landed nobility in Hungary was thus accelerated by the Mongol incursion.

Batu did not abandon any of his conquests east of Hungary, and from 1242 onwards the Mongols began to establish effective control over the Russian lands. Thus began the rule of the "Golden Horde" that would control Russia until the late fifteenth century.

It is often suggested that it was just the chance of the death of Ögedei and the return home of the Mongols to resolve the succession that prevented their conquest of the rest of Europe. While such speculation is always fascinating, it is also worth noting a couple of important considerations. While the way in which western European chroniclers recorded the destruction of Hungary implied that the catastrophe was seen as an event affecting somebody else, on the other hand central and western Europe was now (relatively) densely populated, able to mobilise, if needed, very large forces, and was also provided with an immense number of strongly walled towns and castles. Against what was only a contingent of the Mongol armies, they would surely have represented a significant barrier. Even more pertinent was the simple question of terrain. Batu went to Hungary because the Carpathian basin was the westernmost edge of the central Asian steppe, and was the only place in Europe where the immense numbers of horses kept by the Mongols could be pastured. They might have ravaged further west, but they could not have remained there.

Friars' report

Pope Innocent IV sent a Dominican delegation to investigate the new threat, and their reports, penned following a two year journey in 1247, gave further evidence of Mongol skills, and confirmed their methods. Noting that Kiev was "reduced to nothing", Carpini stated that the Mongols, alongside their other weapons, carried "ropes for dragging machines". This hints at a method of operation otherwise not indicated by the sources or a judgement on the practicality of transport by a high-speed army. But his analysis of how the Mongols conducted sieges was accurate in all respects: rapid encirclement with a barricade of some sort, constant bombard-

Esztergom, E Nagy, 'Rapport préliminaire des fouilles d'Esztergom', *Acta Academiae Scientarum Hungaricae*, 28, 1971, especially 194–6.

[1] P Engel, *The realm of St Stephen, A history of medieval Hungary*, 2001, 101–4; Fugedi, *Castle and Society*, 50–60.

ment by stone-throwers and archery to allow the defence no rest, but with the attackers divided into shifts so as always to be fresh. Other weapons included the use of Greek Fire, which may in fact refer to the Chinese fire weapons the Mongols had adopted rather than the Byzantine weapon known to Carpini. Simon of St. Quentin gave detailed descriptions of how this involved the use of liquefied fat including that made from human corpses, which, flying over the defences and striking the houses beyond, could not be extinguished. If the place stood on a river, the Mongols would block or divert the stream to cause a flood, and they were also capable of tunnelling. After surrender, the craftsmen would be separated out and the remainder of the people massacred or enslaved. To resist them, the friars' obvious recommendations included maintaining a sufficient supply of stones and arrows, to prevent the enemy establishing their own machines within range.[1]

The Mongol attacks continue, and war in the Caucasus and Asia Minor

The armies now mobilised by Mongol rulers were large enough to undertake more than one conquest at a time. Following the crushing of Kwarezm, another force had as its target the lands to the south and west. However, it is worth noting at this point that despite the Mongol victories in central Asia, Muslim rulers continued to rule large territories, and the Mongols did not succeed either in removing Djelal ed-din for a decade, or in making any significant inroads into the Muslim sultanate of Delhi. The rival principalities further west were continually in conflict: the Christian rulers of Georgia and Armenia, the Muslim Seljuk Turks in Asia Minor, along with the other Muslim rulers to south and east. The fragment of the Byzantine empire that remained at Trebizond on the Black Sea was also involved. The Armenian state in Cilicia had become strengthened under strong rulers, Prince Leon achieving recognition as king at Tarsus in 1198. His mountain realm was reinforced with new or more powerful defences on all sides. His arch enemy was the sultan at Iconium (Konya), 'Izz al-Dīn (Kay Kawus I, 1211–20), who had attacked his own brother within the fortress of Ankara in 1211, where despite deploying engines the siege lasted a whole year. 'Alā' al-Dīn (Kay Qubādh I, sultan from 1220–37) attacked Armenia with a full range of siege equipment in 1218/19. At the siege of Kargin, his artillery made great breaches in the walls leading to a successful assault.[2] In 1223, the sultan fell out with the new Greek "emperor" Andronikos of Trebizond and mobilised a large army to attack the city, built in the shape of a parallelogram between sea and mountain, with a powerful citadel and stone walls that had been repaired on the order of Alexius Comnenus (the original work dating back to Justinian). The work, however, was not complete. Stone throwing artillery was employed on both sides. There was

[1] Carpini, Giovanni di Plano, *The story of the Mongols whom we call Tartars (Historia Mongolorum)*, trans. Hildinger, 1996, 68 (Kiev), 72 (ropes), 76–8 (siege methods), 92 (recommendations); Simon de St Quentin, *Histoire des Tartares*, ed. J Richard, 1965, 44–5.

[2] H Duda, ed., *Die Seltschükengrschichte des Ibn Bibi*, 1959, 44–6 (Ankara), 72; G Dedeyan, trans., *La chronique attribuée au connetable Smbat*, 1980, 92; *Vahram's Chronicle of the Armenian Kingdom in Cilicia*, trans. C Neumann, 1831, 43–4.

fierce fighting, assaults and sorties, but the resistance, and severe weather conditions, persuaded the sultan to make a new peace with the Greeks and leave.[1]

The continued warfare of sultan 'Alā' al-Dīn renders more examples of the state of knowledge and practice of siege warfare in Asia Minor in the time before the Mongol conquest. Concerned at the weakness of his key cities, the sultan ordered new walls with towers and gates built at Iconium (Konya), his largest city, and at Sebastea (Sivas), the next largest, hiring the best architects available for the work. It was specified that the work involved creating stronger foundations and higher curtain walls, and the chronicle clearly states that the work was for defence.[2] There is some confirmation in the remains of some of these works. The new citadel built over the Byzantine predecessor at ancient Ephesus is dated to this time: it had strong mortared stonework in its curtain, and 15 projecting towers, and a single entrance.[3]

The sophisticated siege techniques of the early thirteenth century were known to the Seljuk rulers too. The emir sent by the sultan against Kâhta in 1226/7 had the "western" stone throwers, four of which he posted against one of the town's gates. Following the defeat of a relief attempt by rebels from Amida (Diyarbakir) the town gave in. The governor of Malatya was sent with 5000 men and stone throwers to take Cerniskezak, a town so well sited on a high ridge that it could not be undermined. Nor did a week-long bombardment of the town gates by stone throwers have any effect, and in the end the place was betrayed, 50 sappers creating a breach after the besiegers had been pointed to the place to attack the wall. The Seljuk empire continued to expand, as the sultan sent a similar army (1231) to take Harran and Raqqa, both old settlements with probably very old defences. Even these resisted for two months, during which time the bombardment caused great damage inside, and some breaches which the captors ordered repaired. The hardest target proved to be Amida (in 1233), a town with walls made of marble blocks. Stone throwers were brought up by camel but winter brought this first attempt to a close. The second attack was more successful: with the help of traitors, and under cover of a bombardment of stones, naphtha throwers were used to burn the town's water gate whereupon it surrendered.[4]

In 1222/3, the Mongols had arrived in Georgia, the Muslims of Derbend allowing them to pass through into the Caucasus. An Armenian source reported that the pass through the Caucasus having been blocked, the Mongols returned across the mountains, being obliged to discard baggage including siege weapons, indicating that alongside their highly mobile mounted forces, they had also brought

[1] J P Fallmerayer, Original-Fragmente, Chroniken, Inschriften und andere Materiale zur Geschichte des Kaiserthums Trapezunt, *Abhandlungen der Historischen Klasse der königlich bayerischen Akademie der Wissenschaften*, IV, 1846, 23–9 (chronicle of the Metropolitan Joseph Lazaropulos), 42–3 (Chronicle of Michael Panaretos).

[2] Ibn Bībī, 110–11. For a modern study, C Cahen. *The formation of Turkey. The Seljukid sultanate*, trans. Holt, 2001. Cahen lists (121) the major towns of the sultanate as Konya, Sivas, Sinope, Kayseri (Caesarea), Antalya, Erzincan, Malatya, Erzerum, Amasya and Aksaray.

[3] J Keil, *Ephesos. Ein Führer durch die Ruinenstatte*, 1964, 40–1.

[4] Ibn Bībī, 122–3 (Kahta), 124–5 (Cerniskezak), 191, 214–6 (Amida).

artillery.[1] Sultan Djelal clearly still had power and resources because, the Mongols having left, he invaded Georgian Azerbaijan in 1225. He blockaded the town of Tabriz, ordering the building of mangonels, rams and ladders, and the place surrendered after one week. He then attacked Tiflis (Tbilisi), which was better defended with a very strong wall. An assault failed, so Djelal resorted to trickery, drawing the garrison out into an unwise sorties then ambushing them, and breaking in with the retreating forces. Some of the garrison made it back into the citadel, which stood behind a river, but the sultan got across in a day and it capitulated.[2] The poor country was devastated, and it was recorded that nothing lived outside of the walled towns and fortified outposts. The existence of simple towers in many villages has been dated back to earlier than this period of instability, offering a little protection for people as refuges against raiding parties or vendettas.[3]

Sultan Djelal's warring came to an end with a new Mongol invasion and his flight and death in 1231. The Mongols returned to northern Iran and Azerbaijan. The city of Rayy, with its powerful ridge-top citadel, was taken, while Isfahan, whose defenders conducted a vigorous and effective resistance for up to two years, was taken after its river had been diverted (probably in 1237).[4] Meanwhile, in 1235 the town of Gandzak, strongly fortified and held, was taken inside three days. The Mongols deployed a great battery of artillery and the rampart was brought down at many points. At Shamkor, the inhabitants thwarted the first attempt at capture by setting alight to materials being used to fill the ditch. The second time, the Mongols built up a ramp as high as the town walls, leading to the inevitable storm and massacre. The area was fertile and offered good grazing ground for the Mongols' horses, and was also well placed for striking in any direction. Ani, the old Armenian capital, was sacked in 1239. Kirakos noted that their artillery batteries were deployed "with perfect skill". Tiflis was reoccupied and restored. At this point, the Georgian ruler, Queen Roussudan (1223–47) took the step of submitting to the Mongols, and was allowed to continue ruling while being obliged to provide soldiers for the nomads' expeditions. These were directed now against the Seljuks.

The sultan had little chance. Having won in the field, the Mongols proceeded to reduce the main towns of Seljuk Asia Minor one by one. Erzerum (Theodosiopolis) fell first, its walls reduced to rubble by the Mongol artillery, after 20 days, and the people were wiped out despite being promised their lives. At Kayseri (Caesarea), they deployed three stone throwers to bombard one tower for 15 days.

[1] Kirakos, whose account ends in 1270; Dulaurier, ed., 'Les Mongols après les historiens arméniens', *Journal Asiatique*, 5th ser., XI, 1858, 199–200.

[2] Mohammed en-Nesawi, 184–204; Brosset, trans., *Histoire de le Géorgie*, I, 485–507; R Blake, R Frye, 'History of the Nation of Archers by Grigor of Akanc", *Harvard Journal of Asiatic Studies*, 12, 1949, 295–7; Kirakos, 202.

[3] P Harrison, 'The fortified villages of Ushguli in the Georgian Caucasus', *Fort*, 25, 1997, describes a range of towers with defensive functions, alongside a castle erected by Queen Tamara (1184–1211).

[4] Kirakos, 213; J A Boyle, The capture of Isfahan by the Mongols, in *The Mongol World Empire*, 1977, 331–6; aerial photographs of the defences of Rayy: E F Schmidt, *Flights over the ancient cities of Iran*, plates 32, 33.

Great breaches became visible and the town was stormed. The citadel here, with walls three metres thick and 12 high, is probably a rebuild of the Seljuk version. Sivas (Sebastea) fell, and it appears the people offered up the keys but were anyway ravaged. When the Mongols laid siege to Iconium/Konya, the sultan had no option but to submit himself to Mongol overlordship. Of all of this conquest, very few details have been handed down.[1]

We do know, however, of the halting of the onslaught of the Mongols as they marched into the former Ghaznavid and Kwarezmian domains of modern Afghanistan. The mountain town of Uk resisted them for 19 months, only falling after disease decimated the population. The advance into Sind came to grief in 1240. Although Lahore, which was unprepared, fell to an attack in which Mongol mangonels destroyed the walls, the captors were unable to remain there and retreated. When they marched into Hindustan, they found Multan abandoned. Then they attacked Uchchah, and managed to create a breach, but the defenders had created a strong defence from mud and water behind the breach, and drove off the attack. As a relief army sent by the sultan of Delhi neared, the Mongols retreated.[2]

Baghdad, Damascus

At this time, the first Christian efforts were made to establish whether an alliance could be made with the Mongols against the Muslims. It is another fascinating but ultimately pointless speculation whether such an alliance could have maintained the survival of Latin states in Palestine: Mongol "help" was always connected with their ultimate aim of world rule.

Meanwhile, the Mongol ruler Hülegü assembled a great force aiming for the destruction of the caliphate itself, and a new source, Raschid el din, who worked in the Mongol court, provides key information. A force of 1000 men "used to operating machines of war, throwing naphtha and firing crossbows" was brought from China. The figure seems more reasonable than the corps of 10,000 previously ascribed to Genghis, and perhaps should be substituted for that earlier figure too.

First on the list was the land held by Rukh ad din (leader of the Ismaili sect known as the Assassins) with their main centre at Alamūt, an enemy that none of its neighbours was likely to rush to help. But the inaccessible terrain made for a hard campaign for the Mongols and their Georgian auxiliaries. The operation was as ever meticulously executed. A force was dug in to besiege Kerdeh-Kouh in 1254, protected on both sides, while the rest of the army attacked Mehrin with

[1] Brosset, *Géorgie*, I, 511–20; Ibn Bībī, 230–1; Grigor, 307–13; Kirakos, 426–31. G de Jephanion, La citadelle Turque de Césarée, in Mélanges d'archéologie Anatolienne, *Mélanges de l'université Saint Joseph (Beyrouth)*, XIII, 1928, 103–6. The walls of Sivas may have been very similar to those of Ankara, but have largely disappeared: Jephanion, 86–8. As much detail is in fact given by the itinerant friar, Simon de St Quentin, 75–6, who claimed that the Turks were "viles et miserati" and that the Franks would have defended Caesarea without fearing the machines or assaults of the Tartars.

[2] Minhaj ud-Din, 1120–5, 1130–4, 1153–5.

engines, and captured it. Other towns fell in the same way, Toun after 12 days, and once again everyone apart from the artisans was massacred. The main towns held out for a long time but were eventually all taken. When Alamūt was attacked (1256), trees were cut down and trimmed to provide the timber to make the mangonels, and relays of porters were established every one sixth of a mile to carry "the heavy poles and pillars to the top of the next hill" (Juvaini). Sites were chosen and the "parts had been easily fitted together" while negotiations for surrender took place, and the first shot smashed a defending machine. When Alamout surrendered, the Mongols broke up the machines and dismounted the gates of the town.[1] It must have been the case that the attack was directed at the lower parts of the complex of Alamūt. The castle proper stood on a very steep rock, with only very limited access. Like other Ismaili castles, it was strongly built, carefully sited to limit the possibilities of attack, and well-provided with intricately-engineered water supplies, and space for large amounts of food. But the strengths of Ismaili fortresses were also their weaknesses. Provided the besiegers could establish a supply route to sustain themselves, there was no one to bring relief, and the end was inevitable. The power of the Mongols enabled them to overcome the obstacles that had prevented generations of neighbouring Muslim rulers from being able to reduce these strongholds, built in groups to bring aid to one another, and requiring a systematic and long-term campaign to reduce them in turn.[2]

In 1258, Baghdad fell to Hülegü, the nominal centre of the Muslim world was devastated, and the caliph from whom all Muslim rulers derived their authority was cruelly killed. The Georgian chronicler recorded the heroic role of his compatriots in sapping under the walls and creating a breach in 12 days. Raschid recorded that a rampart was raised all around the vast city and stone throwers set up against selected towers, including one called the "Persian Tower", and walls to make a breach, which rapidly occurred. Because the area was barren of stones, they had to be transported from the mountains, while palm trees were cut down and shaped into missiles as well. The defence by the caliph's mainly Turkish troops was vigorous. They fought also from boats on the river, and the Mongols deployed machines hurling pots of naphtha at them. After two weeks of bombardment, the "Persian Tower" collapsed, and the city was taken by escalade.[3] The evidence suggests that the vast sprawling metropolis was partly surrounded by a semicircular kiln-burnt brick wall built originally in the eleventh century, and a ditch also brick-lined, and possibly not in the best of repair, as it suffered from regular flood damage. The original great round city founded by al Manṣūr had long since disappeared, although it may be that its remains constituted the citadel on the

[1] Juvaini, 629–32; Raschid el din, *Histoire des Mongols de la Perse,* trans. Quatremère, 1836, 133, 171–215; Brosset, *Géorgie*, 1, 530; Kirakos, 484.

[2] P Willey, *Eagle's Nest. Ismaili Castles in Iran and Syria,* 2005, 106–14 (and see also 264–5). Invaluable for describing in many cases for the first time these mighty fortresses, the author provides a counterbalance to the otherwise universally hostile coverage of the Ismailis to be found in the writings of their enemies, although he seems to bend the stick rather far in the opposite direction.

[3] Raschid, 281–93; Brosset, 1, 549; Grigor, 333.

west bank of the Tigris mentioned by Raschid. The decisive factor in the Mongol conquest may have been their success in crossing the river Tigris, enabling them to pick their points of attack along a perimeter of immense length (18,000 paces according to a fourteenth century source).[1]

The fall of Baghdad was merely the prelude to further advances, initially to the north and then westwards through the fertile crescent. Irbil refused to surrender, and the heat and insalubrity of the climate caused the Mongols heavy losses and the temporary abandonment of the siege after six months. Leaving the Syrian desert on their left, the conquering horde descended upon Syria itself from a direction from which the area had not been threatened since the age of the Sasanians. En route, they summoned their client kingdoms of Armenia and Georgia to provide additional forces. Leaving detachments to maintain the sieges of cities too strongly held to be taken immediately, such as Mīyafariqīn (Martyropolis),[2] Hülegü took others such as Harran, where the town surrendered but the citadel had to be stormed. This was a small roughly rectangular building (about 30 by 100 metres), with four polygonal corner towers and a square-towered gatehouse in one face, that did not detain the Mongols for long.[3] Hülegü then arrived before the city of Aleppo, that had successfully resisted the Latins throughout, but which now fell within forty days, the citadel being stormed (26 January 1260). This mighty defence standing more than thirty metres high on its steep-sided partly natural, partly man-made hillock, protected by a paved glacis rising from a massive ditch and topped with a powerful towered curtain, had inspired the citizens of Aleppo to stand against the invader. The Mongols first isolated the city with a trench, then set up their artillery to bombard it and their miners to sap the wall. After a week, they stormed through a breach in the eastern walls despite their thickness and pillaged the town, before turning their attention to the citadel. This, along with the town walls, had been modernised by the Ayyubids in the first decades of the thirteenth century, and provided with a substantial twin-towered gate with no fewer than five ninety degree turns. None of this availed against the Mongols, who massacred everyone inside, although the precise means of capture that they deployed to overcome the defences are not known.[4]

The defenders of the recently constructed citadel of Damascus decided not to test out their new walls against so ruthless an opponent, and surrendered the chief city of Syria to the Mongols. The chronicler al-Makīn, who was there, vividly

[1] Lestrange, *Baghdad during the Abbasid Caliphate*, 264–83, 342; Ibn Jubayr, ed. Broadhurst, *Travels*, 234. He had visited in 1183/5.

[2] Having resisted an initial attack with artillery, the place held out until hunger enfeebled the garrison and allowed the Mongols to enter and massacre everyone still alive. It is also recorded that the Mongols diverted the river here: Kirakos, 491.

[3] D S Rice, 'Studies in Medieval Harran', *Anatolian Studies*, 11, 1950, 36–48, plan of citadel at 98.

[4] Raschid, 333–37; Grigor, 335, 349; Dedeyan, *Smbat*, 104–5; Al-Makīn ibn al-'Amīd, *Chronique des Ayyubides*, trans. Edde, Micheau, 1994, 98–104; Burns, *Monuments*, 33 (Aleppo). The outer defences of the gate that are now such a prominent feature of the citadel date from after the capture of 1260. The city walls have disappeared.

described the panic of the people, and of the city's rulers, who fled.[1] The very survival of Muslim rule in the Middle East now hung by a thread, but a new hero was about to reverse the unbroken run of Mongol victory. In 1260, the Mamluk[2] rulers of Egypt destroyed the Mongol army left to control Syria by Hülegü, who was consolidating his new empire based on Iran, known as the Ilkhanate, at the battle of 'Ain Jālūt, which probably deserves recognition as one of the few really "decisive" battles of history. The lost cities were then swiftly recovered, and on the way back to Egypt, Baybars completed his rise to the sultanate. Meanwhile, it is necessary to retrace the course of events further west, in the lands never reached by the descendants of Genghis Khan.

Ayyubid disarray, war on Cyprus, and more crusading kings: Frederick II, Louis IX

With the Ayyubid princes engaged in internecine struggles, the Latin states were clinging to the coast free from threat. Jerusalem, however, remained out of their direct control until the remarkable "crusade" of the emperor Frederick II who arrived in Acre in 1228. This was the only crusade to have been undertaken with its leader under excommunication by the pope, who had lost patience with the numerous postponements of the venture. It was also the only campaign to have been bloodless, and finally, it was the only crusade since the first to succeed in restoring Latin control over a disarmed Jerusalem.

Frederick next attempted to make good the claim to Cyprus which had been accepted by his father Henry VI. However, during the infancy of the Lusignan ruler, heir of the man who had been given the island when Richard I seized it from the Byzantines, the rulers of Beirut were exercising the regency, and Frederick and John d'Ibelin had fallen out badly over questions of their respective rights. In July 1229, the emperor's lieutenants on Cyprus were driven from Nicosia by the d'Ibelin forces and took refuge in the castles of St Hilarion, Kyrenia and Kantara in the north of the island. The Lombards holding the harbour-side fortress at Kyrenia, a roughly quadrangular fortress built by the Lusignans around 1210 around an earlier Byzantine fort, agreed to surrender if not relieved. Scholars disagree about the date of the Frankish additions to Kyrenia, some suggesting that the greatly strengthened curtain wall with its powerfully-projecting towers well provided with arrow loops mainly date from much later in the thirteenth century. On the other hand, the length of the sieges of 1229 and 1232 would be very hard to explain if Kyrenia had then possessed only its thin Byzantine walls and towers. Extra protection had been provided, possibly in the late eleventh or even in the twelfth century in the form of much thicker walls in front of the old curtain, with

[1] Al Makīn ibn-al'Amīd, 111–15.

[2] The Mamluks had the status of slave soldiers and were often Turkish in origin. Baybars was a Crimean Turk who had entered the service of the Egyptian sultan. On his remarkable story see P Thorau, trans. Holt, *The Lion of Egypt, Sultan Baybars I*, 1987/92.

solid pentagonal bastions, on the south front, but there were no such additions on the long west side facing the town where the gate stood.[1]

John laid siege to St Hilarion, standing in a very strong mountain location, but its large garrison would soon become hungry. Here, as at Kyrenia, the Cypriot Franks had reused a Byzantine fortress, but this time had thrown up inner defences within a quite weak outer circuit of walls and towers barely a metre thick, although once again there is uncertainty as to whether much or any of these inner defences existed at this date. While blockading St Hilarion, d'Ibelin went to Kantara to observe a "great trebuchet" constructed by Anceau de Brie. Here, the trebuchet battered down the wall erected to face the only flat ground from which the castle could be approached, but the rock on which the defences stood was too powerful to be scaled, and eventually surrender terms were agreed.[2] It sounds as if the lesser outer line of wall may have been breached, but the very thick inner wall behind it, standing high above the ground from which an attack had to be made, had defied the attack, and indeed it is difficult today to see how stone-throwing artillery could have been mounted to hit the inner wall with any effect.

In 1231, imperial troops made for Beirut itself, where the bishop surrendered the town "timidly", but the castle held out. This fortress, of which only the foundations have been uncovered, was a quadrangular site in the corner of the city adjacent to the sea, rebuilt since its destruction by Saladin. The traveller Wilbrand of Oldenburg left a description. One side was covered by the sea, the other by a deep ditch, while the double wall was made of great stones reinforced by iron rods (and still visible classical columns), and protected "against damage by machines" by strong towers.[3] There was, however, a shortage of troops. The imperial army pressed a close siege, having many sailors and engineers and all the material they needed. The ditch ("one of the finest in the world" according to Philip of Novara) was captured, enabling the besiegers to begin mining under the protection of a massive screen, and covered by siege towers based on stone from which they could cause great damage to the castle. When relief arrived from Cyprus at Christmas 1231 the defenders were delighted, "for the castle was so mined.. it was falling in pieces" (Philip). Reinforcements were sent into the castle and they countermined, destroying the imperial mine and regaining control of the ditch. After a long and now fruitless operation, the imperial forces burnt their engines and abandoned the siege. A new thirteenth century castle had successfully resisted

[1] A Megaw, The arts in Cyprus: military architecture, in K Sutton, *A history of the Crusades, IV: art and architecture*, ed. H Hazard, 1977, 199–202. But the Department of Antiquities guide argues for a date of 1208–10 for much of the work.

[2] Philip de Novara, *The wars of Frederick II against the Ibelins in Syria and Cyprus*, ed./trans. J L de la Monte, 1936, 97–107. According to the editor, Philip was an adviser to John of Ibelin. He hailed from Novara in Lombardy and had served at Damietta in 1218, moving to Cyprus by 1226, and was a participant in what he describes. His account is therefore scarcely unbiased, but it is the only one. The text was reconstructed from its incorporation in the fourteenth century *Gestes des Chiprois*. Francesco d'Amadi, *Chronique*, ed. Mas Latrie, 1891, 143–6, largely follows Philip.

[3] Wilbrandi de Aldenborch, *Peregrinatio*, ed. J Laurent, 1859, 8–9.

everything a powerful emperor could throw at it, but ultimately owed its survival to seaborne relief.[1]

In 1232, an imperial army occupied Kantara, Buffavento and St. Hilarion castles and Famagusta, an important city possibly protected by twelfth century walls, and a (Byzantine) harbour-side tower. Only St. Hilarion resisted, and was relieved by a Cypriot army which also reoccupied the other castles, defeating the imperialists in the pass below St. Hilarion. Kyrenia was besieged. It was held by 50 Lombard knights and 1000 foot, including crossbowmen and sailors, and included "many good makers of engines" who built trebuchets and mangonels to defend the castle. The siege was pressed by land and sea, and many engines were built, including mangonels and "great trebuchets", two siege towers and many cats. A fierce battle was waged across the ditch as the defenders tried to burn the towers. A captured traitor was hurled back against the castle wall by a trebuchet. The defences were greatly weakened by the bombardment, but assaults were unsuccessful and surrender only came after nearly a year, in April 1233.[2] The castles of Cyprus, with the exception of Kyrenia, relied on their remote and inaccessible mountain top locations, but they also had effective walls at their weaker points with excellent provision of flanking towers and numerous well-placed arrow slits. Their siting made it difficult to deploy either artillery or mining to much effect. Nonetheless, they changed hands several times in the four years of conflict, mainly because their defenders were rarely as determined as the defences were strong.

The wars of the Ayyubids, including with the Sultan of Rum (1234–6),[3] enabled the Franks, were assisted by the arrival of crusaders led by Thibaut count of Champagne (1239), to rebuild the Tower of David in Jerusalem. It had been left destroyed following the truce agreed with al Malik, who now marched against the city, bombarded it with artillery and secured its capture after 21 days, the Tower of David surrendering six days later. It was again dismantled, miners being used as the most effective way of demolishing the massive stonework of the ancient tower.[4]

Complex shifts in alliances mark the next period. In 1240, the crusader army sought to rebuild the wrecked castle of Ascalon, a move abandoned after a contingent was defeated at Gaza. A truce was made under which the castle of Beaufort was returned to the Latins. Its garrison, however, had other ideas, and had to be besieged by the sultan, who ordered up the artillery from Damascus and forced its Muslim garrison to surrender at his mercy. The interrupted reconstruction at Ascalon was then resumed.[5] In 1243, John d'Ibelin recovered Tyre from Frederick's supporters, breaking into the town through a postern on the sea side, then

1 Philip de Novara, 121–37.

2 Philip de Novara, 141–8 (1231), 156–68 (1232); Francesco d'Amadi, 147–82. On the castles see E Fiene, *St Hilarion, Buffavento, Kantara, Bergburgen in Nordcypern*, 1992; and Megaw, 197 (Famagusta), 204–5 (brief, on St Hilarion and Kantara). Megaw suggests that the inner Frankish defences at St Hilarion may have existed by 1228.

3 Al Makīn ibn al-'Amīd, 49–50.

4 Makrizi, 10, 323–4; William of Tyre (Rothelin continuation), published as *Crusader Syria in the thirteenth century*, trans. J Shirley, 1999, 40.

5 William of Tyre (Rothelin), 41–57.

besieging the castle closely with many engines. The powerful fortress resisted all attacks, but gave in when the commander's brother was captured and threatened with hanging.[1] But the same year, the Egyptians allied with the Kwarazmian army, available to the highest bidder since the death of Djelal ed-din, against the Syrians, and the horsemen from central Asia swept south, not daring to attack Damascus with its powerful defences, but instead seizing and ransacking Jerusalem in 1244. Not long after, having joined with their Egyptian paymasters to besiege Latin Gaza, they were instrumental in winning a crushing victory over a joint Latin-Syrian army at the battle of La Forbie, where Baybars played a leading part, and once again the crusader states were under serious potential threat. First, however, Damascus was attacked with many stone throwers, the projectiles killing many people and leading to the surrender of the city to the Egyptians. The sultan continued his campaign to reunite the old Ayyubid realm under his authority with a siege of Homs, where a battery of 14 large stone throwers maintained an intense bombardment of the old city. The weapons were described as "devilish mangonels". At great expense, labour had been hired to transport the artillery from Damascus to Homs, and eventually a peace was negotiated.[2]

Then in 1247, the Muslim armies turned their attention to the reconstructed fortress of Ascalon. Unable to prevent the Latins from keeping the place supplied by sea, the besiegers turned the wreckage of their own ships to advantage by reusing the timbers for mantlets and the masts for the arms for stone throwers. Simultaneously, a mine was driven under the hill, and continuous relays of attacks exhausted the defenders. The castle fell as attackers used their mine to obtain entry into the fortifications, a rare case of the tunnel being built for such a purpose. Thus, as the chronicler noted mournfully, all the castles repaired through Count Thibaut's crusade were lost again within a few years except for Saphet (for which, see below).[3]

Two years after the sultan's victory, the crusade of Louis IX of France arrived in Egypt. The target was once again Damietta, but this time the powerful fortress city was abandoned by its guards and people, shamefully according to the Muslim accounts. Louis spent much time making the city even stronger, one chronicler commenting that this was "all folly, we could well have managed with less." Sultan al Malik aṣ-Ṣālih, although mortally sick, led an army to Mansourah to block the crusader advance up the Nile, occupying the same encampment as his father had in 1219. In November 1249, King Louis marched upstream from Damietta, after hearing of the sultan's death, and set up camp opposite, whence he arrayed stone throwers to bombard the Egyptian camp and ships ("although I never heard that ours had done much damage" — Joinville, who was a participant). The French efforts to build a causeway across the river in the face of a battery of 16 stone

[1] Philip de Novara, 183.

[2] Makrizi, 10, 365 (Damascus); 11, 198–9 (Homs). Makrizi's sources are identified as the contemporary al Wasil, who used the original Mongol term Karaokha (Karabogha) for the most powerful of the weapons, perhaps suggesting a connection through the Kwarezmians with the recent events of central Asia.

[3] William of Tyre (Eracles continuation), in Shirley, work cited above, 135–6.

throwers were thwarted, mainly by the immense difficulty of the task itself, as well as by fire bombs hurled by catapult that succeeded in burning up the crusaders' wooden works including two wooden towers and a "covered way" providing protected access to the works on the river bank.

There was apparently heated discussion in the crusader camp on what to do. The Master of the Temple advised patience, and the construction of engines to destroy the walls and houses of Mansourah. But the count of Artois was for a direct assault and on 20 February 1250, the French troops, having found a way to ford the Tannis branch of the river, made a surprise attack on Mansourah, and broke into the town. After initial success, they were repelled by a disciplined Mamluk contingent and most of them killed in the narrow streets. Crusader accounts talk of the gate being closed on them with a portcullis and of being swept by crossbows and ballistas, and also suggest that they were lured into a trap deliberately, something which the Muslim accounts do not bear out. Fleeing survivors were assailed with Greek Fire (perhaps naphtha) as they tried to re-cross the river. The Egyptian defences on the river had been taken in the initial onslaught and the 14 stone throwers were dismantled so that their timber could be reused for the causeway, but the outcome of the whole crusade was to make this irrelevant. Defeat at Mansourah was followed by catastrophe as the Egyptian fleet managed to cut the crusaders' communications with their base at Damietta and enforce the surrender of the disease and hunger-stricken survivors of the army, although in exchange for the return of Damietta, Louis was released.[1] The new sultan ordered the dismantling of the fortifications to prevent the risk of fighting for it a third time.

Louis made himself useful to the Latin cause as over the coming three years he spent money and man-hours modernising the defences of Caesarea, Jaffa and Sidon. At Jaffa, he added a new line of walls around the old castle; at Sidon, he completed the unfinished circuit of town walls, with numerous projecting towers, and reconstructed the castle on the line of the wall on the landward side (now entirely ruined). The sea castle, on its island 100 metres from the shore, had proved its value in holding off a Muslim attack the previous year, when the town was ransacked and 2000 people massacred, but it was too small on its own to offer refuge to the citizens.[2]

[1] J de Joinville, *Life of St Louis,* trans. Shaw, 1963, 213–17 (a participant); Makrizi, II, 204–27; Jamāl ad-Din ibn Wasil, Mufarrij al-kurūb fi Akhbār Banu Ayyūb, excerpts trans. in F Gabrieli, *Arab historians of the crusades,* 1964, 285–99; The assault on Mansourah, *Excerpta Historica,* 1831, an account written in old French by William de Longespee, earl of Salisbury, a participant, 69–84; William of Tyre (Rothelin), 88–99, 111 (dismantling of Damietta).

[2] Joinville, 282, 295, 305, 310. On Sidon's sea castle, which although only built for the first time in 1228 and abandoned in 1291, has a building history difficult to unravel, see Kennedy, *Crusader Castles,* 122–4. The castle in 1253 probably comprised three closely placed square towers facing the land, later provided (by the Templars) with extra shooting galleries. The French in 1253 were probably responsible for rendering the two outer towers more massive.

New Levantine fortifications

The form . . .

King Louis' work was to the same modern standards as the other rebuilding of castles in the Latin states during the first half of the thirteenth century, during which time some of the most spectacular medieval fortifications ever built were created. Often being the last significant developments on these sites, they are the easiest to study. Almost all of the new work was carried out by the religious orders, aided by significant donations either of money, men or time, from visiting crusaders such as Louis, or Thibaut of Champagne. The cost to build, maintain and garrison a modern castle was far beyond the reach of the barons of outremer.

The latest architectural styles were applied to the interior buildings as well, with large and decorative residential quarters, halls and chapels, alongside the other essential buildings: kitchens, stables, workshops, and, everywhere, large cisterns for the collection of water. It is not surprising that there should be parallels with monastic building, and because of the relatively large size of the population of these Templar, Hospitaller and Teutonic castles, the extent of the domestic provision is much greater than in most comparable western castles. Much survives of these immense buildings in Crac des Chevaliers in particular. But also notable was the unprecedented transformation of the defensive features of these castles, the several key elements of which reflected a common approach to the new conditions of the thirteenth century across the Christian–Muslim frontier. The cross-cultural influences are now impossible to pin down, but the descriptions already given of the work at Cairo, Damascus and Aleppo on the one hand, and on the other of the new constructions of Richard the Lionheart and Philip Augustus and their successors, clearly demonstrate the lines of development, even if it is impossible to be certain as to which came first.[1]

One conscious development of this period, for the Latins, was where possible to abandon lowland sites and withdraw to more easily defensible hill tops, where only one face was the likely target of a possible attack by trebuchet or mining. Of the great castles, the Hospitaller Crac des Chevaliers (Plate 3) and Margat (al Marqab) (Plate 2) and the Templar Saphet (Safed) and Safita fit this description. Both in these conditions, and in conditions where it was not possible, such as the seaside city of Tortosa (Tartus) or Chastel Pelerin (Pilgrim's Castle, 'Athlit), another step regularly taken was to build concentric defences, with a higher inner curtain wall overlooking a lower outer wall. Having two lines of walls naturally extended the defensive life of a fortification and represented a considerable change from older two or three-bailey castles, even those as powerful as Château Gaillard. What was lost was the possibility of using the outer enclosure for residential or related purposes, as it was now too narrow to be anything other than a means of protected circulation. What was gained was that an attacker could be resisted simultaneously from two lines of defence, and mining of the main defences could

[1] See my *History of the early medieval siege*, 300–308.

be rendered a much slower process. Such were the thirteenth century developments at Crac, Margat, Chastel Pelerin, Saphet and Tortosa.

Protecting the walls against sapping, mining and straightforward escalade remained a major consideration, and for this regular flanking towers became standard, either circular or rectangular in shape, but always well provided with ranks of arrow loops, the scientific siting and design of which represented a great advance on previous crude loops. Direct protection of the foot of the wall was now provided for by frequent use of machicolations, such as those remaining on the outer wall at Crac. It has been noticed that their design, a projecting, roofed stone box on corbels with holes below and an arrow slit in the front face, is identical with that over the gate of the citadel of Aleppo of the same date, a direct parallel the cause of which is now impossible to explain. (The continuous machicolations on some towers at Crac undoubtedly date from work after the fall of the castle in 1271). Shooting galleries within the walls (such as those of probably thirteenth century date at Kyrenia, Margat and the sea castle of Sidon) added further defensive possibilities. At Tortosa, the great donjon standing on the later curtain wall was provided with ground level external galleries for the same purpose.

Another common feature was the adoption of the "bent entrance" already common in Byzantine and Muslim fortifications, with spectacular examples at Crac, with its long dog-legged covered entrance, and Margat. Donjons became much less of a feature except where, as at Safita, a great tower was of such size as to offer substantial residential space. Instead, in the new defences one of the mural towers would often be larger than the others and provided with accommodation suitable for a Master of the order, as at Crac, Margat and at the Teutonic Knights' castle of Montfort. The latter offers a different design much closer to twelfth century German models.[1]

The most notable feature was one of scale. The new defences dwarfed anything that had gone before. Chastel Pelerin was thrown up in a remarkably short time by a crusader force in 1218, reusing ancient Phoenician stones found on site and taking advantage of a spit of land projecting into the sea. The outer wall alone is six metres thick and sixteen high. On the inner wall were two massive towers 34 metres high and the defences facing an attacker were provided with artfully contrived galleries and covered entrances. At Tortosa, where fragments remain within existing buildings, the inner wall stood 25.5 metres high. The outer wall at Crac, which appears to date wholly from the thirteenth century, was nine metres high, but rising massively above it the interior wall with its three immense towers facing the south is characterised by its vastly greater size (25 metres high), and the visibly dominant feature of an immense glacis from which the towers appear to grow and project majestically. The glacis may have been designed to help protect against earthquake damage, but certainly it also serves to provide an absolutely forbidding obstacle up to 25 metres thick. A water-filled basin in front of the glacis doubles as a moat and a cistern. At Margat, the circular donjon tower at the most vulnerable point of the inner defences is 24 metres high and 20 in diameter, similar to

[1] Kennedy, 129–31, plan and cross section at 130.

those of Philip Augustus, but with walls ten metres thick in places. In many of the castles, the design combined with the size of the works to create substantial platforms which may have been used for the deployment of stone-throwers.[1]

The castle of Saphet can be reconstructed not only from archaeological investigations and nineteenth century engravings, but also from a contemporary tract written by or on behalf of a bishop of Marseilles trying to raise donations for the continued maintenance of the Templar fortress. They set about constructing the new castle from 1240 during a truce. Assisted by a large number of "workmen and slaves", and considerably benefited by the discovery of a spring within the site to provide water, the work went ahead with great speed. On his return twenty years later, the bishop found a castle on a spur with sheer sides, and inner and outer walls. The ditches were 15 metres deep and 13 wide, the inner ramparts 44 metres high, the outer 22, and over three metres thick at the top. There were seven towers each 48 metres high and 22 wide, being 4.5 metres thick at the top. The fortress was provided with shooting galleries above the ditches and numerous positions for archers, crossbowmen and artillery. The quality of the masonry was "exquisite". Archaeological enquiry has largely borne out the description. One feature not mentioned by the account was a large circular donjon, which may possibly have housed the Templar chapel. The bishop was shamelessly fund-raising to support the work he had inspired. The regular garrison was 50 knights, 30 sergeants, 50 Turcopoles and 300 crossbowmen along with 1200 others. 12,000 mule loads of barley and corn were needed each year.[2]

... and the function

The tract stated that Saphet was essential in protecting the territory around it from the Muslims, and additionally enabled its garrison to control, and exact tribute from, the Saracen lands facing it, thereby also denying this sustenance to the sultan. Its garrison prevented raids on Acre, and this allowed for the settlement of the countryside and for agricultural production. As a result, a substantial town with (allegedly) 10,000 inhabitants and a market, had grown up within its lee, and these people helped defray the cost of the castle too. Ibn Furāt later described it as "an obstruction in the throat of Syria", a phrase that eloquently summarises the castle's multiple roles.[3] Similar functions were performed by the other inland castles. Crac des Chevaliers, which seems to have had a permanent garrison including 60 knights (and no doubt a proportionate number of sergeants and crossbowmen)

[1] Kennedy, 120–79, covers all the castles mentioned; Müller-Wiener offers excellent short summaries and plans. P Deschamps, *Les Châteaux des Croisés en Terre Sainte*, provides detailed analysis and description. Wilbrand of Oldenburg, 11–12.

[2] De constructione castri Saphet, translated as appendix to Kennedy, *Crusader Castles*, 191–99; R Huygens, *De constructione castri Saphet*, 1981, analyses the tract. D Pringle, 'Reconstructing the castle of Safad', *Palestine Exploration Quarterly*, 1985, 139–48 (a review of Huygens).

[3] U and M Lyons, *Ayyubids, Mamluks and Crusaders. Selections from the Tarīkh ad-Duwal wa'l Mulūk of ibn al-Furāt*, 1971, 11, 89. Ibn Furāt was a fourteenth century writer, collecting earlier sources.

controlled, through its garrison, a wide and prosperous countryside, and by the same process was able to exact tribute from neighbouring Muslim territory and deny it to the sultan. Margat also had a town adjacent to it, separated from the castle by a deep rock-cut ditch, with its own, substantially weaker, defences. The primary function of these castles was to exercise lordship over the surrounding countryside, and to prevent the other side from doing so. But in order to exercise lordship, the castles needed to be built to the strongest specifications, so that they remained in Latin hands. During the first two thirds of the century, while the Muslim regimes were disunited, their strength was sufficient to deter attack.

Although small numbers could ensure the protection of the place in the passive sense — Margat had a nightly guard of four knights and 28 watchmen (Wilbrand) — large garrisons were needed to carry out the range of functions, and workmen for repairs, so the cost was high. The religious knights were a small minority of each garrison. Wilbrand of Oldenburg said that Crac could sustain 2000 fighting men. By comparison with the number of occupants regularly to be found in even a major castle in western Europe, these numbers were great and the cost commensurate.[1]

Frederick II at home

Emperor Frederick II was well aware of the position he occupied, and like his grandfather Barbarossa was not backward in presenting the imperial dignity to the world. He was, as well, remarkably well educated for the age, and had a broad range of interests that gave rise to all kinds of anachronistic nonsense being written about him in subsequent years.

At the beginning of his reign, still supported by the popes, Frederick had to establish his rule in his own inheritance, in particular in the kingdom of Sicily. On the mainland, the emperor ordered the construction of new castles at Gaeta, Aversa, Naples and Foggia in 1223, and ordered other castles in the area destroyed by the Edict of Capua. His target, which would have been familiar to other monarchs trying to restore authority after periods of royal minorities, was the castle built outside of private demesne since 1189, the alternative being to surrender them to the king. Whether the places he confiscated were really in strategic locations, as has been suggested, is perhaps another matter, but it is quite likely that such considerations did play some part in imperial calculations at times when the kingdom seemed threatened.

In 1229, in the spirit of Christian unity, no doubt, with the emperor abroad on his crusade, the pope who had excommunicated him sent an army against his lands, and the people of Gaeta rose up and destroyed the new imperial castle. The invading forces were however held up by a castle they could not take with their machines, and the return of Frederick to Apulia restored the position. Imperial building continued. In 1230, large numbers of workers were gathered to begin the

[1] For these two paragraphs: De constructione Castri Saphet (Kennedy, 197–8); Wilbrand of Oldenburg, 11, 12.

digging of ditches and the erection of walls and towers at San Germano, and in 1233 the order went out for the construction of new castles at Trani, Bari, Naples and Brindisi. In the following years, the continuing problems at Gaeta led to the construction of an imperial tower on the city's bridge, and then an order for the destruction of all bar four of the urban towers.[1] Meanwhile further north, the incessant warfare between the Lombard cities had continued, the Bolognese in 1229 conveying 12 mangonels and "trabucos" on carts to attack the Parmese, while the trebuchet becomes regularly added to the list of weapons being deployed by the city armies over the coming years.[2]

In 1237, Frederick was engaged in resumed hostilities with the northern cities led by Milan. The accounts of the chronicles are helpfully supplemented with surviving records from the imperial court. The final outcome suggests that the difficulties faced by all outside forces, however powerful, in subduing the large, wealthy and strongly fortified network of north Italian cities, especially when they were allied together and backed by the papacy, had not become any easier since the times of Barbarossa. Frederick's son Manfred deployed siege artillery including mangonels and two trebuchets at the castle of Monte Clario, forcing its rapid surrender. But in 1238 Brescia, which may have been chosen as a weaker target than Milan, proved too hard all the same, despite the emperor using a massive array of siege equipment. The Brescians used their own artillery to destroy the emperor's siege towers leading eventually to the raising of the siege. Clearly, a well-resourced Italian city, with its own supply of modern military skills, could hold off an imperial army for long enough to persuade it of the futility of the exercise. Here, the story is told that one of the imperial engineers, a Spaniard named Calamandrino, was captured, and advised the Brescians on the construction of the artillery which proved so effective, but the resemblance of this tale to so many similar ones, and the improbability that the Brescians were unaware of trebuchet technology, must make one question it.[3] The next year, Bologna informed Milan of its preparations to resist imperial attack with multiple lines of defence, powerful additional wooden works on top of the walls, and the organisation of supplies of timber "suitable for making trebuchets".[4] At the same time, Frederick was ordering his own side's fortifications to be put into a state of defence, for example instructing Andrea di Cicola to reinforce the castles of Monte Cassino and Rocca Janula by making a blida (that is, a trebuchet) "for their stronger defence", and early in 1240, sending orders to transport another blida with its associated equipment from its castle to join the siege of the Rocca Alberici.[5] Later that year, Frederick scored his only significant successes in taking defended cities (as opposed to castles), when he secured Ravenna inside six days (August 1240). His troops had crossed

1 Ryccardi de Sancto Germano, Chronica, *MGH SS*, XIX, 343–73.

2 Annales Placentini Guelfi, 448–9, 451, 455.

3 Salimbene de Adam, *Cronica*, ed. Scalia, 1966, I, 132, 135. Abulafia, 308–9.

4 E Winkelmann (ed.), *Acta imperii inedita seculi XIII et XIV. Urkunden und Briefen zur Geschichte des Kaiserreichs und des Königreichs Sicilien*, 1885, I, 526–7 (no. 660).

5 J Huillard-Breholles, H de Albertis de Luynes (eds.), *Historia Diplomatica Friderici Secundi*, 1852–61, V (i), 443 and V (i), 932–3.

the marshes and rivers without hesitation, storming the suburbs, whereupon the townsfolk had yielded in exchange for imperial mercy. Frederick had planned to continue his offensive by attacking Bologna, but decided that it was necessary first to secure Faenza, which he did not want to leave "intact in his rear."[1] The progress of this siege can be followed from his letters. He had a large force, with a diverse array of siege machinery, which was called into play when the town refused to surrender. Despite his optimism that the place would fall to his aggressive attack with engines, this did not happen, and the advent of winter delayed victory until the following spring, when, with the walls shattered by his bombardment and with his underground tunnels penetrating into the city, surrender was agreed in April (1241).[2]

Normal service was resumed with the siege of Viterbo in 1243. Here, an imperial palace was constructed inside the city walls, and a garrison was placed in the city. Cardinal Rainer, however, organised with papal supporters to launch a coup, bringing knights from outside, and after a fight, driving Frederick's supporters, who apparently numbered about 400, back into the castle Ercolo, that stood on a height within the south-western point of the roughly triangular city (8 September 1243). The account given by the imperial lieutenant, Count Simon, tells of the efforts made by the citizens and their troops to take the castle, which they blocked off, then set up stone throwers on top of the city's towers to bombard with stones and incendiary devices, and also began mining. The emperor, taken by surprise, assembled forces and marched to relieve his supporters in Viterbo, and led an immediate assault three days after he arrived in person on 8 October, which forced its way across the defenders' palisades but was driven back from the walls, including by women hurling down missiles. Therefore he had to establish a regular siege. The city was protected by traditional defences with large square towers and powerful walls where fast-flowing rivers did not provide sufficient protection (as they did on the south). The imperial camp was necessarily extensive, heavily protected, and immensely costly, and the sight of the emperor tied down and unable to reach his own supporters, besieged in their turn, was not helping the cause of the Hohenstaufen elsewhere. So another assault was ordered. An immense array of wheeled siege towers was constructed from which to attack the walls, along with extendable wheeled ladders able to reach from the bottom of the ditch as high as the stockade protecting the ditch, and another very tall wheeled battering ram. As this mighty array was moved up, incendiaries naturally played a vital part. As fires were started on both sides, the wind proved decisive, shifting suddenly and burning 12 of Frederick's 13 towers (10 November). As the year turned and the imperial forces were forced to suffer in their temporary quarters in mid winter, the gallant defenders of the castle had to yield and the fortress was destroyed (January 1244). Frederick had failed.[3] From these accounts, it is clear that the emperor had an effectively organised system for conducting sieges and constructing a large

1 *Historia Diplomatica*, v(ii), 1029, 1050–52.

2 *Historia Diplomatica*, v(ii), 1041–43, 1050–52, 1113–14.

3 Winkelmann, I, 549–53; *Historia Diplomatica*, VI, 125–44. A thorough account is given in E Winkelmann, *Kaiser Friedrichs II Kampf um Viterbo*, 1886, 278–304. For the

and diverse array of equipment. But in using it, he was no more able to overcome the problems posed by the political and geographical situation in Italy than his predecessors.[1]

Meanwhile in 1240, the emperor resumed the construction or repair of castles in the kingdom of Sicily, with work at Bari, Trani, Foggia and Goia de Colle, and the beginning of the building of the remarkable Castel del Monte in Apulia.

Frederick II's castles

The castles of the *regno* of Sicily and southern Italy had been built by the Norman conquerors and were now quite obsolete in terms of style and strength. With the relentless hostility of the papacy and memories of hostile interventions, it would not have been surprising if Frederick had looked to replace the old donjon- and curtain-walled castles, such as those of Melfi and Monte St Angelo, with something more modern. At the same time, a new emperor needed something distinctive to demonstrate his personal authority, and besides, the Norman castles were certainly not palatial, whereas the self-styled *stupor mundi* definitely needed suitable places for the imperial court.

There was however nothing revolutionary about Frederick's programme of building or the way in which the castles were used or maintained. He established a new office of *Provisor castrorum*, one for each district, responsible for maintaining the site. That for the Abruzzi was in charge of providing for 31 castles, that for the Beneventano 85, and in the Capitanata and Basilicata, no fewer than 109. The castles themselves continued to be divided between royal servants and feudal lords holding them from the king. The responsibility for carrying out repairs lay with the local communities. The catchment area for Bari was villages within 16 to 30 kilometres, for Canossa a slightly smaller area, and for Castel del Monte no less than 85 kms. Sadly, the documentary evidence from which this information is drawn is incomplete, and gives no information about the garrisons of these castles.[2] It would be reasonable to assume, on the evidence of royal castles further west, that most of the time the number of occupants was minimal, sufficient to exercise the governmental or domestic functions assigned to it at the time.

Frederick's new castles were characterised by a style all of their own, and the geometrical patterns to which his new works subscribed were quite novel. Generally, the castles were rectangular, and symmetrical, often divisible into quadrants, distinguishing them from the rectangular corner-towered castles being built in France at the same time. Such were the designs of Castel Maniace at Syracuse, Prato, Castel Ursino at Catania, and Augusta. Massive rectangular towers much

defences, A Scrattoli, *Viterbo nei suoi monumenti*, 1988, 33–42; E Bentivoglio, S Valtieri, *Guida a Viterbo*, 1982, 7, 17.

[1] There is a study of Frederick's army in G Amatuccio, *Mirabiliter pugnaverunt. L'esercito del Regno di Sicilia al tempo di Federico II*, 2003, using the above sources, but the edition is marred by numerous misprints and wrong attribution of sources.

[2] The above from H Houben, *Mezzogiorno normanno-svevo. Monastere, castelli, ebrei ,musulmani*, 1996, 168–75.

higher than the curtain stood at each corner, with a gatehouse in the middle of one side. The shape allowed for an equally symmetrical design of the inner courtyard, surrounded by palatial buildings, with elegant if austere two stories of rooms facing inwards, and a grand staircase giving access to the upper level. The imperial insignia still surviving in some of the stonework testifies to the message of *Roman empire reconstituted* being delivered.

The new castle at Trani, now much restored, reveals a plain exterior of rusticated masonry with almost windowless towers of massive thickness, very similar in appearance to the enormous towers added to the former Norman castle at Bari, while similar work was done at Barletta and in the new castle at Brindisi. Powerful towers were added to the Norman walls of Melfi, used by Frederick as his treasury. New towers were added to Monte St Angelo, another favourite residence. The famous Castel del Monte appears to have been an elegant hunting lodge. The strictly octagonal design has intrigued generations of writers, but the interior is quite plain and accommodation would have been rather cramped. Started late in Frederick's reign it may well have been of considerably greater interest to art historians than to its originator. In Germany, the castle of Lahr was constructed as a regular rectangle with rusticated masonry by Frederick's German deputy Woelfflin, on the emperor's orders (around 1220). On flat ground with a deep, wet moat, Lahr also possessed a rectangular Bergfried inside. Traditional castles with a Bergfried continued to be built, as at Neipperg in Swabia, and Brauneck in Franconia.

The full range of functions of a thirteenth century castle is revealed across the building work in Italy of Frederick II. There can be no doubt of the symbolic purpose of much of the architecture, but many of the castles also served as administrative centres and palatial residences for the imperial family and their guests. At the same time, the defences of some of the castles were genuinely powerful, and the intricate measures taken to protect the entrance at Trani, for example, speak of serious defensive intentions, and the combined roles of such castles were as evident in such buildings as they were in the great castles of the Middle East.[1]

Elsewhere in central Europe, the new styles followed by Philip Augustus and Frederick II did not find much imitation. For the lesser nobility, this was as much a matter of what they could afford, and what their local masons could accomplish, as it was of a conscious decision. Even the greater lords might choose to follow traditional paths, such as Godesburg, built in 1210 by the archbishop of Cologne, where there is a two-ward castle with the main accommodation built around the courtyard of the irregular inner ward, itself completely lacking towers, and surrounding a circular Bergfried. The improvements that took place in the Rhineland in the early thirteenth century largely consisted of replacing wooden buildings with stone, rather than replacing old castle designs with new. Other

[1] S Mola, *Itinerario federiciano in Puglia*, 1994, throughout; S Mola, ed., *Castel del Monte*, 1991; S Mola, *Trani: il castello* (guide book, 2000); A Haseloff, *Die Bauten der Hohenstaufen in Unteritalien*, 1920, for descriptions of the individual sites; W Hotz, *Pfalzen und Burgen der Stauferzeit*, 1981, 174–8, 200–1, 325 (Bitonto), for the examples from Germany.

parts of the German kingdom did see the first adoption of flanking towers in new or improved castles, and the use of screen walls and arrow slits to improve their defensive capabilities, as at Grafenstein (Rhineland-Pfalz) and Lichtenfels (Baden-Württemburg).[1] In the lands of the Babenbergs (Austria), Przemyslids (Bohemia) or the Zähringer (south-west Germany), influence from returning crusaders was confined to the names of castles, and only limited additional construction.[2] In Bohemia, stone castle building itself only dated from the twelfth century and was inevitably heavily influenced by traditions of building in eastern Germany. However, influence from French designs has been argued in residential round towers such as at Olomouc Breclau (Moravia), dated 1215–35, and Spiš (Slovakia), well constructed with four metre thick walls but without the internal vaulting found in Philip Augustus' castles. Others, such as the royal hunting castle of Křivoklát, with beaks attached to one side, bear a striking resemblance to buildings further west. Bitov, dated to 1225, has been compared to La Roche Guyon and intriguingly linked with French masons sent to Bohemia to build a palace in that year. Altogether, six Bohemian castles are said to follow French design, but their construction indicates that the intention behind the design was not properly understood by local builders. The square-towered form and the rusticated masonry that distinguished Frederick's buildings is to be found in rather more sites, such as the town castles of Kadan, founded in 1261, and Pisek, 1260. In Austria, the castle of Wiener Neustadt built from 1260 was square, with four square corner towers, and a similar design was used at Vienna (1275). Meanwhile, the royal building programme included a walled new town beneath the castle at Prague by 1250.[3] There was likely to have been a great deal of cross-influence among the higher nobility, not always shared, or understood, by the masons who had to carry out the work.

The Baltic: arrival of the Teutonic Knights

Yet another style of castle-building was to arrive in lands that came to be dominated by Germans as a result of the first stages of the creation of what would become, eventually, a dominion running the whole length of the shore of the eastern Baltic from the river Oder to the Gulf of Finland: the unique realm of the Teutonic Knights.[4] This future was unknown to the first groups of German monk-knights under their leader Hermann of Salza, a close associate of Frederick II, when they accompanied a Hohenstaufen princess to Spain to marry the king of

[1] C Meckseper in *Burgen in Mitteleuropa, ein Handbuch*, I, 91–8; H Hinz, in *Kirche und Burg in der Archäologie des Rheinlandes*, 1962, 137, 144–5 (Godesburg).

[2] H Hofrichter in *Burgen in Mitteleuropa*, I, 108.

[3] T Durdik, *Kastellburgen des 13 Jahrhunderts in Mitteleuropa*, 1994, 13–15, 28 (Tyrov), 45, 110–15, 140–58, 217–19 (Austria), 231 (plans); T Durdik, 'French influence on Bohemian castle architecture of the thirteenth century', *Fortress*, 15, 1992, 18–28; V Humel, Research in Prague, in Austin, Alcock, *Baltic to Black Sea*, 268–78.

[4] The classic modern study is E Christiansen, *The Northern Crusades*, 1980, but there are a large number of immensely detailed German studies of the Teutonic Knights that contain a treasury of information.

Castile, where they were granted lands on which they built the castles of La Mota and Tiedra in 1222, followed by a third castle at Higares east of Toledo in 1231. The first was their headquarters, and had a large tower behind triple walls. Tiedra was built in local style, a large square donjon inside an irregular curtain, while the last had a free-standing quadrangular tower. At the same time, other Teutonic Knights had found a home in Hungary, where they undertook the construction of "castles of stone against the Cumans" in 1225, before being swept aside in the Mongol deluge.[1]

It was 1231 when the Knights were invited by the Polish prince Conrad of Mazovia to join him in a crusade against the heathen Prussians, and they were installed at Thorn (Toruń) on the Vistula (named after Toron in Palestine). Rapidly they began to establish settlements along this river towards the Baltic, towns created with German town law as their foundation, but earth and timber palisades were needed to secure them against the hostility of the native population. Sometimes captured Prussian earthworks were reused or strengthened, as at Alt Christburg, or the promontory fort at Engelsburg. The well-armoured and experienced knights did not need special skills to capture the Prussians' earthworks, the most serious efforts recorded being made by escalade, no doubt under cover of archers and crossbowmen. But their generally small numbers made them vulnerable. In 1239, the Prussians assaulted the new work at Balga, but were shot down by the brothers.

When Swantopulk led a great revolt against the new crusaders, the beleaguered knights relied on their defences to survive: outlying defences were overrun, and the castle of Sardewicz was attacked by the Prussians "with instruments of war, arrows and by other means" for five weeks. Christburg was taken by Swantopulk by storm.

The castles created by the Order were clearly vital to the maintenance of their domain, just as the determination of the Prussians to maintain their traditional way of life made Teutonic rule insecure for many decades.[2] Here, as in Livonia, was one of Europe's most massive and sustained castle-building programmes, and the resources mobilised must have been a great and continuing call on the Order's finances. All the sources are from the Christian side, but the view of the native Baltic communities is indicated by their more or less permanent state of resistance to those promising salvation but in fact bringing relentless destruction to everyone who refused conversion, and submission. The inhospitable terrain of eastern Prussia and of the Lithuanian/Livonian marches helped prolong the agony.

[1] N von Holst, *Der Deutsche Ritterorden und seine Bauten*, 1981, 61–4 (Spain), 76 (Hungary). For illustrations of the Knights' many castles, A Winnig, *Der Deutsche Ritterorden und seine Burgen*, 1956. Now also see S Turnbull, *Crusader Castles of the Teutonic Knights*, two vols (2003–4), for superbly illustrated summaries of the castles in (respectively) Prussia, and Latvia and Livonia.

[2] Peter von Dusburg, Chronik der Preussenlandes, ed/trans. Scholz, Wojtecki, in Buchner, Schmale, *AQ*, xxv, 1984, 1–16, 26–42. The author wrote early in the fourteenth century. Długoss, *Annales seu Chronici inditi regni Poloniae*, ed. Dabrowski, iii, 261, 280 (Elbing). Holst, 80–91. *Livländische Reimchronik*, lines 9934–994, talks of the force taken to construct the castle of Heilsburg including two *bliden* and 300 men who were left there.

From 1240, bricks were used to construct the commanderies (a local authority in the Order's structure) at Thorn and Elbing (Elbląg): it is calculated that it took three million bricks to construct such a standardised fortress-headquarters, using bricks of standard size, and that eight ovens would take six years to complete the task. The design, where started anew, was based on the ancient Roman *castellum*, but built as a convent-castle, a quadrangular site enclosing cloister buildings, chapel and towers, whether of earth and timber or of brick. Balga (Veseloje) and Elbing incorporated right-angled turns in their gates, indicating their eastern (or Castilian) roots. Königsberg (Kaliningrad) was founded in 1255, then moved to a new site and provided with "two walls and nine towers of stone". The Prussians once again displayed knowledge of how to take fortifications when they revolted again in 1260. They used three (unnamed) machines against the bishop's castle of Heilsburg (Mamonova), although the garrison abandoned it because of hunger, and they pressed Königsberg closely. In 1263 the defenders also abandoned the castle of Weissenburg after the Prussians blockaded it for three years, constructing three machines with which they pressed a close attack. In an attack by the Prussians and their Lithuanian allies on the castle of Wehlau, they deployed a stone throwing machine, but withdrew disheartened when an arrow fired by the defenders struck a Lithuanian who was standing on the machine to repair it, and pinned his hand to it.[1]

Events in Livonia

Eventually, knights from Riga stormed Memel (Klaipeda) (1252) and established a direct link between the Teutonic Knights in Prussia and the Christian settlers in Livonia (whose "crusade" had been underway since 1185). There had been several-way conflicts involving Scandinavian settlers from Denmark and Sweden, bishops, native Estonians and Livonians and the Russian principalities of Novgorod and Polotsk at the turn of the century.[2] The monk-knights arrived with the establishment of the Livonian Sword-Brothers in 1202, and Bishop Meinhard[3] had set about his mission with the fort on the island of Holm attacked in 1206. From the description, this first construction was of timber, since it was set alight by fire-arrows propelled by crossbows or ballistas. Subsequently, the fort was rebuilt in stone, remains having been found of a simple quadrangle 40 by 34 metres, the walls being more than three metres thick, with two doors protected by large draw-bars, and bordering the river Narva on one side. The castle seems to have been designed as a refuge. In the same year, the bishop's fort at Üxküll (Ikšķile) successfully resisted attack as well: this appears to have comprised a simple tower, eight metres square, and adjacent to the church. It was Livonia's first stone building.

Riga was established by the bishop and the new Order from 1202, the centre of the new town being the bishop's palace and the castle of the knights, built

1 Peter von Dusburg, 72, 94–5, 117–22; Holst, 93.

2 Described in my *History of the Early Medieval Siege*.

3 Actually a monk from Holstein appointed Bishop of Livonia by the empire-building archbishop of Bremen.

according to the quadrangular design. The town wall seems to have been built simultaneously, as the castle stood on its line. Stone-throwers were recorded in an attack on an Estonian fort (Fellin) in 1211. They had never seen such a machine before and agreed both to surrender, and to convert to Christianity. The Order began to build a stone castle on a hill at Segewold in 1210, and completed it within two years. Other stone buildings followed, many of them featuring strong round towers with apparent functions as watchtowers and for storage purposes (as at Treiden, from 1214, and Ascheraden, by 1234).[1]

In 1222, King Valdemar of Denmark began to build a castle on the island of Oesel (tentatively identified by Tuulse as a rectangular ramparted site some 45 by 42 metres). The natives attacked it, and had by now learnt from their foes the use of siege equipment. We are told (by Henry of Livonia) that they constructed 17 "paterells" and bombarded the fort, and because there was no shelter inside, killed many of the occupants who then surrendered. The Oeselians then passed their technical knowledge onto the Estonians, who succeeded in an attack on Dorpat (Tartu), and in 1228 the people of Courland overthrew and destroyed the crusader castle at Dünamunde (Daugavriva), setting a pattern of struggle in Courland lasting years, in which the Knights would seek to establish themselves by building castles and the Cours would resist by destroying them when an opportunity arose. In 1230, the town of Reval was founded. The Danes had thrown up an earthwork on the site of a native fort in 1219, which had been destroyed by the Oeselians. In 1227, the Order took the place from the Danes and began a stone castle with walls, towers and a deep ditch, only for the Danes to reoccupy it and this time maintain their control.[2]

In 1237 the Livonian Order merged with the Teutonic Knights and thereafter a common pattern of castle building developed, the quadrangular convent-style castle becoming common and clearly having a function as an administrative headquarters and residence for the monk-knights. Lacking projecting towers and therefore flanking, from a defensive viewpoint they were out of date, although the sheer scale and strength of the construction, and the provision of outworks, did remedy some of the defensive shortcomings. The only difference between the castles of the two provinces was that because of the nature of the available materials, limestone was preferred to brick in Livonia. Local taxes, gifts from Germany, and peasant labour services had to be mobilised for the building of the Livonian castles, and they tended to be generally smaller than their Prussian counterparts, defence against raids from the south being helped as much by a devastated frontier zone as by stone walls, while relations with the Russians to the east were, for the most part, and at least at the start, friendly, although by mid-century conflict there was also commonplace.[3] However, the chronicles confirm that the role of the Order's castles regularly included sheltering people from the warfare of devas-

[1] Tuulse, *Burgen*, 40–9.

[2] Tuulse, 57–64; Hermanni de Wartberge, Chronicon Livoniae, ed. Strehlke, *Scriptores Rerum Prussicarum*, 1863, II, 29 (a fourteenth century source).

[3] *Chronicle of Henry of Livonia*, ed. Brundage, 88, 104–7, 126–8, 205–7 (Oesel), 215; Hermanni de Wartberge, II, 29–35; Holst, 130–44; A Tuulse, *Die Burgen in Estland und*

tation and abduction that was common to all sides in this grim and interminable conflict, and very rarely did they fail in this purpose.

Some trends are clear even from these first decades of the establishment of the Teutonic Knights in the Baltic. The technical advantages that the newcomers possessed (including access to sea transport) were substantial, and included more advanced weaponry, siege techniques, and the experience of building strong fortifications. Within a short time, the Prussians, Lithuanians, Letts, Oeselians, Estonians and others had learnt how to overrun rapidly-built earthwork and timber defences that differed little from their own, and had mastered the art of constructing siege engines when the target was tougher. But their resources did not stretch to constructing their own stone defences. The local materials were unsuitable, they did not have stonemasons at hand, and their social organisation did not encourage such a process. Their defences were refuges, and they could be built and rebuilt very quickly. Defence of their territory and of their rulers' lordship did not stand or fall with the holding or capture of buildings. Their armies melted into the wastelands when defeated, and did not trust in ramparts. The same resource problem afflicted the newcomers, who had to import the craftsmen to build their fortresses and churches, local labour being useless for anything other than labouring.

Poland under the Piast princelings

The days when Poland had been a powerful and unified kingdom had ended in the twelfth century with the division of the land amongst the princes of the royal family, theoretically meant to recognise Cracow as the senior but in fact rapidly descending into civil strife. The duke-princes in their small provinces had limited resources, so a substantial programme of castle-building was not to be expected, and as the princes continued to hold (again at least in theory) a monopoly of fortification, and the sole right of appointment of the counts and castellans around whose castles or *grod* (compare the Russian and Hungarian systems) local administration was meant to be based, building by the nobility was also constrained.[1] But many monasteries were fortified with ramparts and ditches, presumably with some degree of princely approval for their likely role as refuges. The convent at Zierzyniec, built on a rock by the Vistula in 1160, was fortified after destruction by the Mongol invasion of 1241. The prince specifically allowed the monastery at Grodzisko to erect defences in 1257. Later, in Kazimierz, the Augustinian house formed a donjon on the town wall.[2] Beside the monasteries, the theoretical control of the dukes over their nobles followed the same path as decentralised monarchies had in the west, with increasing feudal relations in the

Lettland, 1942, 22–33; W Urban, The organisation and defence of the Livonian frontier in the thirteenth century, *Speculum*, XLVIII, 1973, 525–31.

[1] N Davies, *God's playground: a history of Poland*, I, 1981, provides a clear and perceptive description of medieval Polish history in English.

[2] S Kolodzieski, 'De Tyniec à Jaroslaw, des études sur les éléments défensives des monastères en Pologne de midi', *Castrum Bene*, 5, 1996, 98–101.

countryside tied in with castle-building by people increasingly low in the chain of nobility. In another throwback to what had happened in Francia in the tenth and eleventh centuries, conical mottes became popular, dating from the thirteenth century. Located at the centre of a lord's estate, these "tower-hills" formed a landowner's residence, but some also explicitly performed military roles if in a frontier region, or less often as a distant watchtower for an older royal castle. A hundred of these sites have been studied by Polish archaeologists.[1]

Castles continued to be the currency of warfare and settlement. Duke Henry of Silesia seized the castrum of Kalisz when he invaded his neighbour's domain (1236). The Pomeranians betrayed a newly built ducal castle to their duke's rival (1244). In 1246, the duke of Silesia invaded Greater Poland and built a new castle on the river Oder, whereupon peace was made and three castles were conveyed to the Silesian. In 1249, a hint is given of what was there when duke Henry's brother was imprisoned in the "tower" of the castle at Legnica, strongly suggesting some kind of donjon. In 1255, the major castle of Nakel, involved in repeated conflict in the early twelfth century, was betrayed to the son of the duke of Pomerania by some of the crossbowmen in it, and had to be besieged by a regular army assembled by the Polish prince. Despite erecting a siege castle against it, he failed in this attempt, but he left the siege castle garrisoned when he withdrew. The year after, while the Poles stormed the Pomeranian castle at Raczens (formerly Reetz in West Prussia), Duke Swantopulk of Pomerania attacked the siege castle facing Nakel. He began by filling the ditch with timber, but the garrison sallied and burnt it. So next he attacked with mantlets and *fundis lapidibus*, but these stone throwers were not enough and the attack failed.

There is enough detail to suggest the central role of castles in the conflicts among the Polish dukes and princes, and that despite their probably not very modern construction, they often proved strong enough to resist attack.[2] The princes had access to modern siege technology, and this chronicle confirms that the duke of Pomerania had stone throwers and other siege equipment. What they lacked was modern defences and resources. Lacking central authority, the country was unable to prevent the establishment of the theocratic state they had invited into the Baltic and its encroachment increasingly on lands traditionally regarded as Polish, or the continuing eastwards advance into Pomerania and Silesia of the German march of Brandenburg.

From Scandinavia to Scotland

The kings of Denmark had been exporting their brick-built castles across the Baltic. Danish royal castle building got underway in a consistent fashion during the twelfth century. The king and the bishops were the chief build-

[1] Summarised by J Zak, 'Mittelalterliche Kegelförmige Burghügel in Polen,' CG, XI, 1983, 289–91.

[2] Annales Capitoli Posnanienis, *MGH SS*, XXIX, 439–59; Annales Sandovigli, *MGH SS*, XXIX, 427–9.

ers, and castles generally featured donjons of baked brick (after the stone supplies had been exhausted). From 1241, the kingdom fell into two decades of civil strife, and royal power did not recover for a long time. Little work was done, therefore, on the existing stone structures, while the sudden appearance of large numbers of mottes, few of which can be dated at all, and certainly few to before the thirteenth century, suggests perhaps that collapse of central government led to lords looking to their own devices, as in Poland at the same time.[1] Danish scholars do not have the evidence to confirm the function of these structures.

In Norway, King Sverre had begun to introduce stone-built castles into a land that had previously not used them. His long-lived successor Hakon IV (1217–63), presided over a period of growing power and prosperity for the kingdom, and continued the trend, enclosing royal palaces with stone curtain walls (Bergen, Trondheim, Tønsberg, Oslo). The castle at Tønsberg was built on a rock of larger size, inside which an inner rectangular fortress was erected. Castles were also built in frontier areas, featuring rectangular designs and a few rectangular towers and, inside, donjons. The Norwegian nobility did not have the resources for such building, but as in Denmark, the church did, the archbishop of Trondheim fortifying his harbour residence, and there are examples of churches being fortified with enclosing walls and corner towers (Gran, thirteenth century).[2]

The Norwegian king's realm included the Orkneys and Shetland, the Isle of Man, much of the Western Isles of Scotland, and the Norse settlements in Iceland. Examples of siege warfare in Hakon's time arose from the conflicts between the Scottish kingdom, itself consolidating under the modernising influence of King Alexander II and his successor Alexander III (1249–86), and the Norwegians, as well as with the local rulers in what is now western Scotland but then and since Viking times, a place of different loyalties. The kingdom of the Scots had been pushing west for some time, and the crown and its nobles were seeking to consolidate their position by founding new castles to serve as centres of power and also the means of defending it. The Clyde estuary had come under Scottish sway in the 1220s, with the Stewarts starting the castle at Dunoon, and royal castles being founded at Tarbert and Eilean Donan — probably simple stone enclosures with a gatehouse. In 1230, the Norwegians sent an army led by Uspak, king of Man, to challenge the Scottish advance. They landed on the island of Bute where they attacked the recently built castle of Rothesay, standing close to the sea. This had a stone curtain ten metres high and three thick, unusually of round shape. The Norsemen launched a direct assault, but the Scottish defenders poured down "boiling pitch", killing many of the attackers. Regrouping, the besiegers provided themselves with wooden shields to protect them from the aerial bombardment,

[1] H Stiesdahl, Types of public and private fortifications in Denmark, in *Danish Medieval History: New Currents*, ed. Skyom-Nielsen, 1981, 211–14; Stiesdahl, 'Die Motten in Danmark', CG, II, 94–9; R A Olsen, 'Danish Medieval Castles at War', CG, IX/X, 223–31; N Liebgott, 'Brick-making and castle-building', CG, XVIII, 109–18.

[2] Ø Ekroll, 'Norwegian Castle-building', CG, XVIII, 67–71; on Bergen, W Douglas Simpson, *The Castle of Bergen and the Bishop's palace, Kirkwall*, 1961. Simpson assigns a substantial place to English influence in the design of the great hall of Bergen.

and began to hew at the masonry "for the stone was soft". After three days they brought the wall down to the ground and captured the castle.[1] Uspak himself died of wounds received when he had been struck by a stone, and a force from Galloway recovered Bute immediately after.

The Norwegian king repeated the attack in 1263, sending a fleet of 70 ships. Rothesay was again assaulted, and this time surrendered. It is recorded that the Norsemen and their islander allies suffered substantial losses, but again the fortress had proven unable to resist an attack launched without the benefit of artillery or other elements of the arsenal of siege warfare. While the Norwegian role was soon ended following Hakon's death at the battle of Largs, the castle at Rothesay was strengthened with the gatehouse containing a portcullis, in place of the simple archway, and four round towers well provided with slits. Other castles in the west of Scotland have been dated to this general period, of which some of the most substantial were Dunstaffnage, seat of the lords of Argyll, Dunollie, and Skipness (Kintyre).[2] The standard features of a thirteenth century castle were demonstrated in western Scotland too, as it was joined, for the first time, to the kingdom. During the century, the feudal forms that had been common in more traditionally feudal countries also became common, with the institution of castle-guard as a duty for royal castles, but by the time of the Norwegian attacks records also show the payment of money wages to crossbowmen and sergeants serving in the garrisons, indicating the inadequacy of the castle-guard system for its purpose. In 1264–6, for example, the earl of Menteith was paid £80 for the service of 120 sergeants at Ayr castle.[3]

England, Wales and France

Wales and the borders

The English kingdom was, for the most part, at peace under Henry III, and the prosperity of the realm was evidenced by the building works including the major castles put up or modernised by order of the king or his greatest barons. The same modernising tendency was present in Wales, in the form of the dominant prince, Llywelyn the Great of Gwynedd. Like the lord Rhys at the end of the previous century, Llywelyn seems to have recognised that with such a powerful neighbour, the Welsh had no choice but to adopt modern practices and unify their scattered tiny principalities — under his rule, of course. Llywelyn, prince from 1201, had played a skilful game during the troubles of King John, and when con-

[1] The Saga of Hacon, in *Icelandic Sagas*, trans. W Dasent, 1894, IV, 153; Chronicle of Lanercost, in A Anderson, *Early Sources of Scottish History*, 1922, 471; for an excellent general introduction, R A Macdonald, *The Kingdom of the Isles: Scotland's western seaboard 1100–1336*, 1997, 84–5, 90; D Pringle, *Rothesay Castle* (guide book), 1995, 2–6.

[2] Saga of Hacon, 350, 395; McDonald, 237–43; Pringle, *Rothesay*, 7; G Stell, *Dunstaffnage and the Castles of Argyll*, 1996.

[3] N H Reid, *Scotland in the reign of Alexander III*, 1990, 133–5, citing the *Exchequer Rolls*, I, 5, 10–12, 18, 24.

flict broke out between Henry III and the successors of William Marshal, earls in Wales of Pembroke and Chepstow, he was not slow to intervene to his advantage.

In 1231 the Welsh prince marched through the borders destroying castles at Radnor, Hay, Brecon, Caerleon, Neath and Kidwelly, then joined up with local forces to storm the key castle at Cardigan, using mangonels to breach the castle walls.[1] In 1233 he besieged Brecon again for a month with *blifyeu a magneleu*, but this time in vain: we must assume that on the earlier occasion, either the castles had been weakly or undefended, or perhaps the account meant that the Welsh burnt and ravaged the towns. But in the same year, Richard Marshal was in revolt and Llywelyn joined with him to take Cardiff, Abergavenny, Penkelly and Blaenllyfni, destroying all of them except Cardiff. The major castle at Carmarthen successfully resisted them for three months. The king's army besieged Marshal's castle at Usk which, lacking provisions, surrendered. The earl marched with his forces to recover it, laid siege to the royal garrison with siege engines, and recaptured it. Usk castle then comprised a recently built but fairly simple stone curtain with four projecting towers and a rectangular donjon, and a simple gate, standing on its high ground above the town. The revolt ended in 1234.[2]

Llywelyn the Great left a more permanent mark on the Welsh landscape than any of his predecessors with the castles he had built. Among the earliest were those at Castell y Bere in mid Wales, and Dolbadarn in Gwynedd. There was a common style to the first two: irregular curtain walls, rectangular two-floor donjons and towers with apsidal ends, or round towers placed without particular defensive purpose, according to the shape of the site. Dolbadarn differed only in having a circular donjon, although it also had an extra storey and a portcullis. The castle at Cricieth, the dating of which has been a cause of much scholarly dispute, may owe its great gatehouse to Llywelyn, which with its strongly projecting D-shaped towers and arrow loops stands in marked contrast to the rest of the inner ward of the two-ward castle, and may date from the 1230s. Llywelyn would have been familiar with the modern twin-towered gatehouses from the new work on the castles of the Marcher lords. Dolwyddelan, also in Gwynedd, was also an early site, being a princely residence. On a rocky outcrop further isolated by a ditch, the original castle comprised a two-storey donjon probably protected by a wooden palisade, the other rectangular tower, the additional storey and the stone curtain all being added rather later.

For a prince with limited resources, these and other castles represented a considerable outlay. It is not known who designed and built the castles, and where the craftsmen were from. They were plainly intended as centres of lordship and, as with any strong stone-built structures held by armed men, they would put up varying degrees of resistance, but they were defensively passive, a trait reinforced by the fact that the greater part of Welsh armies was spearmen, archers and the

[1] *ByT(P)*, 102; *ByT(H)*, 224–31; *ByS*, 229; *Annales Cambriae*, 78.

[2] *ByT(P)*, 102–3; *ByT(H)*, 231–3; *ByS*, 229–31; *Annales Cambriae*, 70; Matthew Paris, Chronica Majora, III, 203, 250–6, 268; Annales Monasterii de Theokesberia, *Annales Monastici*, I, ed. Luard, 1864, 79, 88, 90 (Usk), 92 (Carmarthen); J K Knight, Usk castle and its affinities, in M Apted et al, *Ancient Monuments and their Interpretation*, 1977, 139–52.

crossbowmen so essential to the active defence of a fortress being rare.[1] Such an analysis can support the theory that defensive purposes were of little importance to the owners. What can be said with certainty is that when the English came to conquer Wales for good, they brought siege trains to ensure that the castles could be taken.

After a substantial period of relative peace, warfare broke out again in the Marches when Llywelyn ap Gruffydd, who had succeeded to the princedom in 1256, took advantage of the civil strife that had broken out in England between Henry III and the baronial party led by Simon de Montfort. In 1257 he subdued most of rival Welsh Powys, except for the castle at Welshpool itself, while in the south Abercorram, Llanstephan, Narberth and Maenclochog fell to Prince Rhys, with Llywelyn himself destroying Trefdraeth and all of the country of Rhos except for the castle of Haverford(west) itself, which held out. In 1260, Builth was taken by surprise although Roger Mortimer's castle held out. Two years later, local Welsh forces captured the old, small stone castle on top of the steep ridge at Cefnllys in Maelyienydd. Roger Mortimer and Humphrey de Bohun assembled a Marcher army to reoccupy the site and began to build a new stone castle. Llywelyn descended on the place with a large Welsh army, dug himself in to surround it, and forced the would-be constructors to surrender (they were allowed to leave unharmed) through hunger.[2] In 1263, as the struggle in England reached its climax, he seized the chance to attack the castles of Dyserth and Deganwy and reduce them to the state of ruin in which they now stand. Dyserth was left such that "no stone stood upon another", Deganwy fell not by assault but through the "bitterness of hunger".[3] However, it is also clear that the prince of Wales had a siege train and was quite able to capture defended castles and towns when he had to.

France

The English king's conviction that he had a right to his father's and grandfather's inheritance across the channel involved inevitable conflict with the new French king Louis IX. Before the latter embarked on his crusading adventures, he had first to secure his own kingdom against Henry III's hostility, and to consolidate the even more recent extension of royal power in the south, obtained on the ashes of the Albigensian heretics. In 1226, Louis VIII, in the last year of his reign, besieged Avignon for three months. The count of St. Pol was killed by a stone from a catapult, alongside some 2,000 royal troops who died in the hard-fought siege. The royal army also deployed trebuchets and petraries, and from the sources they seem

[1] For the castles, P Davis, *Castles of the Welsh Princes*; R Avent, *Castles of the Princes of Gwynedd*; R Avent, *Dowyddelan Castle, Dolbardarn Castle, Castell y Bere* (2004) and the same, *Criccieth Castle* (1998), for the most recent guide books.

[2] *Annales Cambriae*, 100; *ByS*, 245; *ByT(P)*, 112; *ByT(H)*, 251. P Remfry, *Castles of Radnorshire*, 1996, 95–8.

[3] *Annales Cambriae*, 101; *ByS*, 247; *ByT(P)*, 113; *ByT(H)*, 253.

to have been effective in demolishing walls and towers. One source also mentions a siege tower as high as the walls, and the use of incendiaries mixed with sulphur.[1]

There was a major English expedition in 1242, directed at Poitou, where the count of La Marche revolted. Louis ordered his engineers to build mangonels and trebuchets as part of an army said to be of 4,000 knights and 20,000 sergeants and crossbowmen: very large numbers, but possible. The Poitevins and Gascons attempted to delay the counter-attack by felling trees and filling in the wells. But Louis' force was too strong. He arrayed his siege train against the castle of Montreuil and the tower of Bergue, both of which fell and were demolished. At Geoffrey de Lusignan's castle of Fontenay, protected in modern style with double walls and high towers, the French army erected a high wooden tower, despite efforts by the defenders to burn it (with Greek Fire according to one chronicle). After "hammering" (Matthew Paris) the place with artillery for fifteen days, they took it by escalade. The garrison had comprised 41 knights and 24 sergeants, who surrendered. Other castles fell too, and miners were used to destroy the walls. Henry III and the English army fled from an encounter with their foes outside the castle of Taillebourg, abandoning their supplies and engines, leading to the surrender of the castle, and then of the city of Saintes.[2] Thoroughly worsted, the English king made a truce with Louis and gave up the expedition, while the local rebels had already, prudently, made their peace. The events confirmed that modern siege technology, backing the large resources available to a centralised kingdom, could overcome the most modern defences, even with a hostile army in the field.

About this time, the last survivors of the Albigensians were being rooted out in the south, and the last efforts of the previous rulers to recover their patrimony defeated. In 1240 Raymond Trencavel, son of the last viscount, entered what had once been his city of Carcassonne. Now, he confronted a royal garrison in the citadel, the old walled city, whose gallo-roman walls had been reinforced by the construction of an outer wall and barbicans. They had to hold on while help was sent by King Louis. The seneschal left a detailed account in a letter to the queen.[3] Raymond's attempt to set up a mangonel against the barbican was thwarted by accurate shooting from the defenders' "very good" Turkish mangonel. The besiegers were able to use the existing ditches and new defences to isolate the citadel, but could not prevent the garrison aiming their stone thrower at them with great effect. So they turned to mining. A mine under the barbican was detected, heard by the defenders. They could not prevent the tunnel being fired, but they were able to build an inner defence of stones. Another mine was driven under a tower,

[1] *Chronique de Guillaume de Puylaurens*, ed. Guizot, 78–9, and see chapter eight; *Grandes Chroniques de France conservés en l'église de St Denis*, 1836, I, 947–8; Guillaume de Nangis, *Chronique Latine*, ed. Gerard, 1843, I, 175; Nicolai de Braia, Carmine de gestis Ludovici VIII regis, *MGH SS*, XXVI, lines 1570–1818: a verse epic modelled on Guillaume le Breton, talks of the siege tower. He was a contemporary, but the "heroic" nature of his account may mean some invention has crept in.

[2] Guillaume de Nangis, 194–5; Matthew Paris, IV, 188–9, 195, 202–21; *Grandes Chroniques*, II, 970 (Champagne), 980–3; Joinville, 188–9 (with little detail of the fighting).

[3] William of Orme's letter is translated in full in J-P Panouille, *Carcassonne in the days of siege*, 1992, 68–71.

which the defence thwarted by countermining and driving them out. Other mines (there were no fewer than seven altogether), started much further back and dug deeper so as to avoid early detection, succeeded in bringing down lengths of the outer wall, but in every case the defenders had detected the digging as it reached the wall, and had just sufficient time to erect inner palisades. The new temporary defences, it was noted, were well provided with arrow loops, and prevented the besiegers from approaching the breach. Several direct assaults were made against the barbican and elsewhere, but each was driven back by the crossbows of the defenders. Hearing of the approach of the relieving royal army, the besiegers melted away. From the account, admittedly one-sided, it may seem that Raymond did not have many experienced artillerists in his service, but plenty of miners.

The new rulers in the south made sure of their position by building powerful fortifications. In Albi the bishop's palace (palais de la Berbie) had been started in 1240 with a hall on the cliff above the river Tarn, and from 1254 took further shape with a gigantic donjon, like the rest of the place all of brick (now restored), with solid half-found turrets acting as buttresses, machicolations and high curtain walls.[1] The complex is seen as a forerunner of the powerful fortress-palaces that reached their peak at Avignon fifty years later. Next to it, the cathedral of St Cécile, begun in the 1280s, was also fortified, and represented an unmistakable affirmation of catholic and royal power against the city below them.

The Reconquista continues: James the Conqueror

One reason for the lack of support provided to the Trencavels to recover their former lands by their hosts in Aragon was that the Spanish kingdoms were involved in continued and successful offensive action against a much weakened al Andalus. In the wake of defeat, the centralised hold of the Almohads collapsed and local dynasties took over. They were unable to mount any effective challenge to the reconquest, instead passively awaiting the next attack. While the Castilian monarchs advanced to the south, the crown of Aragon conquered the Mediterranean coastline and the Balearic islands, and left in the chronicle written by King James himself a first-hand account indispensable for the evidence of siege techniques of the time, even if one might wonder if its author overstated his own direct role.[2]

James succeeded to the crown of Aragon and county of Barcelona at a very young age in 1213 and had first to secure his position against local baronial resistance. He was compelled from childhood to witness, and master, siege warfare skills. The castle of Alvaro surrendered after two days' attack using a *foneval* (Catalan for *fundibalus*, a generic term for lever artillery). The same weapon was then set up against the castle of Licana, and recorded a remarkable rate of shoot-

[1] Mesqui, *Châteaux forts*, 16–17, for a brief description.

[2] *Chronicle of James I King of Aragon*, trans. J Forster, two vols, 1883; also see P E Chevedden, The artillery of King James I the Conqueror, in Chevedden, Kagay, Podilla, *Iberia and the Mediterranean World*, 1996, II.

ing: 500 overnight, 1000 during the day. The result was the smashing of the walls and storm of the castle. At the tougher target of Albarracin, a walled town whose garrison included 150 knights, the Aragonese king (then just 11 years old) deployed a mangonel and a *fonevol*, but the siege was abandoned after two months because of the treachery of many of James' men, who allowed supplies into the place. Albarracin was built on a high, narrow ridge and the strength of its location, and its geographical isolation, had deterred many attackers, rather than the strength of the walls.[1]

Mallorca

In 1229, a major expedition was launched against the Balearic islands, carried in hundreds of ships. The ships carried "many engines and skilled artificers" (Desclot). Mallorca was the primary target, the base of the Muslim king of the islands. Coming ashore safely in Palomera (7 September), and defeating the defending army, James approached the city, and sent for a trebuchet and a mangonel, and prepared two other trebuchets and two *algarradas*. The sailors from Marseilles offered to manufacture a trebuchet "in our fashion", using their spars and altogether no fewer than 20 engines were used in the Aragonese camp, while the defenders also used two trebuchets and 14 other stone throwers. The king said that one of these was "the best he had ever seen", shooting over the camp and beyond six rows of tents, but his trebuchet reached even further. Other siege equipment included several strong timber mantlets on wheels with earth further protecting the roof, beneath the shelter of which mining was undertaken once it had been pushed up to the edge of the moat. Knights themselves carried the stones for the artillery, which destroyed two of the defenders'. The camp was well protected by its own ramparts and a moat, just as well since substantial Muslim forces remained outside the city. They tried to cut off the besiegers' water supply, but were heavily defeated, and their dead leader's head was cut off and catapulted into the city. The mining caused most damage, bringing down the outer walls and collapsing several towers, despite efforts by the defenders to countermine. On one occasion it had been the opening of an air hole that had revealed the location to the defenders, and allowed them to countermine effectively. The siege was seriously hindered by appalling weather, with rainwater filling the ditches and mud preventing the movement of equipment. But they persevered. During December, the outer wall was brought down, mobile wooden defences were brought up to the ditch and crossbowmen cleared the walls opposite of defenders. Recognising the position as hopeless, the Muslim ruler offered to surrender, but asked for good terms. The besiegers decided that they were so close to victory, as the defenders were so weakened, that they could reject the terms: they really wanted the right to sack the city, and in a dawn assault on 31 December they stormed in. Much of the

[1] *Chronicle of James I*, 26–31; Almagro Goubea, El sistema defensiva de Albarracin, *Arquelogia Medieval Española*, II Congreso 1987, II, 71–81, plan at 76.

population fled to the countryside but the king and his family were captured and the island passed under Aragonese/Catalan rule.[1]

In a sensible move, the right of those Muslims who wished to remain on the island was recognised, helping no doubt to ensure that Minorca surrendered without resistance, although the conqueror had to take steps to secure his new possessions, the risk of a counterattack from Tunis (with which the previous ruler had been connected) being ever present. Ibiza remained Muslim until 1235, when the archbishop of Tarragona landed unopposed on the island, and approached the city. He deployed mangonels, which were "less powerful", against the town, and trebuchets against the citadel. The collapse of the outer wall under bombardment enabled a second phase, of mining, to begin, whereupon surrender followed.[2]

Valencia, Seville, Murcia

Fresh from this triumph, James turned his attention to his southern neighbour, Valencia, and embarked on a strategically thought-out plan of progressive conquest. The first significant target was Burriana, in 1233, where once again trebuchets and mangonels were constructed for bombardment. The king's account tells that Nicoloso, who had built the trebuchet at Mallorca, offered to construct a wheeled siege tower. The king told that his council had never heard of such a thing, but eventually agreed, so the engineer was given the materials and men for the undertaking. The tower's design was in the end rendered useless, because although great efforts were made to provide a smooth approach to the walls, with wooden sleepers, greased ropes and pulleys, and knights with their shields covering the teams of men pulling on them, the great tower became stuck half way. Naturally, the defenders had moved a large battery of their stone-throwers into position to bombard it, and it was now wrecked beyond repair, as it could not be pulled back out of range. The obviously frustrated James noted "thereafter we would not employ again that device", but it is also interesting that a weapon that had played so major a role in the Middle East was now unknown to the Aragonese barons: the portable structures used to enable crossbowmen to get close to the walls of Mallorca had clearly been a different kind of engine. The siege continued with mining into the ditch, and the demolition of a tower by the mangonels, whereupon the Muslim commander surrendered on generous terms after a siege of two months.[3]

Meanwhile, James' Castilian neighbours were also pushing forward their boundaries. The important castle of Salvatierra had been captured by King Fernando in 1226, "with little labour", and he had then taken the castle of Capilla, although it was well supplied with men, provisions and machines of war, when

[1] *Chronicle of James I*, I, 112–73. There is much complementary detail in the non-contemporary Desclot, *Chronicle of the reign of King Pedro III of Aragon*, trans. Critchlow, 1928, I, 87–120. Also noted in the *Cronica Latina de los reyes de Castilla*, ed. Brea, 1999, 79, the victory was noticed across Europe, including by Matthew Paris, *Chronica Majora*, III, 305.

[2] *Chronicle of James I*, 219–21.

[3] *Chronicle of James I*, 259–83.

relief had not come from Seville. Further west, Elvas opened its gates to the king of Portugal. A siege of Jaén in 1230 failed because of the large size of the garrison, but in 1233 Ubeda surrendered to Fernando, while the master of the order of Calatrava took the surrender of Trujillo. A succession of further captures followed, although for the most part the sources offer no details of the event to enable a reconstruction of anything other than the outcome.[1] In 1236, Córdoba, formerly the centre of one of the great civilisations of its age, fell to the Castilians. Clearly, there was no stomach for resistance. A force sent by the king of Granada was blocked at Ecija. Some Christians in the city had secretly contacted Fernando offering to assist an attack, and entry was effected by ladder, with the sleeping guards being overwhelmed and the gates opened. The "mother of all cities of al Andalus", with walls "of sublime height" had ceased to be a Muslim city, and although its central role had long before moved elsewhere, its loss must have been a devastating moral and psychological blow.[2]

The fall of Córdoba was soon followed by the loss of Valencia. In 1238, James laid siege to the seaside city, pressing it very closely from the suburbs outside the walls. The king described how he decided where to deploy the artillery such that it could harm the enemy, but was at sufficient distance that any attempt to disrupt the bombardment would leave the sortie very vulnerable. To help, a trebuchet and two mangonels came by sea from Tortosa. Mantlets were pushed up to the edge of the wet moat and materials were thrown in to allow attackers to cross and storm an outwork. Elsewhere around the circuit, two of the king's barons, without his permission, launched an attack on one of the towers and were repelled. Recognising the psychological impact of this reverse, James switched his assault to this point, and eventually stormed the tower. For a month the town was battered, but was strongly defended. Hunger began to set in, and perhaps the key moment was the arrival of a fleet from Tunis which, observing the strength of James' dispositions, decided not to attempt to land, and sailed away. Valencia, now without hope of relief, negotiated surrender in September.[3] The conquests continued in the years following this significant victory for the Aragonese. In 1244, the hill top castle of Jativa south of Valencia was besieged for six months. The skills of the besiegers made no impact on the defences, but this was because of the site, the (now restored) castle being comprised of a single towered curtain, but in the end the place was taken by surrender on generous terms.[4]

The culmination of the Christian monarchs' drive was the conquest of Seville itself, while the Aragonese were devastating Murcia. Seville proved to be a very long and difficult siege lasting nearly a year and a half. It involved the deployment of a fleet on the river and the armies of the Spanish kingdoms and the military orders. The Muslim chronicles record that the attackers having ravaged the countryside and surrounded the city, it was bombarded by artillery, but that hunger

[1] *Cronica Latina de los reyes de Castilla*, 70–77, 83, 87; *Annales Toledanos*, II, 409–10.

[2] *Cronica Latina de los reyes de Castilla*, 98–101; *Primera Cronica General de España*, II, 729–40; *Annales Toledanos*, II, 409–10.

[3] *Chronicle of James I*, I, 368–84; Desclot, *Pedro III*, I, 132–6.

[4] Desclot, I, 139.

was the cause of surrender, with the Almohads of North Africa doing nothing to attempt relief, being entirely occupied with matters close to home as a new dynasty, the Hafsids, took power in Tunis.[1] At Seville, there was a vigorous defence, and the besiegers had to get across the rivers on which the city stood, and face attempts by the Muslim troops to resist them outside the walls. A bridge of boats across the river Guadalquivir connected the city with its suburb of Triana. In May 1248, the Castilians tried to break through it, despite a strong response from mangonels, and trebuchets operating from the Torre d'Oro (the polygonal *torre albarrana* still standing by the river, although then it must have had a flat roof), and one ship achieved this. Fernando then concentrated his attack on Triana, trying to break through by mining, but the defenders located the mine and destroyed it. Next, artillery was set up, but failed to breach the strong walls. So the king, occupying the river with his ships, left hunger to do its work on both sides of the water until surrender was negotiated on 13 January 1249.[2]

Murcia finally capitulated to James in 1266, through hunger, when the people ejected the king of Granada's governor,[3] leaving the Muslims restricted to the kingdom of Granada. The Nasrid ruler, al Abmar, secured the frontier zones of his still substantial and wealthy if much smaller kingdom, with natural mountain defences, and the occupation and if necessary reconstruction of many old castles. It has been noted that this time, the Muslims modernised by borrowing from the style of the Christian builders to the north, as at Alcalá la Real.[4]

Codification of castle customs and siege technology

Across Europe during the thirteenth century, consolidation of royal power continued. The centralisation of mechanisms of government was strongest in Capetian France and Plantagenet England, but also notable in the strengthened monarchies of Castile, Aragon, and the Norman and then Hohenstaufen realms of southern Italy. Surviving from this time are a number of writings reflecting the concerns of the new monarchies to validate their position as against those who would assert noble privilege or communal liberties. The genre of political tracts now appearing were designed to establish a firm theoretical support for the rulers. In the process, they necessarily made reference to the position of fortifications.

Ptolemy of Lucca wrote his *De regimine Principum* in the late thirteenth century. Classical allusions abounded. He quoted Vitruvius and Vegetius on where to site cities and castles. Kings needed an abundance of wealth to repair and establish

[1] Zarkechi, *Chronique des Almohades et des Hafcides*, 29–38, records how the new regime took power in 1229, strengthened the walls of Tunis, captured Constantine, Bougie and Mahdia (al Mahdiyya), and in 1242 with a great army forced entry into Tlemcen.

[2] *Primera Cronica General*, 750–68; Ibn Idhari, reprinted in Melville and Ubayydi, III, 145–9.

[3] *Chronicle of James I*, I, 560–3; Desclot, 185–7.

[4] A Bazzana, Forteresses du royaume nasride de Grenade, CG, XI, 1982, 29–43.

defences to prevent enemy invasions. Biblical King David was called in evidence for the need to built fortifications. Their roles were laid out: being a safe place for the ruler, preserving the ruler's reputation, inspiring obedience, defence against attack, and preserving the treasury. He pointed out that diggers, carpenters and stone cutters needed to be under military discipline because of their critical role in erecting defences. Clearly, for Ptolemy, there was no distinction between the defensive, administrative and symbolic purposes of fortifications.[1]

Ptolemy used as an authority the power of the king of Castile, in whose country the nobility had "no forts other than by the king's will", and this constitutional state of affairs was confirmed in the compilation of Castilian law known as the *Siete Partidas*, ordered by King Alfonso x in the second half of the century. These laws included detailed references to the duties of the people to defend the country's fortresses, which belonged to the crown, and assigned specific responsibilities to people to ensure that fortifications were guarded and provisioned: loss of the castle through neglect amounted to treason. It recognised that much of the time, only a porter would be employed there. Royal *alcaides* (governors) were specifically charged with ensuring that their castles were provided with watchmen and crossbowmen, victuals, hand-mills, timber and coals for ovens, etc. Their duties included patrolling the ground outside the walls, to prevent surprise attacks, and in the event of a siege, the commanders were called upon to ensure that the garrison was not dismayed by the enemy preparing engines or digging mines. Much attention was paid to the respective rights and duties of lords who held castles in fiefs, and of the occasions when the king was entitled to take them back into his own hands. In a list of the weaponry that must be stored in royal arsenals, especially in those near the frontiers, alongside crossbows, bows, slings, mantlets and cats and many iron tools, were "engines which throw stones by counterweight". The trebuchet clearly deserved a specific mention.[2]

The conditions prevailing in Germany in the thirteenth century were reflected in the compilation of customary laws known as the Saxon Mirror (*Sachsenspiegel*) of Eike von Repgow. No one, it said, may erect a castle, walls or towers in a town or village without the authority of a judge, while even such local authority was not required to build a structure with no more than two stories above ground, with a palisade. Crenellations were not however permitted. The restriction is similar to those stated many years before by the *Leges Henrici Primi* in the Anglo-Norman kingdom. If a castle was condemned — and the laws state that the structure itself was a legal entity that could be found guilty if it harboured thieves or stolen goods — then peasants may be called to level the ditch and ramparts. Substantial detail was given of the rights and duties of castles held as fiefs from another lord, including the customary duty of castle-guard. Frederick ii, for all his centralisation in his Sicilian kingdom, had been forced to grant virtual autonomy to his German

[1] Ptolemy of Lucca, *On the government of rulers. De regimine Principum*, trans. J M Blythe, 1997, 104–8, 119, 129–30, 286.

[2] Ptolemy of Lucca, 212; *Las Siete Partidas del rey Don Alfonso el Sabio*, ed. Lopez, 1847, 2nd partida, 169–70, 173–5, 179–82, 197 f, 226–7, 277.

princely subjects, but it would seem that localisation of power including control of castles had already become an established fact long before it was codified.[1]

The writing down of laws coincided with a rebirth (in the west) of the practise of writing down advice for princes, something that had a long history in Byzantium and the Muslim world. Such texts often contained advice on siege warfare, usually appealing to the classical tradition of Vegetius and Vitruvius as authority. If they may be suspect in terms of their actual influence, it is nonetheless apparent that they must reflect to some degree what their authors saw as current best practise. In Spain, the *Poridat de las Poridades* from mid-century stressed the importance of having artillery to hurl great stones to break walls, and also the value of large wall-mounted crossbows (or ballistas) called *ballestas de torno* against which there was no defence.[2]

Giles of Rome addressed his advice to the son of Philip IV of France (1270–85), and covered siege matters in detail. His authorities included, again, Vegetius, but his advice was based clearly on current methods. Stressing that the best time to carry out a siege was in the summer, before the harvest was collected, he advised the prince to try to divert water courses, not in the way of the Mongols to assist a direct attack, but to cut off the supply to the besieged. If launching an attack, he recommended digging entrenchments to protect oneself, assailing the walls with ballistas, bows and stone throwers. Scaling ladders were discussed, but the best method was mining. Where the site of the fortress made mining impracticable, the chief attacking role was taken by stone throwing machines. Of these, the largest was the trebuchet, operated by a counterweight. By the later thirteenth century, there were different varieties of the machine: the counterweight could be fixed, mobile or composite. It should be made up of sand, stones, lead or other heavy material. This made for uniform throwing. The range could be altered by changing the length of the sling or the weight of the missile. The second type was called *biffa*, the third *tripontum*, while the last of the four types of stone thrower was propelled by hand-pulled ropes rather than a counterweight. Thus for Giles, the old manually-propelled stone thrower was the least of the types of artillery. The writer provided much practical advice on other siege matters, too. The use of rams and cats and other types of protective screens was carefully explained. The height of a wall should be calculated by using the shadow it cast, a rather less complex approach than that of the Byzantine theorists. In defence, incendiaries hurled by stone throwers or ballistas was recommended as the best way to resist attack by machines.[3] This tract gives a very good picture of the accepted best practice of siege warfare of the thirteenth century.

The inclusion of the Scandinavian realms in the common world of European states under Sverri and Hakon also found an echo in the production of advice to princes, with the *Speculum Regale*, probably written by a clerical courtier around

[1] Repgow, *The Saxon Mirror*, trans. M Dobozy, 1999, 114–8, 133–4. The law of Frederick II, Constitutio in favorem Principum, is in *MGH Legum*, II, ed. Weiland, 1876, 212.

[2] Seudo Aristoteles, *Poridat de las poridades*, ed. L A Kasten, 1957, 57.

[3] Aegidius Romani, Libellus de re militari veterum auctores praesertim medii aevi, ed. S Hahn, *Collectio Monumentorum*, 1724, I, 21–3, 44–63.

1240. In chapters advising on how to attack and defend castles, the king was told he would need trebuchets: powerful ones to hurl large stones against the wall, to determine whether they could resist such "violent blows", and smaller ones to hurl missiles over the walls to demolish the buildings within. If the wall was too strong, iron-headed battering rams should be used, and cats and siege towers were also mentioned. In defence, the wall should be reinforced with earth and timber against the bombardment, while mines could be used by the defender to get under and destroy the engines. Reference to a device that "breathes forth flames and fire" suggests the writer had seen, or heard about, Greek Fire.[1] The accounts of Norwegian siege warfare gave little hint that any of this armoury was actually available or deployed at this time, but it would have been very little needed against the walls it confronted. It is however apparent that the writer had seen or heard about the equipment, and may have been informed about it by Norwegians returning from overseas (such as crusaders).

The tradition of Muslim writing on poliorcetics also continued, although many texts have not survived. That of Sayh al-Harawi, written sometime between 1192 and 1215, stressed the importance of paying the wages of the workmen, the masons, the ballista operators, the naphtha throwers and sappers. In a list of weapons to be held, crossbows and the large mechanical versions held pride of place, alongside sapping machines, grapnels, timber to build ballistas and mangonels, hemp cords (etc). The generic term for the artillery does not help determine whether the trebuchet had arrived at the time of writing. The advice on attacking fortresses was commonplace, and reflected the real practice of the Muslim commanders, so it is entirely possible that their practical experience had been based on some learning from the books as part of their princely education.[2]

Conclusions

Asia, and large parts of the Middle East, had been transformed by the Mongol conquests, although after 1260 they would no longer be invincible. They had absorbed the highest levels of siege warfare skills in very little time, profiting from the sophistication of the civilisations (Chinese and Muslim) that they had overwhelmed. Elsewhere, the Christian and Muslim powers had continued to fight each other, and amongst themselves, as if nothing had happened. Here, as has been shown, trends begun in previous decades were deepened, including the construction by rulers (or religious orders) of unprecedentedly strong fortifications. Their attack or defence became less of an issue in their design in many cases as governmental power was consolidated, and their other roles as centres of administration, residences and symbolic statements were enhanced. But that said, sufficient examples have been provided to show that castles old and new could still be called upon as defences, while walled towns also continued to feature

[1] *The King's Mirror — Speculum Regale, Konungs Skuggsa*, trans. M Larson, 1917, 220–26.

[2] J Sourdet-Thomine, 'Les conseils du Sayh Al-Harawi à un prince ayyubide', *Bulletin d'Études Orientales*, XVII, 1962, 227–8, 234–6.

in siege warfare, as invaders understood the significance of conquering centres of trade and craft. The effect of the trebuchet seems to have continued to influence defensive thinking in many sites.

Alongside, the process of consolidation of royal government noticed before continued, the codification of laws and practices discussed here representing a formalisation of existing practice and ideas, including restating traditional laws of royal authority over (some kinds of) fortifications. Early forms of nation-state were now were more clearly marked in England, France, Spain and Scandinavia. In Germany, this took the form of multiple lesser powers consolidating their own realms. By the time of Frederick II, the imperial dream was already just that.

2 Attack and defence in the late thirteenth century (ca. 1260–1320)

The period between the middle of the thirteenth century and the middle of the fourteenth centuries represented the end of a long period of continuous expansion of wealth and population. In western and central Europe, the fourteenth century was one of calamity, with the arrival of the Black Death preceded by years of awful famine. Massive social change, and beginnings of challenges to the social order, would be among the consequences.

In 1260 this all lay ahead. The Byzantines were about to recover Constantinople from the Latin emperors, but not to recover the glory of the east Roman empire. The Staufen dynasty was about to be extinguished. North of the Alps a long interregnum in the German kingdom would consolidate the already substantial independent power of numerous princes, bishops and city-states and leave the empire as a token of great esteem but only such power as the individual ruler's own resources would permit. In contrast, the French kingdom under St Louis continued as a rich and powerfully centralised state, soon to host the popes in Avignon. The Plantagenets found in Edward I a king to extinguish the independence of Wales, and to come very close to doing so in Scotland. In eastern Europe, a strong new political force began to emerge in the still pagan state of Lithuania, but the dead hand of Mongol overlordship still lay over most of the Russian lands. At the other end of the Asian continent, Mongol rule was consolidated in China, but resisted in Japan, and in these struggles is found the first evidence for the military use of gunpowder.

In Iberia, the Muslim kings of Granada consolidated their small realm and continued to relate to the power struggles that characterised the emerging dynasties of North Africa. Many of the crusades launched during this age were to be directed against the papacy's Christian enemies. The new Mamluk rulers of Egypt and Syria were about to remove the Latins from the Middle East altogether, a struggle that tested their ability to overcome massive modern fortifications to the full. Even the chastened Muslim sultans of Rum were able to exercise their power, and in about 1265 mobilised a fleet carrying mangonels, under cover of which they landed to force their way into the Black Sea city of Sinope, taken back from them by the Byzantine Manuel I of Trebizond some eleven years before.[1]

[1] M Nystazapoulou, 'La dernière réconquête de Sinope par les Grecs de Trebizonde (1264–5)', *Revue des Études Byzantines*, 22, 1964, 241–9, reconstructs these events from fragmentary references in Greek and Arabic chronicles. It had been continued Mongol

The end of Outremer

The crusader states had always been aliens in someone else's country, and shortage of manpower was a critical problem. Genoa, Pisa, and Venice were always well represented in the ports, because of their trading interests, but otherwise the rulers relied on transient crusaders for sufficient numbers ever to take the offensive, and for the sustenance of their defences. The military orders, especially the Templars and Hospitallers, held the great castles, especially those inland, and took responsibility for their upkeep and modernisation, and their garrisons, which enabled them to control the surrounding country and its produce. While the coastal cities still held by the Christians were strongly fortified, sometimes with the help of crusaders like Louis IX, the monk-knights holding the inland castles blocked the way to the coast. The greater numbers and resources of the Mamluk rulers could only be brought to bear through a slow process of reducing these mighty strongholds.

Sultan Baybars

The first concern of the victor of the battle of 'Ain Jalūt was with the Mongols, but another attack on Egypt from overseas could not be ignored. Baybars immediately looked to restoring his defences, reconstructing the fortress of the island of Rawda, then ordering the rebuilding of a long line of fortresses that had succumbed to the Mongol onslaught on Syria, including Baalbek, Shaizar and Bosra. Ditches were cleared out, the walls were broadened, equipment and garrisons were put inside. A diploma from the caliph requested him to "make the care of these fortresses your principal concern".[1] The continued Mongol threat would be partly offset by the relations that the sultans were able to establish with the Golden Horde when their khan Berke converted to Islam in 1263, but conflict recurred often with the Persian horde: in 1262, they attacked Mosul with a battery of 25 siege engines, and the city, without the means of defence and lacking provisions, fell and suffered a dreadful sack and massacre lasting nine days.[2]

But from the outset, the sultan viewed his main task as being to send the Latins back over the seas from whence they had come. The Arabic sources, although not contemporary, are based on first hand evidence and provide unusually rich detail of the many sieges offering a first rate guide to the available technology and methods.

First, he had to secure the routes towards the coast by taking the remaining

intervention in the area that had allowed the Byzantines to exploit a dynastic crisis in Iconium (Konya) to retake the city, originally recovered from the Latin empire in 1205 then lost to Iconium in 1214.

[1] Ibn 'Abd az-Zahir, Sirat al Malik az-Zahir (Life of Baybars), in S F Sadeque, *Baybars I of Egypt*, 1956, 115, 117, 130.

[2] Makrizi, I, 181. The contemporary Arabic sources for this period can be found summarised in Michaud, *Bibliothèque des Croisades*, 1829.

Latin strongholds in the south. Then in 1263 he launched a probing attack towards Acre. Although it appears that he initiated an attempt at mining, he pulled back, and instead directed his army towards Ayyubid-held Kerak castle. The opportunistic nature of his attack on Acre was now made clear, since he did not have a siege train with him: from Kerak he summoned from Egypt and Damascus stone-cutters, masons and carpenters who began to prepare an arsenal of equipment to take the beleaguered town. The sultan invited the (Muslim) occupants to surrender on good terms, and they leapt at the chance. Baybars wanted Kerak as a forward base of his own. He installed a new garrison in the citadel and stored arms and equipment there, and to improve its defences he had the ditch deepened.[1]

1265 saw crucial developments on both fronts for Baybars. Hearing that a new Mongol attack was taking place against al-Bira, the sultan, at Gaza, mobilised his forces for its relief. His preparations had served well: the town was now solidly protected and well supplied, and the besiegers departed, abandoning their 17 stone throwers. Ever ready for sieges, the Mongols had also prepared iron pegs with which they had planned to scale the walls. The sultan meanwhile ordered his emirs to construct both Maghribi and Frankish mangonels, for which the timbers were already cut and prepared, and four large machines were constructed. Other machines, masons and labourers were summoned, while soldiers constructed scaling ladders. His biography says that he required his troops to maintain complete equipment including tools for stone cutting and carpentry. Baybars descended suddenly on Caesarea on 27 February, encircled it, and launched an immediate assault. His troops used iron pegs and other paraphernalia to allow them to get across the ditch, while others succeeded in burning the wooden gates. The inhabitants, taken by surprise, retreated to the citadel, against which Baybars now deployed the mangonels and ordered a siege tower erected (the only time this is recorded in this period). The massive walls held together by granite columns protected effectively against mines, but the defenders of the town so expensively and recently fortified by Louis IX surrendered, and the place was destroyed.[2]

He arrived at Arsuf (Arsaaf) on 3 March. The castle standing above the town here had been built by John d'Ibelin as recently as 1241, and had a pentagonal shape backed against the sea. The inner ward was protected by four D-shaped towers (two of them creating a gatehouse similar to the style of contemporary western Europe), with strong walls backing onto an integral donjon. Outside was a concentric outer wall comprising three half-round towers. In front of this was a dry ditch up to 30 metres wide and 16 deep.[3] Baybars ordered two ditches dug, lined with timber, to cover the approach to the citadel. This became the focus of the fighting, with the garrison first trying to burn the woodwork, then seeking to

[1] *Life of Baybars*, 172–9; Makrizi, 199, 205–8; *L'Estoire de Eracles, continuation de l'Estoire de Guillaume de Sur*, 447.

[2] *Life of Baybars*, 238–9; Makrizi, II, 6–7; U and M Lyons, *Ayyubids, Mamluks and Crusaders. Selections from the Tarīkh al-Duwal wa'l Malik of Ibn Furāt*, 1971, II, 69–71; Dedeyan, *Smbat*, 114 (al-Bira).

[3] From the excavations reported by I Roll, B Arubas, 'Le Château d'Arsur: forteresse côtiere pentagonale du type concentrique du milieu du XIIIe siècle', *BM*, 164, 2006, 67–79.

drive a counter-mine under the approaches, which involved tunnelling under their moat. They achieved this feat, and fired the mine using bellows. Thwarted, Baybars ordered a new start, and this time used his trenches to provide cover for his troops bringing material to fill in the moat. Meanwhile, one Karmun Aghe built an engine that shot seven arrows together, obviously a multiple ballista. Eventually, the bombardment from the stone throwers proved decisive, creating breaches in the wall and allowing the attackers to storm across the ditch and capture the town. The defenders of the citadel surrendered at once for their lives. They had not been short of men: Christian sources confirm that there had been 90 Hospitaller knights along with 1000 other troops.[1] The range of skills deployed here demonstrated that even strong defences and willing defenders would fall to a skilful besieger provided with modern siege skills and weapons. The stone throwers responsible for the breaches must, presumably, have been trebuchets and may be the same "four large machines" recorded earlier.

In 1266 the sultan attacked Saphet, arriving on 13 June. The siege equipment came from the store at Damascus, whence it was brought by hand and by oxen, since the camels could not manage the mangonels, arriving on 26 June. The accounts tell of massive bombardment from the artillery, artificers hurling naphtha bombs, and mining, and it was the latter that was decisive in breaching the outer wall despite attempts at a countermine on 13 July. The outer defences finally fell on 20 July. Baybars immediately began to mine the citadel, whereupon the garrison surrendered. What followed next remains murky: according to Muslim accounts, the Christians breached the terms on which they had been granted their lives, and were slaughtered. Once again, Baybars wanted the castle, and ordered it strengthened still further.[2]

The sultan now turned his attention to the north, launching a massive attack on the Armenian kingdom in Cilicia, whose king, Hetoum, had aided the Mongols at al-Bira. This invasion was abortive, but the following year it would be repeated, the city of Sis taken and sacked, but the citadel resisted successfully. Apparently, Baybars had decided that this was not the time to attempt a substantial siege.[3]

Beaufort, Antioch, Crac des Chevaliers

Baybars was proving himself an extremely able commander, and clearly based his success on thorough preparation. Several key settlements and fortresses in the south and east of the crusader states had quickly fallen, despite their strength. In 1268, following a rapid capture of the town of Jaffa, "by treason" on 8 March, it was the turn of Beaufort, standing high on a sheer rock above the river Litani, where defensive works had been constructed in several phases of occupation since the twelfth century. Beginning on 4 April, when the beams for the artillery arrived and a battery of 26 was quickly emplaced, the besiegers used the recently

[1] Al Furāt, 74–6; Makrizi, 9–10 (follows al Furat); Eracles, 450.

[2] Al Furāt, 92–5; Makrizi, 11, 28–31, 41 (repairs); *Gestes des Chiprois*, 179–80.

[3] Dedeyan, *Smbat*, 115–18; *Vahram's Chronicle of the Armenian Kingdom in Cilicia*. Trans. Neumann, 50–1.

constructed outworks as a base to bombard the rest. Meanwhile, the sultan sought to ease his path by distributing false news by letters, hurled inside by machine. The garrison surrendered on 15 April.[1] Making a surprise change of direction, the Muslim troops arrived at Antioch on 18 May, where the vast size of the city enabled them to find weakly defended areas and to take it by escalade on the side of mount Silpius within a few days, many of the people retiring to the citadel which surrendered on terms, thus avoiding the appalling sack and massacre that took place in the city below.[2] Virtually the whole land then reverted to Muslim rule, the once great city becoming a backwater. In 1270 Baybars organised a major raid on Crac des Chevaliers. His purpose was not yet to test the massive defences, but to prepare the ground: his force consumed the surrounding meadows and prevented thereby the re-provisioning of the castle.

He returned in January 1271. Safita (Chastel Blanc) surrendered at once, despite having a garrison of, apparently, 700. On 21 February, the sultan arrived outside Crac. By 14 March, the artillery was in place, and on 21 March, the *bashura* was stormed. Nine days later, a tower in the main wall was breached, forcing the defenders back into the inner ward. The artillery was deployed against this target, but was not needed: Baybars resorted to cunning, sending the commander a fake letter from the Master of the Hospital ordering him to surrender the castle, which he did on 7 April. This is al Furāt's account, and it and others have been studied to try to determine exactly what happened to bring about the fall of this, the most imposing and powerful of all the Latin castles. There was only one approach, from the south, where the now disappeared town may have been walled, in which case the first capture would have been of this, its wall perhaps representing the vague term bashura, elsewhere suggesting outer wall or barbican. This would have brought the attacker up to the main ditch, 10 metres deep, facing a strong wall with two projecting machicolated circular towers. From here, the wall was mined as well as bombarded, and by some combination of these a breach was created very quickly. The source suggests that the south west tower may also have been brought down. The attackers now confronted the massive glacis at the base of the immensely strong inner ward, behind its wet moat, with its equally powerful D-shaped towers, and tiers of defences. While the eventual fall of the place was now inevitable, given the shortage of provisions, the ruse gave the defence an honourable way to save their lives, and Baybars a chance to avoid the heavy losses that would be associated with continuing the assault. Baybars also wanted this castle in working order: the tower he had built to repair the outer wall stands to show where the breach was made. If the dates are correct, Crac had fallen in six weeks, and even that was quite long compared with many of Baybars' other victories.[3]

The sultan's successes continued unabated, other castles falling rapidly as soon

[1] Al Furāt, 110–12; Makrizi, II, 51; Eracles, 455; *Gestes des Chiprois*, 190; Kennedy, *Crusader Castles*, 41–4, 127–8 (description of the works).

[2] Al Furāt, 121–2; Makrizi, II, 58; Dedeyan, *Smbat*, 121; *Grigor of Akanc'*, 371.

[3] Al Furāt, 139, 143–5 (also translated in Gabrieli, *Arab Historians*, 316–19); Makrizi, II, 78–85; and see D J C King, 'The taking of Le Krak des Chevaliers in 1271', *Antiquity*, 90, 1949, 83–92; P Thorau, *The Lion of Egypt*, 266. The Christian sources give no details.

as the artillery was arrayed. His troops had to clear a path through the mountains to bring the weapons to 'Akkār, 40 kilometres from Tripoli, but this Hospitaller castle, with its curtain wall and donjon and deep ditch on a steep ridge, capitulated on 11 May.[1] Baybars now marched south via Damascus to take the Teutonic castle of Montfort. The attack was launched up the gentler slope from the western side, facing several lines of walls. The outermost defences fell on 11 June after the catapults had set to work, and next day the outer bailey wall (*bashura*) was breached, leading to rapid surrender. The ruins of the castle offer silent witness to the progress of the attack. Deep cracks in the walls of the inner bailey indicated where the final mining took place, while heaps of round stone balls from 25 to 43 cms in diameter were found inside the castle.[2]

Tirelessly, the sultan looked not only to capture the fortresses, but also to repair those he judged of use. At 'Akkār, he organised for a catapult to be loosed so that he could measure exactly where the new ditch should be dug. But it should not be thought that Baybars had been able to devote his sole attention to the Latin states. In 1270, Louis IX led a large army to besiege Tunis. The sultan raised Bedouin troops to assist the Tunisian ruler. The city's walls were repaired by adding timber reinforcements and restoring the crenellations, and a deep ditch dug. The crusade ended when Louis died and his troops were bought off after six months, leaving behind baggage including 90 *mandjaniki*. In 1271, the Mongols returned to attack al-Bira again with a battery of artillery, and the sultan, whose governor held out in the town, sent a successful relief force. The following year, hearing rumours of another crusader force being prepared to attack Alexandria, a battery of 100 stone-throwers was deployed along the city's walls.[3] The sultan died in 1277, with the crusader state reduced to a small coastal rump despite the efforts of popes, Mongols and saintly French kings and frequent visits from other European leaders (such as the future Edward I of England).

The conquest completed: Qalā'ūn and Ashraf: Margat to Acre

A breach of the ten year truce with the Latins arranged in 1280 by the garrison of Margat enabled the new sultan Qalā'ūn to attack this castle in 1285. His biographer highlights Qalā'ūn's conquest as a great victory, as the castle had always thwarted his predecessors. Arms, siege engines, naphtha and "flame-throwing tubes" were transported to the site on its high ridge overlooking the Mediterranean. The destination having been kept secret, surprise was achieved. The sultan had also recruited "a number of experts on the art of siege" (Tashrif). On 17 April,

[1] Müller-Wiener, *Castles*, 50 and plate 32, for 'Akkār.

[2] B Dean, 'A crusader fortress in Palestine', *Bulletin of the Metropolitan Museum of Art, New York*, 1927, 15, 38; Burchard, *Description of the Holy Land*, trans. Stewart, 22, confirms it was in ruins in about 1280.

[3] Al Furāt, 141–2 (Tunis); 148–59; Ibn Khaldūn, II, 362–69; Makrizi, II, 103, 110, 113. Tunis remained a seat of power and its arsenal a source of weaponry, including, in 1310, a device able to propel three shots at once: Zarkashī, *Chronique des Almohades et des Hafsides*, 83.

the artillery was arrayed, and mining began. There were three of the huge Frankish type, three of the lighter "Karabughas" and four "devils". The only approachable side, the south, was initially protected by a round tower on the outer wall (the *Tour de l'Esperance*), but the inner defences, at this point represented by the immense donjon, reared up behind. The defenders had artillery of their own, probably standing on the strong, wide roofs of the buildings of the inner enceinte, but these were broken up by shots from the Frankish type trebuchets. The garrison repaired some of them, and used them in turn to destroy some of the sultan's machines. Meanwhile the mining continued, and one of these was fired on 25 May. However, the Tour de l'Esperance collapsed into the breach, killing many attackers, and blocking the breach itself. Other mines, however, succeeded in penetrating right under the ditch and beneath other towers, and the garrison asked for terms. Keen to prevent further damage to "so fine a fortress", the sultan agreed. The point of the first attack is still clear to see, a rebuilt outer tower occupying the spot of that destroyed by the mine.[1]

Tripoli was the target four years later. Apart from four galleys sent to its aid from Cyprus, the defenders were without relief against the siege that began on 17 March. The walls were of strong masonry, protected by a ditch, but there were weak points against which the sultan concentrated his attack with a relentless bombardment with stone throwers. Numerous archers also covered attacks by mining and sapping, driving the defending crossbowmen from their places, and allowing a storming on 27 April. The battery had comprised 19 machines and no fewer than 1,500 stone cutters and artificers had been permanently employed servicing them.[2] The battle had been entirely around the city wall, the old castle of St Gilles outside having fallen into disuse, but a new city was now built beneath the former count's fortress originally constructed to besiege old Tripoli.

Qalāūn was preparing to deliver the final blow against Acre when he died, and it was his son and successor al-Ashraf Khalīl who conducted the siege of the heavily fortified seaport in 1291. A double row of walls, deep ditches, and separate towers each held by the religious orders or various Christian nations, ran for 1900 metres from sea to sea, with a dog-leg projecting inland, the vulnerable point of which was protected by the recently constructed circular King's Tower standing in front of the older *Tour Maudite* (cursed tower), which had also been the main focus of the attack during the Crusader siege of 1191. The Lebanese forests provided the best timber to construct the siege engines, some of them of great size. The writer Abu 'l Fida was with the contingent from Hama travelling via Crac des Chevaliers (now once again Hisn al Akrad), whence they picked up a giant catapult, nicknamed "The Victorious", which was dismantled to be carried in a hundred carts: the writer had charge of one. The array of artillery was possibly the largest ever employed in a single siege outside of China (72 are mentioned in one source), and included probably 15 of the heaviest trebuchets (the "Frankish")

[1] Tashrif, Life of Kalaoun, quoted extensively in Michaud, *Bibliographie*, 561, and translated in Gabrieli, *Arab Historians*, 334–8; Makrizi, III, 86; *Gestes des Chiprois*, 218.

[2] Michaud, citing Life of Kalaoun and Abu 'l Fidā; Makrizi, III, 102–3; *Gestes des Chiprois*, 236–8.

able to project stones weighing above 45 kgs. The battle at Acre was furious, with several nocturnal sorties designed to destroy Muslim artillery, which then had to be repaired. The garrison used their ships to attack the exposed seaward flanks of the Muslims with ballistas and stone throwers, until that carrying the mangonels was wrecked. Despite the vigour of the defence, the sultan had large forces and was confident of ultimate victory. Breaches were opened in the outer wall. The outer wall of the new King's Tower broke away and slid into the ditch as a result of mining, the defenders erecting a wooden barricade to prevent entry by this route. The Muslims used their mounted troops to relay large numbers of stones forward to the foot soldiers who used the sacks of stone to create an approach to the breach. Meanwhile, fear was beginning to take hold among the defenders, and an exodus by sea began. At dawn on 8 May there was a general assault against the entire circuit, the defenders fleeing to ships in the port, while the Templars, Hospitallers, and Teutonic Knights, their whole raison d'être now ceasing to exist, fought to the end from their individual towers in the city, some of which had to be brought down by mining.[1] The Muslim chroniclers noted with satisfaction that Acre had returned to their control on the anniversary of the date that it had fallen to the Third Crusade in 1191. With Acre gone, the defenders of Beirut, Sidon, Tyre and Tortosa gave up their heavily defended cities, along with the impregnable Pilgrim's Castle ('Athlit) without a further blow, many departing for Cyprus.

This was the end for the crusader states, although the fortified island of Arwad at Tortosa would not be taken until the sultan organised galleys to attack in 1303, the garrison of 120 knights and 500 archers being overrun. The kings of Cyprus would launch frequent raids on the mainland, and the end of Outremer did not mean the end of fighting for the sultans. In 1292, al Ashraf captured the castle of Qal'at al-Rūm, on the Euphrates, from the Armenians. This old fortress stood high above the Euphrates and one of its tributaries, surrounded on three sides by water, with a strong rock-cut ditch on the accessible, southern, side, but also had a much gentler slope and flat land before the river was reached on the northern side. After substantial preparations, the outer circuit was taken. Twenty machines were deployed overlooking the southern side, and another 15 to the north, something that must have required considerable skill to permit them to cope with the gradient in front of them. Mining was begun, but an unusual means of assault proved crucial: chains were somehow attached to the crenellations, the other end secured to the ground, up which the besiegers climbed to mount the wall tops. Engines were then arrayed against the citadel which surrendered. In the same year, it was recorded that the work of rebuilding the citadel of Aleppo was finally com-

[1] Makrizi, III, 125–7; Abu 'l Fidā, *Annales Moslemici*, ed. Reiske, 1789, IV, 95–99 (translated in Gabrieli, *Arab Historians*,344–46); Abu'l Mahasin, translated in Gabrieli, 347–50; *Gestes des Chiprois*, 244–50, also in Amadi, *Chronique*, 220–24. The siege is recounted from Arabic sources in D Little, The fall of 'Akka in 690/1291, *Studies in Islamic History*, 1986, 165–78. For a plan reconstructing the defences, Müller-Wiener, *Castles*, 73; Burchard, 9; Ludolph von Suchem, *Decription*, trans. Stewart, 50–1.

pleted, 33 years after its destruction by Hülegü.[1] The new work boasted a most impressive and heavily defended entrance over the vast ditch (Plate 1).

Mongols and Mamluks at Damascus 1299–1300

In 1299/1300, the Mongols returned, defeated the disunited forces of the sultanate and occupied the undefended city of Damascus. Although the local viceroy had given orders for preparations to be made when the invasion was rumoured, it seems that this did not lead to a readiness to defend the city. Alongside the compilations of later Muslim historians, there survive the contemporary and eye witness accounts of al-Yunini and Adh-Dhahabi, allowing a vivid picture of the horror for the citizens caught up in the military struggle, along with the confusion and conflicting statements of fact that give a real feel for the atmosphere of fear and rumour that must have pervaded the ordinary people during this period. It is for this, rather than for any decisive military event, that it is worth looking at this rare evidence. The city having opened its gates, the garrison in the citadel resisted, shooting with artillery on the invaders outside, and the accounts all confirm that heavy artillery was in use. The Mongol commander had houses burnt down to clear space for the siege, ordered the people to cut down trees outside the city to provide the timber for siege engines — clearly they had not brought a siege train with them, but had sufficient engineers — and ordered a stone thrower set up in the nearby Ommayad mosque. The investment of the citadel was in place by 10 February 1300, and citizens were terrified lest the Mongols order them to take on the extremely dangerous task of filling in the moat. But matters did not get so far. The garrison sortied and destroyed the machine, only for it to be rebuilt and the mosque desecrated. According to one account, a soldier managed to slip into the building at night and kill the machinist. Perhaps as a result of hearing of the approach of relief, the Mongols suddenly withdrew (March), and the defenders were able to come out, and destroy the abandoned engines by chopping up their timberwork. They also may have begun to dismantle (or alternatively, according to the source, to further strengthen) their own temporary defences, which sound as if they were hoardings added to the citadel walls. Rumours of the return of the Mongols in April and September caused widespread panic, but the sultan organised a relief army from Cairo.[2] There followed the trial and execution of Damascenes who had collaborated with the invaders, and, somewhat belatedly, an instruction that the citizens should learn archery.[3]

[1] Makrizi,, III, 139ff; IV, 62, 152–68 (Damascus), 196 (Arwad); Abu'l Fidā, V, 103–5 (Qalaat al Rum); *Gestes des Chiprois*, 309–10; Amadi, 238–9; A Stewart, Qal'at Al-Rum/Hromgla/Rumkale and the Mamluk siege of 691 AH/1292 CE, in H Kennedy (ed.), *Muslim Military Architecture in Greater Syria*, 2006, 269–77, describes the site which has been transformed by the creation of a reservoir, turning the fortress into an island.

[2] Makrizi, IV, 153–68.

[3] Abi Fada'il, Histoire des Sultans Mamlouks, ed./trans. Blochet, *Patrologia Orientalis*, 12, ed. Graffin, Naum, 1919, XIV, 1920, 656–58; Li Guo, Early Mamluk Syrian Historiography. Al Yunini's Dhayl Mir'at al-zaman, 1998, I, 129–63; J Somogyi, Adh-Dhahabi's record

A high point of siegecraft

None of the mighty fortifications of the crusader states had withstood the forces and skills mobilised against them by the Mamluk sultans. During each period when there were campaigns against the Christian fortresses, several more were taken. In contrast to the often slow sieges of previous eras, victory was achieved in never more than a few weeks from the time the investment began. There was a combination of reasons. It clearly made a great difference that the Muslim armies could draw on the resources of the whole region, while the Latin forces were too few to raise a siege once it was formed or attempt any kind of relief operation, even on the coast. One consequence was that once it was clear that the besiegers had broken through the first line of defence, the final outcome was not in doubt, so most garrisons negotiated at that point. The psychological element involved appears also to have been very significant. Despite the addition of a strong ideological (religious) element to the conflict, often reported by the chroniclers of both sides, we can set against the sometimes desperate defences carried out even when the position was hopeless the frequent almost hasty negotiated surrenders of garrisons of more than adequate size still holding massive fortifications, including those with access to the sea. It was always the case that capture of a fortress without terms being negotiated placed the occupants entirely at the victor's mercy, and the sense of vengeance in the Muslim accounts of the killings of, in particular, captured Templars, is very obvious. But the sultans generally kept their word to surrendering garrisons, a wise approach that could only encourage others to negotiate. This approach seems not to have been opportunistically varied according to whether the sultan wanted to keep the captured fortress. The contrast with the fortresses held by the sultan's men against the Mongols, an opponent no less fierce, determined and skilful, is telling. Al-Bira and the citadel of Damascus held out, greatly helped by the prospect of a relief army arriving.

The material factors included the presence of an extremely powerful siege train, with the necessary organisational infrastructure. Several references indicate the presence of royal arsenals at Damascus and Cairo, and once Crac des Chevaliers and other castles had been captured, they became forward bases for the storage of artillery. Even so, the logistics of transporting such equipment were extremely demanding, as Abu 'l Feda's story confirms. The organisational skills of the Mamluks were obviously as impressive as their abilities at the siege itself. But once arrived, which sometimes took many weeks, the sheer speed with which the massive defences were broken reflected the skill of the attacker: few siege operations lasted more than six weeks, and many were quicker. In this short space of time, an enormous array of artillery had to be assembled, mounted and deployed, and the approaches to the defence made secure for mining to begin through the careful choice of where to attack, the digging of trenches and the creation of timber shelter protected against fire. The effectiveness of the artillery was clearly important,

of the destruction of Damascus by the Mongols in 699–700/1299–1301, in *Ignace Goldziher Memorial*, 1, 1948, 361–86.

too, especially the "Frankish mangonel", the trebuchet, with its ability to crack open walls as well as to destroy defending machines, while lighter engines cleared defenders away from the point of attack. The importance of the latter, identified in particular with the "Karabugha", is explicitly indicated by one Christian account.[1] Large numbers of people were called for, to transport the engines to the fortress, to assemble them, to collect and prepare stones for shooting. Rarely are the engines of the defence mentioned, but it would seem that although they undoubtedly were used to some effect, they were generally outnumbered, and, to use a later phrase, outgunned. It is likely that among the first targets for the besiegers was the artillery of the defence, the destruction of which would make the attack so much easier.

Mining played the dominant role in breaching walls and towers, followed and sometimes prepared together with bombardment. There are no references to battering rams or siege towers or other forms of attack, common in previous times. They had been superseded. Stronger walls and ditches of great depth and width made them redundant, while the effectiveness of artillery made cumbrous and inevitably slow-moving wooden contrivances extremely vulnerable. What had not changed, however, was that one of the resources of the attackers must always have been a substantial corps of experts for the machines, and engineers who understood how to mine beneath ditches that were now wider and deeper than ever before. Where mining was not possible, as at Caesarea, for example, the besiegers could call on other methods with the same results. Crossing the ditch remained the first objective, and more sophisticated means of doing so had now been achieved. Baybar's biographer also made it clear that not just the specialists had siege skills, with references to the ordinary soldiers expected to construct attack ladders.

What also emerges was a clear appreciation of the strategic function of the captured crusader fortresses. Those on the coast were usually destroyed by their conquerors, concerned that Christian control of the sea would permit them to secure a fortified base. The inland fortresses were repaired and improved. Particular attention seems to have been paid to the point at which a breach had been made. These castles continued to be centres of administration but they also acted as a defence or refuge for the surrounding community, a place both for the collection of taxes and for the maintenance of control. On the coast, during the fourteenth century, the Mamluks built a line of towers, at intervals of one kilometre, around the port of the new city by Tripoli at al Mina. The surviving tower is a substantial quadrangular building of considerable but passive strength, the walls (windowless until the top storey) reinforced with old columns and machicolations at least over the door. Suited perhaps as a refuge against raiders arriving by sea, they were a stark contrast to the fortified cities used by the previous owners of the land. Indeed, the ridge top former Crusader castle at Tripoli was provided with a powerful new barrier wall with internal galleries and projecting, stone-vaulted rectangular towers covering a bent-entrance gateway, a few metres in front of the

[1] *Gestes des Chiprois*, 244, at Acre.

old Latin donjon on the side of the approach.[1] A small new fort was also built in the middle of the island of Arwad, quadrangular with round corner towers, intended it would seem as the base for a garrison.[2] The sultans could not know that the crusaders would not return.

Byzantium restored

While Outremer was being extinguished, conflict was also frequent in the former Mediterranean lands of the Byzantine empire. The Byzantine ruling class had re-established itself in geographically distinct new states: one in Epirus, one at Trebizond on the Black Sea, while the "imperial capital" was relocated to Nicaea (Iznik). Separate dynasties set themselves up in each place, and were as much in conflict with each other as with the Latins. It was undoubtedly fortunate for them that the Seljuk Turks of Rum (with their capital at Iconium) were weaker than before. A Bulgarian state had been recreated to the north and was sometimes in conflict with the Greeks, sometimes with the Latins. Michael Paleologus had been "despot" of Thessaloniki since 1222, and had repaired and improved the rather old and much battered Byzantine defences of the citadel with new towers, taller and stronger. His successors continued to restore the defences of what had been the empire's second city, with new defences for the harbour, and a low outer wall between the city and the citadel for the first time (1316).[3]

This jumble of enfeebled fragments of a once powerful empire was now about to be reconstituted by the chances of dynastic survival and the accidents of battle. Theodore Laskaris succeeded his father, Emperor Michael, in 1254, and continued his father's policy of truces with the Turks to allow him to operate in Thrace, but he died young just four years later, a chronicler interestingly describing the fatal illness as being like an attack by siege engines against his head.[4] Michael Paleologus fought Despot Michael of Epirus. The Despot allied himself with the Latin princes of the Peloponnese and Achaia but was defeated at the decisive battle of Pelagonia (1259). At one blow, most of the Byzantine empire was reunited in the hands of the Paleologi, and in exchange for their release, the Latin princes of the Morea had to cede control of Monemvasia, the Maina and Sparta/Mystra to the new Byzantine ruler. Epirus continued independent.

The new emperor wanted his capital back. In 1260, Selymbria fell easily. Then he tried to attack Constantinople by way of Galata, on the opposite bank of the Golden Horn, and set up stone throwers. The chronicler suggests he was familiar with such weapons from long use, knowing how to deploy them to inflict the most

[1] Müller-Wiener, 42–3. There is no description of al Mina in this study, but a mention in connection with the building of the new city.

[2] Z Hanna, *Castles and archaeological sites in Tartous,* 19–26.

[3] Tafrali, *Topographie,* 45–6.

[4] George Pachymeres, De Michaele Palaeologus, in *Migne, PG,* 143, cols. 550–2; Nikephorous Gregoras, Historia Rhomaike, trans. as *Rhomaische Geschichte,* van Dieten, 1973, 91; also in *Migne, PG,* 148, 187. Gregoras wrote in the mid-14th century.

damage. But he was unable to make any impact, and the particular importance of defenders, unseen behind their arrow slits, shooting down his troops as they tried an attack with ladders on the places weakened by bombardment was stressed by the account. Instead, he set up fortifications around the city whose garrisons kept the Latins in and began to deprive the city of supplies. In 1261, the great city of Constantine with its triple walls that had defied so many attackers fell almost without a fight to a small Byzantine force, working in collusion with the Genoese (keen to supplant their Venetian rivals' privileged place in Constantinople).[1] The successor states to the "Fourth Crusade", composed of a small alien layer of landowners and their followers, were incapable of mobilising any significant force to prevent the collapse of their regime. The Byzantine force was itself very small. Byzantine armies now were largely composed of hired foreign troops, including many Turks, fighting in their own units: the old centralised imperial army was a thing of the distant past, the "emperors" of the successor Byzantine states having had no means to sustain such a burden. What had happened to the siege trains and the employed experts? The expertise must have been available in the forces they hired. Builders and masons were always available. No doubt aware of how Constantinople had first been lost, Michael immediately ordered repairs to the city's sea walls, improving the wall tops and the number of turrets, and allocating enough troops to the garrison. When in 1273 he learnt of the impending threat from Charles of Anjou, alongside a major diplomatic campaign to divert the attack, the emperor took care to ensure that the city was protected with sufficient numbers of weapons including ballistas and stone throwers, for constructing which artificers were employed, while architects were enlisted to once again review the walls and in particular to strengthen the defences of the harbour.[2]

Fortifications played a central part in subsequent campaigns as in 1271 Michael recovered Euboea and the islands of Lemnos, Chios and Rhodes[3] then in 1275, fought against Epirus, capturing "fortlets" (*oppidula*) with engines, but failing to take the fortress of Neopatras, which was relieved by trickery.[4] Renewed hostility with the Turks from 1278 led the emperor to order the building of a line of forts along with river Sangarias, to provide a "shield" for Bithynia, while there was an attempt to reconstruct a more substantial city at Tralles north of the river Meander (Menderes) "to serve as a bulwark". But after four year's work, beginning with the circuit of walls, the Turks besieged it and took it: after a blockade they mined under the wall protected by a shelter.[5] A survey of the remains datable to this time finds that fortifications generally were built on hill tops and followed the line of the ridge, with walls of local stone mixed with brick, and circular or pentagonal towers. The more substantial defences of Magnesia, a capital for the Lascarid

[1] George Pachymeres, cols. 578–83; Nikephorous Gregoras, ed. Dieten, 101–8. The Venetian Marino Sanuto Torselli protected the last Latin emperor's honour by claiming the Greek army was vast: Istoria del Regno di Romania, in Hopf, *Chroniques Gréco-romanes*, 115.

[2] George Pachymeres, cols. 627, 817–18.

[3] Eracles, 461; Nikephorous Gregoras, 112–13; Marino Sanudo, 123.

[4] Nikephorous Gregoras, 121–2.

[5] George Pachymeres, cols. 929–34; Nikephorous Gregoras, 135–6.

dynasty before the reconquest of Constantinople, were formed of two circuits of walls, with the same building styles. These were not sophisticated constructions, but represented the resources available to their builders, and also their function as, primarily, a bulwark against raiders rather than invaders. On the coast, at the major commercial centre of Smyrna (İzmir), the Byzantines have been attributed the building of a curtain wall and small quadrilateral circular-towered citadel separating the much larger "old castle" on its hill top from the walled city below. With walls more than three metres thick, this was a major construction reflecting the importance of the city.[1] The eastern Mediterranean, a hub of international seaborne trade was also an attractive target for private enterprise in the form of piracy, some of it on a grand scale. Corsairs launched an attack on the island of Suda in 1286, and brought with them siege engines including a trebuchet to assail the castle, leading to a Venetian relief expedition from Euboea.[2]

In 1303, Emperor Andronikos II (1282–1320) hired a Catalan-led mercenary army led by Roger de Flor that had found itself unemployed, and the professional corps proved more than a match for the Turks. But the new general became a very expensive nuisance, and was murdered by the Greeks, whereupon the Catalan Company, as it became known, became an independent political force of great military strength in the region. They made their way to Greece where they set up their own principality, having captured Thebes and Athens from their rulers with ease. They were adept at all forms of warfare. They had quickly restored Constantinople's Asia Minor frontier, although few details are handed down of the siege operations. The westerners were frustrated at a long siege of Magnesia, where they deployed artillery, towers and many other implements, but were not supplied with the necessary back-up by the Byzantines. In Thrace in 1308, they captured the town of Madytus, having decided to starve it out rather than attempt an assault, although they had also arrayed siege engines.[3] Their historian and, later, general Muntaner recorded regular deployment of trebuchets, and also considerable acumen. After a long siege, and several failed assaults, a town near Gallipoli was taken (1304) by a surprise escalade launched at the height of the hot afternoon.[4] This powerful force of professionals was in effect a fully equipped army. It had now become a state, too, in its own right.

Rhodes

The Greek revival was undermined critically by a shortage of resources to carry out military operations on so many fronts. The empire largely depended on allied foreign powers (such as Genoa) for ships. In 1306, the now homeless Knights Hospitaller, dwelling uneasily in Lusignan Cyprus, launched an invasion of Rhodes,

[1] C Foss, 'Late Byzantine fortifications in Lydia', *Jahrbuch der Oesterreichische Byzantinistik*, 28, 1979, 297–318; Müller-Wiener, 'Die Stadtbefestigungen von Izmir, Sigacik und Candarli', *Istanbuler Mitteilungen*, 12, 1962, 70–2, 82–3.

[2] Marino Sanudo, 113–14.

[3] George Pachymeres, De Andronico Palaeologus, *Migne, PG*, 144, cols. 481–3, 634.

[4] Nikephorous Gregoras, 175–95; *Chronicle of Muntaner*, 528–9.

where the commander had set up an independent pirate state. The island, fertile and prosperous, had a good harbour and lay on trade routes to the Muslim world. It had already changed hands more than once between Genoa (backed by Byzantium) and Venice, and may have been invaded by the Turks too. The Hospitallers under a new Grand Master, Fulk de Villaret, took a siege train with them. The chronology of events is irretrievably confused. What was described as "an old castle", already "ruined" (Pheraclos) was rapidly taken with the engines, but an attack on Rhodes city five days later was repelled. There had been an earthquake only six years previously that had (reportedly) caused massive damage on the island, but at least some of the defences must have been repaired, it seems. The castle of Philaremos was then taken when a disaffected Greek showed the invaders a way in, and the Turkish garrison of 300 was slaughtered. But conquest of the whole island proved a long, and for the Order very expensive, affair. Investment of the city of Rhodes continued for months, with Andronikos sending occasional seaborne relief. Eventually, its surrender was somehow brought about when a Genoese ship with more supplies was captured in Cyprus and the commander, under threat of death, managed to persuade Rhodes to a peaceful capitulation (mid-1308). The capture may have been timely in preserving the Hospital from the violent suppression then being visited upon the Templars. The Master of the Hospital then began the work that would become one of the strongest fortresses of the region, with an important outpost on the mainland at Bodrum.[1]

1308 proved to be a bad year all round for Byzantium. In addition to losing Rhodes and seeing the Catalan company begin to establish its separate authority, the Turks made several significant gains too. Ephesus fell to the Turks after a long blockade, and another well-fortified town, called Tricoccia, and described as an "outwork" of Nicaea, was subject to direct attack. Strong and well-garrisoned though it was, the Turks were able to fight their way to the edge of the ditch. This, deep and wide, was the defenders' main hope. But the attackers found a novel way of overcoming the obstacle. Their engineers constructed what must have been a remarkable contraption, resembling an "X" in shape, with a bridge suspended from it, the whole protected from destruction by earth, timber, and stones. This vast machine somehow enabled the Turks to cross the ditch and reach up to the top of the walls, across which they stormed.[2] From the description it must have been a defended frame with its top as high as the ramparts.

Italy, Sicily, the Angevins and the Aragonese

When Emperor Frederick II died in 1250, and his son Conrad IV soon after, the Pope seized the chance to finally dispose of the Staufen dynasty. Conrad's son was an infant, and only Frederick's illegitimate son Manfred

[1] D'Amadi, 257–9; *Gestes des Chiprois*, 320–22; George Pachymeres, cols. 431 (earthquake), 697. A careful reconstruction of events is in A Luttrell, The Hospitallers at Rhodes, in Hazard, Sutton (eds.), *History of the Crusades*,III, 280–9.

[2] George Pachymeres, col. 699.

stood in the way of driving the family from Italy for good. Having declared the disinheritance of the Staufen, the Pope needed another candidate for the crown of Sicily, and he offered it to Charles of Anjou, brother of King Louis IX of France.

The consequence was a long period of unrest and violence, because Manfred had no intention of resigning the rich lands he now ruled without a struggle. Manfred recovered Barletta from a papal-inspired revolt: through a hail of arrows, he led his troops to storm the wooden outworks and break down the city gates. On its surrender, he had the walls levelled. In 1252, when Conrad returned from Germany, he and Manfred faced the most serious revolt, in Naples and Capua. They besieged Naples from the side of Mount Vesuvius and also by sea, arrayed machines around and sapped and mined under the wall and ditch, and either this attack, or hunger according to a second source, led to the city's surrender. Its walls too were destroyed.[1]

With the sudden death of Conrad from illness (May 1254), Manfred was careful to place reliable garrisons in Capua and San Germano to protect against an invasion from the direction of Rome. He had to suppress further revolts, in 1256 using his Arab ("Saracen") troops from Lucera to take Foggia, who got into the town where the rampart was very low but were forced to withdraw on the approach of a papal army. At Oria, the defenders were clearly advised by people with experience of resisting sieges. When a wall was brought down by mining, they had erected another inside. When the prince erected some kind of wooden tower to overtop the wall, it was burnt down. Leaving the town closely guarded, Manfred trapped the papal army in Foggia, where its very size told against it. Provisions quickly ran out and the papal legate had no choice but to seek terms.[2]

In 1264, the Romans recovered the town of Sutri which had been taken over by imperial supporters and then besieged the castle of Vico. They erected machines and prepared to mine the castle, Pope Urban IV appealing for funds to pay for the construction of machines to enable the castle to be taken. In fact, it was relieved.[3] Meanwhile, the future king had been making his preparations to secure his realm, and Manfred had been preparing his defences against the imminent invasion by a powerful French army, including preparing siege engines. French forces marching overland had to defeat pro-imperial supporters on the way, and took a number of castles by storm. In 1266 Charles of Anjou, two weeks after his coronation as king of Sicily, set out to secure his new kingdom. His army stormed the castle of Rocca d'Arco (20 January) then attacked the castle of San Germano, a major part of Manfred's defensive arrangements. Charles had an adequate number of trebuchets, but they were not needed owing to the "pusillanimity" of the Italian defenders. The French troops had managed to force their way inside, including through

[1] Jamsilla, De rebus gestis Frederici II imperatoris ejusque filiorum Conradi et Manfredi, in G de Re, *Cronisti e Scrittori Napolitani,* II, 112–17; Saba Malaspina, Rerum Sicularum Historia, *MGH SS,* XXXV, 1999, 96; Thomae Tusci, Gesta Imperatorem et Pontificum, *MGH SS,* XXII, 1872, 516.

[2] Jamsilla, 152–64, 176, 190–2; Saba Malaspina, 108; Bartholomei de Neocastro, Historia Sicula, ed. Paladino, *RIS,* XIII/3, 1922, 5.

[3] Saba Malaspina, 141; *MGH Epistolae Saeculi,* XIII, 1894, III, 604.

a postern, and the defenders fled.[1] Here, the question of morale had proven to be decisive. The area then fell quickly into the new king's hands, Capua and Naples submitting. Manfred was then killed at the battle of Benevento, leaving Charles free to take over his new realm. It was a brief respite. The following year, Conrad's young son, Conradin, had gathered support in Germany and from among his family's former subjects and had returned to the south. In Sicily, various towns rose in revolt for him, but Torone resisted. It was attacked with machines and by mining, but was forced to surrender when the water was lost from its cistern.

At the battle of Tagliacozzo in 1268, Conradin was defeated,subsequently captured, and executed by Charles. He still had to fight to defeat the remaining supporters of the old order. The large castle and palace at Lucera had been developed by Frederick and was the base for his Arab mercenaries. Charles besieged it closely and attacked it with the array of machines, but it was hunger in the end that brought its surrender (August 1269). He had summoned his "barons and feudatories" from across his kingdom and specifically called for people skilled in carpentry and the construction of siege engines, and for supplies of tools and workers for the purpose of digging mines, although this was not how the castle was taken.[2]

Charles spent large sums repairing and reinforcing other castles, having sent out an expert, Pierre d'Angicourt, to survey the existing buildings, for example at Manfredonia, Villanova and Brindisi, providing higher and thicker walls. The reconstitution of the archives of the Angevin chancellery at Naples has provided documentary evidence of the workings of an efficient late medieval government, not dissimilar to the collections of English government materials of this century. The records show not just payments for work at royal castles, but also sometimes very detailed instructions to the overseer. Payment for major works at Melfi was documented in 1279, alongside other work at Canosa. At Manfredonia in the same year, the king ordered his lieutenant to find six suitable masters to oversee the stone cutters working at the castle. At the same time, important work was being carried out to modernise the castle at Barletta, and the instructions here to d'Angicourt specified exactly the dimensions of the new round tower, the ditch and walls.[3] From 1279, he organised the construction of the new castle (Castelnuovo) at Naples. This was built by a French master, Pierre de Chaule, on current French lines: a quadrangle with powerful circular corner towers, said to have been modelled on the large new castle built at Angers itself,[4] and like Angers, it would soon be provided with a magnificent palace inside the walls. From the outset of his reign to the end, Charles showed a close interest in the castles he had inherited

[1] Saba Malaspina, 161–3; Villani, *Chronicle*, trans. Selfe, 208–9; Andreae Ungari, Descriptio victoriae a Karolo Provinciae comite reportate, *MGH SS*, XXVI, 566–80.

[2] R Filangieri, ed., *Registri della Cancellaria Angioina*, 1950–, I, 201 (16 February 1269).

[3] *Registri*, XXII, 74 (Barletta), 75 (Melfi, Canossa), 162 (Manfredonia).

[4] Saba Malaspina, 193–225; Thomae Tusci, 521–3; Bartholomei de Neocastro, 7–9; see also J Dunbabin, *Charles I of Anjou*, 1998, 161, 175, 211. On Naples, L Santoro, 'The fortified royal palace of the Aragonese and Angevin sovereign. The Castelnuovo in Naples', *IBI Bulletin*, 50, 1997, 93–4.

by conquest. At the beginning, he had carried out an inquest into the state of the kingdom including the castles.[1]

Sicilian Vespers, Berat, and Messina 1282

Charles's financial exactions, many of them generated by the needs of the building programme, and strong resentment by the Sicilian nobility at the arrogance of the new French rulers, were partly behind the rebellion of March 30, 1282 known as the Sicilian Vespers, which quickly saw the new regime lose control of most of the island, Messina joining the revolt on 9 April. The Angevin garrison bottled up inside the castle there negotiated a surrender.

King Charles was at the time planning attacks on the Byzantine state across the Adriatic. His forces were engaged in a siege of Berat and he had organised a powerful fleet and army with a siege train and all manner of modern, and not so modern, weaponry. The government records included orders to send "experts in the construction of machines" and another the despatch of the means to produce "Greek Fire" *(igne greco)* stored in his castle of Avellone. But the expedition was a failure, with the Byzantine army winning a battle against the Angevin forces in April 1281.[2] Gregoras made the shrewd psychological point that weapons, gold and siege engines alone were not enough for victory, bearing out the indication in the records that much of Charles's army, assembled from disparate places (like the Byzantine), was not a highly-motivated fighting machine.

Now Charles had to divert his army to try to recapture Messina. He sent his forces by sea to Catania in June and embarked on a major siege. He found the defenders numerous, and he suffered many casualties from crossbowmen, so he ordered the construction of trebuchets. The citizens were not intimidated, and constructed machines of their own to shoot at the Angevin artillery. They had to pay particular attention to the defence of their harbour, protected by a chain. After a number of failed attacks on different parts of the city's land walls, in which the role of women in hurling down stones on the assailants was noted, the Angevin fleet tried to break into the harbour on 14 September. The defenders arrayed ships to defend the chain, six of them carrying stone throwers, and the attack was repulsed. The defence also used an inflammable liquid made of pitch and oil as an incendiary to defend their walls, burning the Angevin scaling ladders when they then tried another assault by land. A "terrible" ballista on the walls fired a projectile which transfixed two of Charles' nobles. With winter looming, Charles was forced to abandon his attack.[3]

[1] Thomae Tusci, 523; *Registri*, II, 69 (1267).

[2] *Registri*, XXII, 64 (28 June 1280), XXIII, 158 (26 August 1280); George Pachymeres, vol. 143, 973; Nikephorous Gregoras, 139. Berat was a strongly walled and towered town and castle with defences mainly dating from the early thirteenth century: M Kiel, *Ottoman Architecture in Albania*, 1990, 53–55; A Baçe, La Ville Fortifiée de Berat (French summary), *Monumentet*, 2, 1971, 59–62.

[3] Saba Malaspina, 313–14, 327; Bartholomei de Neocastro, 20–8 (an eyewitness); *Registri* ,XXVI, 81 (29 September 1282).

The Sicilians offered the crown to Peter III of Aragon, who was married to Constance of Hohenstaufen and therefore represented continuity with the dynasty now extinguished in the male line. Peter brought an army to the island. The struggle begun by the papacy's determination to remove the Staufen now spread, more than thirty years after it had begun, to embrace the whole of the western Mediterranean, and the consequences for the southern Italian kingdom, the fabled wealth of which had attracted so much envy and interest, were awful. The Iberian peninsula now became drawn into a wider war. A significant role in such a war, naturally enough, was played by sea power, and the Aragonese and Sicilians were fortunate to have the services of an outstanding admiral in Roger de Lauria. His skills lay not just with ships, but in using ships to influence land warfare too. Among his conquests was the island of Malta, taken by a small force of soldiers equipped with four trebuchets.[1]

Castile, Navarre and Aragon

Christian Spain had faced ultimately unsuccessful invasion from the new Merinid dynasty of North Africa, and although castles were taken, no cities were captured (1275).[2] Nonetheless, they continued to be a player in the affairs of the peninsula.

Granada still represented a substantial power, with a large population swelled by Muslim refugees from the reconquista, and now benefiting from some natural land barriers, increasingly and heavily fortified with renovated and new physical defences in the form of refuges for the peasantry against raiders, genuine fortresses containing royal garrisons, and major fortified cities. More effective than all of these, however, was the willingness of the rulers to pay a substantial tribute to the Christian kings in return for a truce. Sometimes, indeed, it involved allying with them against fellow Muslims. This was the situation in 1278, when Alfonso X and the king of Granada began major preparations for an attack on Algeciras, which was held by the ruler of Morocco, and Malaga, the Granadan city that had been betrayed to him. While Granada set about recovering Malaga, the Castilians organised a large land army to take Algeciras, along with the Granadans, who were to withdraw (February 1279), although they returned later. Waterborne transport was essential and a large fleet was organised along with many siege weapons to be taken by ship, thus avoiding the expense and difficulty of hauling them overland. In October 1278 the expedition was ready and the town was blockaded by land and sea. The engines (including *trabucos*) began a 24-hour-a-day bombardment of the walls and the town beyond, shooting from the north side across the river Miel. The city was not weak, having been walled from the ninth century, refurbished by the Almoravids (1086) and further strengthened with the addition of an octagonal torre albarrana in the south east corner under the Almohads. With a perimeter of 1,480 metres, the wall, made of stone and lime, possessed square towers every 25 metres, and a low outer wall close to the inner. The gates had bent entrances,

[1] *Chronicle of Muntaner*, 243. The chronicle was compiled about 1325/8.

[2] Ibn Khaldūn, *Berbères*, IV, 88, 110–12, 114.

and inside was a (now disappeared) citadel. Two sides were covered by rivers, and the eastern was on the sea. So Algeciras represented a strongly defended and hard to approach target, if not at this point particularly "modern". The Merinids had begun to make still further improvements when the siege took place. It was strong enough, certainly, to decide the besiegers to turn to a blockade, and this lasted ten months. The decisive factor, however, was that the king of Castile had omitted to keep up payments to the fleet, and the ruler of Tunis sent ships that attacked the Christian-Granadan fleet and dispersed or captured it. With the town now able to be supplied by sea, Prince Peter had to abandon the siege. It was recorded however that he removed his siege engines rather than burn them.[1] Following the siege, the Merinids decided to construct a new city next door to the old, fortifying the site from which Alfonso had bombarded the old city. The new enclosure was twice the area of the old, with walls eight metres high and numerous towers eight metres wide and projecting six, just 20 metres apart. The whole, very well built, was further protected with a concentric outer wall and a ditch eight metres wide. This construction was, remarkably, ready by 1285.[2] In 1299 however, the Granadans stood by while the Christians besieged the Merinid garrison of Tarifa. Attacked from land by siege engines and by ships from the sea, the defenders negotiated a surrender through hunger after four months.[3]

The survival of a detailed account of the strife in the small north Iberian kingdom of Navarre provides further fascinating information about contemporary siege warfare. Thibaut II had been one of many to perish on Louis IX's crusade to Tunis, leaving his tiny realm without an adult ruler, and his queen the object of baronial conspiracies. The walled city of Pamplona had been granted privileges in comparison with its suburbs, including that they would not be fortified against one another, but the inhabitants resolved to build siege engines, and constructed "mangonels, algaradas and trebuchets" to hurl great stones against the two suburbs. Never before, wrote the chronicler, had a town arrayed a trebuchet against another town.[4] The threatened bourgeois decided to build their own machines but had to send for carpenters and engineers from Gascony. Meanwhile, they invited King Philip III to appoint a governor, and Eustace de Beaumarchais was sent. When he arrived in Pamplona, he ordered the destruction of the machines but the citizens refused, and drove him out of the city. The machines in the bourgs, however, were dismounted by engineers. There followed an invasion from Castile and the governor had to defend himself in the bourg. The chronicle details minutely the arrangements for the defence of each tower. A large *algarada* was

[1] *Cronica de Alfonso* X, ed. Gonzalez Jimenez, 2001, 195–204; defences of Algeciras: A Torremacha Silvia, Las Fortificaciónes medievales de Algeciras, *Congreso Internacional: El Estrecho de Gibraltar,* 1987, II, 354–8, 369, 372–77. The full record of the siege is examined by C Chamberlin, "The King sent little relief": the Castilian siege of Algeciras 1278–79, in D Kagay, L Villalon, *Crusaders, Condottiere and Cannon,* 2003, 193–207.

[2] Torremacha Silva, 362–3, 369–70, 378–9.

[3] Ibn Khaldūn, IV, 131–2; Abu-l Hasan 'Ali Ben Abd Allah ibn Abi Ze' Fesano, *Annales regum Mauretaniae,* ed. Tornberg, 1845, 336, says the town was betrayed.

[4] G Anelier, *Histoire de la guerre de Navarre en 1276 et 1277,* ed. Michel, 1856, 47, 57. The algarada seems to have been a smaller stone thrower (see chapter eight).

installed in front of the church of St. Nicholas with an expert to direct it, another at St. Cernin. One of the trebuchets "had no equal", and needed a team of 20 just to lower the arm. Carpenters were assigned to each machine, presumably for on the spot repairs and adjustments.[1]

The fighting began, and seems to have consisted initially of an artillery duel with stones hurled back and forth, houses destroyed. An urgent plea for help to Paris reported that the Navarrese barons were attacking with trebuchets throwing stones weighing more than three quintals, and relief was promised. There was a particular struggle around a mill used as an arsenal. The governor had it strengthened with earthworks, and also raised timber towers in which he installed ballistas. Stones fell like rain, smashing towers and ramparts. There was an attempt at mining, but the defenders managed to drive the diggers out of the mine. A cat prepared to cover sapping was set alight. The top of one of the towers was smashed by bombardment, and in repairing it Beaumarchais had the bottom filled with earth to solidify it. Some 50 of the large stone missiles landed each day, and were collected and reused by the defenders. The bombardment was destroying houses, hoarding and the upper stages of the walls and Beaumarchais was turning to preparing church towers for the defence when news arrived of the approach of a French relief force.[2] Although, then, nowhere was captured, this remarkable event demonstrates many elements of siege warfare in the late thirteenth century, carried out not by kings, but by local barons and townspeople. The barons had access to people to build artillery, the bourgeois had to import it. Large weapons (trebuchets) were particularly destructive, smaller ones were effective against the tops of walls and ordinary houses.

In 1283, as part of the now regional war that had broken out, King Philip III of France sent an army of 10,000 across the Pyrenees. They attacked the castle of Ull, south east of Pamplona, held by an Aragonese garrison of just 30 to protect against precisely such an attack. The French managed to storm the outworks and the "barbican", probably signifying an outer bailey, and drove the defenders back into the donjon. They then dug a mine, leading to half of the donjon collapsing, and because of the collapse they still needed ladders to break into the remainder.[3]

The main confrontation with the French king was to come two years later. Meanwhile, the king having had to make concessions to the nobility and big cities of his realm to allow him to finance his Sicilian adventure, an expedition had to be arranged in 1284 to attack the rebellious lord holding the inaccessible fortress of Albarracin. The king deployed engines and crossbows all round, concentrating his attack where he could according to the very difficult terrain, but he could make little progress, and his attempted assaults by ladder were driven off with stones and arrows. By September, he had decided to continue the siege through the winter, and ordered the construction of shelters to protect his forces. The defence, themselves now short of supplies, opened negotiations. Continuing to follow old conventions of "feudal" behaviour, they sought the permission of their lord

[1] Anelier, 125, 163–5; Guillaume de Nangis, I, 244.

[2] Anelier, 173–253.

[3] Desclot, II, 125–27.

(prudently not inside) and then surrendered.[1] These two examples illustrate the fate of two fortifications, neither of them particularly modern. One, approachable, fell to assault aided by engines, the other, inaccessible, had to be blockaded into submission, a slow and costly enterprise that would not have succeeded had there been anything to divert the besiegers.

1285

King Peter prepared his defences as the French onslaught loomed. As the population fled in front of King Philip's advance, a strategic decision was taken to try to defend Gerona. Provisions were laid in and a garrison organised under command of the viscount of Cardona, consisting of 110 armoured cavalry and 2,500 footmen, including 600 Muslim crossbowmen from Valencia. Hoardings were constructed on the city's "ancient" walls, and the outworks in front of the old town wall were demolished, an unusual step presumably taken to avoid them providing shelter to the attacker. Everything else outside the walls was pulled down for the same reason, except for the church of St Felix, which was fortified. Many castles in the surrounding country were garrisoned for the king, with the effect of making it dangerous for the enemy to forage for supplies. The main French force arrived outside Gerona on 28 June.

Meanwhile the king was at Barcelona, and had to take steps also to protect the major city of Catalonia against possible attack. The city had expanded enormously beyond its old Roman walls and James I had begun to build new walls, although they would have been far from complete at this date.[2] On the land side, he ordered a widening of the ditch, and he also had constructed a substantial earthwork and timber defence on the seaward side, with wooden towers every forty paces, and two stone throwers between each tower. The French meanwhile assailed Gerona. It was midsummer, and the heat and flies, and the harassment of their supply lines, caused great discomfort. They had with them seven or eight engines (according to Desclot) but these were making no impact on the reinforced defences so they ordered up miners. By chance, the "master craftsmen" found a spot where they could dig easily. They prepared a wide and deep mine supported with heavy timber props. The defenders were able to locate it, but were very concerned, as the old stone wall they were defending was built without mortar. However, they were aware of the correct defence, and when the French miners collapsed tunnel and wall, they found another hurriedly thrown up behind. So now they constructed heavily protected wooden cats, covered in hides, to cover an approach to the wall. Overnight, a sally by 500 men carrying oil-filled vessels and torches burnt the cats, with some of the king's craftsmen still inside them. The French next resorted to the construction of other types of shelter (wrongly described as siege towers — from their function it is clear they were not) to cover them while they sapped the base of the wall. Now the crossbowmen came into their own. Using extremely

[1] Desclot, II, 155–58.

[2] For plans of the city, Bisson, *The Medieval Crown of Aragon*, 1986, 77–8.

powerful two-footed devices mounted on strong wooden platforms secured by stones, they caused heavy losses and forced the foe to keep their heads down. An attempted escalade was defeated by the shots of the defenders, too. Finally, their supplies exhausted, the defenders asked for a parley. They were granted a respite to consult the king. In the meantime, the Aragonese fleet arrived from Sicily and worsted the French ships, but the time allowed expired so the town was surrendered (September). Unable to consider advancing any further, the French left a garrison and withdrew, whereupon the Aragonese marched back up the coast recovering their losses. The great detail of this siege, albeit all provided by one side, nonetheless gives a fairly clear example of what happened when two monarchs mobilised against one another. The king of Aragon was able to strengthen an old city wall so much that a strong and well-led garrison was able to repel many strong attacks by artillery, mining and escalade. However, the king of France was able to press the blockade so tightly that, despite many difficulties of his own produced by being in a hostile land, he was ultimately able to achieve capture.[1]

Continued war in Italy

The French invasion over, and all its main protagonists dying within a short time of each other, it was left to their heirs to continue the struggle. King James of Sicily was wrongly informed that the city of Gaeta would open its gates to him. When he arrived (1287), he found this was not true and sat down to besiege it. Engines were erected as usual, and a massive bombardment was launched: roofs were smashed by the trebuchets, buildings destroyed, walls damaged so that parts fell, and towers tottered. However, a large relieving army including French troops arrived to besiege the besiegers.[2] A settlement was arranged whereby the division of the kingdom into Sicily and Naples was recognised. But war resumed, the popes encouraging the Angevins to try to restore their kingdom against the excommunicated Aragonese. In 1291, Robert of Naples besieged Messina, in which there were few provisions. Attacks by sea were thwarted by the garrison which included Muntaner, who noted that the re-supplying of the large city after Robert pulled back was futile, because the troops escorting the provisions consumed most of them on the way. In the end, a fleet from Syracuse brought in wheat, and Robert raised the siege next day. In 1300, the Neapolitan-Angevin army landed at Termini and tried to capture Syracuse for four months, but half the army succumbed to disease. In 1314, Robert attacked again, landing at Palermo and besieging Trapani, the two sides exchanging stones from catapults. Meanwhile, Muntaner was holding the Aragonese stronghold on the African coast at Djerba. Robert sent a fleet of 60 galleys with four trebuchets against him, so he set up his own mangonels and trebuchets, made sure the cisterns were full and had numerous jars filled with water, but was not in the end tested as the fleet was recalled.[3]

[1] Desclot, II, 290–358; *Chronicle of Muntaner*, 309–15; Bartholomei de Neocastro, 71–2, showing the awareness of the Sicilians of the regional character of the new war.

[2] Bartholomei de Neocastro, 105–7; Muntaner, 402.

[3] Muntaner, 475, 594, 618–20 (Djerba); Marino Sanudo, 170.

North Africa

THE STRUGGLE FOR POWER between the Almohads, Merinids and Hafsids in North Africa revolved around control of the walled coastal and inland cities. Most surrendered to blockade. The Almohad dynasty in the Maghreb was finally extinguished when the Merinids took Tinmelal after a long siege in 1275,[1] but conflict continued regularly with the Ouadite tribe with its chief city at Tlemcen. Although there is evidence of the apparently leading role of the Berbers, either on their own or through their relations with more settled powers, this did not mean that some sieges were not pressed with the full weight of current technology and resources, the cities of Mauretania being assailed on several occasions with batteries of siege engines and artillery brought along by the sultans.[2]

The conflicts within the Muslim regimes continued. One of the bloodier episodes in the Merinid conquest of the Maghreb was the capture of Sidjilmessa in 1274. This Ouadite town put up a strong resistance. The besiegers deployed catapults, ballistas and "engines hurling fire", a possible reference to gunpowder which we will examine below. It took a year for the catapults to demolish the wall sufficient to launch an assault, which then succeeded (September 1274). In 1299, the Merinids attacked Tlemcen, the capital of the Ouadite dynasty. The (much later) account says that having driven the defenders back inside the town, the besiegers sat down to a long blockade. They surrounded the town with a circumvallation and ditch, and built gates in it which were heavily protected. While the siege continued, they raided and reduced smaller enemy towns, and they built a whole siege town with a palace and mosque protected by walls and towers. The blockade lasted for an extraordinary one hundred months and was ended when the sultan was assassinated (May 1307). If this is true, it must have been extremely ineffective in preventing the town obtaining provisions. A more successful attack took place in 1335, when it sounds as if the new sultan reoccupied the previous works including the town, but this time pressed a more direct attack. Catapults and other instruments were arrayed behind the ditch, and siege towers were built opposite the town's towers, from which archers drove the defenders from the wall tops. This enabled the engines to be brought closer, and it appears that it was their role that was decisive in leading to a successful storm (May 1337).[3] Tlemcen was clearly an extremely strong town.

Meanwhile further east, in 1304/5, the Hafsid ruler Abū Bekr captured Constantine after a surreptitious escalade by a small group who opened a gate. In 1306/7, the sultan failed, despite a long siege, to take Algiers. In alliance with the Merinids at Bougie, during the sultan's long struggle with the Ouadite tribe, the latter tried in vain to take Bougie (1325) and built a new fortified settlement to blockade Constantine (1327/8). In 1334, Gafsa (Gabès) revolted and was recap-

[1] Ibn Khaldun, IV, 82.

[2] Abu-l Hasan Fesano, 332–33.

[3] Ibn Khaldūn, IV, 69–70 (Sidjilmessa), 143, 169 (1299–1307), 220–23 (1335–37); Abu'l Hasan Fesano, 340–43.

tured when the population capitulated, after the sultan had attacked with artillery. The main cause of the surrender though was that the besiegers began to destroy the surrounding orchards and palm trees on which the town's subsistence relied. The same year, as part of his defences, the sultan completed construction of a new tower at Ras el Tabiya, which had cost altogether 50,000 dinars, a sum large enough to be noticed in the chronicle. In 1348, finally, the Merinid garrison in the massive citadel of Tunis was besieged there by the population, who themselves deployed artillery. But the walls and garrison of the citadel were strong enough to resist and the attack ended when the sultan sent relief from outside.[1] Meanwhile, an account of the attack on a castle by Ismāʿīl I, king of Granada, in 1324, illustrates once more how siege artillery could be powerful enough on its own to achieve success. The castle (near Baeza) had thick stone walls, so the king called on his largest engines, with which he hurled in both naphtha bombs, and great iron balls. One of these projectiles demolished the top of a tower, leading the occupants to surrender.[2]

Sardinia

In 1323, King James II's son and heir Prince Alfonso led an expedition from Barcelona against Pisan-held Sardinia. Muntaner and Alfonso's son, his successor Peter III, told how he besieged Iglesias on the west coast of the island, having landed his artillery some ten miles away, and conveyed two trebuchets and other equipment there. Landing on 13 June, the army arrived at Iglesias two weeks later and began the siege. Following a failed assault on 20 July, Alfonso decided he would not succeed by this method and resolved instead on a blockade, and the place surrendered after more than seven months. By then disease and hunger had cruelly reduced the inhabitants — the conquerors found the place without any provisions at all — and half the besieging army had also succumbed to sickness during the unusually wet winter, a reminder that long blockades could be as hazardous to the attacker as more direct forms of attack. During the siege, the chronicle notes that a friar was killed by a crossbow bolt while in a wooden siege tower he had built, not the first time we have noticed the religious demonstrating such secular skills. Alfonso then assailed Cagliari, the chief town, in February 1324. This was a significant obstacle, and the prince reverted to more old-fashioned means of attack by constructing a fort and settlement to enforce a blockade. His engines duelled with those in the town, but in the end a major seaborne Pisan relief attempt was defeated, and Cagliari surrendered (June 1324).[3]

[1] Zarkachi, 106–7 (Gafsa), Ibn Khaldūn, II, 398 (Mahdiyya), 425, 426, 443, 464–5; III, 2 (Gafsa), 35–6 (Tunis).

[2] Makkari, II, 352.

[3] Muntaner, 662 f; Pere III (Pedro IV), *Chronicle*, ed. Hillgarth, 1980, 146 f.

England, Wales, Scotland

In England, the monarchy had to defeat internal opposition again. Just as at the beginning of his long reign, Henry III had to suppress opponents who took refuge behind castle walls, following a civil war which had revolved around the taking and holding of castles, so too after his son Edward had defeated and killed the rebel leader Simon de Montfort, the now elderly king had to set about another immense military operation to end the war.

The conflict between the king intent on his royal privileges and a baronial party determined to champion their own rights had broken out into open war in 1264. In that year, the royal army laid siege to the town of Northampton, held by Simon de Montfort the younger. The defenders were few in number, and the attackers broke in through the town walls by sapping. The defenders in the castle surrendered.[1] Meanwhile, the barons captured the royal castle at Gloucester, held by a very small garrison unable to prevent the attackers from breaking through the gates from the town. The governor took refuge in the castle's old donjon but its door was also broken down with iron tools. Simon himself, earl of Leicester, took the old motte castle of Warwick. At Easter, the baronial forces attacked Rochester, held by the earl of Arundel. Siege engines were delivered by water from London. Incendiary devices using fat and lard were prepared for an attack on the bridge, while others attacked from the side of the town with ballistas, succeeding in breaking down the gate to the bailey, whereupon the defenders retreated to the donjon, just as they had in the siege of 1215, and engines were again deployed, along with an attempt to mine, perhaps remembering the success of this method during King John's attack. With the king's army successfully engaged in extending control in the midlands, Simon returned to London. The earl's own main base was at the castle of Kenilworth, which he had devoted attention to preparing for defence if needed, leaving it well supplied with throwing engines.[2]

Siege of Kenilworth

Following the death of the earl at the battle of Evesham (1265), his followers continued to resist. A group of Montfortians held out in the isle of Axholme, but the main centre was the earl's castle at Kenilworth. In the ensuing siege, with little prospect of relief, and facing the resources of an entire kingdom, Kenilworth emulated Bedford in 1224 in holding out against Henry III in a prolonged resistance. The event was well documented both in the chronicles and in government records.

De Montfort had carried out much building at the old castle dominated by its Norman donjon. Some of this work had been grand domestic quarters fitting for a great earl, but alongside were powerful defences too, and particularly critical were the water defences, including a wide lake covering the western and southern

[1] William de Rishanger, *Chronicle*, ed. Halliwell, 1840, 23–4; *Cronica Walteri de Gyseburne*, ed. Rothwell, 1957, 189.

[2] Rishanger, 24–6.

sides, while other less substantial water works covered the other sides. These probably dated from King John's time, but now created significant problems for his son. A strong curtain with projecting towers completed the defences. From June to the end of December 1266 the castle held out. The king erected numerous siege engines (11 in one source), which cause much serious damage, although they could only be deployed against limited sectors of the castle because of the width of the lake. This also enabled the defenders, likewise provided with engines, to shoot back with some effect. The attackers resorted to building a wooden siege tower, from which crossbowmen could fire into the castle, but this was a relatively easy target for the defenders' artillery. An attack by boat across the lake was also considered. The water defences deprived the attack of its most potent weapon, the mine, and kept the artillery from creating a breach. In the end, the large number of people being sheltered in the castle told against it, as supplies were finished, and they negotiated surrender. The records show the king instructing his sheriff of Gloucester to send engines to Warwick in preparation for the siege, and subsequent demands for the materials needed to sustain the engines. One order was for horsehair for the engines: this was the first evidence of the reappearance in England of torsion weapons in the shape of the springald (for which see below). The cost was high: in October 1267, the king was authorising payment to the sheriff of Warwickshire for his expenses in "taking down the king's engines" at Kenilworth. As late as 1271, he was still paying off the debts from the siege. The sheriff of Gloucester reckoned £15 for seven beams, pendants and iron pegs and their carriage to Kenilworth; another cost was that of transporting stones for ammunition, while the siege tower mentioned (*berefridum*) involved a further payment for the timbers transported by the sheriff of Worcester.[1]

Wales

Llywelyn had taken advantage of his bigger neighbour's dissensions to consolidate his dominant position in Wales, to pull down English castles in the north and in the middle marches, and to build his own castles. These included a stone castle at Dolforwyn, just a few miles from the new castle erected by Henry III at Montgomery, and the cause of dissension with England. Against the power and ambition of Edward I who succeeded to the throne in 1272, however, neither conventional nor the traditional Welsh guerrilla war could hold out. In 1277 and again in 1281–2, several powerful English armies, each of them much stronger than anything the Welsh could field, would march into the Welsh heartlands simultaneously to destroy the principality and, after the revolt of Prince Dafyd, the whole princely family of Gwynedd. Whereas so many times in the past two hundred years, kings had been unable to do anything about the roadless mountains

[1] Annales de Oseney et Chronicon de Thomae Wykes, *Annales Monastici*, ed. Luard, IV, 191–95; *Flores Historiarum* (Matthew of Westminster), ed. Luard, III, 8–11; Rishanger, 45, 51–8; *Cronica Walteri de Gyseburne*, 204; *Calendar of Liberate Rolls Henry III*, V, 209, 216, 227, 230, 235, 294; VI, 40, 115, 168, 173. On the castle itself, M W Thompson, *Kenilworth Castle* (guide book), 2nd ed., 1991.

to which the Welsh withdrew, Edward had worked out how to overcome this defensive strength. Snowdonia offered a refuge more impregnable than any castle, but without food it must fall.[1] This called for a massive effort to surround the whole area and to seize Anglesey whence food came. Edward's investment in the resources needed to achieve this outcome was rewarded with victory, although at enormous cost.

There were few sieges. In 1277, the English besieged Dolforwyn, with its associated (probably palisaded) town, which surrendered after just nine days for lack of water. The archaeological evidence indicates that it was not however a passive blockade. The curtain wall and both the north-east angle tower and the main square tower had been substantially repaired by the English after capture, and inside the castle were found many catapult balls made from volcanic conglomerate quarried nearby, or sandstone. The new occupants carried out work that included strengthening the castle's defensive capabilities, including (critically) constructing a cistern, widening the ditch and digging an outer ditch.[2]

Edward's follow up to his victory was to begin a programme of constructing castles. Existing buildings were included. An order of June 1277 instructed his captain of munitions to retain in his hands the castles of Dynevor, Carreg Cennen and Llandovery in the south, and to ensure their proper provision. In the north and centre, facing Gwynedd, much more substantial work was called for, and the new castles of Flint, Rhuddlan and Aberystwyth were started.[3] These were immense works, and a great expense. It also meant that none of the castles was complete when, in 1281, Llywelyn's brother Dafyd spotted an opportunity and launched a major Welsh uprising which took the English by surprise. The old castle at Hawarden was seized in a bold move and in the following year the town and the new castle of Aberystwyth were stormed and burnt.[4] King Edward had to plan a new conquest, which he did with thoroughness. Dealing first with the north, English forces drove Dafyd from his bases at Denbigh, Dinas Bran and Hope (Caergwrle) castles. The latter was abandoned, damaged, and was repaired by the English, a large force being needed to protect the workmen (many hundreds of them called from the English shires) as they cleared the well, pulled down an unsafe tower and made the place secure. The nearby Welsh castle at Ewloe seems to have been abandoned too, then Hawarden was too as Dafyd pulled back westwards.[5] Denbigh seems not to have detained Edward long, the Welsh knowing

[1] J Marshall-Cornwall, 'The military geography of the Welsh marches', *Geographical Magazine*, xxx, 1957, 1–12; G Jones, 'The defences of Gwynedd in the Thirteenth Century', *Transactions of the Caernarfonshire Historical Society*, 30, 1969, 29–39; D J C King, 'Defence of Wales 1067–1283', *AC*, CXXVI, 1977, 1–14.

[2] *ByT(P)*, 119; *ByT(H)*, 265; L Butler, 'The siege of Dolforwyn Castle in 1277', CG, XIX, 2000, 25–6, and *Medieval Archaeology*, XLV, 2001, 378–9. See also L Butler, J Knight, *Dolforwyn Castle, Montgomery Castle* (guide book), 2004.

[3] *ByT(P)*, 118; *ByT(H)*, 267; *ByS*, 255; *Calendar of Patent Rolls 5 Edward I*, 212; *Flores Historiarum*, 49–56.

[4] *ByT(P)*, 120; *ByT(H)*, 269; ByS, 257.

[5] All from the Government records used in the work of J Morris, *Welsh Wars of Edward I*, 161–63.

that surrounded castles would be a death trap for them, and building work was undertaken to make it a secure base. The attack on Snowdonia itself could now take place, as another English force had seized Anglesey again and constructed a bridge of boats across the Menai Strait. But the massacre of troops from that island who had launched an undisciplined attack on the mainland delayed the final invasion. Llywelyn was killed after a battle at Orewin bridge near Builth (December 1282), a better fate than that served on the captured Dafyd, a traitor in Edward's eyes. The castles of Snowdonia had been rapidly taken, Dolwyddelan perhaps having resisted briefly (January 1283), Dolbardarn falling in May.[1]

Welsh revolts

While Edward set about consolidating his final conquest with yet more new castles, with the start of work at Conwy, Caernarfon, and Harlech, the Welsh had not yet finally given up. In the south, Rhys ap Maredudd had remained loyal to the king while the house of Gwynedd was destroyed, but in 1287, after a long period of disputes with the English justiciar, he rebelled. This led to a siege of his castle at Dryslwyn. Starting out as a fairly simple hill top Welsh castle of a circular donjon and irregular stone-built ward, the castle had been significantly developed by the prince with a grand new hall and other domestic buildings, and two outer wards reaching along the ridge top, all sheltering a small town on the slope below. Edward being absent did not prevent an efficient military operation by the English to crush the revolt. An army of many thousands was assembled, some of whom quickly recovered the castles seized by the rebels (Dinefwr, Carreg Cennen, Llandovery), while most were collected in the flat plain around the rock on which Dryslwyn stands. State records show some of what happened. Sappers and engineers were brought from England, and £14 was spent on hides, timber, rope and lead for the construction of a trebuchet. Twenty quarrymen and four carters were employed to provide the ammunition. Given the location of the castle, the weapon must have been deployed on the town side. The decisive weapon was mining, stated by the chronicles and confirmed by accounts of payments paid to workers for removing stones from the collapsed chapel tower. If this was where the mine was driven, it says much for the skill of the engineers, because this tower stands on top of a very long and steep slope. Indeed, it is suggested that many men were actually killed by a premature fall in the mine before it was fired. Dryslwyn anyway fell within about two weeks of the siege starting.[2]

Another prince, Madog, launched a revolt in the north in 1294/5. The recently founded new town and castle of Caernarfon was a target and it was sacked, and he also attacked Castell y Bere. When English troops gathering for transport to Gascony arrived in Wales instead, the revolt was rapidly put down. Caernarfon castle was clearly not finished when the revolt began, and probably had no de-

[1] Trivet, *Annales*, 256–59; Morris, 190.

[2] Government records quoted in Morris, 212–13; *Annales Cambriae*, 109; *ByT(P)*, 121; *ByS*, 261; Trivet, 266; *Flores Historiarum*, 66. S Rees, C Caple, *Dinefwr Castle, Dryslwyn Castle* (guide book), 1999, especially 16.

fences except a ditch on the town side, although the town walls were complete. The circuit of town walls was probably rebuilt rapidly after the collapse of the uprising, by September 1295, while at that time the walls of the castle were recorded as being raised to a height of between 12 and 24 feet (three to six metres). A substantial force of foot soldiers, archers and crossbowmen were employed to protect the workers against further Welsh attacks, as was another force at Cricieth, where work was proceeding on strengthening the former Welsh castle. In 1295, work began on the great pile that would be Beaumaris, on Anglesey. By August, an enormous building effort involving 400 masons, 200 carters, 30 smiths and carpenters and more than 1000 other workers had managed to raise walls to between 20 and 28 feet (6.1 to 8.6 metres), ten of the towers on the outer circuit were under construction along with four on the inner and the four great gates already installed could by then be locked at night.[1]

Late thirteenth century castles in Wales

An immense literature has developed around the castles built in Wales around the time of the conquest. A few comments here will suffice. The common features of all of them were tall, thick curtain walls with strongly projecting towers, and an outer wall and ditches. The origin of the concentric style that reached its perfection on the level site of Beaumaris, and on the otherwise much different cliff top at Harlech, can be found at Caerphilly, begun in around 1270 by Gilbert de Clare, earl of Glamorgan to establish his control in formerly Welsh Senghenydd. Llywelyn twice interrupted the construction by force, but within seven years the massive new castle was ready. Here the immense water defences tell of a direct connection with the recently concluded siege of Kenilworth. But water alone did not provide the main strength of the new castle. Inside the lake with its towering screen walls, a strong quadrangular outer curtain marched closely around a towering inner circuit with powerful corner towers and gatehouses at opposite ends comprising twin towers enclosing a gate passage well provided with portcullises and gates. This scientific arrangement of low outer walls overlooked by tall inner defences was not exactly original, since the walls of Constantinople had the same pattern, and nearer home the walls of Henry II at Dover and of Henry III and Edward I at the Tower of London, but Caerphilly's concentric defences were designed as such from the start.[2]

The castles begun after 1277 at Flint, Rhuddlan and Aberystwyth were different. Flint's circular donjon at one corner of the quadrangular enclosure, may have owed its origin to the recently built *Tour de Constance* at Aigues-Mortes on the French Mediterranean coast, which Edward had seen, although the design was now common enough. The other two were lozenge-shaped, with low outer walls but very strong inner defences, and the chief accommodation transferred to the

[1] Prestwich, A new account of the Welsh campaign of 1294/5, *Welsh History Review*, VI, 1972, 89–90 (the Hagnaby Chronicle); Morris, 267–8, Trivet, 282; *Cronica Walteri de Gysburne*, 251.

[2] C N Johns, *Caerphilly Castle* (guide book), new edition by D Renn, 1997.

gatehouses. Conwy and Caernarfon were again different. Conwy relied on a single enclosure of immense strength standing on a rock, with barbicans protecting gates at each end. Caernarfon involved a flat seaside location and shared with Conwy the plan of two wards end to end, but was intended to have two gigantic gatehouses. Among the powerful polygonal wall towers was one of exceptional size and design, the Eagle tower, with walls five metres thick. Built on the site of the motte castle briefly occupied by Norman invaders two centuries before, Caernarfon was truly a royal building, although the original plan for the interior was never completed (Plates 4, 5). The designer and overseer of much of the royal building was the former employee of Count Philip of Savoy, known as James of St George, and the influence of Savoyard castle designs has been argued for the castles in Wales, although not without some dispute.

The castles were designed to overawe the conquered people: as instruments of war they were far more powerful than the likely resources of any possible enemy would require. The ample provision for active defence at all these castle, both from the wall head and from intramural galleries connecting rows of arrow loops, is evidence of a defensive role too, while the arrangement of loops at Caernarfon, where multiple shooting positions were placed to use single loops, was another sophistication of purely defensive purpose. The gatehouses were enormous buildings, all provided with an unprecedented succession of defences for the entrance with rows of gates, portcullises and machicolations. At the same time, they were also all provided with access to the sea so that they could survive blockade or be used as a base for offensive operations. Also of great strength were the defences of the attached English-settled towns at Conwy and Caernarfon, provided with strong projecting towers at frequent intervals. The immense expenditure involved in their construction meant that the later castles were never finished, building work carrying on in increasingly desultory fashion for some forty years.[1]

Scotland

Edward's intervention in Scotland failed to grasp that the people were no longer willing to submit to whoever happened to be made king, especially if he was a helpless puppet of a foreign force. While engaging in traditional war, Edward was invincible, but against a guerilla struggle, his great armies were unable to hold down what they had occupied, and unlike in Wales, he could not find sufficient resources to control the whole of Scotland, no matter how many castles he held. In fact, the cost of the effort in Wales made it inconceivable that he could embark on a similar process of constructing new fortresses. Castles did, nonetheless, play a part, and would have played a larger one if the English had had their way.

Edward's first armed attempt to impose his overlordship on the Scottish realm

[1] Allen Brown, Colvin, Taylor, *The History of the King's Works*, I, 293–395 (also published separately in the name of A Taylor). The internal arrangement of buildings at Caernarfon is discussed by J Mathieu, A new interpretation of Caernarfon Castle, *Medieval Life*, 9, 1998, 11–13; Mesqui, *Châteaux forts*, 15 (Aigues Mortes); A Taylor, Master Bertram, Ingeniator Regis, *Studies in medieval history*, 1989, 302.

came in 1296 when he invaded to impose his will on John Balliol, his own candidate as king in 1292. Early in that year, John Comyn had invaded the western march with his Galwegians and attacked Carlisle. They burnt the suburbs and tried to burn the gates of the city, but the wind changed and thwarted them. The citizens flocked to the walls, men and women alike, and hurled stones and missiles at the Scots, who withdrew.[1]

Edward himself came up the east coast, taking Berwick in 1296. This was achieved by escalade of defences consisting of an embankment, while a fleet stood off the harbour, and was followed by a horrendous massacre clearly designed to cower the Scots, something in which it transparently failed. The king then strengthened his new base with a ditch 25 metres wide and 12 deep. Following a victory at Dunbar, that castle on its headland site also fell to siege. The wall was breached by undermining according to one chronicler.[2] A revolt led by William Wallace broke out the following year. In his 1298 campaign of revenge, Edward seized the castle at Stirling and began to strengthen it, as he recognised its key role in allowing troops occupying it to secure the whole region. The castles at St Andrews and Lochmaben were taken easily. That year, the Scots attacked the English garrisons left at Roxburgh, Edinburgh, Stirling and Berwick. One account says that the attackers of Roxburgh were thwarted by English forces mobilised from Northumberland who approached silently by night, killed the engineers servicing the Scottish engines, and disabled the stone-throwing artillery. Once again the king returned to the north from his business in Gascony, and defeated Wallace at Falkirk.[3] Once again, the king believed he had ended matters in Scotland, only for resistance to continue as guerilla warfare, and for Robert the Bruce eventually to win the internal struggle for the crown and launch his own new struggle. Castles continued to serve as the base for English garrisons: that of Edinburgh in 1300 comprised, alongside 18 crossbowmen and 46 archers, two carpenters and two smiths, and Thomas the engineer and his boy.[4] Berwick served as the arsenal, its stores in 1298 including "100 pounds of lead in one lump, and another great piece of brass of 10 stone". In the castle were four engines with cords, and on the walls three springalds "with all their furniture except three cords". The weights were presumably counterweights for trebuchets.[5]

[1] Trivet, 342–3; *Chronicon de Lanercost*, 162.

[2] Trivet, 344; *Flores Historiarum*, 96–8; *Chronica Walteri de Gyseburne*, 274–9; John of Fordun, *Chronicle of the Scottish Nation*, ed. Skene, 315–16; *Chronicon de Lanercost*, 162, 176–77; J Bain (ed.), *Calendar of Documents relating to Scotland preserved in the PRO*, II, 249 (works at Berwick); Thomas Gray, *Scalachronica*, trans. Maxwell, 15.

[3] Lanercost, 179, 191; Trivet, 312–14; Gray, 20 (Roxburgh).

[4] Bain, II, 289.

[5] *Documents illustrative of the History of Scotland from the death of Alexander III to the accession of Robert Bruce*, ed. Stevenson, 1870, 323–25.

Caerlaverock and Bothwell

In 1300, Edward's army besieged Caerlaverock, just ten miles from English-held Lochmaben. The event was described in a remarkable verse history. The triangular stone castle was not as strong as the chronicle suggests, but it was protected by a ditch. The defenders had crossbows and springalds and the attackers had the full panoply of a royal army. Initially, the bombardment was by smaller missiles, though these were still capable of beating "helmets to powder" and shields "in pieces". Brother Robert was in charge of a stone thrower called "robinet", which shot without pause. Once he had completed the construction of three other large weapons, the garrison (60 men inside the small castle) had nowhere to hide from the devastating bombardment of stones, and as the walls began to crumble around them they were forced to surrender.[1] In 1301, the king besieged Bothwell, one of the most recent castles built in Scotland, by the powerful Moray family. It is likely that what Edward faced was the immense circular donjon and some of the defences of the inner ward, the whole impressive original design never being completed. A great army of nearly 7,000 men was engaged, including 20 miners and other artisans, especially carpenters. A tall siege tower (*le berefroy*) was built in Glasgow and slowly conveyed the eight miles to Bothwell, where a new bridge had to be constructed to enable its arrival, and the records show the arrival of artillery as well, two "good engines" being shipped via Berwick. Wages were paid to 12 Scottish carpenters for operating a stone thrower at 3*d* per day, while the master in charge was paid twice that. Within a month, the castle had surrendered.[2]

Stirling 1304

In 1303, the Scots attacked the English in Stirling. The place on its high hill held a garrison of 60 archers who only surrendered when provisions ran out. The now elderly Edward came back to Scotland. The Scots defended Brechin, confident in its thick walls, solid enough to resist siege engines, but it was a missile from a stone thrower that killed the governor and surrender followed.[3] In 1304, the latest siege of Stirling began. Edward ordered the building of large engines and the defenders were also able to shoot back with their own. The chroniclers recorded the array of machines the king prepared, including enormous trebuchets, and their existence is confirmed in the state records. All the lead was stripped from St Andrews for the engines (no doubt for the counterweights). Two engines, one called Segrave and the other Vernay, were brought from Brechin, while Robinet, perhaps the same engine that had served at Caerlaverock, was also present. The same record (6 April 1304) also noted the provision of 1,384 stones as ammunition. Other records indicated the presence of 13 other engines, each with their own name. Crossbow bolts were ordered in their thousands. Springalds featured too on both

[1] *The Siege of Caerlaverock*, ed. Nicolas, 61–87.

[2] Bain, II, 318–19; IV, 449–52; *History of the King's Works*, I, 412; C Tabraham, *Bothwell Castle* (guide book), 6–7.

[3] *Flores Historiarum*, 113–14.

sides. The bombardment smashed the buildings inside the castle and shattered the walls. Threatened by assault, for which Edward was preparing the ladders and other instrument of assault, the garrison surrendered after three months and were spared,[1] but only after the king had forced them to receive blows from his new engine called the War Wolf, presumably a large trebuchet. Construction of this engine had occupied five master carpenters, 50 carpenters and four pages.[2] The relative value of the work was indicated by the large sum of £40 paid to Thomas Greenfield for his work on War Wolf, while John de la Mullier was paid £20 for making "Greek Fire" and throwing it into the castle "for burning the houses". This is a rare reference to the use of incendiaries in the west.[3] The total cost for the forces engaged in the siege was an enormous £8,881 *7s 8d.*

Bruce's crowning as Scottish king in 1306 led the English once again in pursuit of the resilient "rebel", and once again the army was too strong to challenge in the field. Operations included the siege of Kildrummy. The castle was a modern thirteenth century building, with a strong double-towered gatehouse at the apex and four other D-shaped or circular towers linked by tall ramparts, but it did not have an outer defence. Sir Neil Bruce held off the English attacks until a member of the household was bribed to start a fire in the hay stored in the Great Hall, which having taken hold swept the interior of the castle, the garrison was driven to the ramparts, where they may have been forced to find shelter in hoardings on the external face. The attackers were able to approach and burn the gate, but were themselves kept out by the ferocity of the blaze inside. Before long, and also now having no supplies, the garrison surrendered and those regarded as traitors were executed.[4]

The springald

THE SPRINGALDS first noted in England at Kenilworth were torsion-powered ballista-type weapons, powerful developments of the crossbow idea, hurling heavy iron bolts with great powers of penetration. Their appearance across Europe over these years represented the first reappearance of torsion power since ancient times.[5] The records for the siege of Stirling included an order for "56 pounds of horsehair for springalds in 2 canvas sacks".[6] The specification of the requirement to keep it in sacks was presumably to protect it from weather condi-

1 Fordun, 329; *Flores Historiarum*, 119–20; Trivet, 337–8; Gray, 25–6.

2 Bain, II,389, 395, 405; IV, 477.

3 Bain, II, 406; IV, 477.

4 John Barbour, *The Bruce*, ed./trans. Duncan, 1997, book IV, lines 58–177. Barbour's late 14th century verse account is not contemporary, but offers many details passed down to his time. C Tabraham, *Kildrummy Castle and Glenbuchat Castle*, (guide book).

5 E Gessler, Der Springulf, ein Mittelalterliches Torsiongeschütz, *Basler Zeitschrift für Geschichte und Altertumkunde*, 20, 1922, 89–203, presents the evidence from Swiss, German and French sources of the 14th and 15th centuries. For other studies and modern accounts of the development and history of this weapon, see chapter 8.

6 Bain, II, 405.

tions that might ruin it as a source of power. Thomas Gray records that his father was hit in the face by a bolt from a springald at the siege of Stirling, but, remarkably, survived.[1] The development and use of the springald is discussed more fully in chapter 8 below.

Fortifications of the Scottish wars, and the Scottish counter-attack

Edward's strategy for holding Scotland involved placing garrisons and siege equipment in fortresses that controlled key routes, while also seeking to protect England from raiding. The various lists of fortresses given in government records include Dumfries, Selkirk, Roxburgh and Berwick on the borders, and Lochmaben, Peebles, Lanark, Carstairs, Edinburgh, Kirkintilloch, Ayr, Bothwell and Dalswinton. Mostly, this meant occupying existing fortifications, although in many cases work was carried out to repair existing walls, or to provide new, larger enclosures. The commonest model consisted of timber palisades shielded by ditches, and the expenditure accounts show this frequently being carried out. The brother Robert who built the engines at the siege of Caerlaverock may be the same Cistercian monk charged with organising the building of the "peel" around the old castle at Dumfries, and with the safe transport of an engine and a springald from Carlisle to Ayr castles in 1306.[2]

A new wooden fortress altogether was built at Linlithgow, where the plan for stone towers was abandoned and instead timber ones were constructed on the promontory site where the palace now stands. This was intended as a major base for armies and much attention was paid to its completion, under the direction of no less a master than James of St George. The work resisted a Scottish attack in 1302, but a comparable defence at Selkirk (around the old motte) was not so fortunate and was taken: work on a tower, a postern "faced with stone", a drawbridge and portcullis was recorded, but the peel was not finished.[3] The most notable exception was the strong new stone castle erected on an old site on the western side at Berwick, an irregular quadrangle with projecting round towers. Little remains apart from the remarkable stepped wall that marches from the castle's south west corner down the steep slope to the river Tweed.[4] Overall, the contrast with the building programme in Wales could scarcely be starker: the limits of Plantagenet power and resources had now been exceeded.

The Scottish leaders could not compete in traditional warfare against the English. Robert Bruce switched the approach, rendering castles indefensible after capture. Most of his successes, too, came from secret escalades often carried out at night, whereupon the fortress would be slighted. Edward II could not conceive

[1] Gray, 26.

[2] Bain, IV, 392.

[3] Bain, II, 332, 339–41; *History of the King's Works*, I, 409–22; D Pringle, *Linlithgow Palace* (guide book), 5.

[4] *History of the King's Works*, II, 563.

of new methods to deal with the resurgent Bruce. When he did venture north, his efforts were not exactly successful. In 1308, a fortification at Forfar was taken by nocturnal escalade, the watch being asleep, and the wall was destroyed.[1] In 1312, English-held Perth fell to a silent escalade, and it was observed that this was the only way to do it because the wall was of stone, and the Scots nearly succeeded in taking the key English post at Berwick by the same method, but a barking dog roused the watch. Linlithgow was taken by a clever ruse, an armed party concealing themselves in a cart which suddenly stopped within the opened gates. Another escalade succeeded at Roxburgh in 1314. The garrison was surprised as it dined, the governor could not hold out in the main tower, and the whole fine stone castle was razed. A large Scottish force attacked Edinburgh castle from two sides. With the small garrison diverted, a company got in by ladder after a daring scramble at night to the highest point of the rock, captured the fortification and again destroyed it. At the point of attack the wall was only four metres high so a ladder could reach the top. Then they laid siege to Stirling, the linchpin of English control, and it was the threat to this remaining garrison that brought the English army north, to catastrophic defeat at Bannockburn.[2]

Carlisle 1315

The victorious Scots raided, generally avoiding the castles. But in 1315, King Robert himself made an attempt at Carlisle, attacking for ten days from 22 July. The governor, Andrew Harcla, had perhaps a few hundred fighting men, reinforced from the surrounding area whence people had sought refuge behind the walls.[3] The attack concentrated on the walled town. The defenders shot arrows and hurled stones from the ramparts and prevented the Scots from attacking the town's three gates. On the fifth day, the Scots erected a stone thrower which caused no appreciable harm to the defences. The garrison had seven or eight springalds, which hurled long darts, and also stone throwers stored in the castle, which between them caused serious casualties among the besiegers. The next step for the Scots was to build a wooden siege tower to reach over the top of the rampart, against which the town's carpenters followed traditional methods of defence by constructing a wooden tower to raise the height of their own defence opposite the siege tower. This was not however called upon, because the Scottish engineers had made their engine too heavy to move in the soggy ground. The next step was an escalade with numerous ladders. However, the ditch covering the wall was too deep to be filled in with available materials, and the wooden bridges the Scots made to cross it were too weak, and collapsed. An attempt at mining was also abandoned

[1] Barbour, book IX, lines 313–25.

[2] Fordun, 338–40; Lanercost, 221–29, Barbour, book IX, lines 336, 370–445 (Perth), book X, lines 85–252 (Linlithgow), 357–476 (Roxburgh), 589–705 (Edinburgh), 810–30 (Stirling).

[3] 50 men at arms, 30 hobelars (light cavalry), 150 archers in 1314, and another 176 spread between the town's three gates, according to Bain, III, 76–7. There were also more than 100 men at arms at Cockermouth Castle.

for similar reasons. For the following three days they tried attacks all round, and as a last resort attempted the same ruse that had succeeded at Edinburgh, making diversionary attacks while James Douglas came from a previously unassaulted side with specially long ladders. The alert garrison however hurled them all back, and the besiegers gave up on hearing of the approach of relief from the south, burning their engines.[1] The king of Scotland, therefore, had access to the same repertoire of equipment as any other medieval ruler, but from the account does not seem to have used it very efficiently. A rather less substantial effort was launched against the key border castle of Norham, which was twice blockaded by Scottish forces. On one occasion, Scottish troops managed to fight their way into the outer ward, but they were unable to penetrate the inner defences where Thomas Gray's men held out in the small enclosure around the tall donjon until the attackers gave up, having tried but failed to damage the outer walls by mining.[2]

Berwick 1318–19

Three years later, Robert achieved a significant breakthrough at the other end of the frontier when his troops succeeded in another escalade at Berwick, assisted by a citizen on whose watch the successful attack was launched, seizing the town and forcing the garrison in the castle to surrender for lack of victuals five days later. King Robert decided to retain the fortification for a change and placed in it a garrison of 500. Among them was a Fleming, John Crab, skilled in designing and making siege equipment. Crab oversaw the production of artillery including springalds, and also "Greek Fire".[3] They then had to resist a sustained attempt by the English to recover the town and castle. The counter-attack was fierce. Engines were shipped north from York, ditchers and other workers were mobilised in their hundreds, garrisons were paid at Newcastle, Alnwick, Warkworth, Bamburgh, Mitford and Prudhoe castles. The army included 120 crossbowmen and 1,520 archers among more than 6000 other troops.[4] An initial direct assault with ladders was driven back, although in places the town walls were so low that the top could be reached with a spear. A simultaneous attempt by sea, where a ship had been provided with a bridge to lower onto the walls, also failed and the vessel was burnt and its engineer captured. On the fifth day came a new attack, using a large wooden "sow" to cover an attempt to sap the wall, while scaffolds were also built to over look the defences. Crab responded by preparing movable cranes, and prepared incendiaries using pitch, tar, flax, oakum and brimstone combined into faggots with dried timber. Meanwhile, the captured engineer was forced to operate a stone thrower, with the third shot from which he hit and disabled the sow, whereupon Crab finished it off by hurling burning faggots. The skilled artillerists also succeeded in defeating another seaborne attack by sinking the boats. The

[1] Lanercost, 230–32; M McCarthy, H Summerson, R Annis, *Carlisle Castle*, 135–7, gives the changing numbers of the garrison over this period.

[2] Gray, 64.

[3] Barbour, book XVII, lines 1–204, 235–51 (Crab).

[4] Bain, III, 124–6.

siege was abandoned after a detachment was defeated in the field, so Berwick returned to Scottish control for a few years yet.[1]

Robert's brother Edward launched an invasion of Ireland and embarked on what proved to be premature attempt to raise a pan-Celtic coalition against a common foe. Edward Bruce was killed in battle in 1318, but the native Irish lords were anyway holding their own against the foreign encroachment. The settlers from England were now building stone castles as centres of their lordship. That at Roscommon had been demolished by the Irish in 1270, in a year when many settler castles were destroyed by the O'Conors in the east of Ireland. When Roscommon was rebuilt in the 1280s, it was rightly compared in design to Harlech, although its walls are rather weaker. A castle of similar design was put up at Ballintober, not far away, around 1300, which was even larger. But for the most part, English lordship was restricted to the coastal cities and the lands in the north and east that were already heavily colonised.[2] The defences of Dublin, it seems, were in a hopeless state of disrepair at this time, and as the Scots neared, the citizens had to take their own steps to pull down shops built against the city walls, and demolish the spire of a church to get the stone to repair the walls. The city had been in legal dispute with Geoffrey de Morton for years, the latter having been granted *muragium* (murage) by the king of England for the purpose of maintaining the walls, but allegedly having diverted it to favours for his friends. Not only was he erecting his own houses on the walls, he was making openings for windows in them too (1307–10). Morton admitted his guilt in 1312, and it seems that the defences at least looked sufficiently strong in 1317 to deter Edward Bruce.[3]

France and the Low Countries

Edward I's forays into France were, generally, expensive failures. Gascony remained the only Plantagenet possession in France and it was there that much of the actual fighting took place. In 1295, Edward received the surrender of Bayonne, town and castle in quick succession, after a fierce assault. The French, led by Charles of Valois, besieged Podensac castle, and then the more important town of Rions on the Garonne and took them after eight days of attacks.[4]

King Philip of France, ruling one of the most powerful and centralised states of Europe, was a force to be reckoned with, but he would face greater difficulties in the north. The populous cities of Flanders had wealthy merchants presiding over fractious factions in walled urban centres inhabited by large numbers of artisans.

1 Fordun, 340–1; *Flores Historiarum*, 188–90; Lanercost, 235; Barbour, book XVII, lines 261–474, 588–826. Barbour's editor suggests that Crab and the captured engineer are the same person, but the text does not seem to bear this out.

2 *Annals of Clonmacnoise*, 247–9; Barry, *Archaeology of Medieval Ireland*, 66–8.

3 J Gilbert, ed., Historical and Municipal Documents of Ireland AD 1172–1320, from the Archives of the City of Dublin, *RS*, 1870, 270–3, 274–5, 277–80. A modern account is S Duffy, *Robert the Bruce's Irish wars. The invasions of Ireland 1306–1329*, 2002.

4 Guillaume de Nangis, 288–9; Trivet, 282.

Philip tried to depose the count and impose direct rule on Flanders. In 1297, the king sent his troops against Bruges, occupied it, and had a deep ditch dug around the city. Over the coming year, earthen ramparts and strong gates were erected there. At the same time, he ordered the construction of defences at Dam, Aardenburg, Ghent, Cassel and other towns, designed as much to overawe as to protect the local people. In 1301, the neighbouring count of Holland came down from the north upon Bruges and had the defences pulled down, and the ditches filled in. In 1302 open war broke out, and the knightly army of France went down to dramatic defeat against the commoners of Flanders at Courtrai, a battle the tactics and outcome of which bore uncanny resemblance to those used by the opposing sides at Bannockburn twelve years later.

The war saw little siege activity: the great towns were very large, while the castles of the countryside were small and weak. In 1302, the Flemish forces had captured a house at Waele which was "partly fortified", and they took the town of Wynendaele after three weeks, but had been unable to storm it because it had walls made of brick. The more substantial castle of Termonde only fell after the battle when it had no hope of relief. The following year, the Flemings attacked Lessines with engines, allowing the garrison to leave, but in August they came up against strong defences at Tournai and although they used engines they had made no progress after six weeks. In 1304, it was noted that the French equipment included "horrible" instruments called springalds, "which hurled bolts which no armour could resist", a description confirming the character of these weapons.[1]

The conflicts in the Low Countries were not restricted to Flanders, and in the east of the region there were complex alliances between the local powers. As an example, in 1288 the duke of Brabant was urged to root out robbers based in the castle of Woerinck (Worringen), and he besieged the place closely. This led to a large relief army being assembled by a wide coalition of the duke's enemies led by the archbishop of Cologne, the count of Gelders and the count of Luxembourg, which was however bloodily defeated at the battle named after the castle, which then fell. Meanwhile, the duke's ally, the count of Jülich, took advantage of the victory to destroy castles belonging to the archbishop all over the Rhineland.[2] Few of the conflicts attained this level of massive violence. In 1298, for example, in the civil strife endemic in the bishopric of Liège, one party attacked and burnt the *turris* (tower) belonging to their rivals, who promptly responded by burning a tower belonging to the first party. At this point the bishop, who was also the local prince, intervened. This warlike prelate was involved in serious conflict again in 1315, when he called out the militia of Liège to destroy a number of castles, at one of which he deployed "machines and mangonels" for three weeks, but without success. He made sure of the security of his own fortifications at the same time at Bouillon, Franchimont and Logne.[3]

[1] All from the *Annales Gandenses*, ed. Johnstone, 1951, 5–8, 15, 26, 35, 49, 55 (springald).

[2] Dynter, E, *Chronica nobilissimorum Ducum Lotharingiae et Brabantiae*, ed. De Ram, 1854, II, 720–22.

[3] *Chronique de Jean de Hocsem*, ed. Kurth, 1927, 98–99, 160–1. The author was a canon of Liège and contemporary of the events he described. On the bishop's castles, P

A more serious but still local conflict erupted in 1319 between the Meuse towns of Dinant and Bouvignes, one being in the county of Namur, the other in Liège. Persuading the bishop to invade and devastate Namur, the Dinanters attacked Bouvignes, setting up a regular siege. Their tactic was to construct a tower of their own from which stone throwers could hurl missiles into the town, smashing the houses, on the other side of the river Meuse, so they must have been large trebuchets to have the range. The following year (1321) the attack was resumed by a force including knights and sergeants and also, apparently, siege engineers, because a ram was used against the wall.[1]

In the county of Holland, the long-reigning Floris v (1256–96) made substantial contributions to the development of the traditional trappings of late medieval countship. The verse chronicle of Melis Stoke told of the count's construction of fortresses to contain his neighbours, the West Frisians. Of these castles, the only survivor is Medemblik. This is a small quadrangular castle inside a wide wet moat, with large projecting round towers at the corners, of which only one tower and one side with the great hall remain. It is however evidence that even in the small and relatively poor county of Holland, castles of normal thirteenth century form were built. Medemblik is singled out by the chronicler as being exceptionally strong, indeed "perfect", but records that this was only one of a number built around 1287. The Hollanders also engaged in siege warfare. In 1304, the new Count John's brother Guy, the bishop of Utrecht, tried to occupy the island of Duiveland. His enemies, supporters of, Guy, count of Namur, came with an army and drove him out. They then sat down to besiege the town of Zierikzee with siege engines. But the town was strong and had been well provisioned by the count of Holland, and the attack was fruitless. The siege would take another turn when aid arrived sent by the French king by sea in the form of Italian troops and ships hired from Sicily,[2] and a great sea battle ensued.

The common form of castles built in most of the Low Countries at this time was a rectangular tower, in the style and shape of a donjon, but often without an outer fortified enclosure. Such towers have been dated to the thirteenth and fourteenth centuries, and their builders were the local nobility. That some of them had some defensive role is indicated by the provision of arrow loops, and by the frequency with which attacks on them feature in the chronicles. Other towers, of similar dimensions, had a more passive role, probably acting as refuges for local populations against passing danger, or as a place strong enough to deter bandits. The one thing that the historical sources do make clear is that they were attacked, and defended.[3]

Hoffsummer et al, Naissance, transformation et abandon de trois places-fortes des environs de Liège: Chèvremont, Franchimont et Logne, CG, XIII, 1987, 63–79.

[1] *Chronique de Jean de Hocsem*, 167–8.

[2] *Rijmkroniek van Melis Stoke*, ed. Brill, book IV, lines 538–45 (Medemblik), 550–63 (other defences), book VIII (Zierikzee); *Annales Gandenses*, 52–3, 56. There is a short guide to Medemblik castle available at the site.

[3] Ubregts, Les tours de Fontaine et de Voraux, et le "donjon" de Wassage, *Bulletin de la Commission Royale des Monuments et des Sites*, 4, 1974, discusses three towers in Liège.

The Empire

The death of Frederick II led to an interregnum. Without even a nominal ruler to answer to, German princes and lords at all levels enjoyed a generation-long opportunity to enrich themselves constrained only their neighbours' efforts to do the same. Finally, in 1273, with the support of the church, Count Rudolf of Habsburg was elected king. Based in a small Swabian lordship, Rudolf may not have seemed to be much of a threat to the greater lords of the empire. They were mistaken. The new king quickly laid his hands on the duchy of Austria, and set about an ambitious programme of recovering imperial lands and possessions lost during the previous years, relying on support from groups of the greater princes, and backing from the many German cities.

Ottokar II, king of Bohemia (1253–78) had been trying to extend his domains into Austria, Styria and Carinthia and resented the new ruler's gathering in of those territories. In 1276, assembling a south German army, Rudolf besieged and took Vienna and forced the Bohemian to recognise his rule. Two years later, Ottokar broke the treaty and attacked again. Joined by Polish and Moravian forces he marched to Brunn to gather forage, then pressing on, attacked the castle of Drossendorf, held for the German king by a "brave knight" called Mizow. Ottokar besieged the castle with his large army, including siege engines, and forced its surrender. He was besieging the castle at Laa when he heard of the arrival of Rudolf's army, and in the ensuing battle of the Marchfeld the German-Hungarian army won a resounding victory. Descriptions of the battle have pointed out that only cavalry seem to have been involved on both sides, but the sources clearly indicate that Ottokar had a siege train, without which he could not hope to conquer the walled towns of Austria.[1] The victor stayed in Vienna, and entrusted the duchy to his son Albert. The area would become the heartland of the Habsburg dynasty.

The family's original holdings, however, were on the upper Rhine, and the limited extent of Rudolf's new powers were demonstrated by the frequent conflicts between him and his near neighbours in this area. These included the bishop of Basle, the margrave of Baden and the expansionist counts of Savoy, and the chronicles record a number of conflicts with these powers, all involving the attack and defence of towns and castles. In 1274, he campaigned with Peter of Savoy to reduce several towns in modern Switzerland, and then took more castles in Alsace that had previously belonged to Swabia. In 1286–7, the king faced a conspiracy of

[1] Ellenhard, Gesta invictissimi domini Rudolfi Romanorum Regis, *MGH SS*, XVII, 124; Chronicon Colmariense, *MGH SS* XVII, 247–8; Annales Otokariani, *MGH SS*, IX, 192; Cronica S. Petri Erfordensis moderna, *MGH SS*, XXX, 415. The castle at Drossendorf formed an irregular enclosure, about 50 by 30 metres, with a Bergfried and one large projecting tower, within the town walls: G Dehio, *Die Kunstdenkmaler Österreichs: Niederösterreich*, 1953 edition, 41–2. Laa an der Thaya was set in one corner of some strong town walls, using these to form one half of the enclosure. The thirteenth century Bergfried was ten metres square with walls 2.5 metres thick: R Büttner, R Madritsch, *Burgen und Schlösser in Niederösterreich. Vom Bisamberg bis Laa/Thaya*, 1987, 175–79. Both castles had been founded in the late twelfth century.

15 local counts forcing him once again to assemble troops and invade: he crushed the rebellion by devastating the lands and castles of one opponent, leading the others to submit. In the same year we read, in the rhyming chronicle of great length attributed to Ottokar, a Styrian *ministerial* serving the family, of Rudolf's attack on the castle of Herbotenstein, where stone throwers maintained a 24-hour bombardment with great stones. The chronicle records, in fact, the regular use of artillery called *tribok, mangen, bliden* and *tummeraer*. The latter is not a name encountered often and appears to represent a variant type or size.[1] In that same year, Albert of Austria had attacked Martinsdorf with all these weapons alongside cats and siege towers (*berefriden*).[2] The Habsburgs could lay their hands on a major arsenal of siege equipment and the people to operate it, even though so far most of their efforts had been directed against minor targets.

Castle warfare in a German province: Thuringia

The role of the king as peacemaker was illustrated by a documented account of events in Thuringia, where the towns, headed by Erfurt, were having problems with local princes. In 1290, Rudolf sent an army into the province which set about the destruction of no fewer than 66 castles from which armed men were plundering the countryside. While there, he also resumed control of an imperial castle from the landgrave. The Habsburg died in 1291 and was followed as king not by his son Albert, but by Adolf of Nassau, another prince with just a small territory to call his own. Adolf waged war in Thuringia in 1295, capturing two castles. Conflict between Albert and Adolf developed in Thuringia and neighbouring areas from 1298, but Adolf was killed, allowing Albert to succeed to the kingship.

Meanwhile Thuringia was not peaceful. In 1303, the city of Erfurt, helped by the neighbouring towns of Nordhausen and Mühlhausen, sent troops to capture and destroy the castle of Hopfgarten (near Weimar), and the following year, along with the landgrave, they took three "strong castles", destroying two which had belonged to the burgrave of Kirchberg, and returning the third to him once peace was agreed and the feud settled. In 1307, the city of Eisenach, suffering from the depredations of the sons of the landgrave, appealed for help to King Albert. The king sent troops, who built siege equipment to attack the Landgrave's castles of Wartburg and Winterstein with assistance from the townspeople, but without much success.[3] In 1309, with Henry VII now king, there was war between Erfurt and the margrave of Meissen. The latter was able to mobilise sufficient forces (several hundred mounted troops) to attack the city. The townspeople bravely

[1] Ellenhard, 125–6; Chronicon Colmariense, 242–3; Ottokar's Österreichische Reimchronik, *MGH*, 1890, lines 31,081 ff. The *Mittelhochdeutsches Handwörterbuch* (1876) defines it as a throwing machine using a revolving movement.

[2] Ottokar, Reimchronik, lines 29, 841 ff.

[3] Wartburg was a strong hill top fortress founded in 1200, formed of two wards, and two Bergfrieds and a splendid palace in the upper (B Ebhardt, *Die Wehrbau Europas im Mittelalter*, 1939, 1, 529–31). Winterstein was smaller, a free-standing Bergfried in an irregular, ditched enclosure, newly built at this time (for plan, F-W Krahe, *Burgen des deutschen Mittelalters Grundriss-Lexikon*, 1996, 673).

organised to set up ballistas and other engines on the walls and repelled the margrave's assault, but things still looked bleak until the king organised local princes to send relief to the beleaguered town. The chance for some revenge came in 1312, when the margrave was captured in a feud with the margrave of Brandenburg. The citizens seized the chance to destroy his castle of Rinkelstein while the abbots of Fulda and Hersfeld destroyed other castles threatening their own lands.[1]

The real position of the German kingdom is illustrated in this example. The elected king was unable to prevent feuding and the complete breakdown of order in a central province of the realm. The German kings had nothing to compare with the ever more effective centralised state bureaucracies developing in other monarchies: they were dependent on the voluntary help of other princes, and those same local princelings could choose to ravage the country with near impunity. The cities had to fend for themselves, and indeed they did so, often with help from other towns, only occasionally with effective aid from the crown. Their targets were regularly the castles of the nobility, not necessarily large fortifications, but defensible residences, as the use of the term *Hus* interchangeably with *castrum* suggests. Sophisticated levels of siege warfare were hardly to be expected, but the access of all parties to the skills to manufacture the equipment was apparent.

Struggle on the Rhine

Keen to expand his own power base, King Albert tried to take over the vacant counties of Holland, Zeeland and Friesland. This brought him into the complex relationships of the low countries which we have already seen. Here, it led the powerful Rhenish bishops, along with the Bavarian ruler, to plot to depose him. Striking hard and in alliance with the towns, Albert marched across Bavaria then besieged Heidelberg in 1301, forcing the count Palatine of the Rhine to make peace. Next, he marched on Bingen, belonging to the archbishop of Mainz, prime among his enemies. Bingen stood on the south bank of the Rhine, with a tributary providing cover on its western side. The old Roman town had been reoccupied and reconstructed before, the latest set of walls dating from around 1200. A castle called the Klopp stood on a hill on the landward side. It was said to be well protected against the threat of mining. A strong wall and deep ditch protected the Mainz side. Five counts headed the garrison.

Albert attacked both by land and by ship on the river. He had a substantial army including more than 2000 knights, but the decisive personnel were the engineers. Under cover of a battery of stone throwers (*bliden* hurling great stones), the men of Alsace constructed two large engines, a "cat" and a "crab" (*cancrum*). Lovingly described by the Colmar chronicler who must have either been a witness, or else received a first hand account, both devices were so strongly roofed that they could resist the defenders' stones, but sufficiently mobile to enable the attackers to approach and cross the ditch and reach the foot of the wall. The "cat" was however

[1] Cronica S. Petri Erfordensis, 421–3, 430, 435–8, 442–3, 445, 475; C Bayer, ed., *Urkundenbuch der Stadt Erfurt*, 1889, 1, 354–5 (expenses in capturing and destroying the castles), 371–2 (end of feud), 385–6 (royal protection).

pushed back into the ditch by a sortie. But the "crab" sheltered an immense battering ram, through which a breach was created in the town's wall, and a tower was also brought down. The defenders fled into the Klopp, from which they negotiated surrender. The archbishop quickly surrendered. The terms of his capitulation were interesting: Bingen and three major castles were surrendered into the king's hands. The victorious king resumed his campaign against the other conspirators (the archbishops of Cologne and Trier) the following year and brought them both to similar capitulations.[1]

New engagement in Italy

Albert was murdered by a nephew in 1308. The next king came from yet another new royal family, Henry count of Luxembourg. Henry VII, being in alliance with the pope, could envisage being the first to make the trip to Italy for his imperial coronation for more than half a century, and originally intended to come as peacemaker between the Guelf and Ghibelline factions disputing control of the cities and in shifting alliances with Robert, the Angevin king of Naples and Frederick, the Aragonese king of Sicily. But it had now been a very long time since a king of Italy (the formal title held historically by the emperors) had been able to command obedience in the peninsula, and it was not long before Henry found himself at war with traditional enemies such as Milan and Florence. In May 1311, after taking an ill-provisioned Cremona, a large German army besieged Brescia. The townsfolk built their own wooden outworks outside the walls to shelter archers, and a laborious siege ensued. Henry sent for his craftsmen and they built shelters and tunnels and prepared assaults. However, it seems that the siege was not pressed very closely despite accounts of vicious fighting: Waleran of Luxembourg was killed by a ballista bolt. There was hunger and disease on both sides, the town surrendered in September and suffered the destruction of its walls. Henry then marched on Milan, which opened its gates, and Genoa. The Florentines' greatly expanded city had long outgrown its defences. An extension of the circuit had already been begun (in 1298). Now in 1310, they set about digging a moat around the whole city, resting on the river Arno. Where the new walls were already under construction, they were hurriedly raised in height. The urgency was necessary, as (rather short-sightedly) the city had pulled down parts of the old circuit before completing the new. Sections of these powerful walls still stand around Florence, and are a fine example of a towered enceinte.

The siege began in September 1312. The Florentines mobilised their people to hold the extensive ramparts, and built more palisades as outworks to protect the moats and ditches. The neighbouring cities including Lucca, Siena, Pistoia,

[1] Chronicon Colmariense, 268–9; Cronica S. Petri Erfordensis, 434; Annales Moguntini, *MGH SS*, XVII, 1861, 2–3; Cronica de gestis principum a tempore Rudolfi usque ad tempus Ludowici Imperatoris, ed. G Leidinger, *MGH SrG, Chronicae Bavariae*, 1918, 54; Ottokar Reimchronik, lines 77,685–739; J Scholl, *Annales Bingenses*, ed. Sander, 134–6 (a history written in 1613); L Greb, C Brunner, *Die Kunstdenkmaler im Volkstaat Hessen: Kreis Bingen*, 1934, 12–13 (and with a sixteenth century plan of the town).

San Gimignano and even the more distant Bologna all sent help, so there was no shortage of armed men inside. Faced by such a strong defence, and apparently unable to block off access to supplies, Henry was helpless and had to abandon his attempt. Henry, who had been crowned emperor by the cardinals in Rome, then found himself at war with Robert of Naples, and he was besieging his enemies in Siena in 1313 when he died of disease (or poison) contracted there. Italy had claimed the life of another German ruler.[1]

In the ensuing struggle to succeed, Louis (Lewis) of Bavaria won out over Frederick of Habsburg. The new king's contribution to Italian affairs included a failed siege of Pisa in 1327 and the election of an anti-pope. At home he continued the imperial alliance with the towns, made peace with the Habsburgs, and in 1338 had the rules changed so that the German electoral princes henceforth could elect not just the king, but also grant the imperial title. Thus the empire with no "state" resources of its own was also no longer "holy" or "Roman". The title was still clearly worth fighting for, but its holders were now rarely able to influence the international political stage.[2]

Poland, the Baltic, Lithuania and Russia.

The Golden Horde did not attempt any further extensions of its direct rule over the states to its west, although it remained as a threatening presence throughout. However, it was easier and more productive to collect tribute from the Russians than destroy their towns. The terrain in which cities like Novgorod stood made them difficult targets anyway. Furthermore, the Horde faced other enemies now. Khan Berke (1255–66) had overseen the conversion to Islam, and was frequently in alliance with the Mamluk sultan against the intervening (Mongol) state of Hülegü. There was little evidence that the Horde changed its traditional social and economic organisation very much. The khans founded a capital, Sarai Berke, on the Volga after 1300, which was constructed by transported forced labour and housed imported crafts and manufactures. The tribute of the Russian principalities paid for the construction. The city was unwalled, although the palace was protected by walls and towers.[3] Ironically, it would be Timur who destroyed this Mongol city, which, with up to 100,000 inhabitants, would have been one of the largest in the world. The location of the Golden Horde's main interests meant that much of their continued campaigning was in the south of the region, in south-eastern Poland and western Russia. In 1259, for example, they sought to take Luck. The town was not fortified but was protected by a very high river. The

[1] *Villani's Chronicle*, trans. Selfe, 397–419; *Dino Campagni's Chronicle of Florence*, trans. Bornstein, 89–97; *Cronica de gestis principum*, ed. Leidinger, 71–6; *La Chronique de Jean de Hocsem*, 129–31.

[2] A good summary of the complex political developments is P Herde, From Adolf of Nassau to Lewis of Bavaria, in the *New Cambridge Medieval History*, VI, ed. Jones, 2000, 515–50.

[3] B Grekov, A Iakouboski, *La Horde d'Or*, trans. Thuret, 1939, 135–45.

inhabitants broke the bridge, the Mongols attacked with a stone thrower. The chronicler records however that divine intervention saved Luck: the wind was so strong that it blew the stones back into the machines and smashed them.

Two years later, the Mongols ordered the Russians to dismantle the defences of a number of towns including Lvov and Luck. Whether this means that the original account was inaccurate, or that defences had been thrown up since 1259, is unclear. However, the defenders of Xolm were equipped with stone throwers and crossbows, so the Mongols left it alone and moved against Lublin then Sudomir. Here they built a wooden stockade to enclose the town and set up engines to maintain a continuous bombardment, and after four days the walls had been brought down. An escalade was launched, the inhabitants fled to the citadel, but this (being perhaps of wood) was burnt, and a massacre followed.[1] There is no suggestion from these accounts that the Mongols had lost any of their ability to capture defended towns in short order, provided that other considerations did not prevent them deciding to concentrate elsewhere.

The great distance of the Horde's central interests from the Baltic, however, provided the space for the rival powers now ensconced there to continue their conflicts. The Polish and Russian leaders who had invited in the German crusaders would have ample opportunity to regret their decision. The local tribes proved resistant both to Christianity enforced by massacre and to foreign rule accompanied by harsh exactions, while the Teutonic Knights (including Livonia after 1237) were as adept at wielding the sword against fellow Christians as against those they had come to convert. The pagan Lithuanians were more able to resist than the Prussians, because of the geographical remoteness of their lands and the wastelands that protected them, while the Prussians, pressed between German and Pole, would eventually succumb, although not without several decades of bloody struggle.

The small Polish princedoms were in frequent conflict too. In 1272, there was a fierce struggle for Gdańsk, "perfidiously" seized by the margrave of Brandenburg. Prince Bolesław attacked the town's castle by land and sea, and set up catapults to bombard the walls. Within a few days they had done their job, demolishing stretches of the defences and allowing the prince to launch a successful assault. The castle seems to have comprised a stone tower within a bailey protected by an earth and timber rampart, scarcely a modern or sophisticated fortress, and it fell quickly.[2] Thirty years later, in conflict with the Lithuanians, the Poles however took several months to recover Lublin through blockade (1303), the city proving a much stronger obstacle than the castle at Gdańsk. The German knights having secured the latter in 1310, were themselves thwarted by a Polish castle at Swecze even though they attacked the "humble" wooden defences with four catapults day and night. As late as the fourteenth century, therefore, wooden fortifications could still resist attack if firmly held. The citadel of Cracow was made of more substantial material, however, being "built entirely of stone" and we have an account of

[1] Perfecky, *The Galician-Volhynian Chronicle*, 75–80.
[2] Długoss, v, 20, 60.

a siege here in 1290 when it was held by another prince's soldiers against Prince Bolesław. The defenders were well supplied with stone throwers and crossbows, specifically including large wall-mounted machines, either ballistas or perhaps springalds.[1] Apparently also very well provisioned, the citadel resisted successfully for a year.

Teutonic Knights

The Teutonic Knights were still, primarily, securing their new realm against the native peoples. What would become their great headquarters convent-castle at Marienburg (Malbork) was founded in 1280. In seeking to establish a firm land bridge to Livonia, attention was paid to Memel (Klaipeda), and the Lithuanian "castles" (as described by the German chroniclers) of the area, such as Bisena, were stormed and burnt in 1283 and 1284. But the difficulty of holding the captured sites, and of preventing the Lithuanians simply returning and restoring the earth and timber forts, was great. In 1313 the knights had once again to attack the fort at Bisena, which they did with siege engines, but this time without success. The Slavs had readily adopted their enemies' siege techniques. In 1315, though, their attack on the Knights at Christmemel with two engines and numerous archers was thwarted and they were compelled to burn their engines in retreating.

The attack and defence of fortresses also featured heavily in the fighting in Livonia. The forts of the Lithuanians were frequent targets, one at Terweten (Tervete) being captured and rebuilt more than once, with the Knights trying to secure the area with a castle of their own called Heiligerburg. The Master had to ensure its security, though, with a substantial garrison of 300, and brought two *bliden* along as well. The Livonian rhyming chronicle described a fierce attack on Terweten by the Slavs in which the latter constructed numerous *ribalde*, probably best considered as ballistas, that were placed close to the ramparts and caused much harm to the knights, who responded with stones from their own artillery. The castle was destroyed and its garrison killed in 1279.[2]

The defence of German-ruled Livonia against the neighbouring (and native) Slavs comprised, mainly, a deserted wasteland to the south, backed up with castles held by small numbers of Teutonic Knights with larger forces of hired foot soldiers and local militia. They were centres of refuge and bases from which Lithuanian raiders could be intercepted. The few large castles tended to be built on promontories in the mainly flat and river-crossed landscape. They were structures designed for their purpose, and financed by donations from crusaders, and local taxes, and constructed by imposing labour services on the local people. The important frontier castle of Dünaburg (Daugavpils) on the river Dvina was erected in the 1270s (according to the Reimchronik) with a system of water defences

[1] Perfecky, *The Galician-Volhynian Chronicle*, 116, translates these as "revolving crossbows", an unhelpful description.

[2] *Livländische Reimchronik*, lines 9,934–10,083; Hermanni de Wartberge, *Chronicon Livoniae*, 49.

exploiting its location. It was attacked, destroyed and rebuilt several times, the rebuild of 1312–13 seeing the erection of substantial stone walls and towers.[1]

Another angle on Livonian affairs is provided by the history of Riga, home to the bishop, the Knights' headquarters, and a growing and prosperous urban population driven to occasional revolts against the crusading warriors in their midst. Such an event took place in 1297 when the townsmen enlisted the aid of the local Slavs to attack the garrison of 60 knights in the quadrangular stone castle. One of the large corner towers named after the windmill that stood below it was destroyed, and six brother knights captured. There was another revolt in 1329, when the Rigans invited the Lithuanians in, but their plans were discovered and thwarted. From this time on, the castle underwent further strengthening and development into the form of convent-castle now being built across Prussia.[2] This story also warns against imposing later nationalistic interpretations of the struggle as being simply one of Christian Germans against pagan Slavs.

Lithuanian Grand Dukes

The evolution of the Lithuanian people as a dominant regional power can only be traced in the contemporary accounts of their neighbours. The process was assisted by the country in which they dwelt, which because of its lakes, rivers and forests could only be accessed by enemies when the water froze. During the thirteenth century, a leading noble family became established as grand dukes over a loose federation of nobles. Apart from a short interlude in the 1250s, the Grand Dukes remained, like their people, pagans. A new dynasty assumed the rule in 1289, and it was the sons of the founder Liutauras, Vytenis (1295–1316) and Gediminas (1316–41) who established the power of the realm. The Lithuanians had copied quickly the military practices of the invaders in order to attack their castles, with considerable success. The dukes built their own defended palace at Pograuden by 1305, and during this time they erected a series of forts along the main rivers of the land, usually facing a Teutonic fortress on the other bank. From the 1320s, a ducal capital became established at Vilna, provided with two castles built on hills and formed of wooden towers (donjons) in stone walled compounds. The growing strength of Lithuania was confirmed by victorious wars against Brest (1315), Galicia-Volhynia (1323), and an alleged attack on Kiev, which fell after a siege of a month according to a much later source: whether true or legendary, the tale is suggestive of the newly expansive policy of the dynasty. In general, however, their western opponents had the equipment and technology to remain their superiors

[1] Tuulse, *Die Burgen in Estland und Lettland,* plan on page 88; *Livländische Reimchronik,* lines 8,213–8,264, for a failed Lithuanian attack using *bliden.* W Urban, 'The organisation of defence of the Lithuanian frontier in the thirteenth century', *Speculum,* XLVIII, 1973, 525–31, for a study of the whole system of defence.

[2] Hermanni de Wartberge, 54, 64; Tuulse, 151–3.

in open combat, so the strategy continued for the most part to be one of raiding and looting.[1]

North Russia

Meanwhile, the Russian cities of the north were facing not only the German knights on the Baltic, but also expansionist drives by the Danes and Swedes, which led to several wars. In these struggles, control of the land was always confirmed by the erection or capture of fortified buildings, either towns or castles. In 1263, the Novgorod prince Iaroslav, brother of the famous Alexander (Nevsky), stormed the German-held castle at Iuriev (Dorpat) despite its having no fewer than three lines of ramparts. The garrison must have been inadequate, because the attackers were not recorded as having siege equipment. Indeed, three years later they borrowed engineers from the prince of Vladimir to help them build "wall-battering" machines. In 1268–71, the Knights waged war on Pskov, but were unable to achieve any success despite bringing a siege train to bear. The city certainly had a strong citadel (*krom,* rendered in a much later account as *arx*) on the high ground, and parallel lines of ramparts, but the town defences were still of earth and timber. At the same time, Prince Dmitri was permitted by the city of Novgorod to build himself a *gorod,* which he walled in stone, an event worthy of notice in the chronicle, while it was in 1302 that the town itself was provided with a stone wall. Excavations showed that this was a work of considerable strength.[2] In general, where resources allowed, there was a general process of rebuilding fortifications in stone in northern Russia from the late thirteenth century onwards, and an increasing use of projecting towers to flank the curtain walls, no doubt encouraged by contact with the foreign influences in the Baltic region.[3]

During the 1290s, the Swedes arrived in the Gulf of Finland and built themselves a fort at Keksholm (1293). In 1300, they came to the river Neva, bringing "very clever craftsmen" including one from Rome and they built a town at the mouth of the river Okhta, making it "indescribably strong" and equipping it with stone-throwers. This did not prevent Prince Andrei of Novgorod capturing it the following year with the forces of that city and Vladimir, the chronicle noting that the "clever constructions were of no avail". Then in 1322/4, Prince Iuri Danilovich led the armies of Novgorod and Pskov to assail the Swedes at Vyborg. Despite using six stone-throwers, and attacking for a month, they failed. In 1325, the

[1] S C Rowell, *Lithuania Ascending: a pagan empire within east-central Europe,* 1994, provides this summary, 49–56, 69–73, 111.

[2] *Chronicle of Novgorod,* 101–8; *Nikonian Chronicle,* III, 42, 44, 51. As noted before, the translations called the engines "battering rams" but it is apparent that these "wall-battering machines" are in fact stone-throwing artillery. Tuulse, 53–4 (Dorpat); M Tikhominov, *The towns of Ancient Russia,* 1959, 249, 256 (Pskov); Kirpicnikov, 'Steinfestungen Nordrusslands aus der Sicht der neuen bautechnisch-archaeologischen Forschungen', *Burgen und Schlosser,* 24, 1983 (Novgorod).

[3] V Kostochkin, *Fortress Architecture of Early Russia,*trans. Friedmann, 1970, with numerous illustrations, although most of the examples would be rebuilt repeatedly over the centuries.

Germans attacked Pskov with artillery, and drove trenches up to the ramparts, attacking for 18 days. The return of the prince from Lithuania broke the siege, and the Russians destroyed the engines.[1]

The absence of detail in many of the accounts does not permit an accurate reconstruction of the numerous sieges that took place. It is apparent that the invaders had access to siege equipment on a regular basis, but they and the Russians were often frustrated by the defences they faced, which would in most instances have been of earth and timber. Where walls were of stone, it was sufficiently unusual to justify notice in the chronicles. The earthwork defences were often of substantial dimensions and solid construction, and the ramparts would actually have provided a more useful resistance to trebuchets than a stone wall, but the wooden palisades, apart from having to be regularly replaced, would have been more vulnerable to destruction. Primarily, though, it was the availability of suitable materials and of local building skills that must have influenced decisions on what to build.

The Mongols in the Far East

The overlordship of the Mongol realm resided with the ruler of the eastern end of the gigantic empire, where in the late thirteenth century Kubilai Khan[2] emerged as supreme ruler, and completed the task of conquering the whole of China. The Sung dynasty had remained in control of a large empire in southern China, and from around 1265 Kubilai devoted his efforts to its conquest. Kubilai declared himself emperor, and established the Yuan dynasty of Chinese rulers. The Sung proved a powerful obstacle, but were eventually vanquished in 1279. The enormous scale of this task was apparent. The Sung did not simply surrender their realm. Cities and towns, of which there was a very large number, continued to represent strongholds of considerable size. The city of Jingjiang (modern Guilin), for example, was recorded as having undergone strengthening no fewer than four times over the fourteen years preceding the Mongol onslaught, and the main elements in the new works were the addition of a lower, brick-built outer wall that ran immediately in front of the main inner wall, and the digging of a new moat on the one side not protected naturally by rivers. Internally, the road layout was altered, the traditional Chinese rectilinear grid plan modified with more obliquely-angled roads and more T-junctions, with the effect of confusing anyone breaching the outer walls and wanting to march against the inner defences wherein were the city's headquarters and administration.[3] A key episode in the conquest was the siege of the twin cities of Hsiang-yang and Fanch'eng on

[1] *Chronicle of Novgorod*, 114, 122; *Nikonian Chronicle*, III, 84, 88, 116–17; Hermanni de Wartberge, 60, gives the opposite version and says that Pskov was taken.

[2] Again, a name with several variant spellings of which this is a commonly recognisable form.

[3] Heng Chye Kiang, *Cities of aristocrats and bureaucrats. The development of medieval Chinese cityscapes*, 1999, 188–9.

the Yangtze river. Their capture would take the whole period from 1268 to 1273, mainly because by having navigable water on three sides, the defenders were regularly able to introduce supplies. The operations have also received much notice because of the presence at Kubilai's court of Marco Polo, who recorded that he assisted the khan by offering to provide the Mongol with "men who are able to construct mangonels which shall cast such great stones that the garrison will never be able to stand them". The Venetian wrote that he had a German and a Nestorian Christian who constructed three trebuchets capable of projecting stones weighing 300 pounds (136 kg) which did such damage to the city that it surrendered.[1]

As a warning about the reliability of self-proclaimed eye witnesses, however, this tale is not borne out by other contemporary accounts. Both Chinese and Persian sources indicate that the weapons were indeed devised by foreigners, but that the engine builders were Muslims, Arabs or Persians. The annals of the Yuan (the Yuan Shih) recorded that the engineer I-ssu-may-ying (a Chinese version of Ismail) made a "great stone thrower. It was economical in power and its range was very long", while a later chapter tells of Kubilai summoning Alau 'd-Din for the task. In 1273, the new engines were set up against the south east corner: "When the machinery went off the noise shook heaven and earth; everything it hit was broken and destroyed, and it sank seven feet into the ground".[2]

Since the Mongols had used batteries of, probably, counterweight trebuchets against their European and Muslim opponents, it seems strange that they did not know how to build such weapons against the Sung. The dynastic history recorded the administration of the army of the Mongol emperors, and a significant element was the confirmation that the force was made up of many nationalities. They were also categorised according to their function, and this included catapult troops (*p'ao chun*) and crossbow troops. There was specific reference to "Muslim catapult operators and military artisans" in 1274, following the capture of Hsiang-yang. In 1279 there was reference to 600 artisans who could construct the *Hui-hui p'ao* (Muslim catapults). Another 1000 catapult soldiers were sent to the frontier with Turkestan in 1280.[3]

The normal Chinese manually-operated stone thrower was in widespread and common use by both the Mongols and Chinese forces. It has already been noted how the Mongols relied on foreign experts to build and operate their artillery. To the extent that Kubilai Khan relied on subject Chinese to provide his siege equipment, it would be no surprise that he did not have counterweight trebuchets. The division of the Mongol empire might also account for his not having the more powerful weapon. Certainly, if either the Chinese had already developed the

[1] *The book of Ser Marco Polo*, trans. Yule, 1903, book II, chapter LXX, 158–60.

[2] These accounts are translated in A C Moule, *Quinsai and other notes on Marco Polo*, 1957, 72–6.

[3] Ch'i-ch'ing Hsaio, *The Military Establishment of the Yuan dynasty*, 1978, 85, 113, 175 (a translation of sections of the Yuan Shih). See also J Needham, China's trebuchets, manned and counterweighted, in *On pre-modern technology and science: studies in honour of Lynn White*, 1976, 108–17, and a masterly study in J Needham, R Yates, *Science and Civilisation in China*, v, part VI, 1994, 204–32, with a table of weights and ranges at 216–17.

counterweight version themselves or else the eastern Mongols had retained their earlier knowledge of it, the several independent accounts of the siege of Hsiang-yang, and the emphasis of the Yuan Shih on the role of the Muslims after 1274, would be incomprehensible. There are two decisive elements in the description of its use that confirm that the new weapon was of a different character from its manually-propelled predecessors. The weight of missile it hurled was far greater than anything that could be thrown in China before, and the range was far greater. Traditional Chinese weapons had a range of not more than 100 metres. With its range of up to three times that amount, and a missile of such size as to cause massive damage to walls or buildings, the counterweight weapon would have been devastating to the besieged, and also untouchable by defending artillery or cross-bows. It would be safer then to conclude that the knowledge and skill was held by engineers and artisans from the Muslims further west, and it had to be imported (or perhaps re-imported) by Kubilai Khan.

Uncertainty of a different kind surrounds the new dynasty's failed efforts to extend its empire to encompass Japan. The Mongols had fought to conquer Korea from the 1220s, using their siege skills. Kubilai completed the subjection in 1258. The Koreans were an essential ally in the extension of the empire, because the Mongols needed their ship-building skills. Japan having proudly rejected Mongol demands for submission, the next step was armed conquest, although Kubilai's counsellors were apparently against the project. Nonetheless, in October 1275 an army of at least 16,000 Mongols, Koreans and Jurchen was conveyed in 900 boats to Tsushima and overwhelmed the defending Japanese. They moved on to invade Iki, driving the Japanese back into their castle. This was destroyed by fire caused by the Mongol bombardment, while the Mongol attack also used the tactic of a human shield of captives. On Kyushu, the old castle (dating back to the seventh century) was hurriedly restored, and ramparts of earth topped with palisades and bulwarks five metres high were constructed along the threatened coast. Fire ships were launched as the Mongol fleet approached, and there was fierce fighting on the beaches at Hakata, with the Mongols apparently having artillery discharging iron balls. A great storm scattered the invading fleet.

Frustrated, Kubilai tried again in 1281. An immense army was prepared for the invasion, over 100,000 men in two armies, planned to first land on Iki, then march on the Japanese capital. When general Fan Wen-hu asked for "Muslim firearm (? — see below) experts" he was told they would not be needed. The immense force drew itself into a circle, containing "huge ships" carrying catapults ready to shoot, and the stones sank many Japanese ships. The land defences had been improved and restored at all possible landing sites. There was again fierce fighting, and again the weather intervened, a hurricane destroying the invaders with the loss of thousands of lives. Kubilai wanted to make a third attempt two years later but was diverted by other concerns.[1]

[1] R Tsunuda, L Carrington Goodrich, *Japan in the Chinese dynastic histories*, 1951, 80–9 (translation of the Hsin Yuan Shih); N Yamada, *Ghenko. The Mongol invasion of Japan*, 1916, 47–64 (Korea), 100–202 (Japan), based on the Yuan Shih and other sources.

Fire weapons and gunpowder artillery

The accounts of the Mongol attacks on Japan have been taken to signify that Kubilai's armies had access to artillery using gunpowder as the propellant force for the missile. A Japanese illustration of the conflict made in 1293 shows a ball of fire hurtling through the sky.[1] This may have been the time when the particular formula for making gunpowder, in which the proportions of the ingredients and especially saltpetre, and the purification of the latter, was critical in determining whether the mixture exploded as opposed to just burning, was found and put to practical use.

The Chinese had made gunpowder of the incendiary type for centuries, and weapons such as fire-lances and crude flame throwers, and gradually improving devices of bamboo to propel the flame, as well as ironclad grenades, had been in widespread use in all the sieges since about the year 1000.

Incendiary bombs were either thrown by hand or sling for close quarter combat, or for greater range were propelled by catapult, just as they were in the west, where the Muslim armies in particular had made frequent use of naphtha for their fire weapons. The description of them as "Greek Fire" by Christian sources was frequent as we have seen, and widespread, ranging from the sieges of the fifth crusade in Egypt to the sieges of Scottish fortresses in the early fourteenth century. In fact, none of these writers would ever have seen the original "Greek Fire". The weapons faced by the Crusaders, and many others further west, were incendiary devices that set their targets alight, and of which the crucial ingredient was naphtha. Ready access to petroleum oil in the Middle East meant that the Muslim armies had no shortage of the main ingredient. Different Arabic accounts exist of how the final mixture was achieved, including the famous treatise dedicated to Saladin by al Tarsusi. Other ingredients include sulphur, urine, wine vinegar, asphalt and so on, but there is no mention of saltpetre, so any suggestion of a connection with gunpowder can be immediately dismissed. Having the mixture, which had to be distilled, was of course only the first stage: from a military viewpoint, the crucial question was how to contain and project it against an enemy. One answer seems to have been by using small terracotta jars as both container and grenade. Examples of such jars have survived, although it cannot be certain that this was their function. With a diameter of 82mm and a height of 95mm, those found at Tripoli in Syria were small enough for manual use. Excavations in the mound covering the citadel at Hama destroyed by the Mongols uncovered a workshop containing ovens, numerous vessels of the same spheroconic shape, bits of asphalt for which no other use is apparent than use in the fiery concoction, burnt limestone and evidence, possibly, of ventilation slots low in the brick wall.[2]

[1] F Fischer-Wieroszowski, 'Kriegerisches Einfall der Mongolen in Japan. Ein Japanische Bildrolle', *Ostasiatische Zeitschrift*, NS 9.XI, 1935, 121–4.

[2] M de Saulcy, 'Note sur des projectiles à main, creus et en terre coite, de fabrication Arabe', *Mémoires de la Société Nationale des Antiquaires de France*, 4th ser., v, 1874, 18–31; P Pentz, 'A medieval workshop for producing "Greek Fire" grenades', *Antiquity*, 62, 1988, 89–92; al Tarsusi, translated in B Lewis, *Islam*, 223; Reinaud, Fave, 'Du feu grégois, des feux

If these were grenades, the question remains as to how they were ignited. It has been suggested that they would explode on impact, or alternatively it has been said that having smashed on the target and covered it in inflammable liquid, this would in turn be ignited. Another method of propulsion that seems to have been used was from a siphon, which brings it closer to the "Greek Fire" weapon. The evidence for this is quite clear too. Not only do some of the treatises refer to the use of "syringes" for propulsion, but one is also indirectly described by the early thirteenth century writer al Jazari. In his "Book of Knowledge", he illustrated a machine to raise water which comprised a cylinder and a pipe fitted to one side and stated "this machine is like the ejectors of naphtha (*zaraqat al-naft*), only larger".[1] Another contemporary, Hassan al Rammah, wrote between 1275 and 1295 not just of "pots thrown by mangonels" but also of "fire lances" and "flying fire", suggesting that there may have been an even greater variety of methods, but even though the same person described how to purify saltpetre, this was in a separate context and did not change the way his treatment of fire weapons regarded them as incendiaries. Sufficient contemporary accounts of the use of such devices in action survive to demonstrate beyond doubt the importance of the weapon in defending stone walls against the attack of wooden siege equipment in the Muslim world. Their use is much rarer in Europe, and the kind of mixture used to make the material in the English attack on Scotland has already been described above.

The Arabic term *naft* would be adopted to convey the meaning of gunpowder. Cannon do not feature regularly in the Middle East until at least the middle of the fourteenth century and in India until the fifteenth. This leaves to explain the already mentioned reference in Ibn Khaldūn's account of the siege of Sidjilmessa in 1274. Amongst the catapults and ballistas appeared a reference to "an engine of fire" propelling iron gravel "from the heart of the piece by means of inflammable powder of which the particular property made effects which rivalled the power of the Creator". The Arabic words describe a "naphtha engine", however, and it should be recalled that the writer lived a century after the event at a time when gunpowder artillery was commonplace. This amazing weapon also had no effect whatever on the siege which lasted a year and was resolved by the traditional weaponry and methods. Although the description suggests the powder was a propellant, it might be another way of describing an incendiary weapon of a novel form. If it was gunpowder in the modern sense, it apparently did not catch on in this area for more than half a century to come.

What would be different about the new form of the powder would be its use not just as an incendiary, but as an explosive propellant, which also required a new proportion of the ingredients, the purification of the saltpetre, and the construction of a weapon sufficiently strong to contain and direct the force of the explosion. These weapons, guns in the modern sense, were unquestionably present in

de guerre et les origines de la poudre à canon chez les Arabes, les Persans et les Chinois', *Journal Asiatique*, 4th ser., 4, 1849, 257–324, translates many of the texts and is still full of interesting insights.

[1] Al Jazari, ibn al Razzazz, *The book of knowledge of ingenious mechanical devices*, trans. D Hill, 1974, 187–8.

the fourteenth century. Someone, somewhere, must have been responsible for the first device. There is evidence of use very much earlier in the Chinese empire. A sculpture in a Buddhist cave temple shows what can only be a small bombard held by a demon (the pot shape remarkably reminiscent of the earliest west European illustration from 1326), and reliably dated by inscription to 1128. It cannot show, of course, what the device thus illustrated was made of.[1] Excavation has uncovered a Chinese bronze tube just 34 cms long and with a bore of 2.6 cms, dated by archaeological evidence to as early as 1288. The next earliest known Chinese weapon dates from 1332. The Japanese descriptions of the invasions talk of iron *phao* causing "light and fire" and shooting out 2–3,000 iron bullets. The barrels used were sometimes made of bamboo. Everywhere, arriving at a conclusion is complicated by the fact that contemporaries naturally used familiar terms to describe weapons, so *p'ao* which traditionally meant catapult would be pressed into use later to describe cannon. Why the twelfth and thirteenth century weapons do not feature in the written history can only be a matter of speculation, as can the question as to why they seem to have disappeared from use, and memory. Perhaps, like Greek Fire, they were the victims of too much state secrecy. Perhaps they did not work very well, and were abandoned.

It is nonetheless quite possible that the Mongol invasion of Japan witnessed the first recorded use of gunpowder in its modern form to propel shot, and that its still mysterious spread westwards after that time began in the third quarter of the thirteenth century.[2]

Experts

THE PERIOD FROM THE MIDDLE of the thirteenth century through to the early decades of the following century was one in which the trends of the preceding fifty or sixty years were fully consolidated. They were, coincidentally, the years in which the new counterweight trebuchet had become established as the dominant siege weapon of choice. A new generation of fortresses, epitomised by the completely new castles founded in Wales, demonstrated the advances in technique that there had been, partly in response to the power of the new siege weaponry. One significant element of the period, then, was that to be modern, defences had to be constructed of such strength that none but the wealthiest could possibly afford to pay for them. Another was the arrival of the engineer as a respected and essential servant of the ruler. This was not a new development. The growing role

[1] Lu Gwei-Djen, J Needham, Phan Chi-Hsing, 'The oldest representation of a bombard', *Technology and Culture*, 29, 1988, 594–604.

[2] J Needham, Hu Ping-yu, Lu Gwei-Djen, Wang Ling, *Science and Civilisation in China*, IV, part VII: Military Technology: the gunpowder epic, 1968, throughout; Wang Ling, 'On the invention and use of gunpowder and firearms in China', *Isis*, 37, 1947, 162–76; L Carrington-Goodrich, Feng Chia Sheng, 'The early development of firearms in China', *Isis*, 36, 1946, 113–21, 251; Ibn Khaldūn, as above; D Ayalon, *Gunpowder and firearms in the Mamluk kingdom*, 1956, 21–2, with another translation of the Khaldūn text.

of such craftsmen under the Angevin kings of England and the Capetian kings of France from the late twelfth century onwards has been described already, assisted by the survival of government records of their jobs and what they were paid. Their importance grew through the thirteenth century.

While it may just be that the absence of evidence from earlier times overstates the importance of such craftsmen during this time, it may also be that greater skills and sophistication were needed to build the new fortresses and to construct the new weapons. The manually operated stone thrower depended on meeting precise proportions, as the surviving manuals about their construction (such as al Tarsusi's) confirm. The counterweight trebuchet made even greater demands on the skill and knowledge of the constructor. As the machines grew larger, with ever heavier counterweights and projectiles, the skills needed to design and build them likewise became more critical. The strains on the engines were now enormous, requiring of their builders a clear knowledge of how to cope with them while ensuring the maximum efficiency of the weapon. The consequences of getting it wrong with enormous counterweights and projectiles weighing perhaps hundreds of pounds were severe. A surviving inventory of the royal arsenal stored at Carcassonne in 1298 reveals the range of equipment now called for. The artillery was both large and small, and new. It included numerous ballistas, quantities of parts for ballistas, mangonels and trebuchets such as windlasses, bearings, levers, triggers, and capstans. There were 560 pieces of lead for counterweights, described according to their weight, of which 48 were of half a quintal (50 pounds) and four of no less than a quintal (100 pounds, 45 kgs). The presence of large numbers of various weights suggests how the counterweight could be modified by adding or removing lead blocks. Among the weapons there were also a large number of springalds.[1] For some engineers such as John Crab mentioned at the siege of Berwick, skills seemed to have also included manufacturing incendiaries, so these persons were not just carpenters with a practical awareness of physics and engineering, but must also have had knowledge of chemistry. How they acquired such knowledge is not stated, but presumably the skills must have been transmitted from father to son or from master to apprentice.

There was a ready market for such valuable abilities, in particular where they were also combined with the skills of the master mason. Master Bertram, described as an *ingeniator,* was involved in North Wales in the 1270s and 1280s, having previously been a builder of royal castles in Gascony. He was recorded as selecting the timber at the Tower of London in 1276 for the siege engines used at the siege of Dolforwyn and seems to have been in charge of the whole team of artisans used in the attack including the miners. In 1282, he was responsible for the bridge of boats from Anglesey to the mainland, while in 1283 he was at the siege of Castell y Bere. The superior position of such master workers was clearly marked by their daily pay: at 9*d* it was 75% of what a knight earned, while ordinary engineers in the Scottish wars earned 6*d*, and masons and carpenters 4*d*. John Crab

[1] G Mot, 'L'arsenal et le parc de matériel à la cité de Carcassonne en 1298', *Annales du Midi,* 68, 1956, 409–15.

was from Flanders but served the Scottish king. The engineer captured during the siege of Berwick was sufficiently valuable to be re-employed by his captors. Other skilled engineers seem to have had clerical connections, including the friar who built the engines at Caerlaverock, and another who was killed inside his own wooden siege construction at Iglesias. The Florentines in 1250 had a lay brother who had been a master engineer in the making of "machines and trebuchets and cats and battering rams." The Swedes arriving in the Gulf of Finland had brought with them "clever craftsmen from Rome", although the Russian chronicler might more likely have meant that they were imported from Europe. The critical role played by Ismail and Ala'uddin in assisting Kubilai Khan defeat the Sung empire, and the subsequent role of "Muslim engineers" in the Yuan dynasty's siege train have been noted.[1]

The new demands of modern fortress building and siege techniques seems to have created an internationally-recognised craft, the masters of which could sell their skills anywhere, and who were fairly safe from the consequences of being on the losing side since their captors were most likely to offer to hire them. There must have been another and larger group of workers at a lower level who could turn out or operate siege equipment when required, whose names are lost to us.

The greater weapons, just as the greater and most modern forms of fortified building, were now of such sophistication and expense that only the greatest could afford them. One of the things that distinguished this period from the previous century was that kings, princes and sultans were certainly among those in this group, but so too were a growing number of cities. Joining the larger cities of Italy, which had been effectively independent for a very long time, were the larger settlements of the Low Countries and Germany, whose de facto independent impact on politics and war stemmed from their situation, their population and their wealth. Even the citizens of small, isolated towns such as Riga have been seen exercising an independent foreign policy. The cities of northern Russia, while providing the base for princes, could also find their own voices. This greater role in the world was reinforced by the simultaneous increase in the cost of warfare. Along with the immense cost of building new walls and equipping arsenals of siege equipment and maintaining a corps of experts, rulers also found it necessary to recruit ever larger armed forces for wages, the traditional feudal duty having always been inadequate. The cost of doing so meant that not even so aggressive a defender of royal prerogative as Edward I could afford his wars without having to secure the consent of his wealthier subjects, in order to raise new taxes. The processes that would lead to the establishment of parliaments or estates were set in place across much of Europe during this period.

Meanwhile, as the century turned, numerous transformations at the levels of the economy, society, politics, war and technology, beckoned.

[1] A Taylor, Master Bertram, Ingeniator Regis, *Studies in Medieval History*, 1989, 289–303; A Freeman, 'Wall breakers and river bridgers', *Journal of British Studies*, X, 1971, 2–15; M Prestwich, *Armies and Warfare in the Middle Ages: the English experience*, 1996, 141; Salimbene de Adam, *Cronica*, I, 572.

3 The fourteenth century: siege warfare at the start of a new age

HISTORIANS DEBATE whether there was a late medieval "military revolution" during the fourteenth century, as states continued to develop, alongside big economic and social change. In Europe, a new regional power emerged in Poland-Lithuania. In the Muslim world, the Ottoman Turks rapidly became a potent force and brought Islam into the Balkans. Russia shook the rule of the Golden Horde for the first time under the leadership of a new city state, Muscovy, while the Mongol Yuan dynasty in China was overthrown and native rule restored with the Ming.

Conflicts in Iberia in the 1360s were part of the wider struggle known as the Hundred Years War. The whole of western Europe became embroiled as direct or proxy participants, and warriors who found themselves temporarily without employers began to market their skills as *condottieri* (contracted soldiers) to the Italian cities. In the course of this century we will note the first uses of gunpowder artillery and consider its impact.

England under Edward II

ENGLAND WAS THE INSTIGATOR of the trouble. The troubled reign of Edward II was marked by military disaster in Scotland, and civil war. The king had repeated problems at home. In 1316, for example, riots in the city of Bristol were sufficiently threatening to require a naval blockade, and an attack on the townspeople from the castle with stone throwers, which in a few days demolished houses and stretches of the wall.[1] In 1321 the king was engaged in full-scale civil war with many of his leading barons. The latter stormed through the marches of Wales seeking to capture the base of Edward's favourite, Hugh Despenser, who had been gifted the vast lordships of Glamorgan. The small castle of St. Briavels in the forest of Dean, a royal centre for the manufacture of such munitions as crossbow bolts, powerfully provided with a large modern gatehouse, and the older castle at Newport both fell easily, as their custodians gave them up (May 1321). In an inquest, it was noted that the earl had lost the castles of Newport, Cardiff, Caerphilly, Llanblethian, Kenfig, Neath, Dryslwyn and Dinefwr among others, and amongst the equipment removed by the captors were engines and springalds.

[1] *Vita Edwardi II*, ed. Denholm-Young, 1957, 73.

The list included some of the most massively fortified castles in the island, so confirms that no effort was made to defend them. The king attacked the water-girt stronghold of Leeds in Kent, which also had modern defences, and again the garrison gave in at once. The chronicler noted that the defenders were harshly treated because "just as no one may build a castle without the king's licence", so it was also wicked to defend a castle against the king in his kingdom. The Patent and Close Rolls showed frequent orders of the king relating to the holding or recapture of castles involved in the struggles, including a raid on the earl of Richmond's castle of Bowes when the attackers made off with 4 tuns of wine along with armour, engines and springalds, and the earl of Lancaster's attack on Tickhill castle. But in reality although castles played their role as defended bases, there were no significant sieges: the nature of the civil war was such that there was no leisure for prolonged investments, and the few (if any) defenders had a better chance of avoiding the hangman if they did not resist.[1]

War of Saint Sardos 1323–5

The English crown also ruled Gascony from Bordeaux as dukes of Guyenne, an area of France linked to the Plantagenets by a common interest in trade, reinforced by the preference of the Gascons for a king hundreds of miles away across the sea. In a conflict erupting in 1323, known as the war of Saint Sardos after its place of origin, Charles IV declared the English king forfeit of his fief. A French army invaded the territory and the struggle was conducted entirely through the defence or capture of fortresses: the English and Gascons did not have the resources to field an army. It can be followed through a series of letters sent to the king of England or his chief minister, the younger Hugh Despenser. A French army marched through Périgord to besiege the important castle of La Réole in August 1324, breaking down the bridge over the river to cut it off from Bordeaux. An initial assault the next day was repelled with heavy loss. La Réole was a fortress on the river Garonne built as a *turris* (donjon) when the river became a frontier at the end of the twelfth century. It had been modernised by the provision of a strong curtain wall with circular corner towers. The adjacent town was walled too, and the river and a ravine provided additional protection on the other three sides. It was no surprise therefore that the immediate attack having failed, the French decided to starve the town and castle into submission.

Shortage of money meant the towns were unprovided with victuals or equipment, although Bordeaux itself had in store different types of crossbows and bolts. Later correspondence dealt in detail with the provision of artillery. Lords and bishops received instructions to assist the transport and construction of engines including springalds. Engineers, arrow-makers, carpenters, masons, ditchers and

[1] *Vita Edwardi II*, 110, 116, 121; *Calendar of Close Rolls 1318–1323*, 542 (8 May 1322), *Calendar of Patent Rolls 16 Edward II*, 76, 108, 254 (Tickhill, Odiham, Bowes); *Flores Historiarum*, 199, 205. For a study of the South Wales campaigns, J Conway Davies, 'The Despenser War in Glamorganshire', *Transactions of the Royal Historical Society*, 3rd ser., IX, 1915, 45–53.

miners were in demand. On one occasion, it was ordered that if, for reason of time, no one could manufacture a sufficient number of springalds and engines, trees suitable to provide the beams for stone throwers should be cut down.

La Réole surrendered at the expiry of a truce in October 1324. A letter to Despenser in November explained that the castle of Sompuy could not resist the French because the walls and gates were broken down, while the castle of Montpesant had fallen because the defenders had no engines. At Bordeaux, however, the suburbs had by now been enclosed with a wall and ditch. By February 1325, the city was in desperate need of ships, men, supplies, engines, springalds and everything else. By the end of March, he was warned of French preparations with *brides, despringales et de berffiers* (trebuchets, springalds and belfries). At that time, nonetheless, it is worth noting that the English treasury was still paying the wages of no fewer than 43 Gascon and English castle garrisons' captains in the province. The war was concluded with a truce but had revealed that even without a field force, a heavily castellated region could prevent conquest if there were determined defenders properly equipped in the major fortifications.[1]

Scotland: the struggle resumed

By actively supporting the claim to the Scottish throne of Edward Balliol, the young Edward III (1327–77) was guaranteeing war with Robert Bruce's successor, David. The war was recorded by a number of contemporary and later historians, including the Liègeois Jean le Bel, whose chronicle provided the source for the more famous Jean Froissart for the period up to 1360. In 1333, the king joined the siege of Berwick already begun by Edward Balliol, and assisted by the same Flemish engineer John Crab who, until his capture, had overseen the defences of the town for the Scots. He deployed stone throwers close against the castle. In March, £40 had been delivered to Richard the Goldsmith to hire carpenters, sawyers, smiths and other workers to construct siege engines, while Thomas of Papenham was paid to secure the extraction of stones from the quarry at Tadcaster, which were then worked into ammunition by masons for a cost of 20 marks. In May, the same Richard was ordered to deliver the two new engines along with 606 stone balls to Hull, for transportation to Berwick by sea. They caused serious damage to the defences, but the garrison resisted until forced to agree to surrender if not relieved within fifteen days. They did so the day after the English had defeated the Scottish army at the battle of Halidon Hill (19 July).[2]

[1] P Chaplais, ed., 'The war of Saint Sardos 1323–25', *Camden 3rd Series*, LXXXVII, 1954, 48–51 (La Réole), 65–6 (engineers etc.), 81 (La Réole), 92 (Sompuy), 93, 136 (Bordeaux), 157 (French siege weapons), 162 (ships), 272–4 (garrisons). J Gardelles, *Les Châteaux de moyen âge dans la France du sud-ouest*, 1972, 155–6 and plate XXVI (La Réole).

[2] *Chronicon de Lanercost*, 269, 274; Gray, *Scalachronica*, 91–6; *Chronique de Jean le Bel*, ed. Viard, Deprez, 1904, 118; Froissart, *Chronique*, ed. Diller, 1991/2, 125. This modern edition uses the Amiens manuscript of the text; *Calendar of Close Rolls Edward III 1333–37*, 24, 35, 38, 48–9. For a study, R Nicholson, The siege of Berwick 1333, *Scottish Historical Review*, 40, 1961, 19–37 and the same, *Edward II and the Scots*, 1965, 119–39. There are later

In the ensuing invasion of Scotland, Edward was careful to reoccupy the key fortresses that had been dismantled, such as Roxburgh, Edinburgh, Stirling and Bothwell. At Dalkeith, Froissart recorded that a tower defended by 36 men well provided with artillery and supplies was stormed, following which the rock at Edinburgh was attacked with two stone throwing engines (1334). In 1336, the same treatment was dealt to Stirling on its rock, and it surrendered. On the site of the ruined castle, the English erected a fort "called a pele" (Lanercost Chronicle). In November of that year, Edward sent an order to speed up the payment already agreed to Nicholas the engineer *(Lenginour)*, who had contracted to construct 30 new springalds, for which were required materials such as "dry timber, iron, hemp and hair". The Scottish counter-attack got underway in 1337: an attempt to retake Stirling and Edinburgh was thwarted by a relief force sent from Carlisle but Bothwell fell again. It had been repaired but there had been no time to restore entirely the effects of earlier demolition. The earls of Salisbury and Arundel attacked Dunbar "with many huge engines", but despite attacking for 22 weeks were resisted by the stalwart countess inside. In 1339, Robert the Steward recovered Perth. In 1341, Edinburgh was captured "cleverly" by William Douglas, and in the following year Roxburgh, not for the first time, fell to a dawn surprise escalade. Once again, all the English-held castles except Berwick had been recovered either by assault, by trickery, or by blockade.[1]

Start of the "Hundred Years' War"

HAVING RE-ESTABLISHED THE REPUTATION of the English monarch as a successful leader of armies with his triumphs against the Scots, Edward III had determined to make war on the French in pursuit of his claim to the throne. From 1337, his diplomats organised an international coalition to support his efforts to overturn the Valois king Philip VI who had succeeded the childless Charles IV in 1328. But the fighting would have to be done in French territory by forces raised and paid by the English. Although the English-language histories of the war usually highlight spectacular victories such as Crécy, Poitiers and Agincourt, the war mainly comprised regular siege and defence of fortresses, alongside devastating marches (*chevauchées*) through enemy territory designed to undermine his willingness and ability to resist, while providing the invader with loot, plunder and ransoms to meet some of the enormous costs of the fighting.

The most serious theatre of war was in Flanders, where Philip issued orders to strengthen the defences of all the cities, especially those along the coasts of Flanders, Picardy and Normandy. Edward's attempt to take the city of Cambrai in 1339

references to the use of "gonnes" at the siege but these are uncorroborated. For a modern study, C Brown, *The Second Scottish War of Independence 1333–63*, covers many of the sieges in detail.

[1] *Chronicon de Lanercost*, 287–98; Fordun, 354–6; Gray, 98–102, 116; Froissart, I, 150–3, 169; *Calendar of Close Rolls Edward III (1333–37)*, 724; *Anonimalle Chronicle*, ed. Galbraith, 1927, 12–14.

was in vain, and in 1340, following his naval victory at Sluis, the king and his local allies tried to take Tournai, attacking by land and from ships on the river Scheldt. Expecting attack, the French had prepared their engines, both springalds and "kanonz", and this or a reference at Cambrai the year before are probably the first evidence of their use in a siege in western Europe if we accept that the later and uncorroborated reference to their use at Berwick was anachronistic. The effort was abandoned in a truce when a large French army arrived at Arras (September 1340).[1]

Froissart recounted that at the siege of the town and castle of Mortagne by the troops of Hainault, the troops from Valenciennes had a very large stone thrower. A master engineer inside the town aimed the town's engine at it. The first shot fell 12 paces behind, the second fell in front, and the third was a direct hit, breaking the attacking engine in two:[2] an unusual example of a description of the operating skills of the engineer. Another traditional siege operation in the Low Countries took place in 1345, when the count of Hainault and Holland, William II and IV, besieged the city of Utrecht, attacking it for six weeks with a battery of ten stone throwers. His large army watched as a bombardment of 1,100 "spherical stones" each weighing two hundred pounds (91 kgs) was hurled on the town. Remarkably, although much damage was done to property, hardly anyone was hurt.[3] Seeing the futility of his attack, the count reached an accord. The ineffectiveness of the bombardment suggests the absence, to these employers, of the kind of skilled engineers available a few years before.

First cannon in the west

The traditional trebuchet and the torsion-powered springald remained the favoured weapons of siege warfare, at the same time that the first explicit references to gunpowder artillery appear in the chronicles. The well-known illustration in the Holkham Hall Milemete manuscript dated to 1326 of a pot-shaped gun firing an arrow indicates the earliest evidenced appearance in the west of a new weapon (Plate 7). The first contemporary documentary references to "gonnes" in England date from 1339, but in Italy from, possibly, 1326 (Florence) and in France from 1338, where the arsenal of Lille, hitherto dealing with springalds and crossbows, started ordering bolts then, from 1342, gunpowder.[4] The fifteenth century chronicle of the Dukes of Brabant wrote of "*guennons*" operated by cannoniers at the siege of the rebellious lord of Faucquemont in 1329, a castle that fell to the more traditional method of mining after nine weeks.[5] The fact that this account, like Froissart's for

[1] Robertus de Avebury, *De Gestis Mirabilibus Edwardi Tertii*, ed. Thompson, 1889, 316–18; Froissart, I, 218–19, 270–3, II, 6, 45; *Chronique de Jean le Bel*, I, 181–202; Adae Murimuth, *Continuatio Chronicarum*, ed. Thompson, 1889, 103, 107, 115.

[2] Froissart, I, 7–8; II, 70–2.

[3] *Chronique de Jean de Hocsem*, 336.

[4] De La Fons-Mellicocq, *De l'artillerie de la ville de Lille au XIVe, XVe et XVIe siècles*, 1854, 4–8. The first explicit reference here to a "canon" dates from 1348.

[5] Dynter, *Chronica nobilissimorum ducum Lotharingiae et Brabantiae*, II, 783–5.

the early decades, was not contemporary raises questions about the reliability of the evidence. But sufficient references exist to confirm that cannon, all of the type shown in the Milemete illustration, throwing large arrows, did appear around this time. In September 1331 a number of local lords made an attack on the north Italian city of Cividale. One version of the city's chronicle talked only of an attempted nocturnal escalade. However, an addition to the manuscript describes the use by the attackers of *vasa,* in the context of weapons propelling missiles, and then says that they *balistabant cum sclopo.*[1] *Sclopus* is a much later word used to describe the arquebus in the sixteenth century, so it may be that the whole text is a later insertion. But the date ties in with the other chronicle references already cited, as the new instrument, still clearly small and of limited effectiveness, having only weak ballistic power, spread rapidly alongside the older artillery.

Of inestimable value in corroborating this documentary evidence was the discovery in 1861 of an actual weapon, called the Loshult gun (preserved in Stockholm military museum), a bronze weapon just 30 cms long, and with a "calibre" of 31 mm, weighing nine kgs, which because of its clear similarity in appearance to the pot-shaped Milemete gun has been assumed to be of the same age. Larger guns of cast iron, preserved in Bern, are thought to be a couple of decades younger in age. Meanwhile, the discovery of actual "gun-arrows" in archaeological investigations at Castle Eltz, dated by relation to a historical event (a feud with the Archbishop of Trier in 1331–36 leading to a siege), seem to confirm again that this was at least one of the types of missile used in these very early examples. The five missiles recovered were between 522 and 642 mm in length, and weighed between approximately 400 and 560 grams.[2]

It has long been traditional to regard these original guns as being of relative feebleness, based in part on their small size, and in part on comparison with later and much more deadly examples. Therefore, it is interesting that all the governments listed seem to have ensured that they had supplies of the weapon within a few years of its first appearance.[3] Nothing is known of their actual performance. The results of tests on replicas are all there are to go on. Tests on a replica of the Loshult gun found that gun-arrows made of wood could reach 500 metres, while lead balls could achieve a range of 1,200 metres. Accuracy was poor, but against lines of men shot would have hit something at 200 metres, and both shot and arrows were able to penetrate a mail shirt, although not plate armour, and if the effective range for achieving this was only 100 metres for the arrow, it was up to

[1] Fragmenta Chronici Forojuliensis, auctore Juliano canonico Cividatensi, *RIS,* XXIV, cols. 1288–89.

[2] G Strickhausen, Bemerkungen zu frühen Feuerwaffen in 14. Jahrhundert, in Wagener and Laß, . . . *Wurfen hin in steine/gröze und niht kleine . . .*, 2006, 47–51; W Tiltmann, Die Eltzer Büchsenpfeile von 1331/3, *Waffen- und Kostumkunde,* 36–7, 1994–95, 120–6, 53–6.

[3] Clephan, 'The ordnance of the 14th and 15th centuries', *Archaeological Journal,* LXVIII, 1911, 50 f; J Partington, *A History of Greek Fire and Gunpowder,* 1960 (new ed. 1999), 97–108; M Beyaert, Quelques problèmes de développement des armes à feu, *Military Studies in medieval Europe,* 1997, 71 f; H Brackenbury, 'Ancient cannon in Medieval Europe', *Proceedings of the Royal Artillery Institution,* IV, 1865 and V, 1867. All discuss the same evidence with particular insights.

600 for the lead shot. Tests by the Gunpowder Research Group using different formulae for the gunpowder achieved considerably greater ranges, up to 1,300 metres at a high elevation, but again confirmed the lack of accuracy or consistency in repeatedly failing to hit a target even at 200 metres.[1] There were also tests made on a replica of the gun shown in the Milemete Manuscript. Using the dimensions of the original illustration produced a larger weapon three feet (900 mms) long, and weighing 900 pounds (410 kgs). The team also assumed that there was a common bore throughout (of 38 mm). With small charges of gunpowder, firings of an arrow 54 inches (1350 mm) long achieved ranges of 150–200 yards, with significant swerving to the left.[2] A much larger number of assumptions had of course been made in executing this test than with the Loshult experiment. Those who invested in these first guns must have appreciated their potential: the great range that could be achieved in the Loshult tests was far in excess of anything else existing.

The war in Brittany

English intervention in Brittany proved to be more effective than the unproductive but very expensive operations in the Low Countries. The death of the duke left two claimants, John of Montfort supported by Edward III, and Charles of Blois, backed by the French. John attacked Brest (1341), strongly fortified with a garrison of 300. An initial attack drove the defenders back to the bridge, protected by a portcullis. Engines were then brought up, and their bombardment covered the approach to the ditch, filled despite the hurling of stones and pots of quicklime from the walls. The garrison surrendered before the final assault. Montfort then exploited dissent among the townsfolk of Rennes to gain admission, and tricked the castle of Hennebont, on the coast, into surrender too. Vannes submitted. Auray gave in after resisting two assaults.[3]

Charles of Blois was given a large French army to reconquer the duchy. He first attacked the castle of Champtoceaux on the border, driving up to the ditch which his troops succeeded in filling, then sapping the walls under cover of cats whereupon the garrison surrendered. Charles followed up with a rapid march across country and the seizure (with the aid of some of the citizens who opened a gate) of the city of Nantes, complete with Montfort inside. His countess continued the struggle, however. As 1342 began, Charles besieged Rennes, which also surrendered while an English relief army was held up by stormy seas. Hoping to make an end of things, Charles laid siege to the countess in the castle of Hennebont, which was well provisioned and garrisoned. The countess oversaw an active defence, and

[1] P V Hansen, 'Casting and firing a replica of the Loshult gun', *Journal of the Royal Ordnance Society*, 14, 2002, 5–16; R D Smith, The Ho experiments, report no. 2 (August 2003), Medieval Gunpowder research group, www.middelaldercentret.dk (website).

[2] R D Smith, 'The reconstruction and firing trials of a replica of a 14th century cannon', *Royal Armouries Yearbook*, 4, 1999, 86–94.

[3] *Chronique de Jean Le Bel*, I, 244–57; *Knighton's Chronicle*, ed. Martin, 1995, 44.

soon enough, Charles left the place blockaded by his Castilian allies, who summoned up 12 large engines from Rennes, and went off to besiege Auray.

Despite the arrival of the English fleet at Hennebont, the siege was not broken off and the bombardment continued, but a sortie under cover of 300 archers enabled the defenders to capture and break up the largest engines and set alight the camp. Other towns were falling to the French side, and Auray itself, starving, fell after ten weeks. Charles left it securely held and stocked, then took Vannes with ease. The English meanwhile failed to take the strong castle at Roche Piriou. A new French siege of Hennebont followed, spearheaded by 15 or 16 large engines. However, troubled by shortage of supplies for the besiegers, Charles agreed to a truce[1] which was confirmed after a campaign by King Edward himself (January 1343).[2]

The war resumes

Fighting was under way again in 1345, with an army led by the earl of Derby being sent to Gascony. Once again, sieges formed the main element of the strategy of both sides. First Derby assailed Bergerac in Périgord with a substantial Anglo-Gascon army, and took it by storm. His archers were so effective that no one could prevent him from filling in the ditches at the weakest point of the wall, whereupon it surrendered. Nothing remains of the defences of the town, which had flanking towers around a central citadel, and as many as forty other towns and castles were then taken. Meanwhile, Derby took Monségur after two weeks of heavy bombardment, shaking the walls with artillery brought from Bordeaux and Bergerac. The town, a recently built *bastide,* had a triangular shape and strong stone wall but without flanking towers. He then tried to recover La Réole. Here, he resorted to an old-fashioned approach, building two belfries three stories high, on wheels. The ditch having been successfully filled, the towers were wheeled up to the wall, and sappers set to work. The town surrendered and the castle followed five weeks later.

In 1346, the duke of Normandy's army besieged Aigullon, held by a large force of 300 men at arms and 600 archers. Standing at the confluence of the Lot and Garonne, the town had separate castles of late thirteenth century date at opposite corners. The French tried to build a bridge across the Garonne. A first effort was fruitless when the garrison sent out boats to destroy it. The French then filled boats of their own with soldiers to protect the carpenters, and despite further English efforts, the bridge was completed in two weeks. Across the river, the French tried a direct attack on the castle gate, but were repelled with very heavy losses. The duke turned to artillery, sending to Toulouse for eight big engines stored there, and building another four, even larger. The garrison were well prepared, however. With effective shelter inside, they were protected from the stones, and their own engineers were very accurate: the two defending engines destroyed

[1] *Chronique de Jean le Bel,* I, 264–340; Froissart, II, 146–235, 249.

[2] *Chronique de Jean le Bel,* II, 19 f; Avebury, 340–44.

six or seven of the attackers'. In a further assault, the chains of the drawbridge were broken and there was a ferocious struggle for the bridge. Next day, French carpenters set about building siege towers mounted on boats to overtop the walls. The garrison waited until the completed contraptions, filled with soldiers, reached mid-river, then shot at them with four engines, whose stones destroyed one of the floating towers, the rest being prudently withdrawn. The besiegers had however ensured that there was no relief by surrounding their camp with a deep ditch of their own. In the end, the duke learnt of the French disaster at Crécy and abandoned the siege.[1]

From Castle Cornet to Caen

Meanwhile Edward III crossed to Normandy via the Channel Islands, where he had re-secured English control of Guernsey, which had been taken in a French raid in 1338. This entailed a substantial operation on all sides. Apart from two stone-built refuges, the only castle on the island was that now named Cornet, commanding the harbour of St. Peter Port. This would appear to have been constructed during the thirteenth century, when the rock on which it stands was linked to the mainland by a causeway, before which it was useless as a refuge. A French force under admiral Behuchet arrived in Spring 1338, but failed to take this castle. However, the fleet returned on 8 September, capturing Castle Cornet as well as the old headland refuge of Jerbourg from their garrisons of 11 men at arms with 50 archers, and 12 archers respectively. The French then attacked Jersey, where they failed to take Gorey castle, and then withdrew, leaving a garrison of one knight and 80 Norman soldiers on Guernsey. Edward ordered the island recaptured in 1340, the French troops withdrawing into the castle in October where they were blockaded. The recapture of Castle Cornet was delayed until 1345, when the king ordered it to be assaulted, the governor having refused to yield. Timber was prepared for scaling ladders, and the professional troops supplemented the islanders themselves. The conquest of the castle happened in August in a three day operation and the garrison was apparently slaughtered. Excavation uncovered evidence of a wooden fighting platform along the curtain that had been destroyed, and clusters of large beach pebbles (weighing 4–5 kgs) that may have been missiles dating from the capture or recapture.[2]

On the mainland, Edward III marched across the Cotentin and attacked William the Conqueror's old city of Caen, with its imposing donjon built under Henry I more than two centuries previously. The town was divided into sections

[1] *Chronique de Jean le Bel*, II, 40–2, 58–63; Froissart, II, 316, 327, 335–9, 364–9. Much of Froissart's text is taken directly from Le Bel; Avebury, 356 f; *Chronicon Galfridi Le Baker de Swynbroke*, ed. Thompson, 1889, 78; *Chronicon Ventense*, ed. Piganeau, 1874, 43–4; Knighton, 52; Chronicon Vazatense, *Archives Historiques du Département de la Gironde*, XV, 1874, 44. Gardelles, *Châteaux*, 83–4 (Aigullon), 98 (Bergerac); L Drouyn, *La Guienne Militaire*, 1865, II, 378–89 (Monségur). For La Réole, see references given previously.

[2] K J Barton, *The Archaeology of Castle Cornet, St Peter Port, Guernsey*, ed. N Jee, 2003, 48, 90; T de Guerin, 'The early history and first siege of Castle Cornet', *Report and Transactions, Guernsey Society of Natural Society and Local Research*, 1904, 337–60.

by a number of rivers. There having been no threat for a very long time, the town's walls may well have simply fallen apart, and the hasty repairs now carried out could only have been of ditches and palisades. Jean de Venette, then a Parisian prior, stated that the town was without walls at all. There were strongly walled abbeys on the perimeter, but when the attack was launched, they were not garrisoned by the force of French men at arms and Genoese crossbowmen in the place: possibly 300 in the castle, and another 1,200 in the town, not counting militia. A summons to surrender having been rejected, Edward ordered an attack on several points, and the river Petit Orne was too low to represent a serious obstacle. Storming such defences as there were, the English penetrated both islands, cutting off the Genoese from the castle, and the commander of the fortified bridge between the outer suburb (the *Ile St Jean*) and the inner (the *grand bourg*) (protected by stockades and a portcullis) surrendered after offering stiff resistance. The defenders of the castle did not stir to aid their comrades in the town, which now suffered a dreadful sack and massacre, but the king could not stay to complete the task and left a force to besiege the castle (31 July 1346). The French obtained revenge when the troops in the castle joined with the surviving citizens to overwhelm the English force.[1]

Events of 1346–7

Following the battle of Crécy, Edward began what proved to a siege of Calais of epic duration. His army diminished by losses, the king had quickly decided not to attempt to storm the port. The walls dated from the thirteenth century when they had been erected by the builder of the innovative castle of Boulogne, Philippe de Hurepel. There was a separate castle with strongly projecting circular towers and a large round donjon behind its own moat, and the whole oblong-shaped fortification was protected by two deep ditches which could be flooded from the sea. The low-lying terrain could also be flooded, such that the town could resist any attack. The garrison were alert to likely risks and prepared plenty of sacks of straw and other material to soften the blows from trebuchets. Additionally, because of the nature of the ground, there was a shortage of firm sites on which to erect the heavy machines. Instead, Edward instituted what turned out to be a very long blockade, protecting his encampment with ditch and palisade. The coastal approach was closed off with a wooden fort provided with archers, cannons and springalds when French forces threatened to bring relief. With the French marching along the coast through the dunes, Edward sent out ships carrying archers, springalds and "bombards", while doubling his garrison in the tower at Sangatte, four miles (6.5 kms) away. The French army still managed to storm that tower, but found the

[1] Avebury, 361 (letter of Michael Northbaugh from the siege); Jean le Bel, II, 70; Froissart follows Le Bel; Adae Murimuth, 203–4 (letter from Burghwash); *Chronicon Galfridi le Baker*, 80; Knighton, 54; *Chronicle of Jean de Venette*, trans. J Birdsall, ed. R Newhall, 1953, 40–1. The event was analysed by H Prentout, 'La prise de Caen par Edouard III 1346', *Mémoires de l'Academie Nationale des Sciences, Arts et Belles Lettres de Caen*, 1904, 5–42, with a useful plan of the city.

English camp too strong to attack, so Calais finally surrendered in August 1347. It was noted that Edward had spent a great treasure on wages, building work and siege engines.[1]

During the siege, Derby raided and seized several castles in the south west (such as Lusignan).[2] Meanwhile, King David of Scotland recovered most of the castles held by the English in his kingdom, some by blockade, some by treachery. The Cumbrian border castle of Liddel Strength was, unusually, stormed after an attack with engines that left a great part of it destroyed within six days: it was still an earthwork protected by timber stockades. The ditches had been filled with timber to enable the assault.[3] The battle of Neville's Cross however ended the Scottish incursion south of the border, though in 1347 the remote Liddesdale castle of Hermitage was forced to surrender.

The 1350s

Despite the dreadful ravages of the Black Death the Anglo-French/Scottish struggle continued after a short truce. In south west France, the future king John tried to recapture St. Jean d'Angély and succeeded because a relief force was unable to cross the river on which the important regional town stood, and the garrison's provisions ran out (August 1350). In 1352, a French force recovered the walled town of Guines, which had been secured by Edward III as an important outpost of his newly acquired foothold in northern France, resorting to a surprise attack at night.

In 1356, Évreux in Normandy fell to King John II (1350–64) and a siege of Breteuil followed. The town was attacked from all sides because it stood in open country, and "large engines" were brought from Rouen to launch a non-stop bombardment. A wooden "scaffold" holding 200 men was also constructed to get the attackers close to the wall, which was eventually destroyed by the defenders. However, after seven weeks of attack, the Navarrese garrison surrendered (August). The French had also been besieging Pont Audemer in Normandy since April when Lancaster arrived and raised the siege at the end of June, the French leaving behind all their engines and *arblastes* (springalds?). Their mines had reached within just over a metre of the town wall, and Lancaster had them refilled before leaving a garrison. The same letter which records this success then tells of how the town of Conches (a small place) was taken by storm, and how Vernueil was also stormed on the third day of the attack, despite its strength.

At the same time, on his destructive *chevauchée* in the south, the Black Prince attacked the castle at Romorantin. The town was undefended but the castle

[1] Jean le Bel, 110–12, 132, 152–67, followed by Froissart, III; *Chronica Galfridi le Baker*, 86, 89–91; Knighton, 64. Allen Brown, Colvin, Taylor, *History of the King's Works*, I, 424 (plan), 444, 446. G Wrottesley, *Crécy and Calais from the Public Records* (1898), shows quite small numbers of gunners and engineers on the payroll, but numerous carpenters, smiths and masons: 148, 203.

[2] Avebury, 370–6.

[3] Avebury, 376; Gray, 116; *Anonimalle Chronicle*, 24–8.

comprised a strong donjon in a bailey. Archers kept the defenders' heads down while the courtyard was stormed, although the first onslaught was driven off. The donjon surrendered after three further days of attacks (30 August – 3 September). Froissart talks of *kanons* projecting "in the manner of Greek Fire" being used in the bailey, others refer to mines and engines. Whether or not the English had cannons, and they are only mentioned in Froissart, it was apparently the effects of fire and mining that induced the surrender. One convincing account not contradicted by the others talks of stone throwers being set up to hurl fire into the castle, setting alight the roofs and compelling surrender.[1]

The Black Prince went on to win the battle of Poitiers and take prisoner King John, eventually setting in train the negotiations that led to the peace of Brétigny (1360) and the cession of large parts of France to Edward. Lancaster, meanwhile, had submitted Rennes to a prolonged blockade of nine months, although he had also deployed engines including a siege tower, and springalds, and a mine. According to the epic verse biography of Bertrand du Guesclin, who was involved in the action, the location of the mine was discovered by the method known to the Byzantines of placing pots such that they would vibrate if near to the underground workings. The direct attacks having failed, it may also be that on this occasion there was no incentive to seek an early capture, and it is possible that his small force (about 2,400 men) was unable either to complete a blockade or to mount a seriously threatening attack. A truce brought the siege to an end. Further north, the count of St. Pol was besieging a Navarrese garrison in St. Valéry, collecting his siege engines from Amiens and Abbeville. The prolonged operation cost Picardy dearly to sustain the army, and the defenders also had many *kanons*, springalds and wall mounted crossbows. A large engine deployed to the east of the town (presumably a trebuchet) began to cause serious damage with its great stone projectiles, but in the end it was lack of provisions that brought about surrender. Two years later, during the siege of Rheims as the English tried to force the French into peace, the English captain Burghersh (whose letters have been quoted in Avebury's account) attacked the castle of the archbishop of Rheims at Cormicy. This was a strong square tower. with thick walls, and Burghersh overcame it with miners hired from the bishop of Liège. When the mine was fired, the tower split in two.[2] The English forces were living off the land. At Easter, Edward's troops marched through Beauce, where it was reported that almost all the monasteries were fortified, and sheltered the collected "provender" of the neighbourhood. These places were either stormed, or else surrendered as soon as siege engines

[1] Jean le Bel, II, 226–7; Froissart, III, 90, 133; Avebury, 463–4; Galfridi Le Baker, 141 (citing a diary of the campaign); Gray, 121; R Crozet, 'Un episode de la guerre de cent ans. Le siège de Romorantin', *RH*, CLIII, 1926, 187–92, reconstructs the now disappeared castle and the events, arguing that Froissart was probably wrong about the cannon. However, we have already seen several examples of earlier use.

[2] Froissart, III, 90, 135, 161, 168, 223–28; Jean le Bel, II, 242–3; *Chroniques des règnes de Jean II et de Charles V*, ed. Delachanal, I, 51–2, 67–9, 110–11, 170; Gray, 148; K Fowler, *The King's Lieutenant. Henry of Grosmont, Duke of Lancaster 1310–61*, 1969, for a scholarly study of the English commander's life and deeds.

were set up against them.[1] The troops were fed, but the effect on the local peasantry can be imagined.

While these military events were taking place, the kingdom was also shocked by the series of massive peasant revolts known as the *jacquerie* from 1358, and the effects of war, hunger and plague were among the causes. With Paris itself in uproar, the French royal chronicles confirm the existence of a major arsenal of artillery held at the Louvre. This was seized by the citizens, who set about blocking existing gates and making new entrances protected by bastilles, and digging new ditches in front of the gates. The "natural order" was quickly and ruthlessly restored. The Île de France continued in a state of tension for fear of the English and their allies, however, and before peace had been made, the French regent and the Parisians, now on the same side, undertook the destruction of monasteries and churches standing just outside many towns, to deny them to an attacker, or alternatively as was happening elsewhere, added ditches and hoarding to stone churches and belfries "as one does castles" (Venette), and laying in supplies of stones and crossbows.[2]

Berwick 1355–6

In late 1355, the Scots made yet another attempt to recapture Berwick. Thomas, earl of Angus, came by ship at night, camped under the walls, and at dawn on 2 November raised ladders at the Cowgate to get inside, overpowering the sentries. Once again the Scots, on this occasion aided by a contingent of French troops, were unable to make any headway against the castle, and although they managed to storm the Douglas Tower, which the gathering English forces outside had intended to use to gain access, they found themselves caught between the castle and the army outside, and with Edward III raising another army to march north, abandoning his immediate plans in France, the few defenders capitulated on the promise of their lives (January 1356). The king had brought with him 120 miners from the Forest of Dean, according to Avebury, in order to allow him to get soldiers inside the town, but perhaps for the traditional purpose of creating a breach.[3]

So far, then, the history of the Anglo-French wars has not thrown up any new developments in techniques of either attack or defence. The standard weapons remained the trebuchet and the springald, with mining a common way to create a breach. Guns were becoming widespread, but as yet of limited power, and therefore limited effect in the siege of stone walls. In fact, early adaptations of fortifications suggest rather that their primary use at this time was as anti-personnel defensive weapons.

[1] Gray, 156.

[2] *Chronicle of Jean de Venette*, 72, 84–5.

[3] Fordun, II, 362–3; Avebury, 450–1; *Anonimalle Chronicle*, 34–5; Walter Bower, *Scotichronicon*, ed. Watt, 1996, VII, 281–3 (written in the 1440s, using Fordun and another source).

The struggle in Iberia

The peace of Brétigny left vast tracts of land and people who identified themselves as French under alien rule. But in the short term, with the French still paying off king John's gigantic ransom, the fighting continued by proxy, with the chief protagonists now engaging each other in Spain, taking sides in a violent struggle between Aragon and Castile

Before the interventions, in 1328, King Alfonso XI of Castile had taken the Muslim town of Teba with its ridge-top castle, after his engines had breached the walls.[1] In 1331, an army from Granada laid siege to Gibraltar. A detachment attacked Castro del Rio near Cordoba, making six breaches in the wall. Reinforcements from Cordoba slipped into the town overnight and worked hard to block up the breaches with doors and other timber from the houses, and the next day these makeshift defences held out. Gibraltar was besieged for three months and then surrendered despite the approach of a Castilian relief army. Alfonso tried to recapture the crucial fortress, deploying six engines against the castle with its quadrangular tower and against the Muslim ships in the harbour, shooting from the higher ground. The chronicle said the tower was destroyed, but clearly not entirely, as the attackers had to bring cats to its base to cover sapping, all the while under ferocious attack from above. With shortage of food afflicting the attackers, and the arrival of a Muslim army from Granada, a truce was concluded. Abu Malek, son of the Sultan, set about restoring the damaged defences. According to a contemporary visitor, this involved the construction of a "huge tower" in place of the "small turret" destroyed by Alfonso's engines (this being the still-standing "Moorish Castle"), and the building of an arsenal and new curtain walls.[2]

The next period witnessed something of a revival of Muslim fortunes in Iberia, and the establishment of the Merinid dynasty in place of the Almohads seemed to encourage a more aggressive approach to the Christian states. First, though, the new sultan had had to capture Tlemcen, something that involved another prodigious siege operation, with a wall and towers built to mirror those of the town, from which catapults and archers maintained a powerful attack. Slowly and methodically, the sultan's troops inched their way closer to the ramparts with their cats under cover of the bombardment, until at last (1 May 1337) an assault succeeded.[3] In September 1340, a North African army besieged Tarifa. Twenty engines were deployed on the landward side, in such a way that the engines inside the town could not hit back at them. The defenders sent messages to Alfonso for help. The Moors constructed a stone wall to cut off the town from the sea, and proceeded to target the tower of Don Juan, constructed of *tapia*, taking advantage

[1] *Cronica del rey don Alfonso XI*, ed. Cayetano Rosell, 226. Teba was handed to the Order of Santiago and a new castle built: Ortíz Echagüe, 14.

[2] Accounts from the opposing sides give quite different details but the same outcome: Ibn Khaldūn, IV, 216–18; *Cronica del rey don Alfonso XI*, 242–53; *Travels of Ibn Battuta*, ed. Gibb, IV, 934–6.

[3] Ibn Khaldūn, IV, 220–23.

of a hillock nearby that rose to two-thirds of its height. Mounting engines there, they set about constructing a wooden tower of their own to overtop the defence. A fierce battle waged back and forth around this tower, with four separate sorties by the Christians, and running repairs to the tower, and with the defenders shooting back the stone shot hurled at them, until in the end a victory by the king saved Tarifa.

The Castilians invaded Granada in 1341 and besieged Alcala (la Real), a strongly defended city. On the top of the hill, an ancient circuit of walls had been strengthened by the addition of a massive *torre albarrana* which created a powerful new gate passage, and two strong towers. The initial attack broke through the defences of the suburbs and captured the outer town. Eight engines were deployed against the towers of the inner defences, in particular against the large tower that contained the water supply. Against this well-constructed tower the artillery made no impression, so miners were used to dig under it, then fire the props and the whole edifice collapsed, leading to the surrender of Alcala. Alfonso ordered the tower rebuilt and the outer walls repaired.[1]

Algeciras 1343–4

The Castilian attack switched to Algeciras, a vital communication between Muslims across the Mediterranean. Control of the sea was gained through the defeat of the Merinid fleet. The joint Castilian-Aragonese army crossed the river Guadalete in August and instituted a close blockade of the town, protecting its own camp with ditches and a wooden rampart. The defenders shot large iron balls and arrows but could not prevent the attack developing. Engines were set up against the wall towers. In December, 20 more engines were built, including two "trebuchets" constructed at Seville by Genoese engineers, described as having two bows that maintained a rapid bombardment against the defending engines, so were presumably actually some kind of multiple crossbow. The fighting continued over the winter, the Castilians advancing against the weakest points of the defence, carefully building and guarding wooden shields for their own artillery. From the outside, the Granadans tried in vain to bring relief, and only small boats were slipping through into the harbour. In January 1343, the king concentrated his artillery at a point near the sea and there was a severe battle around the gate tower at this point, with the Christians driving forward either a ditch or a mine (the term *cava* could imply either), the Moorish defenders protecting their walls with screens and sallying out to destroy the mine. The besiegers suffered heavy losses from arrows, engines and iron balls from the defenders' *truenos*, a word strongly suggesting that the Muslims were using cannon. The conflict continued for months, with the king switching his point of attack to other sides of the fortress, building new siege works and initiating new mines or saps, and cost immense sums: only aid from Aragon and France enabled him to continue operations. Finally, when a truce with

[1] *Cronica del rey don Alfonso XI*, 331–5; Ibn Khaldun, IV, 234. R Collins, *Spain*, 46–7; Ortíz Echagüe, 15, 20,21.

Granada expired in March 1344, the place, still too strong to storm, yielded at last.[1] The siege is notable for the successful resistance of the garrison to every form of attack available to a king, except for gunpowder artillery, which may in fact have been available to the defence, in which case this is the first use recorded in Iberia.

During the same time, the Aragonese king had come into conflict with his relative, King James of Mallorca. During May and June 1343, Peter III/IV of Catalonia and Aragon besieged his opponent in the town of Argilers, bringing his supplies and siege engines by sea. The first engine set up caused severe damage. The second, a two-drum trebuchet from Barcelona, and the third, an even larger engine from Valencia, were extremely threatening. The garrison tried to use their own machine to target the new attacking engines, but put too much on the counterweight so that the arm flew straight up and wrecked the engine. As the Aragonese moved in close for the assault, with their cats and other devices to shield their artillery, the town surrendered (2 June). The attack then turned to the castle of Pujol. Artillery created a breach and a battering ram was being set up when it surrendered (9 June). Peter then moved on to Collioure, arriving on 15 June and immediately arranged his engines. The Barcelona weapon was assembled on the hill overlooking the tower outside the tower of Puig Musart. The other engines were carefully located to attack the town. Under the bombardment the walls rapidly crumbled and the external tower was stormed, a second tower (with a Genoese garrison) was captured, and then the town's suburbs fell and the castle capitulated (22 June). With the rapid surrender of other castles, James quickly submitted.[2]

The chronicle noted the precautions taken by the king to ensure the functioning of his defences, when it records how he had the cisterns at Murviedro cleaned out, and the walls repaired before refilling them with water brought laboriously on the backs of animals.[3]

War between Aragon and Castile 1356–65

King Peter of Castile (1350–69) achieved the distinction rare for a fourteenth century ruler of being so exceptionally violent as to earn the title of the "Cruel", but he also had the military skills needed in his dangerous occupation. In 1357 he began war with Aragon by taking the town of Tarazona, which was only weakly walled and garrisoned. Perhaps the Aragonese had believed it was not at risk, being inland. Other frontier castles fell as quickly before a truce was negotiated. In 1359, Peter of Aragon's siege of Alcala (de Moncayo) was abandoned because of harsh winter weather, but in the spring the Castilians attacked the Catalan coast, the heart of the kingdom. The Catalans organised to protect Barcelona against the

[1] *Cronica del rey don Alfonso XI*, 343–89; Ibn Khaldūn, IV, 235–6. W Giese, Waffengeschichtliche und terminologische Aufschlüsse aus katalanischen literarischen Denkmalern, in *Homenatge a Antonio Rubio i Lluch*, I, for a survey of Catalan literary evidence for military equipment in use at the time.

[2] Hillgarth, trans., *Pere III of Catalonia (Pedro IV of Aragon) Chronicle*, 321–48.

[3] Hillgarth, *Pere III*, 422.

enemy fleet, mounting four large (two-drum) trebuchets and drawing up twelve ships in line across the harbour, protected by shore based artillery. One chronicle says they also deployed numerous bombards, and the king's own account says his artillery hit the wooden castles erected on a Castilian galley, destroying a mast and killing a man: clearly not yet a very large weapon. Thwarted by the firepower of the defence, the Castilians sailed off to try their fortune at Ibiza and besieged the town with their engines, but abandoned the effort when the Aragonese-Catalan fleet followed them.[1] These accounts allow us to confirm the arrival of gunpowder artillery in Aragon by 1360, and that the weapons were still small, and of more use to the defence than the attack.

In 1362, the major but ancient inland fortress of Calatayud was taken by the Castilians, surrendering because there was no prospect of relief, and many neighbouring castles also yielded. The most detailed account says that the king deployed many engines and also constructed powerful siege works all round, pressing the attack close to the walls. However, he decided that he would prefer to have the place undamaged so agreed to a forty day ceasefire, at the end of which Calatayud surrendered without further fighting. The Aragonese response, unable to compete in the field, was to place garrisons in many of the remaining castles and towns, such as the powerful inland fortress at Albarracin previously described, so Peter the Cruel tried a bold new direction, marching overland towards Valencia, taking Teruel by storm and occupying Alicante (1364). This turned out to be the high point of his victories, however, as the following year the Aragonese began to recover their losses. Murviedro, held by a strong Castilian garrison of 600 men at arms plus archers and crossbowmen, surrendered after a long blockade because the Castilian king could not trust his nobles, many of whose families were victims of his cruelty, not to switch sides in battle.[2]

Foreign involvement

Peter of Aragon now enlisted the active support of the French commander, Bertrand du Guesclin, who collected a force of veterans now unemployed in France and came across the mountains to turn the scales. A coalition assembled to support the claim of Peter the Cruel's half-brother Henry (of Trastamara) to the throne of Castile and led by Bertrand, Castile was invaded. As castles and towns fell easily into his hands, powerful Castilian arsenals were found to have been laid in at captured fortifications such as Calatayud and Teruel, with arms, engines and trebuchets listed. They also found that much work had been done to repair and modernise the defences themselves. Towers and walls had been strengthened, ditches deepened, more wells and cisterns provided. But as ever, without willing defenders and facing a rival king, the walls and arms alone were useless. Peter

[1] Hillgarth, *Pere III*, 523–26; Lopez, *Cronica del rey d. Pedro I*, ed. Macchi, trans. Steuncu, 1985, 119–23, 145–9; Lopez de Ayala, 422, 495–6, refers to *truenos* at Barcelona.

[2] Lopez de Ayala, 525–35; Lopez, *Cronica*, 197–205. A Gutierrez de Velasco, 'Las fortalezas aragonesas ante le gran ofensiva castellana en la Guerra de los dos Pedros', *Cuadernos de la Historia*, ed. Zurita, 12–13, 1961, 10–38, examines the role of the castle in this conflict.

the Cruel was unable to hold his crown unaided, but he enlisted help even more formidable than that obtained by his rival: the Black Prince brought a seasoned army of veterans from Aquitaine to Spain, and again reversed the balance with a decisive victory at Najera (1367).[1] The victory did not end the war. The next year, Henry's army besieged Peter's supporters in León, building *bastidas*[2] and pressing a close attack. The proximity of a monastery building to the defences enabled his troops to make the defences opposite untenable and bring about surrender. Henry marched against Toledo. The ancient city in the bend of the river Tagus was powerfully provided with walls and towers, and fortified bridges crossed the river. Henry had to construct his own bridge to enable him to complete the blockade. One of the wall towers was betrayed to the besiegers but retaken by the defenders. The tower on the far side of the St Martin's bridge was mined from the shelter of a *bastida*, but the defenders, aware of what was happening, constructed a wall of tapia behind it. But with this also under threat, they resorted to destroying the bridge itself by removing stones from the central arch. The attack was continuing but without success when Peter the Cruel died in an attempt at relief[3] leaving Henry II the undisputed monarch.

His neighbours did not fail to take advantage of the long civil war. The Merinids recovered Algeciras, while Henry was engaged in a substantial war with the new king of Portugal, Ferdinand I (1367–83). A North African fleet brought an army across from Ceuta in 1368 to back up an attack from Granada. While the fleet blockaded the harbour, the Granadan ruler Ibn el Ahmer deployed siege engines by land, and in a few days the defenders, knowing relief to be impossible, capitulated.[4]

When the Castilians invaded Portugal in 1382, Ferdinand protected the harbour of Lisbon by placing his galleys facing the sea, mounting "guns and engines" (*trõos e engenhos*), while his army was similarly provided, specifically with *bombardas*. The Anglo-Portuguese invasion of Castile that year found the army able to take frontier castles by storm, after blasting in the gates and swarming across ditches and through breached walls (as at Lobón, held by garrison of 60, and Cortijo, held by a more substantial force of 230). In 1384, the Castilian sieges of Almada and Lisbon (the first successful, the second not) were each conducted by long blockades, the defences or the size of the garrison sufficient to deter assault. In the final stage of the fighting, in 1387, two Castilian fortresses suffered different fates. The Anglo-Portuguese forces were powerless in front of the small castle of Benavente de Campos. The besiegers camped outside crossbow range, because the defenders had "no other kind of sling or engine capable of endangering them", but King João I (1385–1433) also had no equipment, so raised the siege after eight

[1] Lopez de Ayala, 570–77; Lopez, 207–11; Hillgarth, *Pere III*, 570–78.

[2] Both in Spain and France, this word seems to have evolved from its original sense as a planned, sometimes fortified, settlement into a term signifying a strong but temporary siegework, such as a siege fort or a strongpoint from which to attack a hostile fortress.

[3] Lopez de Ayala, 580–84 for a detailed account.

[4] Ibn Khaldūn, IV, 380–1; Cronica del rey d. Enrique II, *Biblioteca de autores Españoles*, 69, 1877, 4.

days. At Valderas, however, the King had a small belfry and an engine (?gun). The town's wall was flimsy (it was of mud brick), and when they saw the Portuguese weaponry prepared to lead the assault, they surrendered. Thus was the presence or absence of siege weaponry decisive against smaller or isolated defences, if they were themselves without such artillery.[1]

Anglo-French hostilities resume

The French, keen to recover lost lands, but now more cautious about meeting Edward's archers in open battle, had found in Bertrand du Guesclin a commander (soon to be made Constable) whose life, while told in the form of a chivalrous epic, also reveals his mastery of contemporary warfare including siege warfare.

Du Guesclin's early victories (before his adventures in Spain) were in Brittany, where in 1363 he recaptured Mantes and Meulan from the Navarrese. At the latter, the French army first fought its way up to and across the palisade surrounding the main defences, then attacked one of the wall towers, but was repelled. Turning to their miners, they drove a tunnel under the defences, soaked it in bacon grease, then fired it. The town and the donjon inside surrendered. At Valognes, he had to besiege the old castle and called up six engines from St. Lô. The castle's location on a rock prevented mining but the garrison surrendered anyway. Carentan also yielded. At Pont de Donne, the defences included a church provided with a donjon. Facing strong resistance, the French turned to their miners. The defence identified the mine through a rattling pot, but, according to the verse account, their countermine proved fatal, as the French used it instead to gain entry.[2] The following year, the duke of Burgundy continued the attack by taking the castle at Macherenville using eight large engines that did significant damage. Meanwhile however, Louis of Navarre marched from the Auvergne to take La Charité-sur-Loire by escalade. Burgundy continued his own march, using his powerful engines to capture several more castles. At Couvray, they smashed the walls and towers of the castle.[3]

With the main field of hostilities displaced to Iberia, there was now a pause of several years in full scale military action in France. In 1370 the Black Prince captured Limoges, during yet another devastating march across south western France. The town was in good defensive order and provided with engines. Having surrounded it, the English army took three weeks to undermine the walls, bringing down a great length once the props had been fired, and allowing easy entry for both horse and foot.[4] The appalling massacre that ensued serves as

[1] F Lopes, *The English in Portugal 1367–87. Extracts from the Chronicles of Dom Fernando and Dom João*, ed./trans. D W Lomax, J Oakley, 1988, 78–9, 82–3, 128–9, 250–1, 254–7.

[2] Cuvelier, *La vie vaillant Bertrand du Guesclin*, ed. Charrière, 1839, lines 3960–4040, 5085 f (Valognes), 5265–5339; Delachanal, *Chroniques*, I, 342.

[3] Froissart, III, 321, 324–8.

[4] Froissart, IV, III.

a powerful reminder of the reality of warfare to set against the image of chivalry so vigorously promoted among the nobility of both sides. The English continued to operate in the region in the coming years, generally successfully. In 1371, the castle of Montpont in Rouergue was taken. Here, the ditch having been filled in first, wooden shelters each containing 20 men at arms and 40 archers were driven up to the wall, enabling sappers to work to create a large breach and leading to the capture of the place. An English force also attacked the famous old castle of Moncontour, and took it by storm after two weeks' bombardment, finding inside a good store of provisions and "artillery".[1]

In 1372, the French recovered the castle. A force marched from La Rochelle and attacked Moncontour, which had a garrison of 60. The castle fell once the ditches had been filled, after just four days. Bertrand du Guesclin had been in command, his biographer suggesting a rather more aggressive English defence than Froissart, with stones and quick lime being hurled down on the attackers who nonetheless broke in after a week-long operation. Du Guesclin had been recovering English-held castles in Maine and Poitou during the sieges just described, relying sometimes on surprise night attacks (Chatellerault), sometimes on direct assault by escalade (Vaux). Du Guesclin's attack on Saint Sever was a full scale operation with a large army. The English had strengthened the town with new walls and a citadel. There was a fierce struggle for the outer wall, with the French using sapping as well as ladders. The crucial event was the creation of holes in the wall by sapping. As a fire was started in the town, a general assault was ordered and the place surrendered.[2]

Having secured the submission of most of Poitou, the Constable transferred his operations to Brittany in 1373, taking the castle of Suscinio after just four days, but constructing blockading forts at Derval, which agreed to surrender if not relieved. Of the whole province, only the powerfully defended port of Brest, and Auray, remained securely in English hands. St. Sauveur-le-Vicomte in the Cotentin fell to the French after a long siege. It was a powerful, modern fortification largely rebuilt while under the control of English governors, with two strongly-walled wards, of which the inner, well preserved, boasted strong walls and projecting circular towers, and a massive, buttressed quadrangular "donjon" built to replace a much smaller circular donjon (dated to about 1200). The French had great stone-throwers that demolished roofs and crenellations, but the garrison (of 300) surrendered when it ran out of victuals rather than through serious damage to the defences.[3] Gradually, the French king's commanders recaptured castles, towns and provinces, much harder to hold than to take in the face of superior numbers. Large numbers of engines were used in the capture of La Réole, Bergerac, St. Foy, Castillon on the Dordogne, and the town then the castle at St. Macaire. Cannon were among the battery deployed to take Duras after three weeks. In this year of 1377, the royal

[1] Froissart, IV, 146.

[2] Cuvelier, lines 18,534–55, 19481 f., 19,672–780 (Moncontour), 19,811–20,390 (Saint Sevère).

[3] Froissart, IV, 321–22; *Anonimalle Chronicle*, 77–8; Mesqui, *Châteaux forts*, 335–6; Anon, *Le vieux château de Saint Sauveur le Vicomte*, local guide book.

chronicle proudly noted, 133 towns and castles were recovered. They included, in Brittany, Auray, which fell to Olivier de Clisson, the new Constable, leaving only Brest holding out, and against this town siege forts were erected.[1]

Continuous warfare had awful consequences for the people. Raiding parties from English garrisons brought devastation to areas even when no large scale military operations were taking place. A detailed study of the area around the town of Saint Flour in the Auvergne, based on the town council's register, shows that it was in a war zone from 1356 to 1391. The town's suburbs, unwalled, were destroyed four times between 1363 and 1366. Eventually they were provided with ditches, walls and barriers on the roads, while the old town walls, dating from the eleventh century, were themselves reinforced in 1377. The destruction of the old walls provoked a riot and royal troops had to be called in to restore order. Raiding English troops did not bring heavy siege equipment, but attacked using iron bars to sap the walls and explosive petards at the gates. Night attacks were also commonplace.[2]

The death of Edward III in 1377 brought his young grandson Richard to the throne, and left royal government facing immense financial burdens to sustain the war, including having to resist French naval attacks on coastal towns, small by comparison with the consequences of the war so far in France, but politically unacceptable and very difficult to resist. The measures taken by the government included levying the poll tax, inducing, in 1381, the peasants' revolt. In 1378 there occurred the great schism in the papacy, shocking the Christian world. Amidst the international upheavals, the Scots tried, yet again, to take Berwick, forty men "of the meaner sort" managing a surprise escalade of the castle. This time it was the townsfolk who cut the bridge to the castle, isolating it and enabling its rapid recapture by a force assembled by the duke of Northumberland. His men attacked from all sides, setting ladders against the wall while also setting fire to the gates, and effecting capture within two hours. In the same year, the duke of Lancaster led an expedition against St. Malo, deploying, it is said, no fewer than 400 cannons around the seaside town, although the defences of that time were much and many times reconstructed by future rulers.[3] There was more trouble for the English in 1384 when Archibald Douglas attacked the English-held fortress of Lochmaben, having heard that it was lacking both men and supplies. In this they seem to have been right, as the governor offered to surrender if not relieved within just eight days, and this date having passed, the place was yielded and the Scots "cast it to the ground" (February). This castle, it appears, had been a strongly-constructed stone fortress erected by the English only twenty years previously, who had created an earth and timber work in front to protect themselves while building went on.[4]

[1] Froissart, IX, 8–9; Cuvelier, lines 22,581–22,608; Delachanel, *Chroniques*, II, 171, 175, 186–92.

[2] M Boudet, 'Assauts, sièges et blocus de Saint Flour par les Anglais (1356–91)', *Revue d'Auvergne*, 1893, 6–24.

[3] Froissart, IX, 25–9; Fordun, 371; *Anonimalle Chronicle*, 125–6.

[4] Bower, VII, 395–7; C Tabraham, *Scottish Castles and fortifications*, 45, for the description.

Plate 1. Aleppo (Syria). The citadel stands high above the city. The walls were strengthened by the Mamluks and later by the Ottoman Turks.

Plate 2. Marqab (Syria). The crusader castle is of two periods: the round donjon dates from 1180–1200, the outer curtain from the thirteenth century.

Plate 3. Crac des Chevaliers (Syria). The Hospitaller castle boasted massive defences inside and out. Sultan Baybar's attack carried first the now deserted settlement area to the right, then mined and breached the outer wall at the point where the rectangular Mamluk tower was subsequently erected.

Plate 4. Caernarfon Castle (Wales). Built under Edward I to assert his power, it also contains formidable defences.

PLATE 5. Caernarfon Castle (Wales). The defences include galleries from which archers could shoot in any direction.

Plate 6. Sis (Turkey). The massive Armenian castle occupied the top of the hill and was defended by powerful walls and strong towers. It was captured in 1375.

Plate 7. The earliest western illustration of a cannon from a sister volume to the well-known Milemete MS of 1326–7, this from a treatise dedicated to King Edward III. The vase-shaped gun will fire a large arrow and is fired by the man behind with a long match to the touch hole. The real weapon must have been tied to the frame. From British Library, Add. MS 47680, fol. 44v.

PLATE 8. This shows English king Henry v besieging Rouen in 1418. Note the wooden siege works and the battery of guns mounted on flat beds, and the prominent gun loops in the walls of the city. British Library, MS Cotton Julius E IV (6), fol. 19v.

PLATE 9. Southampton (England). God's House tower projects a long way from the curtain wall, constructed in 1417 to mount guns that could fire in all directions.

Plate 10. Southampton (England). Gun positions in God's House tower, for floor-mounted weapons.

Plate 11. Strasbourg (France). Massive towers were built (between 1220 and 1334) to guard the point where the river enters the city.

Plate 12. Luxembourg. The town wall was extended in the fourteenth century under Emperor Wenceslas and comprised tall walls and massive gates such as the Dinsel Gate.

Plate 13. Metz (France). The *Porte des Alemans* (named after the Teutonic Knights) has inner towers built in 1240, and an outer defence erected in 1445 when provision was made for guns, and loops were inserted into the older towers as well.

PLATE 14. Tábor (Czech Republic). The pentagonal three-storey Žižka bastion built in front of the curtain in the 1420s was a formidable and precocious artillery defence.

Plate 15. Tábor (Czech Republic). The Hussites turned the Kotnov tower of the former castle into a circular gun tower with walls ten feet thick, and provided with gun-firing positions at several levels and outlets for smoke.

PLATE 16. Raglan Castle (Wales). The new-built fifteenth century baronial castle was provided generously with gun-firing positions. There is still no explanation for the grooves cut into either side of the embrasure.

Continued Anglo-French conflict in the 1380s: Flanders and Brittany

The often-rebellious cities of Flanders provided further opportunities for the English to intervene in their neighbour's business. In 1379, Ghent revolted against Count Louis de Male, and laid siege to the count's supporters in Oudenarde. Their cannons threatened to set fire to the city and the inhabitants covered their houses with earth to protect them. At the same time, other Flemings attacked the castle at Dendermonde, mounting their cannons on boats from which they fired large missiles.[1]

Charles v died in 1380 and was succeeded by his son Charles vi. Conflict continued to brew in Flanders. The count of Flanders threatened Ghent using cannon firing quarrels in that year. In 1382, the Flemish leader van Artevelde renewed the attack on Oudenarde. He had a large quantity of the new weapon, but decided not to risk assaulting the large town because of the likely losses. However, his blockade was accompanied by substantial weaponry. A vast contraption called a "sheep", more than six metres wide, six metres high and thirteen metres long was erected from which a bombardment was maintained. One of the bombards was so large that when fired it could be heard five leagues away. Its missiles were enormous quarrels. The Flemish army was destroyed at the battle of Roosbeke and van Artevelde killed, when Charles vi brought his army to the count's aid.[2]

The following year witnessed more strife, as the bishop of Norwich, Hugh Despenser, led a "crusade" to Flanders (the name being one of the effects of recognising different popes). The expedition began promisingly for the crusaders, as they captured Gravelines, a town protected by wooden palisades and held only by sailors. The defenders withdrew to the security of a monastery with deep ditches, but that also fell to assault. Other towns along the coast submitted, then the bishop decided to besiege the large city of Ypres. His attack involved what would be a major effort to fill in the ditch and make possible the undermining of the walls, but he was unable to cut down and transport enough trees for this task before a substantial French army came to the city's relief. The royal troops recovered the castle of Dringham from its English captors and then attacked Bergues, held by a large Anglo-Flemish force. Pressing their attack from three sides, the French broke in and the defenders fled. Closing in all the time, the French then attacked Bourbourg, firing cannon into the town while preparing to attack the ditch and palisade. The bishop had arrayed a large trebuchet and a "great gun" called "Canterbury" against the tower above one of Ypres' gates. However, disease was ravaging his army and there were frequent desertions. A moveable bridge built to allow an assault was destroyed by stones fired from the defenders' cannons and with the advance of the relief force he abandoned his efforts, leaving the towns to make their peace with their king.[3] Generally, the defences of the Flemish towns were not

[1] Froissart, IX, 196–8; Delachanal, II, 365–6.

[2] Froissart, X, 246–8.

[3] Froissart, XI, 96, 121; Knighton, 326–8; Delachanal, III, 55–61; *Chronique du religieux de Saint Denys*, I, 269 f; Thomas Walsingham, *Historia Anglicana*, ed. Riley, 1864, 96–9.

strong, but the growing role during this time of gunpowder artillery for attack, as well as in its by now traditional role in defence, is evident from the sources.

The records of Bruges provide considerable evidence for the standard measures regarded as necessary by the late fourteenth century. Defeat at Roosbeke had led to emergency measures for the protection of the city, while the acceptance of French victory then left the citizens fearing the English. Two types of activity followed 1382: the purchase of large supplies of artillery, and major works on the walls. For a year there was non-stop work on the ramparts, continuing in all weathers, the ice being broken up when necessary. Brigades of labourers were employed to widen and deepen the ditch, while brick and wood were purchased for the construction of bastions on the perimeter wall, themselves given an outer protection of wooden palisades. The Ghent and St. Croix gates were among six gates transformed into self-supporting fortresses, surrounded by wet moats and equipped with drawbridges and portcullises, and roofed in tile specifically to protect them against fire arrows (*vierpylen*). Extra defences were constructed at the two points where the city's access to the sea pierced the circuit, where bridges and underground passages were built to provide a continuous passage. Outside the wall, trees and houses were pulled down to leave a clear field of fire for the defenders and the large quantity of gunpowder artillery deployed all round the walls.[1] There are also detailed references to the artillery. Hundreds of guns (*bussen*) of various sizes were ordered made, thousands of missiles, hundreds of pounds weight of gunpowder. Guns were generally small, and were described by their missiles as "stone" or "lead" shooters. Craftsmen were paid to mount the guns on carts (*ribaudequins*) or on wheeled carriages. Since the Flemish cities had been amongst the first to adopt gunpowder artillery (with references in these accounts that could signify such weapons dating to the 1330s), the preponderance of gunpowder weapons in 1383 is no surprise. What is perhaps a little surprising is the disappearance from the evidence of the traditional catapult artillery. The springalds and *bliden* were a key part of the city's armoury in the 1340s, but while other contemporary military leaders continued to use them, the government of Bruges relied now on the new propellant (alongside crossbowmen and archers).[2] The resources available to even a great industrial and trading city such as Bruges were stretched to the limit by the scale of the defensive preparations, no less than three quarters of the city's entire budget for 1383 being expended on defence works.[3]

The English continued to hold Brest, where the powerful town defences and the adjacent castle were strongly garrisoned and could not be reduced while the harbour remained open. In 1386, the duke of Brittany renewed operations against it. On the landward side, he ordered two forts constructed, using massive timbers including the masts of ships, from which stone-throwers kept up a bombardment of the defenders, who responded in kind. The *bastides* were of fine quality, protected with their own ditches and towers. One chronicle said that the *bastide*

[1] L Gilliodts-van Severen, *Inventaire des Archives de la ville de Bruges*, 1875, III, 22–28, 33–4.

[2] Gilliodts-van Severen, III, 18–21, 31, 39 (1339), 58–9 (springalds, bliden).

[3] Gilliodts-van Severen, III, 35f.

called the "dovehouse" had a rampart three metres thick and seven towers, and was completed in 19 days with a thousand workmen toiling under the protection of 10,000 soldiers. Completed, it was fully provided with supplies and weapons including "gunnis" and 300 men. It so happened that the duke of Lancaster was on his way to Spain with an army, and stopped off in Brest. The besieging army withdrew but left garrisons in the forts. They resisted frontal attack for two days, but were not able to resist mining, which brought down one of the towers of one of them. The survivors of the garrison surrendered.

A remarkably similar set of events took place the next year. The duke of Brittany resumed his siege and built a new timber fort on the sea front, and two stone forts on the landward. The earl of Arundel came with a fleet and launched an attack from the waterfront and destroyed one fort, and left Brest well supplied. Henry Percy then paid another visit, again destroyed some of the siege forts, and re-employed one of them for the benefit of the defenders. Nothing remains of the works, but fragments of the duke of Brittany's accounts confirm the chronicles, showing payments to workmen for building the forts, and to cannoniers operating within them during the summer of 1387, totalling the large sum of more than 25,000 livres.[1]

Coastal attacks — and defences

While English armies ravaged French lands and towns for forty years, the ability of the French to strike back was limited to two avenues. Their efforts to encourage Scottish invasions of northern England had not led to any significant diversion of resources. The other approach comprised seaborne attacks on the towns of southern England. It had the great advantage of being impossible to avert and impossible to prevent from having an impact, it being beyond the resources of a medieval state to fortify and garrison every port on such a long coastline. Anglo-French warfare in and across the Channel had by now a long history, but security was very expensive, and generally not sustained during times of peace.

From 1338, when Southampton was laid waste, its castle apparently providing no assistance whatever, to further destruction wrought on the Isle of Wight (1340 and 1377), Hastings (1377), Portsmouth (1338, 1342 and 1370), Rye (1339 and 1377), Winchelsea (1359 and 1380) and elsewhere, the risk of attack was borne out by actual events. Additionally, in 1386 the French prepared an invasion of southern England from Sluis. This never happened, but it appears from English sources that the preparations included a prefabricated wooden palisade 3000 paces long and 20 feet (six metres) high, provided with towers 30 feet (10.7 metres) high, to be erected presumably to create a defensible base on landing. The invasion force also conveyed "stone-throwing machines, guns (*gunnis*) and powder". The captured section of palisade was instead used to provide a defence for the town of Sandwich.[2] Given its proximity to the embarkation port, it is no surprise that the

[1] Froissart, XII, 182, 305–7; Knighton, 342–5, 390–1, 400–1; Walsingham, 143, 155; A de la Borderie, Le siège de Brest 1387, *Revue de Bretagne, de Vendée et d'Anjou*, 1, 1889, 198–202.

[2] Walsingham, 1, 147; Knighton, 348.

city of Bruges should have been called upon to provide much of the material. The accounts of the city show the paying of four artillery "masters", the commissioning of hundreds of pounds weight of gunpowder, lead balls, and all the tools and items needed for the planned invasion.[1]

Many of the town walls erected in the fourteenth century were provided with embrasures for small cannons. There was also work done on castles, although there is debate as to the real purpose of several of them. On the Isle of Wight, the old royal castle of Carisbrooke had work done by 1341 that consisted of the construction of a new gatehouse with a drawbridge and portcullis, but there was no reference at this date to providing for artillery. This followed later, with the insertion of gun loops ascribed to the 1380s.[2]

Southampton and other towns

In Southampton, the reconstruction in an intense building campaign in 1378–88 of the old motte and bailey fortress used the existing site but provided it with a massive round tower within a round curtain on top of the old motte, and probably also saw the rebuilding of several of the gates, and the insertion of gun ports in the old hall which stood against, and was part, of the town wall on the quayside. Henry Yevele, the noted engineer in royal service, was employed here, and the new tower was ready for occupation within just three years. Other measures indicating the seriousness of the defensive intentions included the appointment of a chaplain, not because of his religious fervour, but rather because he was an expert in the "management of artillery". Simultaneously, work was taking place on the town walls, which had 29 towers and seven gates. Some of these walls dated from the previous century but substantial parts were either constructed or rebuilt at this time, and gun ports were frequently inserted, alongside provision of mountings for more traditional weapons including crossbows, springalds and "magnels". The seaward-facing western walls in particular have been dated to the mid fourteenth century onwards. The circular holes in the new walls put up in the arcade on this side (it is thought between 1370 and 1380) were 5" (127 mm) in diameter, clearly for the use of small guns. The remarkable projecting "Gods House" tower, a three-storey quadrangular building covering one end of the quay (see Plates 9–10), and the "Catchcold" tower had been designed to mount a number of guns, and had certainly been completed by the early fifteenth century. Both of these had the capacity to carry heavy guns, on the roof or inside the tower, using large rectangular apertures presumably protected with wooden mantlets. Other parts of Southampton's renewed defences confirm that the role of flanking for the artillery defences was an integral part of the planning. Saunders also suggests that "Gods House" tower was the earliest purpose-built gun tower in England.[3]

[1] Gilliodts-van Severen, III, 97–100. Yet more guns were commissioned in 1388 and 1389, at great expense for the metal and powder. With the truce of 1389, the city withdrew its now large arsenal into storage in the town's treasury and weigh house: III, 113–16.

[2] Colvin et al, *King's Works*, II, 592–3.

[3] J Oxley, *Excavations at Southampton Castle*, 1986, 114–15; J Hodgson, *Southampton*

The squat rectangular tower with projecting corner turrets known as the "Ypres" tower at Rye seems to have been constructed during times of anxiety under Edward I, but not to have done anything to assist the town when it was ransacked under his grandson. The townsmen's appeal for permission to raise murage in 1369 stated that they were still unenclosed, but they were burnt again in 1377, so the walls may have only been raised after that date. The Landgate is a magnificent structure with twin towers nearly 15 metres high, with a portcullis. The fragments of curtain wall however suggest it was crudely built, though at least five metres high. At Winchelsea, where Edward I had initiated the construction of a new town because of changes in the coastline, there were also frequent references to town defences, but no evidence of anything apart from a ditch and bank. There were short stretches of stone wall (but less than a metre thick) adjacent to the three stone gates, one at least of which had a portcullis. It would be hard to justify, however, that any of these measures were expected to resist a serious attack, although no doubt they might have deterred a small raiding party.[1]

There were varying experiences at other towns, too. Chichester, which had a full circuit of Roman walls still six metres high, spent money on repairing them, and there was a ditch. Canterbury began the major task of reconstructing its line of Roman walls in 1378, and the work was largely finished by 1402. Henry Yevele was also engaged in this work. The magnificent West Gate, nearly 20 metres high, almost certainly dates from this time, with gunports on each floor, arrayed in staggered fashion (as Renn noticed) on three floors, drawbridge and portcullis. This is a very early example of careful planning of gun positions for maximum coverage. Many of the town wall's 21 towers were also provided with gun ports. A gate tower very similar to the West Gate was added at the same time to the old castle at Saltwood, and keyhole gun loops are to be found in one of the towers of the outer curtain at the same castle. The west gate at Winchester was similarly provided for in 1390. At Portsmouth, however, despite its own experience of being sacked, the townspeople actually discouraged the work of fortification for fear of the cost and effect on trade, and it seems as if no substantial works were carried out, while at Sandwich, the fortuitous provision of a captured palisade added strength to the existing works, although it is unclear what defences actually existed by this date, with much of the present remains added in the following century.[2]

Castle (guide book); Colvin et al, *King's Works*, II, 840–44; Turner, *Town Defences*, 165–76, a detailed account and description of the town walls; A Saunders, The defences of Southampton in the later Middle Ages, in L Burgess, (ed.,), *The Southampton Terrier of 1454*, 1976, 21–8. In the sixteenth century the new "donjon" was the "glorie of the castell", and Leland also indicated that the town wall was preceded by a double ditch (Leland, *Itinerary*, ed. Smith, I, 278); Harvey, *English medieval Architects*, 361.

[1] D Renn, 'The castles of Rye and Winchelsea', *Archaeological Journal*, 136, 1979; G Bagley, K Clark, *The story of the Ypres tower and Rye museum* (guide leaflet); Turner, *Town defences*, 159, 176.

[2] Turner, *Town defences*, 148, 158; Canterbury Archaeological Trust, *The West Gate* (guide leaflet); Harvey, *Architects*, 361 and the same, *Henry Yevele, Life of an English Architect*, 35–6; D F Renn, 'The Southampton Arcade', *Medieval Archaeology*, VIII, 1964, 228–9 (West gate).

Other new works of the age also made provision for artillery. The function of the brick-built "Cow Tower" that stands outside the line of city walls at Norwich more than 15 metres high and nearly two metres thick has been the subject of debate. It seems certain that it was constructed in the late fourteenth century (accounts survive for 1397/8), and that it was provided with (or for) small guns.[1] As it stands in a bend of the river Wensum, the suggestion that it was designed as a blockhouse to resist a waterborne attack is credible. In the defences of the East Anglian city itself, possibly very early gun loops have been identified in the town wall.[2]

Royal and baronial castles

Other new buildings with similar provision for guns were the baronial Cooling castle (Kent) and Bodiam castle (Sussex). Cooling, although provided with gun loops, was not strongly built. Bodiam, started in 1385, stands in an artificial lake and has all the appearance of a serious defensive work, but close examination has suggested that this was all for show. The new royal castle at Queenborough was ordered by Edward III in 1360. Surviving plans and descriptions show that it consisted of a large circular inner tower, surrounded by an inner circular wall with six projecting towers, in which ward was the principal accommodation, while this was further surrounded by a circular outer wall containing the main gate and another gate on the opposite side, inside a moat. Records show that it was provided with artillery for its defence. This was the only completely new royal castle built in England in the whole century, but how much its design owed to defensive considerations is impossible to confirm. It was intended as a residence for the queen. The name of the royal Master Henry Yevele has been linked with the design of Queenborough and the later work at Cooling, Canterbury West Gate and also with the grand and powerful new gatehouse at the archbishop's Kent castle of Saltwood.[3]

A number of other older fortresses in southern England also had new loops inserted to allow the firing of small cannons. Generally, they consisted of round holes often added at the bottom of existing arrow loops, to create the appearance of a keyhole, although at the new castle of Cooling they were simply circular. The gun would have been mounted either on the floor, or else on a trestle. The very earliest in England has been attributed not to a royal castle, but to a monastic establishment: Quarr abbey on the Isle of Wight. Royal authority was sought to

[1] A Saunders, 'The Cow Tower, Norwich: an East Anglian bastille?', *Medieval Archaeology*, XXIX, 1985; Ayers, Smith, Tillyard, 'The Cow Tower, Norwich: a detailed survey and partial reinterpretation', *Medieval Archaeology*, XXXII, 1988.

[2] B Ayers, *English Heritage book of Norwich*, 1994, 65.

[3] Colvin et al, *King's Works*, II, 793–804 (Queenborough); Harvey, *Architects*, 359. Among the plentiful literature, B H St John O'Neil, *Castles and Cannon: a study of early artillery fortifications in England*, 1960, set the standard; review by A Dufty, *Archaeological Journal*, 119, 1962, 368–72; J Kenyon, 'Early artillery fortification in England and Wales', *Archaeological Journal*, 138, 1981, updates O'Neil; Harvey, *Yevele*, 25–6, 41.

build a curtain wall to enclose the precinct in which the loops occur, in 1365. It proved sufficient to deter the next French attack in 1377.[1]

Neither town nor castle could be defended without a garrison or weapons. Dover castle received 12 guns in 1385, eight more in 1386 and a further six in 1387. Commissioners of Array were several times ordered to select men for service and in 1360 the sheriff of Wiltshire was ordered to array local men to defend the castles of Old Sarum and Marlborough, both far inland. Local knights who absented themselves from the danger area were instructed to return. The feudal levies as well as the households of the great lords and bishops and abbots were called upon for service too. The Isle of Wight had its own militia of perhaps 200 fighting men. The coastal towns had the authority to array their own citizenry, and Portsmouth, Southampton and Yarmouth also had a royal "keeper" and garrison.[2]

While it is possible that towns were willing to undertake the immense work and expense of building a circuit of walls as an example of civic pride, the refusal of some towns to carry out the work, despite governmental pressure to do so, suggests that not all adopted this approach. In the 1380s, when so much of the building was begun, it should also be recalled that the peasant's revolt was a recent event, in which the archbishop of Canterbury had been a prominent victim. In the debate on the function of Bodiam castle, where the case for its defences being a sham is strong, it has been suggested that the obtaining of the licence to crenellate from the crown to protect against French raids was an elaborate charade. If so, its purpose was to enhance the social status of its holder rather than create a military fortification. That said, even a sham fortress might have appeared strong enough to deter raiders, or rebels, if it had some armed men inside it, so Edward Dalyngrigge, the old soldier who built it and who would certainly have known how feeble it would have been against a real threat, might still have slept secure behind its walls and lake as well as being able to impress the neighbours.[3]

One area where the works of the English crown were certainly not intended for show was the newly conquered zone around Calais. The defences of this port were already very strong, but the water-logged country, also bereft of timber and stone, all of which had to be expensively imported, involved the English government in non-stop expenditure just to maintain the waterworks, towers, walls and ditches. Following the capture of Guines, major work was undertaken on stone castles there and also at other nearby sites (Oye, Marck, Audruicq, Hammes), creating

[1] D Renn, 'The earliest gun port in Britain?', *Archaeological Journal*, 125, 1968, 301–2.

[2] Kenyon, *Early artillery fortifications*, 209 (Dover); J Alban, English coastal defence: some fourteenth century modifications within the system, in R Griffiths, *Patronage, The Crown and the provinces in Later Medieval England*, 1981, 57–77; M Hughes, The fourteenth century French raids on Hampshire and the Isle of Wight, in Curry, Hughes, (eds.), *Arms, Armies and Fortifications of the Hundred Years War*, 1994, 123–41 and the sources cited there.

[3] C Coulson, Some analysis of the castle of Bodiam, East Sussex, in C Harper-Bill, R Harvey (eds.), *Medieval Knighthood*, 1992, demolished the case for "defence"; N Saul, 'Bodiam Castle', *History Today*, 45, 1995, adopts the same approach; but others have argued for defence having a secondary role.

a line of defences around Calais itself. As well as the garrisons, permanent teams of workmen had to be retained to maintain the defences.[1]

The building of fortifications in France

In France the sources have indicated that the most usual addition to the defensive strength of a fortress was the provision of outworks of earth and of timber palisades, representing a means of impeding the advance to the ditch, from which a sustained attack could be mounted, especially by sap or mine, or by escalade under the cover of intense archery. The commonest substantial improvement appears also to have been the deepening and widening of the ditch, again with the primary function of making mining more difficult. But not everywhere in the French realm was subject to such worries, and indeed after the peace of Brétigny in particular, the crown itself undertook some substantial new building.

Charles v was responsible for some spectacular new work. In Paris, this involved new palaces away from the populace, who had proved so threatening in 1358, but also involved creating the Bastille Saint Antoine, turning a twin-towered gatehouse into a veritable urban fortress. This fortress provided the king with a secure gate between Paris and Vincennes. Here, the king turned an existing lightly-fortified manor into a new palace-fortress of gigantic proportions. The centrepiece was a rectangular donjon of no fewer than six storeys (52 metres), designed as royal apartments, but protected with its own curtain inside a ditch. Following its completion by about 1370, a rectangular outer curtain was provided, with angle towers, three gate towers and intermediate towers on the long face opposite the donjon complex. These rectangular towers were 40 metres high with walls three metres thick. Containing a wealth of sculpture, the whole has been described as a residential tower within "a fortified envelope with the character of a symbol of power".[2] These signs of defensive intent included a massive talus at the foot of the donjon, and around the top of the curtain and towers a continuous machicolation.

This style of building seems to have been characteristic of this period, whether intended for defence as well as show, or mainly for show. The papal palace at Avignon, of which the "old" section was built in the 1340s and 1350s, certainly intended to keep its occupants secure, with its massive external curtain and towers, originally provided with barbicans, arrow loops and machicolations. The walls are three metres thick and the tower that served as the donjon is 52 metres high. The architect came from Paris. At the same time the town was provided with a brand new wall to replace that destroyed by Louis viii, liberally provided with towers and machicolations, but of modest height. Sharing some of the papal palace's characteristics was the castle erected at Pierrefonds a little later (the 1390s) by the brother of Charles vi, which had a deep rock-cut ditch, a tall curtain with a talus, towers 38 metres high, and the circuit capped all round with machicolations.[3]

[1] Colvin et al, *King's Works*, i, 423–56.

[2] Mesqui, *Châteaux forts*, 107 (Bastille), 407–11 (Vincennes); Fino, *Forteresses*, 1977, 475–9.

[3] Mesqui, *Châteaux forts*, 42 (Avignon), 292–3 (Pierrefonds); Fino, *Forteresses*, 338–42

Among the various features that characterised these new constructions (as well as plenty of others, less substantial) is an emphasis on height, on even taller towers, on strong foundations (the defensive value of which would lie in the greater resistance to mining or sapping), and on machicolations. To the extent that these developments had a military rationale, they had the effect of minimising the effect of trebuchet bombardment and close quarter assault. But none of the examples given incorporated provision for guns.

Low Countries

The patchwork of small principalities east and north of French Flanders was a land where princely power (often exercised by a bishop who could not, at least in theory, establish a dynasty) was weak and nobles were substantially independent, each operating from their own castles. Towns were walled and their inhabitants often represented a competing power to that of, for example in Liège, the bishop and chapter. The princely families also feuded amongst themselves.

In 1328, John, duke of Brabant, was at war with the archbishop of Cologne and his powerful supporters, but the duke's target was the town and castle of Faucquemont across the river Meuse, held by an obstreperous nobleman with a garrison, provisions and engines. The castle stood on a high rock behind double walls and ditches, with the town below. Taking advice, the duke dammed the local river and flooded the town. With his allies unable to raise a relief force, the lord of Faucquemont submitted. The next year, however, he broke the treaty and refortified the castle. The duke brought an army including crossbowmen, archers, ditchers and miners, as well as *canoniers*, and pressed a close siege. Mines were however met with countermines, and it took nine weeks to induce surrender.[1] While it is technically possible that the duke of Brabant had cannon in 1329, it is not corroborated anywhere, and it is more likely that this was something added later to the account.

In 1346, as part of the interminable factional conflict between the "Waraux" and the "Arans", the people of Liège and Huy besieged the bishop's castle of Clermont and destroyed it, then attacked Hermale with several mangonels, the garrison of 40 capitulating for their lives. In 1347, a papal nuncio arrived to assist the bishop, while the Liègeois lay siege to Argenteau. In July, while the episcopal party was gathering its supporters, the citizens assaulted the castle with large stones from the mangonels. They also projected molten metal and quicklime in earthenware pots, in the form of incendiaries. In the end it was mining that achieved the result, with a stretch of the curtain wall falling and the defenders retreating to the donjon where, after a vigorous resistance, they recognised the inevitable and surrendered. The castle was then razed. A second account added that bombards were present

(Avignon), 446–9 (Pierrefonds). For the armoury of defensive artillery provided at Avignon, see Rathgen and Schäfer (p. 146 n. 1 below).

[1] Dynter, *Chronica nobilissimorum Ducum Lotharingiae et Brabantiae*, II, 780–1, 783–5.

for the attack on Hermale.[1] It seems that the craftsmen of Liège knew how to build and operate siege equipment, and could call on the services of miners, and in all the circumstances it would be surprising if they did not yet have guns, although the reference to this did not come from an eye witness.

In 1363, and again in 1381/2, Wenceslas (emperor 1378–1400), in his capacity as duke of Brabant, besieged Louvain to force the city's commune to his will, on the second occasion the operation lasting more than seven weeks. In 1380, there was war between the bishop of Liège and one of his feudatories, who seized the town of Bouillon, but was unable to capture the powerful castle. The bishop's response was the destruction by burning of two of the rebel's own castles. In 1387, the bishop's castle of Franchimont was burnt down by accident, and rebuilt, an operation which gives another insight into the building styles of the late fourteenth century. Franchimont was a ridge top castle dating from the eleventh century, comprising a simple stone-walled ward and a rectangular donjon at one end. Study of the remains suggests that this new work was a strong reinforcement of the donjon on its exposed side, with two semicircular turrets joined by a beak. The solid walls doubled the strength of the structure on this side and can only have been intended as a means of defence. At the other end of the castle, a new barbican was added to cover the entrance.[2] Meanwhile, an immense new wall had been erected around the city of Brussels. Taking more than twenty years to construct and finished in 1379, the new circuit with its seven large gates was protected by a wide ditch, wet at lower levels, with the river Senne controlled by sluices. It dwarfed the original circuit, parts of which still stand, and was immensely expensive.[3]

In 1388, the duchess of Brabant gathered her host to lay siege to the small town of Grave, on the Meuse. The contingent from the cities stormed the *boluwert* erected in front of the gates, showing once more how this form of additional earth and timber fort had become a regular addition to fixed defences. The defenders withdrew to the drawbridge. The attackers turned to their artillery, using bombards and *veuglaires*. But they did not press the attack. Later, she attacked a castle at Gasebeck, in which its lord had left provisions, a garrison with artillery, and his wife. Although the men of Brussels had cannon and *veuglaires*, it was nonetheless resolved to summon miners from Liège. But the castle surrendered before they were needed.[4] These sources thus make it clear that gunpowder weapons had begun to take over the main attacking role by this date, but the weapons concerned were not yet powerful enough to create breaches in the walls on their own.

[1] *Chronique de Jean de Hocsem*, 336, 352, 358–9; Jean d'Outremeuse, Chronique abregée, *Chroniques Liègoises*, II, 170–2, 179.

[2] Chronique Latine de Jean de Stavelot, *Chroniques Liègoises*, I, 70, 81–3, 89 (Franchimont); Dynter, III, 571, 606; Jean d'Outremeuse, 215; P Hofsummer, *Le Château de Franchimont*, (guide book), 1997, especially 12–15.

[3] A Henne, A Wauters, *Histoire de la Ville de Bruxelles*, 1845, I, 114–16, 119–121 (illustrations), 151 (plan).

[4] Dynter, III, 616–17, 623–4; Jean de Stavelot, 91–2.

The Empire and Italy: Lewis of Bavaria and Charles iv

The fourteenth century emperors included some ambitious and energetic rulers who may not have managed to restore imperial authority, but at least took the opportunity to strengthen their own dynasties. The death of Frederick of Habsburg left Lewis of Bavaria unchallenged on the throne, but he was followed by another dynasty, represented by Charles of Luxembourg, heir to the throne of Bohemia (which he inherited on the death of his father at Crécy in 1346), and grandson of Emperor Henry vii. Lewis proved an astute politician at home but in Italy, where he marched in 1327 in the hope of an imperial coronation, he found himself stuck before the walls of Pavia. During the same period, the Florentines, champions of the Guelf party in the north, besieged Lucca (October 1329), and built a stockade and huts all round. The arrival of relief from the king of Naples forced them to abandon the attempt. Much later (1341), when balances had shifted, the Pisans besieged Lucca, and completely blockaded it with a ditch and stockade, noted by the usually hostile chronicler as "marvellous work in little time". The Florentines, having changed alliances themselves, tried to raise the siege but found the countervallation so effective that they could not.[1]

The war of Metz 1324–5

Illustrative of some of the siege techniques of the period during when gunpowder artillery first appeared was the remarkable war in which the imperial city of Metz successfully withstood the onslaught of no fewer than four of the neighbouring princes in 1324/5. King John of Bohemia, through his family connection with the archbishop of Trier and the duchy of Luxembourg, the duke of Lorraine and the count of Bar, joined later by the bishop of Verdun, declared war on the prosperous independent city of Metz, in an apparently unprovoked attack. The city stands on the confluence of the rivers Moselle and Seille and was protected by originally Gallo-Roman walls, which had been substantially reinforced, although little now remains of this period except for the thirteenth century "German gate" (Porte des Alemans: Plate 13), named for a house of the Teutonic Knights, and comprising a powerful twin-towered gatehouse with internal defences making altogether a very substantial fortification which, if indicative of the rest of the circuit in 1324, suggested that the town was well protected. The city's wealth also enabled it to hire 700 mounted soldiers from neighbouring lords including the count of Saarbrücken, while it could also mobilise a similar number of native knights and a large number of foot soldiers. A compilation of chronicles permits the course of events to be followed in day-by-day detail, while also sowing considerable confusion to later historians by suggesting a very precocious involvement of firearms.

[1] G Villani, *Croniche Fiorentine,* ed. Porta, 1990, trans. Selfe 1906, (Porta ed., ii, 650, iii, 255–8, 282).

The hosts of the four princes marched against the lands of the city in the late summer of 1324 causing massive destruction to the previously peaceful countryside. Monasteries were not exempted from the destruction. A bridge of boats had to be built to get across the river Seille. The citizens meanwhile acquired a boat which they turned into a floating arsenal equipped with springalds (*espingoles*) and crossbowmen, and used to great effect against the enemy as they tried to camp too close to the water's edge. The attackers got into and destroyed some of the city's outlying villages, but were repelled from other of the suburbs, suggesting that the walls had been extended. The monastery of the Holy Cross was attacked but successfully defended. The city, bristling with armed soldiers and well equipped with springalds, ballistas and crossbowmen holding a powerful set of ramparts, was obviously far too daunting for the princes' army to envisage taking by storm, so they withdrew from the immediate area in October. The city government, aware that this was a pause rather than the end of hostilities, appointed a commission of seven to strengthen the defences. The work carried out on their instructions was described in detail. It included repairs and maintenance to the circuit of walls and towers, and the provision of springalds and crossbows *à tour* around the circuit. The previous fighting having illustrated where the city might be vulnerable, they ordered the demolition of houses and gardens that stood too close to the walls, for which the owners were awarded compensation, and the creation of new ditches 25 metres wide and wooden palisades to double the defensive line. The precautions were necessary because the raiding and counter-raids continued over the winter, with the city's hired troops carrying out devastating assaults on the enemy's lands. Finally, in March 1325, peace was concluded.

One of the chronicle compilations talks not of springalds, but of the citizens deploying firearms: *coulevrines, serpentines, canons*. However, internal evidence (the text in describing the abandonment of the city by its bishop refers to the elevation of the county of Savoy to a dukedom, which took place in 1416) demonstrates that this chronicle was written much later when such terms were common, so the reference may be discarded as anachronistic. The other chronicle, of Jacques Dex, seems to have a common source (even the words of the two are often identical), but makes no mention of what would surely have been a noteworthy happening. What the story does demonstrate is the strength of an imperial city with pre-gunpowder age defences which had been brought up to date, and ably and adequately defended with professional soldiers.[1]

Charles IV

Meanwhile, the young future Charles IV of Bohemia and the empire also had his problems with the politics of the Italian peninsula, vividly recorded in the book described as his autobiography. In 1332, he had to assemble an army to resist at-

[1] G Wolfram (ed.), *Die Metzer Chronik des Jacque Dex über die Kaiser und Könige aus dem Luxemburges Hause*, 1906, a long verse chronicle, pages 101–205, which does not refer to such artillery; and J Huguenin, *Les chroniques de la ville de Metz*, 1838, 42–63, with the references to firearms. C-L Salch, *Atlas des Villes*, 86–9.

tacks from a "secret alliance" of enemies including Robert of Naples, the Visconti rulers of Milan, the rulers of Verona and Ferrara, and the republic of Florence. Successfully preventing them gaining any of his loyal cities, and throwing up a fort of wood with a ditch ten miles from Lucca on Monte Carlo, Charles withdrew but returned to the offensive in 1333 alongside his father, to bring relief to the castle of Pavia, which was being held against those of the town. This turned into a stand-off, as the Bohemian army could bring succour to the castle, but could not capture the town which had constructed a ditch and outworks to separate it from the castle.[1] Once again, it was proving very difficult even for large modern armies to make headway against populous and well-defended Italian cities. In 1337, however, a large coalition including Venice, Florence and Milan laid siege to Padua with an immense host. The town surrendered through hunger after a siege of three months: the weakness of populous cities being, of course, the problem of keeping the people fed if the besieger was strong enough to prevent all contact with the outside world.[2] There is no reference in any of these accounts to the deployment of gunpowder artillery, although that does not mean that it was unavailable. It certainly seems that guns were already a regular element of the armoury of sieges in Italy by 1340, as examples drawn from the Vatican archives have long established.

Papal States

Between 1340 and 1350, papal forces were engaged in several sieges of towns and castles in central Italy, and they deployed a wide array of weaponry. Against Terni (October 1340) the main ammunition comprised thousands of quarrels, for ballistas, but they also paid for two "tubes . . . or bombards of iron" costing all of four florins, while the following year, another Master was paid for working a trebuchet. In 1350, a Florentine engineer was paid six florins for obtaining and operating a battery of four trebuchets, but a larger battery was used against Salverolo in the same year. This required various payments for "cannons, bullets and iron hammers for bombards", timber for trebuchets, powder for the bombards, construction of mangonels, manufacture of stone projectiles and of iron bullets (*ballotas*), and assorted animal hair, cords and wax and grease for the "mangonels". The information together suggests that the various materials were readily available from providers in central Italian cities (Imola, Modena and Bologna were among those cited as the source of the different items). The "mangonels" cited were also clearly torsion-powered weapons and it might be concluded that they were in fact springalds, since this otherwise popular weapon does not appear in the inventory. The editors of the source calculated from the information given about the weight of the shot for the bombards that their diameter was just over 4 cms, confirming that these were definitely small devices, as would be consistent with the other information

[1] E Hillenbrand, ed/trans., *Vita Caroli Quarti. Die Autobiographie Karl IV*, 1979, 92–9, 100–102.

[2] Hillenbrand, *Vita*, 140–1.

about these early guns.[1] The trebuchet remained the main strike weapon for the commanders of the papal armies.

Florence

It was also the weapon of choice for the wealthy communal government of Florence at this time, despite the precocious reference to the city's use of firearms thirty years before. The republic was regularly at war with its neighbours. In 1350, Florence fought the Sienese for control of Pistoia. The town defences were augmented by wooden hoarding on the walls and by outworks of palisades of wood and stone. Cauldrons of boiling water and of quick lime were readied to make the foot of the town wall an unwelcoming place for an attacker. The Florentines organised to build cats and "wooden castles" and induced the defenders to negotiate.[2] Both defence and attack were relying on traditional methods.

Later the same year, a Milanese army prepared to attack the Florentine town of La Scarperia, apparently optimistic about their chances because the town itself was unwalled, and the castle was old. The townsfolk and local peasantry demonstrated great loyalty to the city and quickly cleared out the old ditch (and appear to have excavated a second) and erected a wooden stockade, while aid was rushed in from Florence. The attack was made with ladders, cats, and siege towers under the cover from crossbows and trebuchets. The ditches were dry, and the assailants made it across the first before being repelled from the second by a strong defence, itself relying on crossbows, lances, and catapult stones to demolish the attackers' siege towers. The assault was called off, no fewer than 64 ladders being left behind at the first ditch. The besiegers switched to attempting to mine under the rampart and drove a mine with four different tunnels into the defences, but the garrison must have included experts in this art, because successful countermines were dug. Two further assaults, one of them at night, were similarly repelled by the powerful and well armed garrison, before the siege was abandoned altogether.[3] Trebuchets continued to provide the main artillery used by the republic, as in the capture of a castle in 1352.[4]

Germany

In 1349 Charles IV secured the crown by defeating the counter-king Gunther at a siege of Eltville on the Rhine. Eltville, described as an "oppidum cum castro", had been built on the Rhine near Mainz, and the conqueror also took and destroyed

[1] B Rathgen, K H Schäfer, 'Feuer- und Fernwaffen beim päpstlichen Heere im 14. Jahrhundert', *Zeitschrift für Historische Waffenkunde*, VII, 1915, 1–4, reproduces and analyses the accounts in the Vatican archives. Contemporary accounts from the papal palace at Avignon show that the fortress was provided with thousands of quarrels for different sizes of ballista and several springalds were specifically commissioned as well (8–9).

[2] *Cronaca di Matteo Villani*, ed. Racheli, 51.

[3] *Cronaca di Matteo Villani*, 65–6; Marchione di Capo Stefano, Cronaca Fiorentina, *RIS*, NS XXX/1, 1903, 238–9.

[4] *Cronaca di Matteo Villani*, 94.

a new fortification built near Speyer by the Count Palatine of the Rhine.[1] The sources give no detail of the operations. However, they do confirm that feuding and conflict between the numerous separate powers within the empire continued unabated, and that these conflicts often involved the attack and defence of fortifications. In 1344, the archbishop of Mainz had ordered a supply of *ignisagitarium videlicet Fürshutzen* sent to Aschaffenburg, and a few years later (it is believed around 1355–60), the construction of a new shield wall 21 metres high and five thick to protect the 39 metre tall Bergfried at the castle of Schwalbach (owned by the count of Katzenellenbogen) suggested an awareness of the need for greater passive defensive strength in this age, in the middle Rhine region of the empire.[2] In 1351, the archbishop of Trier organised the capture of Montclair castle, held by a recalcitrant lord,[3] and later that same year Albert, duke of Austria mobilised many of the south German bishops and cities to besiege Zürich and bring it to a temporary submission. In the spring of the following year the conflict here resumed, with Albert bringing a large army into (what is now) Switzerland and assaulting the town of Zug with "machines, cats and fire-throwers (*iaculis igneis*)", leading to its surrender. Zürich was next, and again submitted, along with the league of towns which it led. It may be that the reference to fire-throwers signifies the presence of guns, or it may refer to older incendiary devices.[4]

The chronicler of these events, a cleric at Strasbourg, went on to recount proudly how his town extended its circuit of walls from 1374 to enclose new suburbs, creating three large new towers at the same time (Plate 11). There is little doubt that urban pride was a major consideration in this expensive operation. But there was also continuous warfare between the city's bishop and neighbouring nobles and rival towns – in 1360 the citizens besieged Schlettstadt, in 1366 they destroyed the castle of Friburg, in 1360 and again in 1380 they attacked Hagenau, in 1368 they destroyed a castle at Rotenburg, in 1373 they were part of an imperial army sent to take and destroy a castle full of robbers at Herlesheim, and in 1382 another at Ochsenstein.[5] It was certainly not inconceivable therefore that the walls might be called upon one day to provide protection too.

Other chronicles record similar patterns of events in other regions of Germany. The cities had the protection of stone walls, although many of those in the south were extremely old, and they had town militias. But for professional soldiers they had to hire mercenary captains, and many of them appointed their own native engineers (*Werkmeister*), though some may have had to hire experts to operate siege equipment. In 1376, Speyer was attacked by troops of their own bishop. The

[1] Chronica Mathiae de Nuwenburg, *MGH SrG* NS IV(2), 1936, 433–34.

[2] O Fink, 'Burg Burgschwalbach und die Einführing der Feuerwaffen im Mittelrheingebiet', *Burgen und Schlösser*, 14(2), 1973, 94–6. Plans and diagrams in Ebhardt, *Wehrbau*, I, 368–9.

[3] Stones that might have been used by the archbishop's blide were found, of around 30cms diameter, in excavation: Wagener and Laß, De Monckler Destructione: Die Belagerung der Burg Montclair 1351 und ihre Burgstelle in ihren Umgebung, in Wagener and Laß, 133–53.

[4] Chronica Mathiae de Nuwenburg, 453, 461, 463–5.

[5] Chronica Mathiae de Nuwenburg, 489, 491–3, 501.

bishop's troops stormed the unwalled suburbs, but were hurled back from the walls of the city proper. The account then says that the bishop deployed his guns, but on this occasion peace was brokered by the Count Palatine.[1] The ancient city of Augsburg had been in regular conflict with the rulers of Württemberg and Bavaria. In 1372, as the princes' forces ravaged the city's hinterland and built themselves some kind of fort, the city found itself extremely ill-prepared, with stretches of the wall being flattened and the ditch in places quite filled up. With their militia mobilised however, and having a battery of 20 iron guns, they seem to have deterred attack. In 1378, the bishop of Würzburg marched against Rothenburg ob der Tauber, but the citizens provided themselves with "good cannon" (*buchsen*) and other equipment and marched out to face the enemy. In 1356, Nürnberg was making use of guns and gunpowder, while the duke of Württemburg had also deployed guns in an attack on Mühldorf in 1364.[2]

The expansion of Savoy

Another regional power was emerging in central Europe: the aggressive and expansionist house of Savoy. The counts of this small landlocked Alpine land produced in the course of the fourteenth century a line of successful and skilful warriors who managed to considerably extend their power at the expense of their neighbours, building on the foundations created by their predecessors who had established a powerful dynastic homeland well provided with modern castles during the previous century. These exploits would soon extend to Asia Minor and the Balkans, but the immediate target of their aggression were the rulers of neighbouring Dauphiné. The surviving history is very biased and not contemporary. The statement that the count of Savoy, Amadeus v (the Great, 1285–1323), deployed bombards in a successful attack on the town and castle of St. Germain d'Amberieu is plainly anachronistic.[3]

Count Edward the Liberal (1323–9) captured the high-sited new castle of Montforcheys with a siege train, after an attack of 13 days. His successor Aymon (1329–43) engaged in a ten-day siege of the castle of Monthouz, held by brigands who were ravaging the country around Geneva. Having deployed siege artillery, the count in person led a successful assault on the eleventh day of the attack. The Dauphinois retook Monthouz by a nocturnal escalade, and the garrison commander retreated to the donjon, but capitulated after four days because he had no provisions. The much-fought over town and castle were then recovered by the count after another siege of three days.[4] The struggle continued, revolving around the attack and defence of castles and towers. During an attack on La Berrière,

1 Lehmann, *Chronica der freyen Reichs Statt Speyr*, 1612, 819–20.

2 Chronik des Burkard Zink (1368–1466), *Die Chroniken des Schwäbischen Städte*, 1866, 5–6, 21 (Rothenburg); Chronik des Hector Mülich (1348–1487), same series, 1892, 10; many of the references are collected in J Würdinger, *Kriegsgeschichte von Bayern, Franken, Pfalz und Schwaben von 1347 bis 1506*, 1868, II.

3 Chroniques de Savoye, in *Monumenta Historiae Patriae*, 1839, II, cols. 227–30.

4 Chroniques de Savoye, II, cols. 234–5, 258–62.

the Dauphinois dug a mine against the inner defences, having stormed the outer ward, and the defenders countermined. The attackers prevailed and once again the garrison retreated to the donjon. This last resort was surrendered when the attackers began to drive a mine under the tower.[1] The same year brought the succession of a new ruler in Savoy, the so-called "Green count" Amadeus VI (1343–83), whose long reign saw the consolidation of Savoyard power.

The Green Count used the full range of siege techniques known to the age. The castle of Sion was taken by an attack using ladders, mantlets, cats, mining and mangonels, the engines being used specifically to create breaches in the walls. At Turbillion, the count adopted a different approach, ordering the construction of a bastide. In his conquest of the town of Guex, the Savoyard deployed everything: bricolles, cats, trebuchets, cannons, bombards (unless the chronicler is simply listing all the weapons known to him for the sake of appearances!). The surrender of the town after five days brought the attack to the walls of the castle, against which Amadeus used mines and artillery. In a counterattack carried out by the dispossessed lord with the aid of the Dauphinois, it is recorded that several castles were taken including Bonnevaux, where Hugh de Guex noticed a light at a window in the castle one night, and ordered a bombardier to direct an incendiary missile through the window, setting fire to the building and burning down the castle.[2]

In 1363, Amadeus brought the marquis of Saluzzo to heel by invading his territory, bringing artillery with him and capturing his castles. Saluzzo submitted to the attack and became part of the count's domains. Amadeus then embarked on his crusading adventure (see below).[3] As the Green Count aged, he left some of the warring to his son, who would succeed as Amadeus VII (the Red Count) in 1383. The latter is recorded during these years immediately before his succession as leading a number of victorious sieges, attacking the castle of Beauregard with engines and bombards before carrying it by escalade, then using bombards to smash the walls of the castle of Montmerle, finally reducing the castle of Charlemont with engines shooting stones and pebbles.[4]

Castle building in Italy

Another reason for noting the actions of the counts of Savoy is that the fortuitous survival of building records enables a study to be made of their castle-building as well as their castle-capturing. As with other expansionist rulers, the Savoyards took care to ensure that they continued to control territories they had taken over, and the castles they built represented the familiar mix of functions: visual statements of the ruler's power and prestige, but also administrative bases and modern fortifications too. The count's castle at Fossano was begun in 1324, and records of the work survived over the following eight years. There was a massive engagement of labourers, but much of the total cost was incurred in transporting bricks and

[1] Chroniques de Savoye, II, cols. 262–66.
[2] Chroniques de Savoye, III, cols. 274–5, 279–82.
[3] Chroniques de Savoye, III, col. 296
[4] Chroniques de Savoye, III, cols. 353–5.

materials to the site. The castle's design was traditional for the region. Set inside a very substantial ditch, and backed against the town wall, the castle comprised an exact quadrangle with a tall, projecting rectangular tower at each corner, all roofed and tiled, and buildings all round the interior courtyard. The main entrance was across the ditch by a bridge from the town, and there was a second entrance into another set of wards that lay outside the town.[1]

A major new castle was constructed in Turin from 1317, at the command of the count, on the site of a thirteenth century castle. It is apparent that the work was done for the count, not for the town, which did not contribute to the cost. The design seems to have been similar however to those at several other comital castles, including those at Ivrea and Biccherassio for which accounts also survive.[2] The castle of Ivrea was built by Amadeus VI from 1358 as a purely defensive structure, of plain appearance and the same basic rectangular plan with tall towers at each corner, a gallery with machicolations running all round, a ditch and drawbridge. The town had contained previous castles, one belonging to the marquis of Montferrat having been demolished by the citizens in 1300. The palace of another local dynasty was pulled down to make space for the new count's castle and the town walls also had to be broken through to complete the plan. Built over four periods between 1357 and 1391, a large workforce was employed, 1000 workers being paid in a single day and craftsmen being gathered from all over northern Italy. The outward-facing wall and towers were constructed first, the interior buildings last.[3]

The fourteenth century was a time of significant construction by ruling families in Italy, as powerful individuals established control over the great cities of northern Italy, suppressing in the process the communal governments that had so successfully thwarted the ambitions of numerous German emperors in the past. The new rulers, seeking to consolidate their authority, had to ensure that their own power base was secure, so frequently set about building a powerful castle in the very city that was the source of their power and title. The Della Scala (Scaliger) who established themselves in Verona have left a clear legacy in their castle-building. In the city, but connected separately to the outside world by a new bridge across the river Adige, they built the (now so named) Castelvecchio between 1354–75. The castle still presents a tall and dominating front towards the city, and originally stood behind a much more substantial ditch than now. Inside, the castle is in two distinct sections. One portion lies within the circuit of town walls and had a function as a garrison area. A double wall and ditch separated this from the second section, lying outside the city wall, and containing the Scaliger residence. The two sections were connected by continuous walls at both ends, and inbetween stood

[1] G Falco, *Sulla costruzione del castello di Fossano (1324–32)*, 1936, 65–76 (the accounts are at pages 79–116); F Conti, *Castelli del Piemonte*, 1975, III, 48–50.

[2] F Monetti, F Ressa, *La costruzione del castello di Torino, oggi Palazzo Madama*, 1982, introduction and reprint of the building accouts (*libro di spese*); but see also the critical comments of A Settia, Un castello a Torino, 1983, reprinted in *Proteggere e dominare*, 1999, 169–86.

[3] G Roddi, 'Note sulla costruzione del castello di Ivrea', *Studi Piemontesi*, XI, 1982, 139–48.

the gate to the river bridge, protected by tall towers at each end. The tall, red-brick built building was truly both a statement of the occupant's power and a real instrument of it.[1] The Scaligeri were also responsible for several powerful castles around lake Garda, of which Sirmione is among the best preserved. Begun in the late thirteenth century and representing a design familiar at that time, and with a wide moat and successive lines of defence represented an effective fortress, that was only strengthened in the fourteenth century as later Scaligeri reinforced its walls and constructed the remarkable fortified dock.[2]

In the same region, the Visconti established their domination in ancient Bergamo with the construction of a new citadel on an existing castle and part of the city wall, from 1355, serving as an administrative and military headquarters with massive rectangular towers, the fortress was another example of contemporary fortress building.[3] The Florentine republic was also devoting resources to the construction of modern fortifications. Facing continued conflict with their opponents and fearful of imperial intervention, the commune determined to construct a "noble castle" in Valdipesa in August 1355, when the ditches were dug. In September, the curtain wall was begun with foundations more than a metre wide, while the towers were no less than 12 *braccia* (feet, ?= 3.7 metres) thick, with internal passages. The towers stood 50 feet (15 metres) apart, and there were two gates. Two of the towers had been completed by the following September, at a total cost to date of no less than 35,000 gold florins. This was not the only expenditure for the city on defences, as a couple of years previously they had begun a "large and noble *rocca*" against the threat from San Gimignano, and had modernised defences at Prato and had castellated a great tower *a moda d'una rocca*.[4]

Hungary: Louis the Great and the siege of Zara, 1345–7

A branch of the dynasty of Anjou ruling Naples inherited the crown of Hungary in 1307. After the long reign of Charles I ended in 1342, his young son Louis, who would earn the title of the "Great", was crowned. He faced a revolt in Dalmatia and Croatia. Calling on the aid of the ruler (Ban) of Bosnia, Louis set out to restore his authority, and found himself colliding with the power of Venice. When he arrived with some 20,000 armoured troops, most of the people of Dalmatia flocked to submit. But the Venetians, who had previously exercised authority on the coast, were furious and sent an army and fleet to recover Zara, control of which meant much to their strategic and commercial interests (as it had previously at the time of the fourth "crusade"). The citizens of Zara appealed to

[1] Verona's Castelvecchio has been carefully restored but its basic shape is as it was intended.

[2] F Conti, V Hybsch, A Vincenti, *I castelli della Lombardia: provincia di Bergamo e Brescia*, 1993, 212–14.

[3] Conti et al, *I castelli*, 40.

[4] *Cronaca di Matteo Villani*, 116 (Prato), 181 (Valdipesa).

their now departed king for help and meanwhile looked to their defences. A chain was stretched across the harbour and ships and artillery prepared to protect it. The Venetians hired a large force of German, French and Italian troops and built a bastide against the town. The anonymous diary compiled by a witness tells of the size and beauty of the construction, provided with ditches and outworks. The Zarans deployed 18 machines on their circuit, some of them targeted at the bastide, and the besiegers responded in kind, constructing two immense trebuchets, one on the beach at each end of the town. The Venetians made all the early running, capturing various of the outlying castles and demoralising the defence, and still no relief came from Hungary as the summer of 1345 turned into autumn. The attack was taking place by sea as well, and the defenders commissioned the construction of five new trebuchets to protect the chain against Venetian ships, and had to resist a fierce assault by galleys with reinforced shelters and carrying wooden towers. By land, the Zaran trebuchets and springalds recorded successes against the Venetian siege works and burnt several siege towers, but the attackers were not dissuaded and kept up their assaults. Spring 1346 brought renewed bombardment from new trebuchets, while cats and other wheeled sheds were readied. An assault on the south side with ladders under the shelter of ballistas and springalds was driven back by a defence including seven catapults.

A key moment occurred in July of the second year: King Louis arrived within seven miles of Zara with a large army and launched a ferocious attack on the Venetian bastide, simultaneously with a sortie prepared by the governor of the city. But the Venetians were very well prepared. They extinguished the fires started by the assailants and, with their own trebuchets maintaining a non-stop bombardment, launched their own counter-attack. This succeeded in overrunning the Zarans' artillery, and it was recorded with dismay how the defenders of the city watched as three cats, seven trebuchets and four springalds were consumed by flames. More important still, Louis, who had suffered great losses in the attack, many of them from bolts fired by ballistas, saw he could not take the bastide and gave up. The Venetians repaired the damage and resumed the siege, and Zara finally capitulated through hunger in February 1347. It had been a siege of epic proportions, involving all manner of weapons by land and sea, and large concentrations of stone-throwing artillery causing much damage, but not sufficient to allow a successful assault.[1]

In Naples, where his brother had been murdered in a disputed succession, Louis' military exploits proved rather more successful. The walled towns of Apulia (Puglia) submitted, and Canosa fell to an escalade led by the king in person. At Aversa, he had to construct engines, but the town actually fell through hunger after three months, the king meanwhile having suffered a wound from a ballista bolt.[2] A further conflict with the Venetians took place in 1356, focussing on the

[1] Obsidione Iadrensis, scriptoris anonymus, *Scriptores Rerum Hungaricarum*, ed. Schwandtner, 1748, III, 667–719; Chronica Hungarorum, inserta simul Chronica Ioannis Archidiaconi de Kukellew, same series, 1746, II, 174–9; Johannes de Thurocz, *Chronica Hungarorum*, I, 167.

[2] Chronica Hungarorum (Kukellew), 180–5; Thurocz, I, 172.

Venetian city of Treviso which was besieged in June. The defenders had been reinforced by boat, and possessed trebuchets, which they used to effect. The Hungarians tried to mine, but found the ground too waterlogged, and the surrounding country proved unable to support them, so the attack was abandoned.[1] An attack on the town of Giadra was more successful, long stretches of the wall being brought down by mining and the Venetians fleeing to the castle by the sea, at which point negotiations to end the war were begun.[2]

Poland

In 1370, Louis of Hungary succeeded to the throne of Poland. The Piast family had converted their land into a kingdom only in 1320, and in 1333 Casimir III had become king. That year, the Mongols had invaded and attacked Lublin, which the chronicler noted was then constructed of wood. It held out. The new ruler set about transforming his capital at Cracow, and erected stone walls around the castra there and at many other places, providing "strong walls . . . high towers and surrounding them with very deep ditches and other outworks" in order to provide refuges for the people. The citadel of Cracow was already protected by strong stone walls in the thirteenth century: what happened from 1320 was a massive rebuilding of the whole area known as the Wawel, containing palace and a royal cathedral within a two-warded modern castle defended by towered curtain walls. The hill top site, separated by water from the city below, was nonetheless integrated into a defensive circuit of stone around the town too. To the west of the capital, Casimir also oversaw the reconstruction of the frontier castle of Olsztyn. Much extended later, the fourteenth century work saw the building of a roughly quadrangular stone curtain on top of a ridge above a separately-walled lower ward of much greater area. Access to the upper castle involved a journey around the foot of an immense circular tower 20 metres high. Another significant royal castle was built on the ruins of a castle destroyed by the Mongols at Będzin. In the inner ward, rose a vast circular tower with walls four metres thick. Only a narrow space was left inside, similar to the design of the German Bergfried. Accommodation was provided in a square donjon with its own (this time square) tower attached with walls of similar thickness. A drawbridge connected the two towers. The inner curtain was 8 to 12 metres high and the outer, which ran concentrically with the inner, and itself inside a deep moat, was five metres high. The castle was built in 1358 and connected to a city where the walls were constructed from 1364. This truly formidable fortress, lacking only sophisticated gate defences, revealed that the builders were well-versed in some ways but had yet to make any provision for cannon.[3]

[1] Raphayni de Caresinis, Chronica, ed. Pasterello, *RIS*, NS XII(II), 1922, 11; *Cronaca di Matteo Villani*, 207.

[2] *Cronaca di Matteo Villani*, 251.

[3] English language summary and plans in J and H Kaufmann, R Jurga, *The Medieval Fortress*, 2001, 255–60; Czarnkow, *Chronicon Polonorum*, II, 625–7. See also S Kolodziejski,

Louis of Hungary maintained the Polish crown until his own death in 1382, though not without opposition: in 1376, Polish lords attacked the castle of Zlothor with an array of engines and instrument. The defence doubtless took advantage of the fact that the castle was a store for the siege train of the crown. However, it fell.[1]

Relentless war in the Baltic region, Lithuania, Poland and Russia

The fourteenth century saw the consolidation of the theocratic realm of the Teutonic Knights around the whole coastline from Gdansk to the Gulf of Finland, the settlement of the interior accompanied by a massive programme of castle-building, and throughout, an utterly brutal and almost continuous war against the Lithuanians, in a struggle that provided an outlet for the crusading zeal of western Christians without the even greater dangers of heading for the Middle East. By the end of it, Lithuania not only remained independent, but emerged modernised and strengthened to play for a while a major role in European politics. The consolidation of a princely dynasty in the course of the thirteenth century struggle led to the establishment of a capital at Vilnius by Grand Duke Gediminas (1315–41), which was mentioned for the first time in 1323. Sited at the confluence of two rivers, the location was favourable for many reasons including security. The wooden forts originally established on neighbouring hills were replaced by brick buildings under Gediminas, and town and castle were protected further by a curtain wall. Inside, a town of crafts grew up, never taken by the crusaders. Elsewhere, the land was dotted with earth and timber hill forts. Not less than 1000 have been identified, some 600 of which were in use during the medieval period, which is a remarkable density. From those that have been studied, it would appear that almost all of the previously existing sites in the north and west of the country had been destroyed during the wars with the Knights, but that during the fourteenth century a distinctive policy of re-establishing wooden forts in what shows on the map as a clear inner and outer ring on the borders between the principality and its aggressive neighbours had occurred. The historical account certainly refers to attacks and defence of fortifications during the "crusades", and it remains likely that the constructions were chiefly designed as places of refuge for the local people.[2]

The annual warfare was described by Hermann of Wartberg, whose chronicle ended in 1378, and several other local accounts, with the knights leading *reise* (expeditions) to burn and destroy the enemy's land during the season when the

'Château et ville en Pologne de midi', *Castrum Bene*, 6, 1999, 139–57, for the towns and cities.

[1] Czarnkow, II, 660, 678.

[2] A Kuncevicius, 'Why did Vilnius become capital of Lithuania?', *Castella Maris Baltici*, II, 1996, 94–6; G Zabiela, 'The end of wooden fortifications in Lithuania', *Castella Maris Baltici*, II, 196, 224–8 (map on 226).

water-laced wastelands permitted their passage. The fortunes of battle were certainly not only one way. In 1330, the Lithuanian prince attacked the bishop's castle of Pilten "with various machines" and ravaged the land. In 1345, while the Livonian Master was campaigning on the island of Oesel, the Lithuanians stormed the castle of Terweten (Tervete), killing eight knights, then went on to assault Mitau, taking the suburbs but being denied by the stone walls of the castle. The next year Mesoten fell too.

Siege of Kauve 1362

In 1362 the Master, Arnold, attacked the Lithuanian frontier castle of Kauve, and entrenched his army around it. By now, Kauve was protected with "high walls" but these were eventually stormed. Wigand of Marburg provided a detailed day-by-day history of this siege. Having managed by his circumvallation (possibly comprising a connected line of redoubts) to prevent the Lithuanians from bringing relief to the place (in which many leading Lithuanians were trapped), the Order's Master turned his attention to the enclosed fortification. The defences were strong (*robusta*) and resisted the atack of the stone-throwers (*bliden*) and their arrows caused many casualties among the attackers. A master carpenter from Marienburg named Marquadus (later named as a "brother") built a " machine or battering ram" ("commonly called a tomeler") by means of which the outworks on one side of the castle were thrown down. A similar machine was constructed on another side by a carpenter (Matthias) from Königsberg, by means of which the outer wall was demolished down to the ditch. As the ditches were filled in, the carpenters began to build new machines equal in height to the castle. Having now got across the ditch, another ram brought the inner wall crashing down. The crusaders and knights now attacked, and set fire to the wooden buildings of the fortress. The people inside tried to flee, but the surrounding army prevented them, and there was a dreadful killing. It is fairly clear that several battering ram-type machines were the vital component of the attack, which had taken more than a month.[1]

In 1369, the Knights began to build a fortress (Godeswerder) on the same site as Kauve. The speed of construction suggests this must have been a timber fort. The Lithuanians did not intend to leave it in peace, and assailed it with no fewer than 18 machines (alternatively "15 machines and 5 *tomeler*"). The defenders held out, a relief force came to their aid and the attackers withdrew, burning "six machines and four other instruments of war". Clearly, the Lithuanian prince could call on the services of masters able to build and operate substantial batteries of siege equipment, although the record of their use here does not suggest they were very effective.[2] However, telling a different story is the account of their capture from the Poles of Vladimir (Galicia). This was described as "an ancient wooden

[1] Hermanni de Wartberge, 81–2; Wigand von Marburg, Chronica nova Prutenica, *Scriptores Rerum Prussicarum*, eds. Hirsch, Toppen, Strehlke, 1863, II, 532–7; Johan von Posilge, Chronik des Landes Preussen, same series, III, 82.

[2] Hermanni de Wartberge, 85, 94–5; Franciscani Thorunsenis, Annales Prussici, *Scriptores Rerum Prussicarum*, III, 82; Johan von Posilge 82.

castle" that King Casimir had ordered strengthening with a stone wall, but the work, begun two years before, was incomplete when the Lithuanians attacked in 1370. On this occasion, they resorted to mining, and brought the wall crashing to the ground.[1] The principality therefore had access to the people with the necessary skills for all kinds of siege warfare by this time. A little earlier (1369), the Knights led by their Grand Master Winrich von Kniprode had attacked the significant Lithuanian centre of Kovno (Kaunas), which was protected by walls and two rivers. In a clearly well prepared operation, the crusaders entrenched all round the town, and bombarded it day and night with stones and burning arrows. The Lithuanians were unable to relieve it and it fell.[2]

The chronicle was also full of references to warfare with the Christian Russians, although Hermann justified the building of castles against them by describing them as schismatics. In 1342, the Knights launched an attack on Izborsk deploying what the Russian chroniclers called "wall-breaking rams" which probably meant either siege equipment in general, or stone-throwing artillery. Although the attack was pressed for ten or eleven days and the town's water supply was cut off, the siege was abandoned for no stated reason.[3] In 1368, they attacked the same target again, surrounding Izborsk with a circumvallation and deploying a battery of engines, but were unsuccessful. Indeed, the Russian version says that the invading army was defeated and the defenders destroyed the engines.[4] In 1377, the Master led an army up the river Duna to attack a Russian castle standing some 18 kms from the Order's fortress at Dünaburg. They attacked for two weeks using four machines, with two other instruments of war called "huke", but without success.[5] The meaning of this word is unknown, though it might signify the arrival of cannon in the region.

Hermann's account also told of castle building or improvement. Dobelen and Terweten were constructed "against the heathen" in 1335–9, in both cases using existing sites. In 1345 a strong castle was erected on Oesel. The next year, the Danes sold their lands on the coast to the Order, bringing Reval (Tallinn), Wesenburg and Narva into their hands. Substantial resources were expended bringing them up to date with strengthened walls, towers and ditches. In 1348, the castles of Grobyn, Doblen and Dünaburg were improved with stone walls and towers. In 1371, the Master added two towers and a high wall to the castle of Reval.[6] The later rebuilding of the major castles would also involve substantial accommodation for the occupants, as at Reval. Narva, very much a frontier site, was provided with a convent-style structure and at this time was protected by a strong wall and the castle's high position. During this period, the fortifications of the settlers' land around the river Duna (Daugava) were also reconstituted. Many of the castles es-

[1] Czarnkow, II, 643–4.

[2] Długoss, V, 312; *Nikonian Chronicle*, III, 207.

[3] Hermanni de Wartberge, 70; *Nikonian Chronicle*, III, 151–2; *Chronicle of Novgorod*, 137.

[4] Hermanni de Wartberge, 91; *Nikonian Chronicle*, III, 206; *Chronicle of Novgorod*, 152.

[5] Hermanni de Wartberge, 113.

[6] Hermanni de Wartberge, 67, 72–4, 76, 99.

tablished during the initial invasion had been laid waste, but the Knights created four new ones around Riga, including Dünamunde, Neuermühlen, where a large square tower was added to a fortified water mill, and Neu Kircholm, built around 1380 in convent form behind a dry ditch and a double wall. At this time, there was still no provision for firearms.[1]

Russia: Moscow challenges the Horde

The Russian principalities continued to suffer the overlordship of the Golden Horde. The primary beneficiaries of this were the Lithuanians, who managed a complicated combination of diplomacy, dynastic marriages and military strategy enabling them to extend their dominion over vast tracts of Russia: Smolensk, Brest, Polotsk, Suzdal, Riazan, Tver, and at some point Kiev, were either subjected or became obedient allies of the Gediminid dynasty in Vilnius. The other beneficiaries were the princes of Moscow, who themselves operated a skilful and effective campaign to extend their power and influence and who by the second half of the century were clearly the leading Russian principality. In 1367, the foundations of the city's stone walls were laid and construction work thereafter was continuous. Although it was unlikely that they had progressed very far by 1368, Moscow held off an attack from Lithuanian prince Olgerd in that year, and three years later resisted another attack lasting eight days. In 1369, a new wooden fortress was constructed at Pereiaslavl, completed apparently in just one year, indicative therefore of a substantial mobilisation of resources. In 1375, Dmitri-Donskoi, Grand Prince of Moscow and Vladimir, attacked Tver, besieging it for four weeks, burning the suburbs, building two bridges across the Volga and deploying siege artillery, and bringing about the city's submission.[2]

In 1380, Dmitri Donskoi won a bloody victory over an army of the Golden Horde at Kulikovo, the first major challenge to the "Mongol yoke" in 140 years. However, the Horde's domination was rapidly restored. Khan Tokhtamish quickly gathered a great army, sending forces ahead to Bulgar (Kazan) to kill or capture any Muscovite merchants to prevent word of his plans reaching the city. He then marched rapidly north and crossed the Volga. The prince of Riazan submitted at once, and his assistance included advice on how "to take the fortress of Moscow", suggesting perhaps that the khan had not himself seen the new defences before. As vengeance loomed there was panic in Moscow. The prince prudently absented himself, and it fell to a prince Ostei to take a grip, preparing for siege and burning the suburbs that offered shelter to the attacker, and relying on the city's stone walls and iron gates. On 23 August 1382, the Mongols rode around the walls, while

[1] Tuulse, *Burgen*, throughout; K Altoa, T Aus, J Tamm, 'Narva Castle', *Castella Maris Baltici*, II, 1996, 14–18; A Caune, 'Steinburgen des 12–16 Jahrhunderts in Dünamunde ungsgebiet', same journal, 19–27. See also now S Turnbull, *Crusader Castles of the Teutonic Knights, (1) Prussia, (2) Latvia and Estonia*, 2003, 2004.

[2] *Nikonian Chronicle*, III, 199, 205 (Moscow), 207 (Pereiaslavl), 210 (Moscow), 225–6 (Tver).

the citizens shot arrows and "threw stones and shot from crossbows and guns" (*tiufiaki*). The first Mongol assault was an attempt at escalade, relying on their massive barrage of arrows to provide cover. The Muscovites had prepared vast kettles of boiling water with which they drove off the attack. A further attack was resisted by crossbows, stone throwers and guns. On the fourth day, the cunning Khan induced an obviously naive Ostei outside to a parley, where, having resisted an offer of money to surrender, the prince and his entourage of senior citizens and churchmen were killed and the city stormed, with the resultant massacre matching traditional Mongol records for ruthless brutality.[1]

Scandinavia

The Scandinavian kingdoms went through periods of dynastic rivalries and weakened monarchies, which reduced the resources available to monarchs to strengthen their rule through the development of modern fortifications. Meanwhile, the nobility had increased their own political power, and it may be that the period from the late thirteenth to the mid fourteenth centuries was when many of the Danish mottes were erected, although rarely can it be proven by dating evidence. Such structures were not militarily strong, but represented a defensible enclosure for the landowner and his household, farm and workshops. That they had some defensive aspects is suggested strongly by the fact that they *were* attacked, and defended.[2] The chronicles recorded successive expeditions within the Danish kingdom and its near neighbours, in which the castles and walled towns were the object of the attack, sometimes successful and sometimes not. King Erik VI (1286–1319) laid siege to the towns of Rostock and Wismar, bringing both to yield, in 1311–12 through hunger. In 1313, he marched into Norway, building new castles (Ulstorp and Linforth) and strengthening old ones. In 1316, he sent a fleet to attack Lund, but failed to take it. In 1318, the king's steward gained possession of a *turris* and a *castrum*, while Swedish rebels successfully seized the castles of Nyköping and Stegeborg. Erik led an army overland to recover Nyköping, attacking it with siege engines that "none could resist".[3]

A period of royal chaos followed Erik's death, marked by short-lived reigns, depositions and short-lived resumptions of the throne, before strong government was restored by Valdemar IV (1340–75), who set about the capture and destruction of many of the nobility's castles. His steward, for example, marched through the province of Laland in 1346 "and besieged fortresses" (Zealand chronicle). Resisting his power to the south were the rulers of Sleswig and Holstein, and

[1] *Nikonian Chronicle*, IV, 1–7.

[2] H Stiesdahl, 'Die Motten in Danmark', CG, II, 94–99; same author, Types of public and private fortifications in Denmark, Skyum-Nielsen ed., *Danish Medieval History: new currents*, 1981, 207–15.

[3] Chronica Sialandie, Continuatio, in Kroman, *Danmarks Middelalderlige Annales*, 1980, 116–18; Annales Ripenses, same volume, 265–66' Chronica Jutensis, same volume, 291.

during the 1350s there was frequent conflict marked by successful sieges. In 1358, one castrum was captured quickly by the royal army, while against another the king surrounded it, then deployed machines including *arietes* and other *vasis bellicis*. This may possibly be no more than a general term to describe equipment, or possibly could be a reference to a *vasa*-type gun. In 1360, Valdemar laid siege to Helsingborg, and King Magnus II of Sweden (1319–64), seeing how well entrenched the Danes were, resigned the important castle and province to him. In 1361, Borgholm on the island of Öland fell to Valdemar, and the next year Wisby on Gotland likewise fell. An alliance was formed of his opponents, led by the city of Lübeck, which brought an army to attack Helsingborg, to which they conveyed 16 throwing engines, and led Valdemar to negotiate with them.[1]

In Sweden, where the dynasty of the Birgers had been in control since the 1250s, there is debate and little certainty over the dating, and purpose, of the various fortifications built after this time, of which little remains. The distinctive tall, strong and free-standing round stone towers (*Kastals*) built along the east coast, are judged to have been constructed at different times, for different purposes. Some were, or became, part of later royal strongholds. Many seem to have been connected with the development of towns, which also became established during this period, often associated with a castle and town wall.[2] The chronicle accounts cited confirm that several of these sites were the targets of attack and counterattack by the rulers of this time.

The 1380s: arrival of guns in Eastern Europe and Russia

In 1389, the Polish historian Długoss recorded, the Lithuanians had reinforced their castle of Vilnius with "bombards, ballistas and apparatus of war transported to Lithuania" (a piece of evidence the usefulness of which is weakened by his previous references to the use of bombards dating back to as early as 1272), while in 1400, the Novgorod chronicle records that the Lithuanians attacked Smolensk with four guns, the first time the term is used in that text. The Russian language sources along with the account from the much later Nikonian chronicle confirmed the first use of firearms by Moscow as being during the attack by the Horde in 1382, using a new word (transcribed as *pushki*) which is distinct from the terms used to describe stone throwers.[3] There is a reference in the German chronicle of Johan von Posilge to *groszin steynbuchsen* in his account of the siege of

[1] Chronica Sialandie, 127, 136–8, 142–3; Annales Scanici, same volume, 72–3; R A Olsen, Danish medieval castles at war, CG, IX/X, 228–31,

[2] See three articles in *Castella Maris Baltici*, II, 1996: K Drake, 'Burg und Stadt in Schweden in der Zeit von 1250 bis 1320', 30–33; C Loren, 'The Round Towers on Sweden's eastern coast', 98–102; T Svennewall, 'Even the tallest tower. Early medieval fortification builders and buildings on Sweden's eastern coast', 202–7.

[3] Długoss, VI, 175; *Chronicle of Novgorod*, 176. The Pskov annals and other sources analysed by A N Kirpicnikov, *The Artillery of Old Russia* (in Russian), 23–4.

Kauve in 1362 which has been cited by some historians as evidence of their presence. But actually the writer, who was a local official of the Teutonic Order, wrote, at a later date, that the attackers did *not* have these weapons, and the very detailed account by Wigand, used above, never mentioned them at all, despite his precise account of the construction of equipment at the siege, which would be strange if they were making their first regional appearance.[1]

Guns had certainly been used by the Teutonic Knights in an attack on Vilnius in 1383, when the assault was abandoned because, among other reasons, they had consumed their supplies of stones and gunpowder.[2] They were used again in their large scale attack on Vilnius in 1390, when they were accompanied by a crusading party led by Henry Bolingbroke, the future king Henry IV of England. The defences of the Lithuanian capital were now substantial. The attackers blasted the walls with "thunderbolts" hurled from their bombards (Długoss), or attacked with *buchszen, bliden und tumeler* (Johan von Posilge) and captured their first objective, but after five weeks fighting they were unable to make any significant progress against the *castrum* and eventually abandoned the siege. By collating the different accounts, it appears that part of Vilnius's defences were either "unwalled" or perhaps more likely protected only by a timber palisade. This was rapidly stormed amid fierce fighting and many thousand Lithuanians killed or captured. The stronger defences, which also included guns among their weaponry, resisted successfully. One of the reasons given for the withdrawal was that the crusaders' powder had all been expended. The crusading English earl, with a party of perhaps some hundreds, certainly travelled in style, with six minstrels amongst his train. Among his arms, however, was very little artillery. The extremely detailed accounts refer only to the purchase of a ballista while passing through Danzig (Gdansk), and to payment to *diversis engynours* at the castle of Vilna.[3] Meanwhile, evidence of the range of weaponry used by the Lithuanians in attack came from the German accounts of the siege of Marienwerder by a combined Lithuanian and Russian army in 1384. The besiegers were provided with engines, rams and incendiaries, and another chronicler specifies *bliden, tomelern unde buchsen*. The allies bombarded continuously while on the other side preventing the Order's Marshal breaking the siege, and Marienwerder (Kwidzyn) fell to the princes Jogaila and Vytautis after six weeks.[4] By 1389, when the duke of Gelderland came crusading and specifically used *groszen buchsen*, and 1390, when the king of Poland besieged Grodno for six weeks with *blyden unde buchszen*, the new weapon was obviously now a major element of the arsenal.[5]

The new technology was not used in eastern Europe and Russia until at least

[1] Johan von Posilge, 82.

[2] Annales Thorunensis, 128.

[3] Długoss, VI, 185–6; Walsingham, 198; L Toulmin Smith (ed.), 'Expeditions to Prussia and the Holy Land made by Henry earl of Derby in the years 1390–1 and 1392–3', *Camden Society*, LII, 1894, in which are also gathered the pertinent excerpts from several Teutonic chronicles, cvi–cx, 94 (ballista), 106 (engineers). The ballista cost 16 "scots" or precisely four shillings and sixpence; Johan von Posilge, 164–7.

[4] Annales Thorunensis, Johan von Posilge, 135.

[5] Johan von Posilge, 156, 163.

thirty years after its arrival further south and west. This cannot have been due to simple ignorance. Not only were there continuous trading relations between all the countries of the continent, but the rulers of central European states belonged to dynasties that also ruled in places where firearms were already in regular use, while the Teutonic state was in continuous contact with the empire, and entertained western crusaders. The Knights organised large scale resources to construct and modernise their castles and fortifications, and the absence of provision for guns in these work need not indicate that they were unaware of their existence. They were too small and their ballistic power was too weak to cause any change in design, while in defence it could be that they were mounted on roofs without any change in the design of firing positions. It must rather be that until more powerful weapons began to be constructed (by the 1380s), rulers did not see the need to establish their own arsenals. Długoss's comment that the Lithuanians imported guns may be significant. Vilnius was the capital and contained the main concentrations of manufacture and artisan workshops in the principality: at the time, therefore, it may be that there was no one in the principality who could be called upon to manufacture them.

However, perhaps the most significant change in the politics of the whole region had already taken place. In 1386, Jadwiga, child heiress to Casimir the Great of Poland, married the long-living Lithuanian ruler Jogaila, who as part of the deal was baptised as Lladislaw-Jagiello (king, 1386–1434), and created a dynastic union of the Polish crown and the now greatly expanded Lithuanian duchy. The Teutonic Knights no longer had any reason to exist.

The Turkish conquests in Asia Minor and the Balkans

In the space left by the collapse of the authority of the Turkish sultanate of Rum after their encounters with the Mongols, new clans, tribes and confederations of tribes of Turkish origin began to fill the power vacuum in Asia Minor. The Byzantine empire was too feeble to attempt to recover its lost Anatolian heartlands. The Turkish statelet that achieved the greatest impact in the first part of the fourteenth century was the emirate of Aydin. To its north, another group of holy warriors (ghazis) following the leadership of Osman and his descendants, were about to initiate the march to empire of the Ottomans. A limited number of Turkish histories (none of them contemporary for the Ottomans) combined with the accounts of hostile Christian writers enable the outline of what they achieved to be composed, though reliable contemporary information and dates are rare for these events.

The emirate of Aydin took shape when Mehmed Beg captured the ancient city of Ephesus in 1308. These Turks may have been semi-nomadic mounted warriors but they already knew how to use stone throwing artillery (Turkish word *manciniq*) and how to besiege stone-walled Byzantine cities. The future hero of this dynasty, Umur, was born at this time, and was later made emir of İzmir (Smyrna).

This vital and substantial seaport fell into Turkish hands after a long period of fighting in around 1328 and would remain the centre for continuous raiding across the Aegean by the Turks, and the target of several attempts at counter-attacks by the Latin states in the area, encouraged by the papacy. Smyrna was heavily fortified and had been modernised during the thirteenth century. The "old castle" stood on a ridge south east of the town and reused ancient Hellenistic and later, (middle) Byzantine defences. The plateau, which was linked at both ends to the town walls, was itself walled with a strong enceinte with many towers. The town end of the enclosure was itself separated to form a quadrangular inner castle (using two of the outer walls for its own enceinte) in the thirteenth century rebuilding. The wall was up to 3.4 metres thick, there were four projecting round towers, and a cistern in the middle. There was a second castle in the harbour, now disappeared, but described as new in 1251.[1] It was the town castle that had been taken by the Turks first, apparently by surprise, but a Genoese force held on in the harbour castle for several years before finally it was given up. Control of the harbour allowed Mehmed, then from 1334 Umar, to equip fleets to raid the islands, and indeed to cross to the Greek mainland. The crews were provided by volunteers engaged in holy war, and the ghazis created much havoc, including, according to their own account, capturing numerous forts in Greece and on Euboea by escalade during the 1330s. The only sophisticated weapon recorded in use was a fire lance (*tufek*).[2]

By now Umur was a significant player, and had made an alliance with the Byzantine ruler John VI Cantacuzenus, to whom he lent his fleet, and alongside whom he besieged Pentharion in Thrace in 1343/4. Here the Turkish troops attacked by sapping at the walls with tools and also used incendiary devices and stone throwers.[3] Vengeance was now in the offing, however. Following a failed Genoese effort to recover Smyrna (1334) a large expedition was put onto Genoese ships and carried numerous crossbowmen and "makers of engines" alongside *tufeks*. The Christian army captured the harbour castle. Umur organised a rapid counterattack. But the high stone walls meant that standard siege technology would have to be used to supplement the courage and enthusiasm of the ghazis, and we hear tell of many stone throwers hurling missiles of considerable weight (the Turkish term used could mean anything between 36 and a more unlikely 108 kgs). One of the volunteers serving with Umur, a Moor, built a remarkable "small mangonel" which he used to hit individual "Franks" and also to damage the boats on which they in turn mounted mangonels in the harbour. After ten days, the crusaders launched a sortie, and although the Muslim writer says it was defeated, he admits it burnt the Turkish stone throwers, which was anyway the purpose. A big effort now went into cutting down trees to rebuild the artillery, and this succeeded in propelling naphtha and destroying another ship which had had a wooden tower built on it.

[1] All from W Müller-Wiener, 'Die Stadtbefestungen von Izmir, Sigacek und Candarli', *Istanbuler Mitteilungen*, 12, 1962, 60–86.

[2] *Le Destan d'Umur Pacha (Dusturnam-i Enveri)*, I Melikoff-Sayar (trans.), 1954, 42 (Smyrna), 56 (tufek), 66–83 (raids). P Lemerle, *L'emirat d'Aydin et l'occident. Récherches sur 'La Geste d'Umur Pacha'*, 1957, is a scholarly study of the text.

[3] *Le Destan d'Umur Pacha*, 97–100.

A stalemate however ensued, with the siege being raised and then re-established and in the fighting around the defences. Umur himself was killed in 1348. The Christian forces, which received reinforcements, were occupying the town and harbour, but were unable to seize the upper castle from the Turks, and the latter could not recapture the heavily defended town. In an intriguing sideline, the Muslim writer describes the Christian *tufeks* as making a noise like "tchat tchat" as they fired.[1] While it is entirely clear that the Turks had nothing more advanced than naphtha, what were the Genoese using?

Further north, Osman Beg established himself in Bithynia. By effective leadership and profitable raiding he built up a ghazi following and gradually carved out a lordship for himself. The area contained several large and once important cities, not least Nicaea, Nicomedia and Bursa, all of them provided with large fortifications, and early successes in the area of Nicaea in 1308 as reported by the Byzantine historian Pachymeres were discussed in the last chapter. The histories of the dynasty confirmed that Bursa was too strong to storm, so Osman instead ordered the construction of two siege forts (blockhouses), both of which were completed in one year, and their garrisons of ghazis seriously harmed the citizens, by preventing them securing the produce of the land. Capitulation duly followed, the governor explaining his defeat as being through hunger. The town became instead the capital of the new regime, now represented by Osman's son Orhan (about 1324). The weakness of the Byzantine empire, unable to organise relief of such a city, was clearly demonstrated, and may have encouraged Orhan to use the same tactic to induce the surrender of Nicaea. Here, again, blockading fortresses were constructed and the townspeople, whose loyalty to Constantinople may well have been doubtful, were driven back inside their walls. A somewhat insubstantial attempt at relief organised by the chastened emperor was apparently ambushed as it landed near Nicomedia, and Nicaea followed Bursa into Ottoman hands (1331).[2] The former Seljuk metropolis, recovered in the first major siege of the first crusade, was now lost forever to the Christian empire. Nicomedia too fell by the same method (1337). While the Ottoman operations so far had revealed tenacity and skill, they had not yet been put to the test of taking a properly defended fortress, and the usually generous terms offered to the besieged populations proved also a considerable inducement to reaching a peaceful surrender settlement.

In the following decades, good fortune smiled on the Ottomans which, combined with their own undoubted skills, allowed their continued expansion. The emirate of Aydin had been put on the defensive by the Latin counter-attack, and fell to a swift Ottoman offensive by 1356. Ankara was secured in the 1360s. By a combination of skilful diplomacy and armed might, the Ottoman realm

1 *Le Destan d'Umur Pacha*, 110–25. The Christians' tufeks are described on 112. The course of events is confirmed by the traveller Ibn Battuta, *Travels*, II, 445–6.

2 Dervish Ahmed (Asik Pasha), trans. by R Kreutel as *Von Hinterzeit zür Hohen Pforte*, 1959, 44–5, 52–4, 67–8; *Die Altosmanischen anonymen Chroniken*,ed/trans. F Giese, II, 13–16, 19–20. The anonymous reverses the order of events. Both were compiled in the fifteenth century. H Inalcik, Osman Ghazi's siege of Nicaea and the battle of Bapheus, in E Zachariadas, *The Ottoman Emirate*, 1993, 77–93, for a study of all the accounts.

gradually expanded across Anatolia, reaching the Mediterranean shores at Antalya in 1380. But the main Ottoman advance at this time was to the north, across the Dardanelles and into Thrace – and eventually, beyond into the Balkans.

The initial step of obtaining a foothold on the European shore was the conquest of Gallipoli, which the new ruler Suleiman besieged closely, attacking day and night from land and sea and eventually capturing it (1354?).[1] Although no detail is given it is a safe conclusion that the Turkish forces that captured this Byzantine castle and then proceeded over the years ahead to take control of much of Thrace were more than capable of mounting an effective siege, even if in the actual event they did not always face much resistance. Although it may have appeared that nothing more than raiding was intended, in fact the Ottomans were going to stay in their new lands, and indeed seem to have encouraged considerable settlement as well as converting captured towns into Muslim cities. In no time, the Ottoman emirate would boast of two distinct armies, one in Anatolia and one in Rumeli (Thrace). The chronology of events is utterly confused, with different dates (if any at all) given by various (non-contemporary) sources. The major city of Adrianople (Edirne) fell into Ottoman hands either in 1361, or 1369, or 1371.[2] Philippi (Plovdiv) in Macedonia followed not long after. The Byzantines were helpless, and the Serb principality to the north proved unable to reverse the tide although it tried. Shortly after their arrival, however, their line of communication was to be broken by the arrival in the Dardanelles of the count of Savoy, leading his own expedition with another objective altogether.

Western intervention

(i) from Italy

INSPIRED BY THE SUCCESSFUL ATTACK on Alexandria described below, Amadeus V really intended to rescue his relative, the Byzantine emperor, from captivity by the Bulgarians. Sailing on Venetian ships via Euboea, he arrived off Gallipoli and found the Turks arrayed against him. The count mounted cannons on his ships which he used to support his army and drove the defenders back behind their walls, which he then proceeded to assault. His first attack was driven back but the next day he returned with mantlets and cats which were driven to the foot of the wall, and used to shelter sappers who worked away at creating a breach under protective fire. The Turks responded with stones and incendiaries but could not prevent a breach being made. The place fell as the garrison abandoned it overnight.[3]

Amadeus returned the castle to Byzantine control and resumed his original

[1] Dervish Ahmed, 75–7; Altosmanischen anonymen Chroniken, 23–4.

[2] I Beldiceanu-Sternheit, 'La conquête d'Adrianople par les Turcs', *Travaux et Mémoires*, I, 1965, argues for 1369; H Inalcik, 'The Conquest of Edirne (1361)', *Archivum Ottomanium*, 3, 1971; E Zachariadou, 'The conquest of Adrianople by the Turks', *Studi Veneziani*, XII, 1970, supports 1369.

[3] Chroniques de Savoye, 306–8. The event is not described in Ottoman sources.

campaign. Proceeding by sea up the Black Sea coast, he seized several coastal castles with little resistance and then encountered a significant Bulgarian stronghold at Mesembria, the once powerful headland fortress that has featured in this history centuries earlier. This time, the Savoyard was in a position to attack by both land and sea, and again mounted cannons on his galleys. Mesembria was unable to resist and was stormed and sacked. With the all-conquering invading army moving on to Varna, the Bulgarians gave up their imperial prisoner (1366).[1] It seems remarkable that a seaborne expedition by a minor Italian prince should be able to humble a kingdom, but the Bulgaria of Emperor Ivan Alexander (1331–71) was a pale reflection of the mighty first Bulgarian empire. The new state had located its capital at Tarnovo, strongly sited in a river bend. Defences were important, with strong square and round towers and defended entrances to a palace complex. However, excavations of the site have shown that it was throughout of unworked stone, and whereas Pliska had covered 50 hectares and Preslav 25, Tarnovo occupied a mere 0.6.[2]

(ii) The crusade against Egypt

King Peter of Cyprus had assembled a multinational fleet and force to attack the Turks at Satalia in 1361, which he had taken by storm. Genoese troops and ships had protected Caffa from an Ottoman onslaught from Sinopoli.[3] Peter's engagement against the Turks led to one of the remarkable events of the mid-century, even though not aimed at the Turks. A contemporary chronicle of his reign gives much detail. With the growing consolidation of Ottoman Turkish power in Anatolia and the triumphant Mamluk regime controlling Egypt, Palestine and Syria, there was a long and potentially hostile coastline. Following Peter's battles with the Turks in 1360 and 1361,[4] he raised support for a new crusade. The Venetians provided ships in 1365 and the armada assembled at Rhodes. Cyprus was turned into an arsenal as great quantities of equipment were prepared, including specifically artillery for attacking and defending towns, slings, engines, *bricolles* and cats were all listed. The target, however, was not the Turkish presence, although it was kept (remarkably) secret until the fleet was at sea. Instead, the great fleet sailed to Alexandria, which they attacked on a Friday, 10 October 1365, on the basis that the Muslims would be at prayer. In this scheme they were disappointed, as there was resistance to their disembarkation on the beaches, but the Hospitallers came ashore further along and took the defenders in the flank, driving them back and allowing the crusader army ashore.

The defenders turned out to defend their vast circuit of walls. Every tower was provided with a large crossbow (or ballista) and large numbers of stone throwers,

[1] Chroniques de Savoye, 308–12.

[2] K Mijatov, 'Das Schloss der Bulgarischen Zaren in Tarnovo', *Acta Archaeologica Academia Scientarum Hungaricae*, XVII, 1965, 77–81.

[3] G de Machaut, *La Prise d'Alexandre*, ed. La Mas Latrie, 1877, 20; M Villani, *Cronica*, 352.

[4] Jean Dardel, Chronique d'Arménie, *RHCA*, II, ed. Dulaurier, 1906, 32.

and the barons advised the king against trying an assault, but a long siege was not an option, so he sought out the weakest points of the long circuit. This proved to be the Gate of the Customs House, which had been locked on both sides by its keeper, so that defenders could not get into to it to organise a direct defence. The first assault had been repelled by stones and bolts and *espingales*. However, the king noticed that there was an undefended stretch of rampart between this and another gate. So while he launched a new attack on the original targets, and managed to set a fire at the foot of the gate, others found a way through the defences using an old latrine shaft (shades of the popular version of the capture of Château Gaillard) and reached the wall top. Meanwhile, as the crusaders fought to hold their foothold on the wall, the gate was finally consumed by fire and collapsed, allowing the invaders in, and the defence collapsed. An appalling sack was then carried out as desperate citizens tried to flee and many were caught and killed in the narrow streets. The next day, Peter was advised by the leading barons of the crusade that he could not possibly hope to hold Alexandria against the gathering Mamluk counter-attack, because its extent was too great to be defended by his troops, and his artillery had been used up in the attack. His followers, laden with their booty, retired to their ships ahead of the Muslim army advancing up the Nile to recover the city.[1] The attack on Alexandria had been well designed, as the governor was absent and the sultan was a minor, but the long-term effect (except on the unfortunate city itself) was negligible. The Egyptian eye-witness records, probably, the first certain reference to the use of cannon by the Mamluk forces: "I saw a cannon (*midfa*) made of copper and lead and fastened by iron chains. A great ball was fired from it".[2]

Indeed, the Cypriot king's next actions were all once more directed at the Turks in Anatolia. In 1366, he sent his admiral to the Gulf of Satalia and tried to take the castle of Candelour, this time in vain although the initial attack had carried the barbican. The next year, the Turks attacked the castle of Gorhigos. This castle consisted of a strong land-based fortification with a narrow approach, and another fortification on a nearby islet, providing a secure line of communication for seaborne relief. The land castle largely comprised work carried out by the Comneni reusing classical stonework. A double circuit of walls, the inner roughly parallel with the outer, was further protected by a substantial ditch. Projecting rectangular towers strengthened the more vulnerable landward-facing northern and eastern walls. Two ships brought hundreds of extra troops to hold this formidable fortress. The Turkish commander set up camp outside and placed two engines to shoot into the castle, most likely on the rising ground to the north. There was heavy fighting and a Turkish assault was repelled. Meanwhile, however, heavy losses were being inflicted by the stones of the engines, and reinforcements had to be summoned. Eventually, the Cypriots fought and won a battle in the open and managed to destroy the engines that had been plaguing them. The siege was raised. Peter showed the effectiveness of sea power again in 1367 when his troops

[1] G de Machaut, 53–108. The eye witness Muslim account agrees on all the main points.

[2] Al-Qalqashandi, quoted in D Ayalon, *Gunpowder and firearms in the Mamluk kingdom*, 1956, 2. The original text has not been translated from Arabic.

ravaged large stretches of the Syrian coast, capturing and ransacking Tripoli along with Laodicea and Tortosa among other towns, none of whose defences saved them.[1] The assassination of the king in 1369 must have been a great relief to his Muslim neighbours, and introduced a period of domestic conflict on Cyprus.

In 1373, the Genoese invaded the island and attacked and seized Famagusta. Early the next year they marched across the island to attack Kyrenia. The royal forces resisted from the castle and repelled the first attempt at an escalade, in which (the chronicler says) 400 Genoese were killed. The defence used trebuchets and incendiaries, and the besiegers were compelled to set themselves further back to avoid losses from this bombardment (3 February 1374). Regrouping, the invaders organised a more sustained bombardment, building siege equipment that specifically included numerous stone throwers, and a trebuchet hurling gigantic rocks that smashed houses and walls. But this intended attack failed to materialise after a convoy of supplies including twenty scaling ladders was intercepted by the Cypriots, and the Genoese withdrew to Nicosia. Returning, they set about making new machines and tried further assaults, which were driven back as well, with heavy losses. On 3 March they tried again, by land and sea. Their arsenal included four engines. One was a wooden tower three storeys high, covering sappers at the base, with crossbowmen on the first level, while the third was built like a *castelletto* and it and the fourth were full of men ready to storm across the wall. This great structure however fell victim to a nocturnal sortie that burnt it, and when another attack on the harbour, this time by two galleys linked together and provided with a tower occupied by crossbowmen was thwarted by the defenders constructing their own high screen to block the view, and by bombardment by Cypriot trebuchets sited to protect the harbour, which according to one source succeeded in smashing the elaborate Genoese structures on the galleys. With this setback, the siege was abandoned.[2] The Genoese however remained in control of Famagusta, which resisted an attack by King Peter II in 1378, and two years later the king ordered the collection of taxes to pay for the building of a citadel at Nicosia. Using Genoese prisoners to excavate the ditches, the work was, remarkably, finished in one year.[3]

Meanwhile, the final destruction of the Armenian kingdom in Cilicia had taken place. Originally an artificial creation of the Byzantine empire to form a solid defence against Muslim advance, the Armenians, as we have seen, had provided their small realm with a large number of very powerful and often inaccessible castles. Defended by brave warriors, they formed a formidable barrier to any invasion. The destruction of the Latin kingdom to the south, however, had brought the

[1] G de Machaut, 121–64 (Gorhigos), 207–15 (Syria). For the land castle: Müller-Wiener, *Castles*, and personal observation.

[2] F Amadi, *Chronique*, I, 439–47 (Famagusta), 460–2, 464–6 (Kyrenia). Amadi was not a contemporary but used earlier sources; Leontios Makheiras, *Recital concerning the sweet land of Cyprus entitled "chronicle"*, ed./trans. R Dawkins, 1932, I, paras 470–87. This very biased source was also a non-contemporary compilation, but provides a day by day account of events.

[3] Amadi, 488–90.

Mamluk empire into close proximity, and if the Turkish statelets to the north did not individually pose a critical threat, the advance of the Ottomans towards the coast was another matter. On top of this, the Armenian ruling families demonstrated the kind of mutual loathing and bloody rivalry that seem to beset doomed dynasties. In 1373, a Mamluk army laid siege to the Armenian capital of Sis (Kosan), a strong hill top site with walled town and castle. When Cypriot envoys from Gorhigos slipped into the town unobserved, the defenders were sufficiently encouraged to sortie and defeat the Syrian forces who raised the siege, which may anyway not have been intended as a full scale attack, judging by the tone of the account. By the time a new king, Leo V, had been crowned in 1374, the Mamluks had secured control of much Armenian territory, while Turkish raiders regularly rode within sight of Sis. The Mamluk Abu Bekr, aided we are told by Armenian traitors, made a further attempt on the city, seeking to deny it access to provisions. Cypriot crossbowmen helped Leo break the blockade. In 1375 however, the Syrian enlisted the aid of Turcoman auxiliaries who began a new blockade of the capital. A large besieging army arrived and launched a preliminary attack. With the arrival of further forces led by the emir of Aleppo, assaults became continuous and the exhausted defenders, unable to hold the circuit of walls, withdrew to the castle, where trebuchets were deployed on the walls. There was only a single approach to the castle (Plate 6), and the platform outside the gate became the site of fierce fighting the very next day. King Leo, himself fighting on the wall with a crossbow, was hit by an iron shot from a bombard. This shot broke several of the king's teeth. Armenians hostile to the king tried to kill him as he recovered in the donjon, but, aided by a few loyal servants including his "master of engines", a Greek crossbowman from Smyrna, he escaped by rope and, rallying his troops, tried to recover the great tower. After four failed assaults, a way in was found, and the door opened to bring about recapture. It was all in vain, however, as traitors opened the gate of the fortress to the besiegers and the last Christian king to rule in Armenia surrendered (April 1375).[1] The account also confirms that the Mamluks were using artillery, but this must have been a very small bombard.

Ottoman advance in the Balkans resumed

During the middle years of the century, the Serb ruler Stefan Dušan (1331–55) took control of large stretches of former imperial territory (Macedonia, Epirus, Thessaly and Thessaloniki) without much fighting. Serres resisted, and was taken in 1345, becoming the local Serb capital. This city, standing in a fertile plain beside the river Strymon and the road connecting Constantinople to its western provinces, had become even more important with the virtual disappearance of the Asian section of the empire. Little can now be reconstructed of its medieval form, but it seems to have had a powerful citadel on higher ground, connected to well-built town walls. Given the great significance of the reign of Stefan Dušan in the history of the Balkans, the absence of primary sources for the events of the period

[1] Jean Dardel, 48, 70–81.

is frustrating. The Byzantines noted the town fell through hunger, after a long siege. Then the Ottomans arrived, the Serbs were defeated at the battle of the river Maritsa (1371), and Serres was reoccupied by the Byzantine Manuel Palaeologus, who set himself up as ruler in Thessaloniki and became emperor in 1382. In the Byzantine civil war that had preceded these events, both sides had relied on outside forces: one on the Venetians, the other on the Ottomans. In 1383, Murad sent an army under his future Grand Vizier to capture the city. Serres fell in September of that year, although details of how are scant, and by November Thessaloniki was under siege. The city's powerful but old walls were held by a Byzantine-paid army and sea communications remained open, but the Turks did not attack, instead resorting to what proved to be a three year-long blockade, at the end of which, resources exhausted, Manuel slipped away and the city gates were opened.[1]

Elsewhere in the western Muslim world

The chronicles of events in the North African states reveal little change in the methods of warfare during the fourteenth century. In 1375/6 the Hafsid sultan had to lay siege to the ancient city of Kairouan, held against him by rebels. The decisive move in the attack was to begin to cut down the city's date trees, whereupon the inhabitants underwent a dramatic switch in allegiance. In 1390, a fleet from Genoa and Barcelona launched an attack on Mahdiyya in order to capture this base for corsairs. They built a wooden wall across the isthmus, and towers for their soldiers to occupy, to cut off the town. Then they constructed a wooden tower as high as the town wall. The defenders resisted strongly and the sultan mobilised an army to bring relief. On the critical day, there was a fierce battle on the outside, while the defenders sortied firing arrows, hurling stones and naphtha, and succeeded in burning down the siege tower. Next day, the Christian army withdrew to sea. There is no reference in the history to anything apart from traditional weapons in use.[2]

An account of the sultans of the Resulid dynasty in Yemen allows us to see how this once thriving kingdom was now developing. In 1324, Umer laid siege to the town of Ta'izz, in which the sultan was residing, camping on the nearby mountain. An engineer was sent down to Aden to bring up a catapult. The timbers were conveyed by sea. Once assembled it hurled a number of rocks but "without any effect whatever" against the mud-brick walls. So a second machine was summoned up, which also bombarded, throwing 40 rocks a day. In the end the attackers abandoned the siege. In 1338, the sultan sent a force of 400 cavalry, 11,000 foot soldiers

[1] The confused chronologies were unravelled in G Dennis, 'The reign of Manuel II Palaeologus in Thessalonica 1382–87', *Orientalia Christiana Analecta*, 159, 1960, 56–78; G Ostrogorsky, 'La prise de Serres par les Turcs', *Byzantion*, 35, 1965, 302–19. On the 1345 siege, Nicephorous Gregoras (*PG* 148), col. 983. On Serres in the early nineteenth century, M Cousinéry, *Voyage dans la Macédonie*, 1831, 157–9. For a recent history, B Ferjančić, *Byzantine and Turkish Serres in XIV century*, 1994 (in Serbo-Croat, English summary, 135–41).

[2] Ibn Khaldūn, III, 93 (Kairouan), 117–19 (Mahdiyya).

and one catapult to capture Dhamār, and the place was stormed. There are several references to repair work carried out on defences. The town of Zābid had works lasting three years done to wall, gates and moat in 1338, while fifty years later a second, outer moat was added. The town was then attacked, the besieger riding around it to study it for weak points. Eventually they assaulted each gate, seeking to sap the walls with shovels under the cover of crossbows. They were repelled and the sultan arrived to bring relief.[1] There is nothing in these accounts to challenge the perception that the Mamluk use of gunpowder in 1365 was definitely amongst the first in the Muslim world.

Venice and the War of Chioggia

By the fourteenth century, Genoa and Venice alone competed for domination of the lucrative trade routes, and the concessions they could wring out of enfeebled governments like that in Constantinople. Genoese and Venetian ships and soldiers were regular participants in the conflicts in the Aegean and Asia Minor, Cyprus and North Africa. The cities' wealth gave them access to the latest technology. The struggle between them was incessant, but often conducted at second hand. In 1378, though, there began a war that unusually was to be fought out within a few kilometres of Venice itself. Not surprisingly, there is a great wealth of contemporary records, which are generally agreed on the main course of events.

Venice's land frontier confronted powerful and jealous neighbours, such as Padua, or the kingdom of Hungary. In a struggle with Padua and Hungary in 1372, the Venetians used their ships to convey siege equipment including a three-storey siege tower (belfry) to attack a Paduan tower at Churan, near to Orgiago. The defenders had bombards and *schiopi*, but the Venetians mounted mangonels on their ships and used them to demolish the tower to the ground: clearly the stone thrower here was the heavy weapon. They threw up a *bastia* in just ten hours. In order to protect his lands, the Carrara lord of Padua ordered the construction of a number of *bastias*. He also led an armed force and no fewer than 4,000 workers to reconstruct the tower demolished by the Venetian catapults.[2] In 1373, the terms of the peace treaty between Venice and Hungary included the destruction of a number of *bastias* and *castra* on the Venetian terra firma, while the signor of Padua took the opportunity to begin the construction of a castle for himself in his city.[3] As we have seen further west, it was a common feature of the warfare of this time to construct forts or bulwarks as field fortifications, serving a number of different purposes. In 1376, in another conflict with Leopold, duke of Austria, Venetian forces drove the Austrians from a fort they had constructed, using *bombardella parva*, which was the first time they had been seen in Italy according

[1] El Khaznejiyy, Aliyyu'ben 'l Hasan, *The Pearl Strings. A history of the Resuliyy dynasty of Yemen*, trans. J Redhouse, ed. E Browne, 1906, II, 15–19, 55, 178–82.

[2] G & B Gatari, Cronica Carrarese 1318–1407, ed. Medin, Tolomei, *RIS* NS, XVII(1), 79–83, 95–6; Andrea de Redusiis, Chronicon Tarvisinum, *RIS*, XIX, cols. 743–44, 748.

[3] Raphayni de Caresinis, 29.

to the chronicler, who was perhaps referring to a specific type of small firearm.[1] A distinctly larger weapon was mentioned shortly after, deployed by the Venetian and Trevisan forces against a new Austrian fort. The physical description given was of a large iron weapon with a smaller powder chamber, firing such large stones that "no kind of wall could resist them, however large". Using the same weapons, the Venetians went on to capture the town of Feltre, and other castles and a fortified church (August 1376).[2]

Genoa put together an international alliance including Hungary, Francisco Carrara of Padua and the Patriarch of Aquileia, and in the first battles in 1378 Genoese fortunes prospered. Venetian ships sought out their enemies at Zara in October, failed to find them, but hit the town and port with bombards and other artillery. Meanwhile, forts were constructed (or repaired and prepared) on the land side. The dismantled castle at Orgiago was rebuilt, and *bastias* were prepared and munitioned at Gorgo, Lugo, Solagna and St. Elaro. In July 1378, a large force of troops from Padua, Forli and Hungary approached the landward frontier and began a ceaseless attack on the town of Mestre (on the mainland opposite Venice) with "a great quantity of bombards and mangonels", and at the same time began to construct a fortified bridge across the water to cut the fort off from Venice. The fortified outpost of S. Lorenzo was targeted with a large bombard that caused immense damage, but an assault launched with three cats on 1 August was repelled by the Venetians, who then set about repairing the defences and raising the stone walls still higher. But the attackers placed a bombard in the bell tower of a church to fire down on the defenders.[3]

In April 1379 the Genoese surprised them and captured 14 Venetian galleys, leaving the city itself exposed to attack. Rapid measures were taken to secure the city, with chains across the waterways, protected by artillery fortifications at each end. Aware of the threat posed by the Genoese fleet, the Venetians also sought to close off access to Chioggia, at the other end of the lagoon, with a large bastion and, by water, a big cog (round ship), both equipped with bombards and crossbowmen. In June, under the direction of a master engineer named Faxuolla, the Paduans captured the castle of Romano by sapping the walls under the shelter of a cat. In August, the Genoese led by Ambrosio Doria suddenly turned their attention to Chioggia, and met up with a large Paduan army. Cutting the road to Lombardy, they pressed the town closely after several days' fierce fighting. The attackers had two big mangonels and many large bombards and they also produced a belfry. The weight of fire from the artillery played a significant part in the dislodging of the Venetian defenders (who were foreign mercenaries) from the bridge and its bastion leading to the city, and the Genoese and Paduans entered Chioggia on 16 August, the defenders fleeing by boat. From this base they were able to organise further attacks on Venice's outer defences, using a combination

1 Redusiis, 753.

2 Redusiis, 754.

3 Caresinis, 29ff; Gatari, 150–1; Daniele di Chinazzo, *Cronica de la Guerra de Veniziani a Zenovesi*, ed. Lazzario, 1958, 33–5; Redusiis, 763–4; Johannis Stellae, Annales Genuenses, *RIS*, XVII, 1107–1110.

of land and sea-based forces, artillery, crossbowmen, and traditional siege equipment. The Paduans had captured and occupied S. Lorenzo and other forts were abandoned, reducing Venetian control to the city itself. It was noted that some of the forts were deliberately given up, and their garrisons and their bombards were transferred to the city, where work on the defences was redoubled.[1]

In this crisis, the Venetian nobility rallied to the city's cause and contributed large sums to build and equip a new fleet, 34 new galleys being provided. At the same time, the Genoese fleet was dispersed, and from being on the defensive, the republic, under the inspirational leadership of its doge Andrea Contarini, was able to retake the initiative. They did so by counter-attacking Chioggia on 22 December. The Venetian army with allied troops from Treviso quickly surrounded the town and threw up siege works, in particular one large *bastia*. There followed a long and bitter struggle around the town, with the Genoese forces inside conducting a brave and aggressive defence but unable to break out, while they were strong enough to prevent the Venetians breaking in. Part of the struggle revolved around a fortified monastery at Brondolo, some three kilometres from the city itself. The doge ordered a fort built against it, provided with "many large bombards" and a large mangonel; the Genoese responded with their own stone thrower, but only had small bombards. The Venetian fort called *La Lova* on the causeway was another point of fierce struggle, defended both by men and guns inside, and from galleys moored close by.

There are numerous references enabling a judgement to be made about the effectiveness of the artillery of both sides. On 25 January 1380, the Genoese commander Piero Doria was killed, his head smashed by a bombard shot. The gun makers of Treviso were apparently highly skilled in their craft. It was noted that the two guns installed in a tower recovered and rebuilt by the Venetians were very large, that from Treviso firing a ball weighing 195 pounds, while one made by the Venetians fired a shot weighing 140 pounds (88.5 and 63.5 kgs respectively). The Trevisan gun on 24 January knocked down a great piece of the wall of the monastery, crushing 24 Genoese. It became customary for the Venetians to shoot off all their guns and three mangonels simultaneously each evening. Galleys brought supplies of ammunition daily. The Genoese, as noted before, only had small bombards, although they had many of them.[2]

On 18 February, amidst a great discharge of weaponry, a Genoese sortie was routed, whereupon Brondolo was abandoned and its stone throwers burned. By March, not only were provisions running very low, so was gunpowder. Although some relief and powder from Padua made it through, the end was only postponed. The blockade was tightened to prevent a repetition of the supply effort, and in May a Genoese fleet was beaten as it attempted to break through to rescue the

[1] Gatari, 170–78; Chinazzo, 52–62; Caresinis, 37–8, Redusiis, 770–73; Stellae, 1111–112; V Lazzarini, 'La presa di Chioggia (16 Agosto 1379)', *Archivio Veneto*, LXXXI, 1952, 53–70, reproduces contemporary letters recording the event, and noting that Chioggia had been thought impregnable.

[2] Gatari, 186–95; Chinazzo, 83–128 (references to large guns on 103); Caresinis, 42–57; Stellae, 1114–117.

trapped army. On 21 June Chioggia was yielded once again to Venetian control.[1] For all the deployment of the most modern weaponry, and the effective use of sea power, it was, in the end, old-fashioned hunger that brought about the surrender of Chioggia. The Venetian victory marked the consolidation of their power.

The conflict between rival factions for control of the city and lands of the north Italian city of Bergamo, meanwhile, provides an illustration of the reality of life in other parts of the peninsula towards the end of the fourteenth century, and of the resources at their disposal. In 1383, for example, the city at its own expense organised the construction of several *bastitas* against the threat of the rival ("Guelf") party. These had considerable forces. In 1398, 3000 men attacked the tower of Albino, held by just 26 men, and used bombards to destroy the tower completely, and everyone in it. The artillery had comprised 11 bombards and a large mangonel. In 1404, they built a fort above the town of Nimbro in which they based a large mangonel which hurled large stones into the place. The following year, a force of 300 Guelfs attacked the castle of Poma di Suixio, in which were the lady of the house with a handful of her household, and deployed mantlets, bombards and ballistas. Not surprisingly perhaps, they forced their way in next day. In October of that year, the "Ghibelline" party attacked the *rocca* of Urniano, using first one and then a second, much larger bombard brought from the citadel of Bergamo, as a result of which the castle surrendered at once. The castle of Tricio also surrendered to the Bergamascan bombards, seeing that it could not resist them. In 1406, however, the defences must have been significantly strengthened, because a rebel force held the castle against a substantial force sent from Milan. Hundreds of workmen were used to construct siege forts, but the whole operation was called off when Lord Galeazo of Mantua was killed by a shot in the head from a *veretonus* and the defenders sortied and burnt down four of the besiegers' forts.[2]

A new stage of gunpowder artillery

RIGHT UP UNTIL THE END OF THE CENTURY, traditional forms of siege weaponry continued to be used alongside the new gunpowder weaponry. Siege towers and battering rams were built and used in the wars of the Teutonic Knights, Poles and Lithuanians in the 1390s. Powerful trebuchets continued to be a weapon of choice by Venetians and Genoese right up to the end of the century. Blockade by siege fort was a favourite tactic of the semi-nomadic Ottoman Turks. Changes in technique of attack seem to have been minor. For the defence, there continued to be a greater reliance on setting up outworks to keep the attackers further away from the curtain wall, often in the form of temporary palisades, and rapidly constructed outlying redoubts. Their use has been observed everywhere from the wars in Brittany to those in the Veneto to those in Lithuania. In general, it amounted to a change in defensive *tactics*. The first line of defence was being

1 Caresinis, 44–47; Gatari, 190–5; Chinazzo, 109–28.

2 All from Chronicon Bergamense Guelpho-Ghibellinorum, *RIS* NS, ed C Capasso, 1940, XVI/II, 20, 79, 142–3, 168–9, 174, 175, 185–88.

deliberately moved away from the massive masonry of brick walls and towers of the main rampart, even where it had been doubled with an outer, concentric wall, another regular improvement undertaken where the resources allowed. In the next period, there would begin, quite suddenly, a change in the design of permanent fortifications too. The biggest change, noticed in the west as a result of the fighting in the hundred years war, was the provision of specifically designed gun ports in towers, to allow defenders to fire small cannon at attackers who were close to. In the fifteenth century, there would be a far more radical change to the nature of defences.

These developments were in response to the increasing size and power of guns in the last quarter of the fourteenth century. Improvements in iron-working made it possible to construct larger and stronger guns and to increase the ballistic force, previously weak for numbers of reasons, including the risk of exploding the gun itself. The quality (fineness) of the gunpowder mix was also being improved. Stone shot, previously the same as that thrown by catapults, was sometimes replaced with iron shot, which had a much greater weight to volume ratio. These improvements together added up to making for rather more effective guns as the century advanced. An example, quite reliably dated to 1375, is the bombard that resided at Boxted in England for centuries. 2.38 metres long and with a bore of 343mm (13.5 inches), it was built of iron staves bound with iron hoops. It had a breech chamber much narrower than the business end, the bore of the powder chamber being 102mm.[1] In appearance, therefore, it was very similar to the large (but not measured) weapon that so impressed the Trevisan chronicler cited above. Accounts of the duchy of Burgundy for 1376/7 show the size of shot had increased to between 50 and 130 pounds (23 to 59 kgs), while the even greater weight of the shot used at Chioggia had clearly merited the attention of the chronicler. Apart from the example given, hardly any weapons themselves actually survive, but there are illustrations in many dated manuscripts that suggest the same trend. While most weapons continued to be small and therefore ineffective against defences, larger weapons were beginning to appear.[2]

One consequence of the construction of significantly larger weapons was, at the other end of the scale, the development of guns designed specifically to be held and operated by a single person. The last quarter of the fourteenth century produced some (quite reliably) dated survivals of the earliest of such types, such as the two pieces found at Tannenberg castle in Hesse, a robbers' haunt destroyed by the neighbouring rulers in 1399. Both are of similar design and dimensions. The one studied most closely is an eight-sided bronze weapon 320 mm long, made up of 154 mm of barrel (with a diameter of 17.3 mm) with a narrower powder cham-

[1] H Blackmore, 'The Boxted Bombard', *Antiquairies Journal*, 67, 1987, 86–96.

[2] R D Smith, Artillery in the 100 Years War, in Curry, Hughes, *Arms, Armies and Fortifications*, 154–6; A Norman, 'Notes on some early representations of guns and ribaudequins', *Journal of the Arms and Armour Society*, VIII/3, 1975, 234–6, reproducing MS illustrations of 1390/1410, 1397, 1400. The evidence from Burgundian sources is analysed in R Smith, K De Vries, *The Artillery of the Dukes of Burgundy 1363–1477*, 2005, 11–18. See also the authorities cited in p. 117 n. 3 above for discussion of other evidence.

ber behind, and beyond that an attachment for a (presumably) wooden handle. In a test firing of a replica, a 28 gram shot of 16.8 mm diameter went through 0.5 mm of sheet iron, and penetrated three to six cms into spruce wood, at a range of 25 metres. Another bronze gun was found in the Baltic area of similar design but smaller (length 191 mms, calibre 20 mm). This one's spike attachment below might mean it was to be fixed to a ship, or placed in a slot in a wall. Another gun recovered from the river Weser (the "Bremen Büchse") is not so easily dated except by comparison, but, not having a separate powder chamber, may be even older. This gun (also bronze) was 385 mms long with a calibre of just 13 mm, and weighed 1.7 kgs. Such weapons could play a very important part in close-quarter fighting, and were of much use to defenders of a fortress.[1]

Far greater developments lay ahead, but by 1400 rulers were now able to call upon large quantities of gunpowder artillery of many sizes and of increasingly devastating effectiveness. The impact on battles would remain limited owing to slow rates of fire and lack of manoeuvrability, but the impact on stone walls would become quite devastating in the new century.

[1] P Galperin, 'Die "Bremen-Büchse"', *Deutsches Waffen-Journal*, 1/1987, 48–9; J von Wlassety, 'Combat anno 1399. Welche Schlussleistung besass die Tannenberg-Büchse?', same journal, 6/1977, 719–21; P Dannecker, J von Wlassety, 'Combat anno 1399. Das Fragment der zweiten Tannenberg-Büchse', same journal, 10/1981, 1444–46; P Post, 'Das gotische Faustrohr in der Leibrustkammer in Stockholm', *Zeitschrift für Historische Waffen- und Kostumkunde*, NF 3, 1930, 140–1. On the castle, J von Hefner, J W Wolf, *Die Burg Tannenberg und ihre Ausgrabungen*, 1850.

4 The age of Timur "the world conqueror": the fourteenth century in the East

The Mongol Empire and the emergence of Timur

We have surveyed the sieges of Europe (and Rus'), north Africa and the middle east during the first decades of the century in which gunpowder weapons appeared, a time marked by social upheavals and marred by plagues and apparently incessant warfare in every country. During this same period, the history of Asia came to be dominated by the extraordinary career of a single man, the creator of the last great nomad empire. The empire of Timur would stretch from the border of China to Asia Minor, from Delhi to the Russian steppe, and never during his life was this remarkable leader ever defeated. At his death, he was preparing an invasion of China, where the Ming dynasty had ruled since 1368, and who can say that the immense forces of Timur would not have repeated the conquests of his predecessors Genghis and Kubilai Khan had he survived a little longer? Timur has been an object of fascination or horror ever since. But his empire, constructed with extraordinary political skill as well as military prowess, would not survive him, as it was based on creating loyalty to him personally.

The Mongol empire, now divided into several khanates, disintegrated as rival local dynasties became established. The Golden Horde had divided into several parts, with the Blue Horde becoming established east of the river Volga, and the regime of the Ilkhanids based in Iran had fallen apart altogether after 1335. The situation was promising for an ambitious and capable leader. The early part of Timur's rise saw his progress from being, basically, a bandit with a small following among the Barlas tribe in that part of central Asia assigned previously to Ghengis's son Chaghadai, to the capture of sole power in that region, with a base at Samarkand, during the 1360s. By 1370 that task had been largely accomplished and the new emir (who had found himself a puppet Khan from the old ruling family to front his rule) could begin his path of wider conquest.[1] During these years, there was little reason for siege warfare. The forces involved were nomads, and the settled populations of the towns and cities either were not involved in the conflicts at all, or else were inclined simply to admit the victors. There was a rare example of a siege in 1366 when an army from Moghulistan (the original Mongol

[1] B Forbes Mainz, *The rise and rule of Tamerlane*, 1989, is an authoritative study of the politics and methods of the competing forces of central Asia.

homeland in the east) invaded Transoxiana and was resisted by the citizens of Samarkand, which "at that time had no citadel", but they were inspired to erect defences consisting of chains across the streets. The attackers did not have the opportunity to reveal any skills, and the shortage of food for the large population would soon have led to surrender, had not disease began to kill off the Mongols' horses and forced them to withdraw.[1]

Timur recognised the importance of maintaining a capital city appropriate to the status he intended to acquire, and quickly set about organising to provide Samarkand with defences. According to the chronicles, and witnessed also by a Spanish ambassador some thirty years later, there were two elements. An earthen rampart and a deep ditch surrounded the city, outside of which there were numerous suburbs, gardens and orchards. Inside, there was a citadel protected by water-filled ravines, in which the central treasury would be secured along with an arsenal.[2] Later, having also secured the cities of Bukhara and Balkh, he ordered a palace built at Kech and also a set of walls (an earthen rampart and a deep ditch crossed by drawbridges). At Bukhara, there was (by 1405) a deep wet moat and walls of sun-dried brick, "for in this country there are no stones for building the masonry of walls" (both descriptions by Clavijo).[3] Ibn Arabshah described the steps taken to fortify the eastern frontiers. Timur sent his grandson Mohammed beyond the river Jaxartes to construct several forts to protect against Moghulistan, the most distant being at Asbara, a whole month's journey from Transoxiana, which was provided with "an impregnable citadel".[4] The purpose of such work, it must be assumed in the absence of other information, must have been to provide secure bases for Timur's garrisons from which the rule of his governors could be effectively maintained.

As with his predecessors, Timur would take advantage of the skills of the craft workers of cities he conquered by uprooting them and sending them to beautify his capital. These would also include armourers and weapons builders. The accounts of his victories are themselves largely silent about who carried out the tasks of capturing the cities and forts, and this information will have to be deduced. But it is safe to say that the digging, woodwork, mining and operation of the artillery almost certainly involved people other than the highly skilled nomad horsemen who formed the core of Timur's armies. The leader himself, however, and his generals, clearly knew how to set about a siege. Since the time of Genghis Khan, the highly cultured and educated Persian-speaking city-dwellers of Iran had provided the civil administration for their Mongol overlords and, perhaps, much else besides.

The first years after the consolidation of Timur's rule in Transoxiana witnessed

[1] Chereffedin Ali, *The history of Timur Bec*, trans. J Darby, 1723 (Sharaf al-Din, *Zafarnama*), 1, 64; Mirza Mohammad Haidar, *A history of the Mongols of Central Asia* (Tarikh-i-Rashidi), ed. Elias, Ross, 1895, 37.

[2] Sharaf al-Din, 1, 134; G de Clavijo, *Embassy to Tamerlane, 1403–1406*, trans. G le Strange, 1928, 287–9.

[3] Sharaf al-Din, 1, 134, 200; Clavijo, 302.

[4] Ahmed ibn Arabshah, *Tamerlane*, trans. Sanders, 1936, 47.

cities such as Isfahan, Sheraz and Kerman taken during the 1370s but there is no detail of how. In 1380, the Khurasanian town called Fordiendge by Sharaf al-Din was attacked by the Mongol forces. This was protected by a wide, wet ditch but this did not save the defenders. Timur's troops spent three days preparing their equipment, then launched their assault on the fourth. Despite the strength of the defences and their supply of stone throwing artillery and of bowmen, the attackers crossed the ditch using rafts, reaching the foot of the wall, and fought their way in from all sides. One must conclude that the attackers must have had overwhelming numbers to achieve this outcome.[1] The following year, the important city of Herat yielded as soon as it had been surrounded. Perhaps the massacre of the defenders during the previous siege had worked its effect on Herat's inhabitants, who were spared.[2] The strongly-sited mountain town of Terdniz, however, perhaps confident in the effectiveness of its high walls and deep ditches, and its provision of stone throwing artillery and stock of supplies, did not surrender. Timur therefore ordered the construction of battering rams, and set his engineers to work. Their first task was to drain the ditch, which was accomplished. Then the miners went to work, bringing down the breastworks and walls. At this point the defence gave in, and the victor spared their lives.[3] In 1383, it was the turn of the important city of Sistan. Once again, Timur's complete mastery of standard siege technique is apparent. Having driven back a spirited sortie by the defenders, the emir decided to protect his encampment and siege works against a repetition by erecting a palisade and digging a ditch. A further attack launched at night by the inhabitants was defeated, and the next day the Turco-Mongol army prepared to storm the walls. They succeeded in breaking down the gates, but the citizens continued to resist and were duly massacred for their pains.[4]

Heading west: Azerbaijan, Georgia, Russia

Timur now felt able to seek new lands to plunder or absorb. Toktamish of the Golden Horde tried to make an alliance with the ruler of Azerbaijan but this just exposed the latter to a deadly attack, the city of Tabriz falling to a siege after just seven days.[5]

From here, Timurid forces marched into Armenia and Georgia, treating the local populations with particular brutality. Timur's thinking seems to have been based on a calculation about how best to achieve the capture, a result that might sometimes be encouraged by the demonstration of clemency, at other times by complete absence of mercy. The Armenian and Georgian chronicles told of forts and towns captured and destroyed along with their inhabitants, and whole

[1] Sharaf al-Din, I, 206–8.

[2] Sharaf al-Din, I, 210.

[3] Sharaf al-Din, I, 225–6.

[4] Sharaf al-Din, I, 240–2.

[5] Ahmed ibn Arabshah, 61–2; Neve, ed., *Exposé des guerres de Tamerlan et de Shah-Rokh dans l'asie occidentale, d'après la chronique arménienne inédite de Thomas de Medzeph*, 1860, 30; and T'ouma Metsobets'i's *History of Tamerlane*, trans. Bedrosian, 1987, 5.

regions utterly devastated, although many inhabitants took refuge in the mountains. There are references to refuges, as well as caves and other inaccessible places, provided for such use. The history of the Georgian and Armenian realms would suggest that refuges for the local people were a commonsense precaution. Timur himself came to Tiflis (Tbilisi), and laid siege to the city in which the Georgian king Bagrat v (1360–95) was penned, with his queen and household, while other nobles dispersed to their own strongholds elsewhere in the country. The city had strong walls and the Georgians were prepared to defend themselves. The invaders were well provided with engineers, as the latter were recorded as preparing engines for the assault, and the town itself was quickly carried. The prince retired to the citadel, and, according to a rather sycophantic Georgian account, organised a successful sortie against the besiegers before, his position hopeless, he negotiated a capitulation. The Georgian chronicle suggests that Timur organised an assault by his troops with equipment to sap breaches in the wall. The king was taken off into captivity, only to reassert his independence when he was released.[1] On this first occasion, the Mongol leader was not planning on staying in the region.

In 1387, he was back, attacking Van, the governor shutting himself in his lakeside mountain top castle and refusing to surrender. The fortress stood at the highest point of the ridge occupied by the apparently then unwalled city, defended by double walls still based on massive ancient masonry. Timur ordered the construction of rams and stone throwers, the first time there has been an explicit mention of his use of these weapons although we can be certain they had been used before. Here, the defence lasted 20 days before hunger brought surrender, and, since the defenders were regarded by Timur as rebels, they were all hurled to their deaths from the mountain they had bravely sought to defend. The same year, the inhabitants of Isfahan also rebelled, and the emir showed them similar treatment, the city fell to assault and 70,000 skulls were collected.[2]

Some of Timur's campaigns were aimed at conquest, others were massive plundering expeditions,so the fate of those who stood in his path was already mainly predetermined. In 1393–4, the emir returned to Mesopotamia. The sultan of Baghdad fled. The strongly fortified town of Takrit (Tikrit) resisted. Timur was advised that it was too strong to be taken by battering rams and engines, the sort of advice usually calculated to be seen as a challenge and to ensure that the attack was particularly ruthless. The outer defences were sapped and stormed, the defenders taking refuge in their rock-sited citadel. It sounds as if a massive battery of stone throwing artillery was arrayed, the bombardment wreaking destruction on the buildings inside, but the crucial element of the assault was the mining. No fewer than 2000 diggers went to work under cover of the bombardment, props were installed and fired bringing down a length of the wall, and despite the brave defenders' efforts to block the breach, the fortress was stormed and the garrison all

[1] Sharaf al-Din, 1, 267–8; Medzpeh, ed. Neve, 17, and ed. Bedrosian, 9–10; Brosset, *Géorgie*, 1, 651–7; Brosset jeune, *Chronique Géorgienne*, 1830, 1.

[2] Sharaf al-Din, 282, 290–2; Medzpeh ed. Neve 45, ed. Bedrosian, 20–1. Van citadel: Sinclair, *Eastern Turkey*, 1, 177–85 (sketch plan at 181). Van was substantially rebuilt under the Ottomans.

killed. The entire operation had taken just over a month (November–December), and the scale of the mining operation was indicative of the effective level of organisation in the victorious army.[1]

Powerful cities such as Diyarbakir, Mosul and Edessa (Urfa) were captured with ease and in some cases devastated, but Mardin was (apparently) taken by surprise. Timur ordered his troops provided with shields and helmets, then took the town by escalade from all sides at once, there presumably being no ditch to prevent the ladders being carried up to the foot of the wall. The citizens retreated to their citadel, whence they either resisted or surrendered according to the source used. Amida, with walls of immense size and strength liberally provided with projecting towers, was rendered vulnerable by the fact that the river Tigris was low enough to be waded across. Timur's troops were able to launch a direct attack with their sappers under cover of archery, and breaches were quickly made through which assaults were launched and the city taken after just three days.[2] From here, attacks were made on Erzincan and Avenik, which were themselves taken and laid waste. The latter was held by the local power, a Turcoman tribe, and once again, having had the outer defences overrun, their ruler sought refuge in a rock-sited citadel and refused an offer of quarter in exchange for surrender. Timur therefore ordered stone throwers set up, and under cover of bombardment reverted to an ancient siege technique by ordering the building of an earth and timber mound to reach higher than the defences. According to Sharaf al-Din, this was achieved within just three days, and then the artillery was moved up the mound and bombarded the town with so many stones "it seemed a mountain had fallen on their heads". Meanwhile, sappers had been working away at the foot of the rampart with success, at which point the enemy submitted, but were massacred anyway.[3]

There was no power able to field an army against the invader who was therefore free to march where he chose. Timur's own army was thoroughly attuned to the techniques of siege warfare, and evidently he had ready access to siege equipment of all kinds, from ladders and mantlets to stone throwing artillery and battering rams. Sapping and mining, however, clearly featured as the most effective methods used.

In 1395, Timur launched a devastating plundering expedition across regions already devastated in his previous attacks, defeating Toktamish at the river Terek. Southern and central Georgia, lying in the emir's path, again faced destruction, its mountainous terrain with its fortified caves, towers and castles being unable to withstand the conqueror's overwhelming strength. The chronicler recorded one instance of Timur's troops assembling ladders that had to be joined together to

[1] Sharaf al-Din, 444–7; ibn Arabshah, 63–4. The use of administrators called *tovachis* to supervise and organise such tasks as the mining at Takrit is revealed in the untranslated *Zafarnama*, quoted in Manz, *Rise and rule*, 173–4.

[2] Sharaf al-Din, 464–6; ibn Arabshah, 65–8; Medzpeh, ed. Neve, 58–9, ed. Bedrosian, 34.

[3] Sharaf al-Din, 475–7; ibn Arabshah, 69.

deal with the height they faced, and of troops ascending a mountain in order to drop down into the castle of Birthwis from above using rope ladders.[1]

Before returning to the Middle East, Timur next brought the benefits of his presence to his neighbours in the Indian subcontinent.

Siege warfare in India before and during Timur's onslaught

India continued to be divided, culturally and politically, between Hindu and Muslim dynasties. Conflict between and among these dynasties was commonplace, added to in the later thirteenth century by struggles between the Muslim rulers and the regimes of Genghis Khan's successors in the west. In due course, Fīrūz Shah (ruling from 1355 to 1388) established a period of rule in Delhi so effective and culturally rich as to attract the attention of many contemporary writers, and it would be the relative decay of the power of the sultanate on his death that would draw the greedy attention of Timur to the area in 1398.

To place the invasion in context, however, we should examine the preceding period too, and are fortunate that a number of sources have been translated covering the history of the region. In 1276, when the sons of the ruler of Jessulmer unwisely ambushed a convoy of tribute on its way to Delhi, the city had to make preparations to resist the expected revenge attack. It was recorded that great stores of grain were laid in, and stones were collected and piled up all around the walls to hurl at any besieger, the elderly and infirm were evacuated to distant parts, and the surrounding countryside laid waste to deny resources to an enemy. The garrison inside amounted to 5000 men, defending 56 bastions, while another force was based outside, to harass or restrict an attacker. The chronicler records fierce fighting as the Delhi troops sought to construct a blockading entrenchment, but also notes that the siege last eight years, which, if accurate, rather implies a blockade than a close siege. The city finally fell in 1297, and its walls were demolished.[2]

Ala'uddin Khaljī, nephew (and murderer) of his predecessor, took over as sultan of Delhi in 1296 (ruling until 1315), but had to overcome the resistance of the garrison of the citadel. This he did by the novel tactic of using his artillery to project bags of gold over the walls, presumably illustrating his wealth while also buying the support of the soldiers. In 1299/1300, he faced a determined opponent in the Hindu Rajputs defending Ranthambhor. It may be that they were particularly optimistic, having seen off a previous Muslim attack in 1290, despite the preparation on that occasion of mangonels, trenches and siege redoubts. The thirteenth century towered wall defending the place followed the contours of the hilltop on which the city stood, around one mile long by half a mile wide (1.6 by 0.8 kms) with the single approach protected by several consecutive gates and a bent-entrance into the town itself. The attackers were led by Nusrat Khan, who

[1] Sharaf al-Din, 506–8; Brosset, I, 674–5, describes the same incident but ascribes it to the Timurid invasion in 1401. Brosset jeune, 2.

[2] Annals of Jessulmer, in Tod, *Annals and Antiquities of Rajasthan*, II, 198–201.

ordered the construction of a siege mound. From the top of this construction, it is told, stone throwers were set up to bombard the defence. Nusrat himself was struck and killed by a stone from the defenders' artillery, specifically described as being of the *maghribi* (translated as "royal western") variety. This brought the sultan to oversee the works in person. Bags were filled with sand to fill up the ditch, redoubts were raised and artillery set up to bombard the defence with stones and naphtha shells (made of earth), which fell like hailstones. Fireworks of some description were employed by both sides, but an accurate interpretation of the nature of these weapons is not possible from the almost incomprehensible translations of the florid original Persian-language manuscripts of the sources, despite some arguments to the contrary. Whatever the truth, there was starvation inside, bringing about the place's capture (July 1301).[1] Ranthambhor's location along with its stone walls had obviously prevented a direct assault. Ala'uddin also successfully besieged the Rajput fortress of Chitor, but here he did so by forming an effective blockade lasting two months, and this proved to more effective than the accompanying bombardment by stone throwers, leading to a mass suicide by the defenders (August 1303). Here, once again, the natural strength of the spacious hilltop site with steep sides had been enhanced with powerful crenellated stone walls and the approach was provided with no fewer than seven defended gates.[2]

In 1308, the sultan sent an army against Siwana, the commander of which fled when the "western mangonels" were set up against it. These were reported as "making a cave in the wall with every shot", but the place was stormed when a siege mound was constructed that reached to the wall top. The following year, the target was Arunepal, a large fortress with walls made of mud, but so solid that "a spear could not pierce (it) . . . and a ball from a mangonel would rebound like a nut which children play with." Both sides had *maghribi* and exchanged shots for several days, the defenders hurling out naphtha from which the Muslims had to protect themselves with shelters. The attack was difficult, and initially the Muslims' stones all "flew high". Eventually, the ditch was filled with earth, and the attackers' stone throwers finally succeeded in battering the wall to such effect that long lengths of it were crumbling into the ditch and no longer offered any protec-

[1] Abdullah Wassa'f, Tazjiyatu 'l Amsar wa Tajriyatu-l Asar, 41 (1296); Amir Khusrau, Tarikh 'l Alai, 75; Zia'd el Din Bami, Tarikh-l Firuz Shah, 146–7, 175. All in Dowson, Elliott, *History of India,* III, 1871. There is a slightly different translation of Amir Khusrau by Mohammed Habib in his collected writings: *Politics and Society during the early medieval period,* ed. K Nizami, 1981, II. See 183–4 for Ranthambhor. The same source is used by I A Khan, employing a third different translation of the original, to support the case for the defenders having access to gunpowder weapons: and recently in 'The role of the Mongols in the introduction of gunpowder weapons and firearms in South Asia', B Buchanan, ed., *Gunpowder: the history of an international technology,* 1996, 36, and in *Journal of Asian History,* 30, 1996, 40. The likelihood of there being gunpowder in use by this date, brought from China by the Mongols, was considerable. But Khan goes beyond what the evidence supports, and weakens his case further by assuming that medieval mining involved planting gunpowder under a wall, a much later technique. Description of the defences: S Toy, *The fortified cities of India,* 1965, 75–8.

[2] Annals of Mewar, in Tod, I, 210–15; Amir Khusrau (trans. Habib), II, 187–9. Description of Chitor: S Toy, *The strongholds of India,* 1957, 84–90.

tion. At this point, the Muslims managed to secure scaling ladders to the walls of three bastions in a nocturnal escalade, and the commander (who had retired to a powerful stone-built citadel inside) agreed to surrender with a great tribute that required convoys of elephants to convey it (including, apparently, the famous Koh-i-noor diamond).[1]

Fortress building in the subcontinent

Both the sultan and the Hindu princes devoted resources to their states' defences. The walls of Delhi and other fortresses were repaired, an arsenal of artillery weapons (*manjaniks, 'arradat*) was strengthened, and experienced engineers were taken into service.[2] The reason given for this work was anxiety about the prospect of attack by Mongol armies, but the marauding invaders seem to have been successfully repelled by the sultanate's very large field forces. However, the work of the sultans in ensuring that their rule was reinforced by such defences continued under the enlightened rule of Fīrūz Shah. Fīrūz ordered the establishment of a fortified city to be called Hisar Firozah, where a fort "of great extent and height" was built of hard brick over a number of years, then surrounded by a ditch. Inside was created a palace, and a large cistern fed by water brought from the rivers Jimma and Sutlej by canal. Fīrūz's building work also took place on his northern frontiers, where one more distant witness recorded the construction of new towns near Ray and Jurjan, and a blocking wall beyond the river Oxus, between Iran and Turkestan.[3] Altogether, the sultan built four new fortified cities, at Firuzabad, Jainpur, Fathebad and the new citadel at Delhi, mentioned above, which was a rectangle with dimensions of one half by one quarter mile (approximately 800 by 400 metres), and featured high battlemented walls and frequent towers. The gateway was further protected by a barbican.[4]

Fīrūz also operated offensively. In 1359, he invaded Bengal and besieged the local sultan, Sikander, who took refuge on the island of Ikdala. Fīrūz blockaded the island from both sides, with batteries of artillery. One of the bastions of the wall suddenly collapsed. The chronicler explains that this was because it was made of mud, and could not sustain the weight of men standing on it. It was not this mishap that brought proceedings to a close, however, but shortage of food, which induced peace negotiations. At the siege of Nagaikat, Fīrūz drove the ruler back into his citadel. Both sides had large number of stone throwers, "so many that the stones clashed in the air". After a six month siege the defenders submitted.[5]

[1] Amir Khusrau, 81–4, and trans. Habib, II, 187–9 (Siwana), 205–13 (Arunepal); Zia'd el Din Barmi, 202. Khusrau offers the most detailed account.

[2] Zia'd ed Din Barmi, 191.

[3] Shams-i Siraj 'Afif, Tarikh-'l Firuz Shah (Dowson, Elliott, III), 299; Annals of Hamzah al-Isfahani, trans. O M Daudgupta, *Journal of the K R Cama Oriental Institute*, 22, 1932, 102.

[4] Percy Brown, *Indian Architecture: Islamic period*, 1968, 16–23; on Delhi and its eight distinct sets of fortifications, see C Asher, Delhi walled: changing boundaries, in J D Tracy, ed., *City Walls. The urban enceinte in global perspective*, 2000, 247–61.

[5] Shams-i Siraj, 308–9, 318–19.

Siege warfare in the subcontinent, then, was on a very large scale, with the rulers having the resources to deploy large amounts of artillery along with their large scale armed forces generally. They must, therefore, have had the requisite corps of engineers. The defences around which they were deployed were themselves of considerable strength, and must have been large to accommodate the numbers of people regularly mentioned in the accounts. Hindu building methods did not however include the vault or the relieving arch, nor did they use mortar, relying on gravity, and precision stone cutting, for the masonry. Architecture also played a very significant symbolic and religious role, with many highly decorative features. At the same time, the accounts leave little doubt that these dramatic walls and towers, provided with regular towers, battlements and wall walks, were also intended to provide a defence when necessary. The remains of the walls of the city of Dabhoi have been studied. On a pilgrim route, it certainly had a major symbolic function. But its stone defences, originally built in the tenth century as a protection for the site against mountain tribes, and restored in the thirteenth against the more substantial Muslim threat, seem to have a serious military function too, and it is noticeable that the outer gates in the double circuit of walls are considerably less decorated than the inner. There were 52 shallow rectangular bastions placed every 60 or so metres, and circular bastions at each corner. The whole wall was crenellated, and rose about four metres (Toy measured it as more than five metres at some points) to the level of the wall walk, which was nearly three metres wide, though it stood over a colonnade beneath. Dabhoi belonged to the Hindu kingdom of Achillawada, before its conquest by the Muslims, and the same kingdom was also responsible for the frontier fortress of Jhinjuwada, dated to the same period, and was very similar to Dabhoi but less ornate.[1]

The Hindu kingdom of Mewar successfully resisted the attentions of the sultanate. The annals specifically record that 84 fortresses were built for this purpose, of which 32 were thrown up by the king who succeeded in 1419. The largest, Kuombhomer, was placed impregnably on a hill top, while the others were located in the passes on the country's western frontier.[2]

Timur's invasion of India

The country invaded by Timur in 1398 was a rich, populous trading empire ruled by a Muslim élite presiding over a Hindu and Buddhist population, in and around the borders of which were many smaller but still substantial non-Muslim dynasties. One of the emir's grandchildren led the attack on Multan, organising a continuous bombardment and two assaults a day. Timur himself followed up, and also sent a force to attack the city of Batnur, described as a powerful

[1] Brown, *Indian Architecture: Buddhist and Hindu periods*, 1971, 65–6, 123–4; J Burgess, H Cousens, *The antiquities of Dabhoi in Gujarat*, 1888, 1–13; J Burgess, Memorandum on the antiquities of Dabhoi, Ahmedebad, Than, Gimar and Dhank, *Archaeological Survey of Western India*, II, 1875, 25–6; Toy, *Strongholds*, 73–9.

[2] Annals of Mewar, ed. Tod, I, 231–2.

and populous city, ruled by a rajah. The invaders attacked the city from two sides at once, forcing the defenders back into the citadel. The besiegers took position against the ditch. After an attempt at deceiving Timur had failed, the assault was renewed, with mining being undertaken to drive passages beneath the wall in order to undermine it from all sides. The garrison defended themselves bravely with stones, arrows and fireworks, but the wall was breached and, remarkably, the capitulating garrison was spared according to one account, and massacred according to another, which seems more in character.[1]

The sultan of Delhi, Maḥmūd II (1394–1413), then challenged Timur in battle near Delhi, his army including 125 armoured elephants carrying, it is said, "throwers of grenades", fireworks, and rockets. Against the military genius of Timur, and the size and ability of his army, this weaponry did not prevail, the sultan fled, and Delhi fell to the conqueror though without a siege, which was fortunate for Timur because it was observed to be "a great city" with too great a circuit even for his vast army (January 1399). The Hindu religion of the majority of the population gave the conqueror the excuse to perpetrate another appalling massacre, while carefully ordering that artisans and master mechanics be spared for transportation to Samarkand, to work on buildings to glorify the ruler. The town of Mirat refused to submit, and Timur sent a corps of 10,000 troops to enforce his control, and while the chronicle describes them as cavalry, the attack they carried out was done by sappers undermining the wall. While this work was in progress, however, the walls were stormed by an escalade with ladders and ropes. The excavations were not allowed to go to waste, being used instead as a quick method of demolishing the captured rampart.[2] Timur did not attempt to remain in India. Devastating though his incursion had been, it had no long term impact.

Return to the west: Mesopotamia, the Caucasus and Syria

The news that Timur was in India may have encouraged the Ottomans to prepare their first great attack on Constantinople — but the respite was brief. Starting a new campaign in the same year, 1399, Timur's forces marched through Azerbaijan and Georgia, again capturing Tiflis with a sequence of events remarkably similar to those of the previous conquest, if the later Georgian chronicle has not confused them. The castle of Zerit, on a ridge, was among many that the Georgians tried to hold, and here the besiegers deployed machines and rams to such effect that it fell to assault in seven days, everyone being slaughtered. The same fate was meted out to many other forts and towers in the unhappy country.[3]

Next (1400) was Sebasteia (Sivas), which was encircled at the beginning of September. The old Seljuk defences were still strong, with walls 20 metres high

[1] Sharaf al-Din, II, 5, 41–4; Malfu'za't-i Timur, in Dowson, 422–4 (a source using material from the Zafarnama).

[2] Ibn Arabshah, 99; Sharaf al-Din, 61–2; Malfu'za't-i Timur, 439–51.

[3] Brosset, I, 666; Sharaf al-Din, II, 141; Ibn Arabshah, 105.

on strong foundations, wet ditches on three sides, and seven gates made of iron. Timur's engineers began to erect a high mound from which stone throwers hurled missiles into the town. After just 18 days, breaches began to appear. The attackers also mined under wall towers on the one side that this form of attack was possible, and propped the pits up with wooden stakes. Once they had been fired and an attackable breach created the garrison expelled their governor and offered to surrender. Having promised not to shed the garrison's blood, Timur had them buried alive instead.[1] From here, Timur was able to challenge the Mamluk dynasty ruling Egypt and Syria, a campaign that took place between autumn 1400 and spring 1401.

Between Malatya and Aleppo, the (Turkish) garrison of the Syrian castle at Behesna had the temerity to shoot a catapult ball at Timur which rolled into his tent. Setting up his own battery of 20 machines, it is said that the first shot hit and destroyed the offending weapon. Treating this as a good omen, the attack was launched, the towers mined (underneath "resembled a honeycomb" – Khandmir) and the place surrendered (7 October 1400). Brushing aside the Mamluk army, Timur then reached Aleppo, entering and sacking the city (11 November). He had with him elephants brought from India which were used to carry wooden structures holding archers and "flingers of wild fire". At Aleppo, the governor had retreated into the massive citadel, and his garrison hurled down this "wild fire" along with sulphur bombs, stones and arrows. Timur organised an attack from all sides and the citadel surrendered, having been undermined inside two or three days. Some sources mention the now familiar towers of human skulls that were created in the aftermath.[2]

A Timurid force was despatched to take Hama. The governor held out in the citadel until Timur himself arrived, at which point, prudently, he surrendered. Homs submitted and so did Baalbek. The sultan meanwhile was preparing to make a stand at Damascus, where field fortifications were being constructed to strengthen the defences of the city and citadel, now nearly two centuries old. On 19 January 1401, the Mamluk army was routed and the sultan retreated to Egypt. On 23 January, Damascus submitted to Timur, but the garrison of the citadel resisted. The Timurid chronicler noted how strongly-built were its walls, and the scope of its ditch. The defence was also well provided with artillery and naphtha bombs. Three raised platforms were created to command the fortress. At the same time, other troops filled in the ditch, enabling mining to begin, under the shelter of Timur's stone throwers which also used fire bombs. The siege revealed the true strength of the Ayyubid fortress as well as the courage of its defenders. The sapping succeeded in creating cracks in the rampart. A mine brought down one of the towers, but the tumbling masonry also killed 80 of the attackers, giving the defence time to block up the breach. Mines were being driven in from all sides, and more were fired. With the defeat of the Mamluk army, there was, of course,

[1] Sharaf al-Din, II, 151–4; ibn Arabshah, 116; Khandmir (Kwandamir), *Habeeb-us-Siyar* (Life of Tamerlane), 30, translation of parts V and VI, 1900, was written about 1475 but contains much detail from the Zafarnama.

[2] Sharaf al-Din, II, 162–4, 174–6; ibn Arabshah, 116, 124–5; Khandmir, 31, 35.

no prospect of relief, and the governor now surrendered (after 43 days of siege), his brave troops being enslaved.[1]

Damascus was sacked, but Timur did not overthrow the Mamluk empire. Instead, he had to return east with his army to recover Baghdad which had revolted. Investing the city on all sides, he used by now familiar techniques. Platforms were raised for his artillery, and miners went to work. The desperate defenders tried to repair the breaches as they were created, but the great stones hurled from the engines were devastating, and on 23 July 1401, in the full heat of midday, the final assault was launched, the defence was overrun, and the victors claimed that 90,000 were killed in the massacre.[2]

Timur's violent suppression of all resistance continued. Georgia, for the third time, had to suffer Timurid invasion, the castle called Tartoum being stormed and razed to the ground, its garrison of 200 being killed, and the garrison holding in the citadel at Arzinjan resisting long enough to negotiate surrender with one of the conqueror's offspring. King George VII (1395–1401) submitted. Ibn Arabshah described how one of the mountain caves in which the hapless populace sought protection was taken. The cave roof was secure against "the thunderbolts of the engines of war", and too high to be reached by ladders. Ever resourceful, the Timurids constructed mantlet-shaped boxes which they then lowered with strong chains from the most accessible point above the cave, their troops using the boxes as a vehicle for archers to clear the mouth of the cave and permit them to storm in.[3]

Meanwhile, it is worth noting, because it was such a rare event, one siege abandoned by Timur. In 1401, on his way to Baghdad, he had attacked Mardin in Mesopotamia. The citizens did not try to defend the city but withdrew to the ridge-top citadel, standing high and approachable only by steep paths. Its location meant that "there was no plain for arraying" catapults, so he turned, as so often before, to his miners. However, the rock on which the fort stood must have been of exceptional toughness, because "the hoes were worn away and the hammers blunted" in failing to penetrate the rock, so he gave up the attempt. No doubt the judgement that the recapture of Baghdad was a more urgent task saved the defenders of Mardin.[4]

Ottoman expansion halted by Timur

Timur turned his attention to the remaining large Muslim power, the Ottoman emirate. By the turn of the century, Ottoman power was firmly established on both sides of the Bosphorous, and the descendants of Osman represented a powerful force in both the Balkans and in Asia Minor. The eastern Roman empire relied almost entirely on the support of rival Italian merchant

[1] Sharaf al-Din, II, 178–97; ibn Arabshah, 142–51; Khandmir, 40–1.

[2] Sharaf al-Din, II, 213–16.

[3] Ibn Arabshah, 203; Sharaf al-Din, II, 240–2; Khandmir, 56; Brosset, I, 670.

[4] Ibn Arabshah, 164.

cities, on mercenary soldiers, and on the passive strength of the ancient walls of Constantinople. Manuel II Palaeologus (1391–1425) had won the civil war to inherit the empire, but humiliatingly paid tribute to the Ottomans.

During the 1390s, the emperor decided to throw off Ottoman overlordship, seeking urgent help from Hungary and the west to guarantee continued survival of the empire. The Ottomans decided that the time had come to capture the capital itself. Bayezid I had succeeded his father Murad I in 1389, and the first years of his reign witnessed an extraordinarily successful period of expansion of Ottoman territory, as Bayezid alternated his personal efforts between the two geographical portions of his realm, consolidating control over Thrace, Bulgaria and Macedonia on the one hand, and overthrowing the minor dynasties of non-Ottoman Turks in eastern Anatolia, where he secured Kastamon and Konya among other strong cities, with very little resistance. In 1394, he began the first serious Ottoman attack on Constantinople, by land and by sea, the emperor having to devote resources to the strengthening and repair of the ancient defences. On land the Turks established batteries of stone throwing engines, a later historian recording (very helpfully) that the Ottomans "did not have cannons at this time". The engines were causing serious damage to the crenellations. On this occasion, news arrived of a Hungarian army marching east, the siege was raised and the engines burned, and the army marched north to Tarnovo to plan the relief of the Ottoman garrison holding out at Nicopolis.[1]

This army was a multinational force raised as a new crusade with the support of both rival popes across Europe. A significant element of the army was made up of French and Burgundian knights, and it was their arrogant rejection of the advice of the Hungarians and Wallachians, who had experience of Ottoman fighting tactics, that was chiefly responsible for the outcome: the crusading army went down to catastrophic defeat near Nicopolis (25 September 1396). The battle had come about as a result of a siege.

In 1394, the Ottomans had recovered the small town of Nicopolis Minor on the north bank of the Danube, and put in a garrison to guard against King Sigismund of Hungary, using a great battery of siege artillery which had reduced much of the town to ruin. The crusading army had assembled at Buda in July 1396. An initial target was the town of Vidin, on the south bank of the Danube, where there was a Turkish garrison in a largely Slav town. The French Marshal de Boucicault described how this was attacked "par grant ordonnance", definitely indicating the presence of guns, but in fact the inhabitants rose up and expelled the garrison. The crusaders left a small garrison (and massacred the prisoners). The town of Racho (Rahova, Orjahovo) underwent a similar experience. Boucicault described how his troops collected trees to construct assault ladders, and had driven the defenders back over the town's bridge to prepare for attack, but that on

[1] Giese, *Altosmanischen Chroniken*, 37; Dervish Ahmed (Asik Pasha), 96–8; 'Auszüge aus Nesri's Geschichte des osmanischen Hauses', ed. Nöldecke, *Zeitschrift der Deutschen morgenländischen Gesellschaft*, xv, 1861, 334, 340, 345. Nesri was an Ottoman official writing at the end of the fifteenth century; M Ducas, Historia Byzantina, *Migne PG*, 157, cols. 807–14.

the second day, with the arrival of the main body of the crusader army, the citizens had driven out their small garrison, which had also been slaughtered.[1] Crusader detachments raided south of the Danube, and Nicopolis Minor was recovered by the Wallachians.

Sigismund's main target was the other Nicopolis, on the south bank, where he arrived on 12 September to rendezvous with a fleet that had come up from the Black Sea. The defences of the city have disappeared, but it seems to have been a town with a stone wall and ditch, with a separate citadel on high ground within the walls. It had been selected as the main Ottoman strongpoint in the region. The king set about deploying "great engines of war" but without making much progress in the 16 days before learning of the approach of Bayezid. Boucicault records that the main attack was by means of two mines, wide enough for three armed men to fight abreast in. The French account suggests that the mines had reached the walls when battle intervened. If this was true, it would imply that the ditch was dry, although being adjacent to the Danube this would have been strange, or that the mines were very deep, in which case they had made remarkably fast progress. Ottoman accounts say that the garrison was in good heart and well supplied, and that a Turkish party had ridden up to a spot opposite the citadel and exchanged morale-boosting words with the governor. After the battle Bayezid secured the city and went on to take Silistria as well before returning south.[2]

Bayezid now tried to complete the task of taking Constantinople. The sultan established a blockade, hoping to starve the city into submission, and he ordered the construction of a fortress on the Bosphorous towards this end, and what is described as an observation tower on the Galata side of the Golden Horn. The siege led to many of the residents fleeing, but clearly the blockade was not total, and year followed year without surrender getting any closer. Marshal Boucicault, having been released from captivity, served at Constantinople in 1399, and his biography recorded how he and the Hospitallers of Rhodes led a sea-borne attack on a Turkish garrison in Anatolia, taking the town by a combination of mining and escalade (the ladders constructed with timbers from their ships), despite fierce resistance in which the Turks set up burning wooden barriers to impede the attack on the breaches.[3] The event was not important enough to register in the Ottoman accounts. Still facing the Byzantine capital, he decided once again to try the direct method of conquest, and was assembling a great arsenal of siege equipment, ladders and (possibly) siege towers (rendered as *elepolei* in an anonymous eyewitness account) when the news of Timur's invasion of Asia Minor caused the second raising of the siege.[4] A Venetian report from Crete stated that Timur's

[1] *Le Livre de faits du bon messier Jehan le Maingre, dit Boucicault,* ed. Lalande, 1985, 93–4, 96–8; J Schiltberger, *Bondage and travels,* trans. Buchan Telfer, Brown, 1879, 1–2.

[2] Dervish Ahmed, 98–101; *Altosmanischen anonymen Chroniken,* 40; Thurocz, 48, 56; Schiltberger, 1–2; *Le Livre de faits . . . ,* 102–3; Nesri, 346–7. D Nicolle, *Nicopolis 1396,* 1999, is an excellent English account of the campaign.

[3] *Livre de faits,* 142–6. The account of this siege bears some striking resemblances to that of Nicopolis, leading one to question whether one or both versions are in fact formulaic inventions.

[4] Dervish Ahmed, 101–7; *Altosmanischen anonymen Chroniken,* 47; P Gautier, ed., 'Un

army was 70,000 strong, not counting those carrying the "instruments suitable (*apta*) for fighting against castles".[1]

The ruler of half the world had decided there was only room for one dominant Muslim leader, and having humbled the sultans of Delhi and of Cairo, he now planned to exert his authority over the sultan who ruled from the rather more humble surroundings of Bursa and Adrianople, taking advantage of the resistance of the recently dispossessed local rulers to the Ottoman regime in the east of Asia Minor. Timur marched into the Ottoman realm and laid siege to Ankara. Here, he faced the combined Turkish armies of Rumeli and Anatolia, and crushed them, leading Bayezid off to humiliating and quickly fatal imprisonment. Ankara fell and there followed the usual massacre. Bursa and Nicaea capitulated quickly, not surprisingly in all the circumstances. In December 1402, Timur took an army to Smyrna (İzmir), where the land fortress was held by the Muslims, while the sea fortress was in the hand of Christian troops following its occupation by the Knights Hospitaller. One source dismissively reported that the great conqueror simply made the necessary arrangements to capture the fortress, and did so. Sharaf al-Din provided more detail. The attack on the sea castle took two weeks. The by-now familiar repertory of machinery was put in place: stone throwers including propellants of fire bombs, battering rams (to attack the walls), and mining. By land, the ditch had to be filled in, while by sea wooden floating platforms were constructed to allow his troops onto the water to complete the encirclement of the fortress. The decisive weapon once again was the mining, the completed excavations being filled with fascines covered in naphtha and then lit. The wall above collapsed, and the male occupants were all massacred.[2] The sea castle, which had never previously been captured, was destroyed, and Timur ordered the land castle strengthened and munitioned before leaving.

The Ottoman state was not permanently damaged by Timur, although the settlement imposed by the victor included the return of a number of captured cities (including Thessaloniki) to the Byzantines. It might be judged that the eastern ruler was chiefly responsible for the continued independence of Byzantium into the fifteenth century, and certainly contributed more to this postponement than the western powers. With Bayezid's death, the dynasty underwent a long period of civil strife, which also contributed to the reduction in the Turkish threat for nearly a generation.

Timur however was now planning a conquest of China, ruled since 1368 by a native Chinese dynasty, the Ming. On his route across his domains, Timur had to attack the fortress of Firuzkuh in Ray. It proved to be his final conquest. Arriving on 31 May 1404, he ordered his stone throwers and fire bombs prepared against the ridge-top fortress, and gave instructions to have the river diverted away from the foot of the castle. The garrison of 300 men capitulated after just

récit inédit du siège de Constantinople par les Turcs 1394–1402', *Revue des Études Byzantines*, 23, 1965, 107–9.

1 G T Dennis, 'Three reports from Crete on the situation in Romania 1401–2', *Studi Veneziani*, XII, 1970, 245.

2 Sharaf al-Din, II, 275–8; ibn Arabshah, 192; Khandmir, 68–9; Nesri, 359–68.

two days.[1] The scale of the operations planned by Timur against such a modest target is indicative both of the resources at his disposal, and his complete command of siege techniques. Timur, who had attained a good age (unlike many of those whose paths had crossed his), died at Otrar in February 1405 on the way to launch the invasion of China, which as a result did not occur as the struggle for the succession began at once. Not much information is recorded about his preparations for what would have been an undertaking of immense complexity. He had ordered the assembling of sufficient supplies for four years' campaigning, but even before reaching the border the severe hardship suffered by his troops as they marched in winter and across deserts was noted.[2] Such a campaign would have involved siege operations against some of the largest cities in the world, so there can also be no doubt that a substantial siege train would have accompanied the invasion.

Siege methods in the east in the late fourteenth century

Our summary of the non-stop successes of the last great nomad conqueror provides an explanation for this extraordinary run of victories. At no point is there any indication that Timur had access to gunpowder artillery. Nor did the Ottoman Turks, a judgement confirmed by later "official" historians of the dynasty. It is inconceivable that either regime was unaware of their existence: the Balkan princes and the Italian naval powers facing the Turks, and the Mamluks facing Timur, already used the technology. However the Serbian princes, who definitely had guns in the 1380s, may only have had small weapons, and that these may have been imported rather than manufactured locally.[3]

Nonetheless, its use in the form of cannon had not spread in the east just yet. This may seem strange since gunpowder as an explosive, and as the propellant for missiles from a tube, had really become established in China by the second half of the thirteenth century. Having been available and used to propel firelances, and to provide the ingredients for bombs that were thrown by catapults for some time, gunpowder somewhere became converted into a substance with the right balance of ingredients to act as the propellant for other missiles rather than the missile itself. But whereas on its arrival in Europe and the middle east (from China), it was very quickly used to make guns and gradually but continuously experimented with and ultimately developed into the qualitatively more destructive artillery of the fifteenth century, in the east the development of gunpowder artillery seems to have stalled. There may have been many cannon barrels sur-

[1] Sharaf al-Din, II, 342–4; Clavijo, 172, passed through a couple of years later, and reported its ruin, though he suggested that it had been very strong.

[2] Ibn Arabshah, 225.

[3] D Petrovic, Firearms in the Balkans on the eve of and after the Ottoman conquests of the fourteenth and fifteenth centuries, in Parry and Yapp, *War, technology and society in the middle east*, 1975, 164–90.

viving from mid-fourteenth century China, while the new Ming dynasty would manufacture them in large numbers for posting to their frontiers early in the following century,[1] but the technology does not seem to have spread very rapidly to their neighbours. Indeed, it does not seem that gunpowder artillery was adopted in India until as late as the 1440s, although here there is an isolated (and disputed, see chapter 8 below) reference as early as 1406 to "guns" (in Bengal), which may mean gunpowder weapons, but perhaps only the already familiar rocket, already imported from the north, and known in India as the *ban*, which took the form of an iron tube about one foot long and an inch in diameter fixed to a bamboo rod ten to twelve feet (3 to 3.6 metres) long. Certainly, as noted, there are sufficient different references to show their appearance between the introduction, courtesy of the Mongols, as Chinese-origin fireworks in the 1250s, and the references to fire arms operated by foreigners in the Bahmani realm in 1368 and Vijayanagar in 1365 which all use terms distinct from those used to describe naphtha weapons.[2]

A number of explanations are possible. Least likely is that Timur and Bayezid, both of them dedicated to lives of military conquest, would spurn a new weapon deliberately. More likely is that they were at that time lacking either the people with the appropriate skills, or the practical ability to manufacture the guns or the powder.

The evidence indicates that the available weaponry and techniques were generally adequate. Many of the defences that Timur had to overcome were quite elderly, but even those of exceptional strength could not resist for long. In the very rare cases where capture was not immediately achieved, it was because the location of the fortress prevented use of the normal techniques, and even many of these apparently impregnable sites were taken by using other approaches, including the patient blockade by siege fort. Timur clearly had a powerful train of stone throwing artillery, operated by experts. Time and again, however, the decisive attack was achieved by mining, once more indicating the presence of experts. One feature of Timur's siege technique that was quite distinct from practice further west was the use from time to time of artificial ramps to create a high platform against a defence. This method, as we have seen, was in use both in China and in the Indian states, and it was employed because rulers disposed of sufficient manpower to make it practical. Against the disadvantage that it must have been time-consuming, and certainly necessarily involved much hard physical labour for the builders, the psychological effect on the besieged must have been truly terrifying, and, unlike other devices designed to neutralise the height of a wall such as a siege tower, the earthen ramp was immune to almost any defence except being undermined.

1 Needham and Yates, *Science and Civilisation*, 5, part 7, 276–90, 307–14.

2 For various interpretations of the same limited evidence, P Gode, 'Use of guns and gunpowder in India from AD 1400 onwards', *Studies in Indian Culture and History*, v, 1960, 2–6; I A Khan, 'Origin and Development of gunpowder technology in India AD 1250–1500', *The Indian Historical Review*, IV, 1977, 20–9; and *Journal of Asian History*, 30, 1996, 27–40; J Sarkar, *The art of war in medieval India*, 1984, 121–30.

The evidence demonstrates not just that Timur employed a corps of specialists to build and operate all this equipment, to plan and execute the mining, and to design the siege platforms. To do all this also called for enormous amounts of manpower, and it was only the availability of large workforces (soldiers drawn from his very large armies, presumably, though he may also, like previous Mongol leaders, have pressed local peasants into such service) that made such forms of attack possible. A western medieval ruler would quite simply have been unable to undertake such immense projects of construction or excavation because they would not have been able to mobilise the resources or the personpower. The conclusion must be that the Timurid army was accompanied by a substantial corps of miners and engineers, and that the regular soldiers provided the experienced muscle. Whether the siege machinery was carried with the army or manufactured on site is not clear. On the one hand, the speed of movement of the armies often recorded implies that they were not accompanied by slow-moving baggage trains, on the other the speed of many of the siege operations suggests that the weaponry did not have to be constructed from scratch.

The fate of the defeated enemy has attracted much attention. While the massacre of inhabitants of cities taken by storm was not unusual anywhere in the medieval world, Timur developed the ruthless slaughter of captured inhabitants or soldiers into a careful act of policy, calculating that his reputation might well speed the surrender of fortresses that would otherwise hinder his progress. A study of the Persian-language accounts confirms that policy lay behind the fate of prisoners. Rarely, but occasionally, we have seen defenders' and residents' lives preserved, but the motive was never charity, rather it was designed to encourage others to submit. A town might have been wanted undamaged. But this was not a uniform picture. Sometimes, the policy called for the wiping out of defenders, or their enslaving, and the sack of towns. One reason would have been the need to find rewards for his soldiers; another would have been that the example of terror might bring about easy captures elsewhere. Sometimes this worked well. At other times, defenders who believed they were doomed anyway might be driven to fight to the end. The most brutal mass slaughters took place when there had been rebellion: such as at Baghdad or Isfahan, when the killing was no doubt designed to demonstrate that no one could challenge the great emir and survive. The towers of human skulls that were raised were a visual symbol of the utterly appalling events that had taken place and were a method inherited from Timur's Mongol predecessors. The discipline of his army meant that when he did grant protection to a surrendered place, it was usually observed, but not always (Delhi). There were occasions when he seemed to enjoy playing fatal games of deceit, such as at Sebasteia (Sivas), the capture of which was observed by the young German captured and enslaved after Nicopolis (and then captured again at Ankara). Here, as we saw, the Turkish garrison were all buried alive after surrendering on the promise that their blood would not be spilled. On other (rare) occasions, Timur's orders horrified even his own blood-stained emirs. When Timur ordered the children of the captured city (apparently 7000 in number) laid out on the ground, and then instructed his troops to ride over them, they objected at the cruelty, so

(Schiltberger says) Timur began the trampling himself and forced his horsemen to complete the act.[1]

Despite this use of psychological warfare, not apparent everyone was prepared to give in without a fight, even when there was no prospect of relief. Timur may have continued the long tradition of ruthless atrocity as a weapon of war, but there was sufficient courage and loyalty in Anatolia, Syria, Georgia, India and elsewhere to force him to fight for his victories by more honourable means.

The rise of Ming China

Kubilai Khan and the Yuan dynasty of Mongol rulers that he established in China succeeding in pushing his boundaries south into Indochina, but failing in his several attempts on Japan. His efforts called for enormous resources, and a by-product of this attempt was a census in 1290, showing more than 58 million people owing tribute — not counting the many who had fled to the mountains and formed the core of the growing tide of rebellion that would eventually unseat the Yuan. By the 1350s, the whole of southern China had successfully broken away from the Yuan emperor Chun-ti, with a number of rebel leaders operating in different provinces. Closest to the seat of power were the Red Bonnets operating in Hunan province. In 1355, Chu-juen-chang, founder of the Ming, took the city of Ho-yuan, but prevented any pillaging and thereby gained much support. Similar peaceful captures took place of Taiping and Nanking (Nanjing).[2] In the following years, there were numerous attacks on fortified towns, counterattacks by the Yuan, and struggles between and among rebel leaders involving. Although some of the available translated Chinese sources were a much later compilation, and must be strongly suspect because of their bias towards the Ming, there also remains an account of the defence by the Yuan of the city of Shao-Hsing, which we will summarise for its valuable insights into the tactics practised in 1359. Generally, the pattern was of siege by blockade, with numbers of places brought to surrender this way — but not always. Shao-Hsing, occupied by a warlord named Lu-Chen owing allegiance to the Yuan, resisted a seven month siege.[3] The writer, Hsu Mien-chih, was a witness.

The account naturally glorifies the personal role of the leader, but in the process gives an excellent understanding of the day to day tactics. The city was powerful, its defences modernised in 1352, and enjoyed the protection of many waterways, that also provided a link with the sea. It had a circumference of 14 kilometres.

[1] J Aubin, 'Comment Tamerlane prenait les villes', *Studia Islamica,* XIX, 1963, 83–113, reviews the evidence and challenges the traditional western account of Timur's barbarity as casual. On the massacre at Sebasteia, Schiltberger, 20.

[2] Tong Kien Kanj Mou, *Histoire générale de la Chine,* trans. J de Moyriac de Mailla, 1779, IX, 614–16.

[3] H Franke (ed.), Die Belagerung von Shao-Hsing im Jahre 1359: ein Beitrag zür Kriegsgeschichte Chinas im Mittelalter, in Franke, *Krieg und Krieger in Chinesischen Mittelalter,* 2003.

Walls were 8.7 metres high and 6.5 metres thick (at the top), and there were five land and six water gates. A wide glacis separated the inner from the outer walls, and bridges allowed the defenders' cavalry to cross over easily. Lu-Chen ordered the checking of the palisades on the outer defences, and the deepening of the wet moats to allow his flotilla of military boats to use them, as reports of the arrival of the enemy were received during January 1359. Stocks of stones were laid in, stone-throwers were constructed, and slits were cut into the palisades.[1] The siege began in March as the besiegers set up and fortified camps on three sides, and launched a first attack against one gate, but were driven back by stone throwers and cannon fire (*huo-t'ang*). Over the following days, the besiegers tried to burn the outer palisades but were held off by fire arrows shot from the ramparts. At the end of the month, a large supply of rice arrived in the city by boat. Following simultaneous attacks on both the south-east and south-west gates (2 April) that failed, they switched their target to the north-west corner (8 April), where they were met outside the walls by the defenders who used the narrowness of the approaches here to drive them off with bows and crossbows. The very next day, 10,000 besiegers across a long front tried to bring up timbers and bamboo to fill in the ditch, but they could not remain there because of the strength of the defenders' fire, including from dozens of guns, from both the wall and from boats in the moat. Four days later, the defenders decided to deploy stone throwers on top of each gate. Lu-Chen did not allow the enemy peace to plan their attacks, leading frequent sallies, including sending substantial forces to capture and hold nearby sites. Fighting continued over the next weeks, never seriously threatening to break into the city, while further supplies were safely received using the Yuan control of the waterways. On 8 June, thousands of attackers got up to the ditch using fences made of woven bamboo for cover and set them up less than a metre from the outer wall, despite vigorous defensive fire. Over the next two days, ferocious attacks were launched from dawn till night under cover of a hail of arrows, fire arrows, stones from catapults and "iron balls". The citizens cooked rice for the troops all day, as they fought from the ramparts, and then overnight Lu-Chen ordered sorties using the boats which wrecked some of the attackers' works. Finally, on 13 June, having suffered severe losses, the besiegers sent away their sick and wounded, then drew up in ranks and marched away. Lu-Chen was careful to reward his troops, but also to order strengthening of the defences where they had been shown to be vulnerable, and in particular, he ordered that the ditches be doubled in width and the breastwork (of the outer wall) raised in height by 1.5 metres.[2] The city would not fall to the Ming until almost the end of the war (1367), when the strengthened defences could not compensate for the overall political and military balance. The sheer scale of the operations whose outline we have given here dwarfs anything happening in the west or the Muslim world at this time. Although the defences of the city were strong, they were not exceptional, and the great length of the circuit called for immense numbers of troops, on both sides. At no point were even the

[1] Franke, Shao-hsing, 151, 212–13 (description of defences).
[2] Franke, Shao-hsing, 157–88.

outer defences breached, and the reasons for this seem to have included the professional skill (and sufficient numbers) of the defenders, their arsenal of missile weapons including (small) cannon, and the foresight and planning of their commander. Even had the rebel army broken in, the citizens had been organised into militias organised in each quarter of the city.

In the relatively few cases where capture was brought about by other means, the only method cited was by digging a tunnel under the walls, not for mining, but to enable the attackers to enter unseen behind the defences. On the one occasion that a city was taken by direct assault at the first attack (in 1368), the capture by escalade was noted as something unusual. It was in that year that the new emperor Chu Yuan-chang and the new dynasty were established in Tatou (Beijing).[1] The account suggests that though from time to time, the Yuan armies took places by assault, this too was rare. The numbers of troops engaged was immense by contemporary and later standards, armies of 100,000 being regularly mentioned. Given the ancient Chinese tradition of organisation of specialist siege corps within their armies, it may seem odd that they seem to have become so relatively ineffective, but the evidence of long blockades points in this direction.

Perhaps this arm had fallen into relative decay, although the official records still listed "myriarchies" of trebuchet operators,[2] while perhaps the rebels did not (at first) have access to the trained personnel. While firearms had become widespread in China, they remained of small size, and they were used more on ships, or as hand weapons in battle, or as weapons for the defence, rather than against walls.[3] The military treatises commissioned under the new dynasty gave a picture of the reorganisation of the engineering and artillery troops that would have awaited Timur had he actually undertaken his intended invasion, although it is stated that their first significant use of gunpowder weapons in sieges dated from after their conquest of Annam (Indochina) in 1407.[4] Having gathered the technology and perhaps the technicians from their conquest, the Ming emperor Yang-le (1403–24) created a "firearms training division" divided into five sections each commanded jointly by a palace eunuch and a general. The guns were termed "supernatural guns", *shen chang, shen pao*.[5] Given previous evidence for the use of gunpowder as a propellant for weapons throughout the fourteenth century, this may seem strange. Perhaps the reality was that guns continued to be relatively small hand-held weapons, while firearms in the form of much more powerful can-

[1] *Histoire générale*, IX, 628–31 (1359), 638–9, 644 (1363), 652, 654–5 (1368), 657.

[2] D Farquhar, *The Government of China under Mongolian rule. A reference guide*, 1990, 257–9, 389 (based on the official records of the Yuan dynasty). The word literally means units of ten thousand, but may also signify the superior commanders.

[3] C Peers, D Sque, *Medieval Chinese armies 1260–1520*, 1992, 19–36, analyses the surviving accounts of the rise of the Ming.

[4] A du Picq, *Histoire d'une citadelle annamite — Bac Ninh*, 1935, 44–7, 50–3, recounts without detail the conquest by the Chinese of Annam by capture of the fortresses of the Red River valley, followed by a successful reconquest by the Annamese which forced the Chinese garrison of Hanoi to surrender in 1427.

[5] Foon Ming Liew, *The treatises on military affairs of the Ming dynastic history 1368–1644*, 2 vols, 1998, I, 51, 95, 103.

non only became recognised as a distinct component of the state's forces with this development.

The vast Chinese empire would have posed an immense challenge to an invader, not least for the number of fortifications: the imperial administration was responsible for overseeing no fewer than 392 garrisons of fortified places, both major and minor, in 1393.[1] Generally, the fortresses of the interior may not in practice have had to be held against opposing regular armies, and the most likely direction from which a foreign invasion could be expected continued to be the north. Throughout the history of China's imperial dynasties, the struggle with the nomads who ruled in this region had been continual apart from the time when the nomads had themselves ruled China. The Great Wall had been the ancient solution, and had been extended and repaired repeatedly over the following centuries. When the Ming overthrew the Yuan, they therefore resurrected the conflict in the north with the Mongols, the Oirats and other nomads. The wall itself had not been maintained.

The Ming solution was a system of defensive towers along the length of the wall, strong enough for local garrisons to hold them against small-scale raiding parties, and able to communicate by signal (and later, cannon-shot) with regional garrisons where larger forces were based and could be mobilised to tackle where the threat had been reported. The Ming established nine military commands behind the wall, each subdivided into areas for garrisons based in a fortress. There were ultimately thousands of these towers, built to a number of standard plans according to the function in a programme lasting well into the sixteenth century. A Persian account of 1419 described one that was designed as a fighting tower, as being square, 20 metres tall and accommodating a garrison of 50, with overhanging corner turrets and crenellations. Others were much smaller, accommodating only small numbers of operators, and with no function apart from signalling. The Ming did not neglect the potential value of the wall itself either, and devoted massive resources to trying to keep it in good repair, and even to extending it where necessary with palisades and earthworks to the border of the Korean peninsula. Much of the surviving structure dates to their time. The value of the system was demonstrated when the Oirats won a victory over the empire in 1449, but were prevented from penetrating into the heartland of China by the system of frontier garrisons.[2]

It was not unusual for a new dynasty to established a new capital, and the Ming set theirs in Peking (Beijing), a city of ancient origins, but began by reducing the enclosure of the previous Yuan walls. By 1437, when it was ordered that a brick facing be added, the circuit was 41 li (around 23 kilometres), and there were nine large and well defended gates with outer barbicans, portcullises and other defensive mechanisms. It seems that the Yuan walls were also heightened and widened, and varied in height between 10.3 and 11.9 metres, with the normal sloping shape,

[1] Foon Ming Liew, I, 368–9.

[2] H Serroys, 'Towers in the northern frontier defences of the Ming', *Ming Studies* 14, 1982, 9–60; A Fang-chih Chen, 'China's northern frontiers', *Yenching Journal of Social Studies*, IV/I, 1948, 15–59.

ranging between 14.8 and 22.8 metres at the base, to 11.3 to 17.6 at the top. Not all cities were so strongly provided with protection.[1]

Accounts of Burma, Siam, Cambodia and Vietnam

The Ming dynasty's consolidation of their use of cannon after their initially successful, but ultimately defeated, invasion of Annam is a reminder that the countries of that region were also substantial powers with well-established civilisations, and that the area could boast a history of great empires such as that of (modern) Kampuchea (Cambodia), the home of immense structures such as those of Angkor Wat. Although access to native sources of their history remains limited, there are from Chinese reports and accounts, and archaeological studies from the time of French colonial rule and after.

In (modern) south-west China, the province of Yunnan was independent, and the city of Yunnan-fou, originally seven centuries before, was rebuilt in 1382, a great undertaking as the circuit of the curtain was 9 *li* (five kilometres), and the wall was ten metres high. In 1388, the Ming sent an army said to have numbered 300,000 to crush the ruler, accompanied by 100 elephants which were used as platforms for launching fire arrows and what the chronicle calls "petards", which appear to have been some kind of gunpowder weapon since "their noise made the mountain valleys tremble". The arrows were launched from ballistas.[2]

Yunnan stood on the borders of (modern) Burma, Laos and Annam (northern Vietnam), and blocked Chinese efforts to extend their power into Indochina and in the direction of Burma, and these attempts generated Chinese reports about these countries. A weapon in common use was a large-scale crossbow or ballista, although its origin has been a source of academic dispute. The weapons were illustrated in bas-relief on stone statuary at Bayon and elsewhere in Kampuchea, and took several forms, including some designed to be carried on the backs of elephants, and others on wheels. Their role in sieges was evidenced from the accounts. They were not illustrated at Angkor Wat, from which the scholars concluded they were an innovation of the twelfth or thirteenth centuries, and their resemblance to known Chinese weapons has been observed. It is certainly possible that smaller hand-operated weapons may have been familiar to the Cambodians, but that larger machines intended for siege warfare were introduced from China.[3]

[1] E Bretschneider, *Récherches archéologiques et historiques sur Pekin et ses environs*, 1879, 25–31; O Siren, *The walls and gates of Peking*, 1924, 15–52; Heng Chye Kiang, *Cities of Aristocrats and Bureaucrats*, 1999, 149.

[2] C Sainson (ed.), *Nan-Tchao Ye-Che. Histoire particulaire du Nan-Tchao. Traduction d'une histoire de l'ancien Yun-Nan*, 1904, 196 (Yunnan-fou), 223. The account was written under the Ming in around 1550, so is not contemporary.

[3] P Mus, 'Études indiennes et indochinoises, III: Les balistes du Bayon', *Bulletin de l'école Française d'extrême orient*, XXIX, 1929, 331–8; G Groslier, *Récherches sur les Cambodgiens d'après les textes et les monuments depuis les premiers siècles de notre ère*, 1921, 90. The picture has been confirmed by M Jacq-Hergouliach, *L'armament et l'organisation de l'armée khmere aux XIIe et XIIIe siècles d'après les bas-reliefs d'Angkor Vat, du Bayon et de Bonteay Chmar,*

There is evidence of substantial fortresses in the region, sometimes being walled and ditched temples. Angkor Thom was three square kilometres, with a double circuit of walls nine metres high with a ditch in between, and with corner towers, all well constructed. A report from a Chinese embassy sent in 1295 reported a capital city 20 *li* (about 10 kms) around, with five gates, and a wide ditch. The walls were of carefully jointed stone blocks, with corner towers. There were no crenellations on the ramparts, but there were powerful bastions. The account reported that the soldiers, who went very lightly equipped because of the climate, were without armour and did not know of the ballista, confirming the above supposition about its being introduced by the Chinese. The text's translator noticed the closeness of the description of the capital to that of Angkor Wat. Elsewhere, the kingdom was divided into 90 provincial governments. Generally, wooden palisades sufficed as ramparts.[1] Examination of other cities shows examples of very large (rectangular) enclosures behind terraced earth ramparts, sometimes doubled, as at Sambor Prei Kuk (possibly the capital of King Isanovarman I). Successive kings appear to have undertaken building on a monumental scale, but this is not to rule out that their walls and gates also served a defensive (as well as a primarily religious, symbolic) function.[2] East of Cambodia, a strong Cham state occupied the territory that is now the central and southern provinces of Vietnam, and a considerable number of powerfully-fortified sites have been identified, usually on the coast or a short distance up rivers in this area, which was not finally lost to Annam until the 1470s. Generally square or rectangular, they possessed ramparts either constructed of brick, as at the small (forty metre square) citadel at Co Luy, or with brick facing of powerful earthworks, as at the citadel of Hue. The fortress at Chau Sa was 400 metres square, with a shallow berm between the rampart and ditch, and angled postern gates at three of the corners, and a main gate in each face. At some sites, such as Thanh Ho, a large fortress with sides of 700 metres behind a wide (30 metre) ditch, there were towers at regular intervals along the rampart (except where it abutted mountainous ground), constructed also of brick.[3] The various remains have not been precisely dated, but it is known that the ancient kingdom of Champa dating back to the fifth century had brick walls with towers. These examples presumably belong to the later period of Cham rule that lasted until the late fifteenth century, and indicate that the structures had similar characteristics to others in the region, with their geometric design, and

1979, summary: 227–8, recording in particular that the state knew of texts from India, but was unable to copy their prescriptions because of the circumstances prevailing in the country.

[1] P Pelliott (ed.), 'Mémoires sur les coutumes du Cambodge par Tcheou Tu-Kouan', *Bulletin de l'école Française de l'extrême orient*, II, 1902, 140–1, 172–3, 175–6.

[2] G Coedes (ed.), J Bosselier, Asie du Sud-Est, I, Le Cambodge, *Manuel d'archéologie d'extrême orient*, 1966, 101–6.

[3] H Parmentier, *Inventaires des Monuments Cams de l'Annan*, 1909–18, I, 137–8 (Thanh Ho), 235, 512–14 (Hue), II, 579 (distribution map). G Maspero, *Le royaume de Champa*, 1928, 27–9.

that they no doubt had a variety of functions. But the frequent warfare suggests that one of these was defence.

Chinese accounts, alongside native histories, also allow the reconstruction of sieges and warfare during this era. The collapse of the ruling dynasty of Pagan (in modern Burma) followed the impact of the Mongol invasion of the Far East, although the regime had been disintegrating for some time already. The royal chronicles for the reigns of the last five monarchs told how in 1255 King Narasihapati faced several rebellions, and sent armies two armies to deal with the rebels. The southern army (infantry, cavalry and war elephants) marched on Dagon (Rangoon). The governor fortified himself in the upper town, but after a week of the royal army pressing forward with siege works, the defenders, running out of water, abandoned their palisade and surrendered.[1] Apparently, the city had been previously unfortified, and there was no mention of any sophisticated weaponry. A similar outcome in different circumstances occurred when later that year the other rebel retreated to a mountain, which he fortified with entrenchments, but submitted through hunger.[2]

The Chinese invaded in 1281 on the orders of Kubilai Khan, but the country still had no significant fortifications. At the town of Nga-caung Khyam, the Burmese resisted the Chinese from behind a palisade and ditch that had been created for the occasion, and held out for three months before being overwhelmed. The king was anxious that his city of Pagan only had low walls, so he had decided to construct new defences, but because of a shortage of bricks or stone (an indication of the source of the problem overall), he would have to obtain these by demolishing temples, something that naturally produced dire predictions of doom that were fulfilled when the king was poisoned and the Chinese walked into Pagan. A Chinese account written around 1321 noted that in the original frontier fighting (in the year 1277), the imperial forces had overcome and destroyed some 300 palisaded posts.[3] After a number of further engagements and collections of tribute, another major invasion took place in 1300, when the Chinese forces penetrated into Burma as far as the city they called Mou-lien tschang (Burmese Myin-Saing, south of modern Mandalay), where there was a "three-walled" town, and the defenders had placed ballistas (*pa'o*) on the walls all around, including "weapons with one arm" and with "three arms". The besiegers erected their own defences, but it was slow work and it was not until 1301 that they completed the circuit. An assault was heavily defeated, losing 500 men to arrows, stones and tree trunks. Having also lost large numbers to disease, the Chinese eventually gave up the attempt and withdrew, to the anger of and punishment by their emperor.[4] Clearly, the Burmese had artillery by 1300.

The survival of royal chronicles from the Cambodian kingdom after 1340 has

[1] E Huber, 'Études indochinoises, v: La fin de la dynastie de Pagan', *Bulletin de l'école Française de l'extrême orient*, IX, 1909, 641.

[2] Huber, Pagan, 642–3.

[3] Huber, Pagan, 664, translating both the Yuan Shih and Houang Yuan tcheng Mien lou ("Notice of the expeditions in Burma under Yuan dynasty").

[4] Huber, Pagan, 679; Mus, Balistes, 339, translating the same text.

allowed the reconstruction of a number of struggles between neighbours at this time, which also cast light on the levels of technology available to the contestants. In 1352 there was war between Cambodia and Siam. The Siamese invaded, and the inhabitants of the suburbs of Angkor Thom took refuge within the walls, which were guarded. The deep ditch and the walls held off the initial attacks, then the besiegers began to prepare what the account calls towers from which to bombard the city with "cannon" (the word used is komphloeung, combining words for ball and fire). Clearly these were not sufficient, as the siege lasted four months (December 1352 – March 1353) until there was first a failed attempt at relief, then a bloodily repelled attempt at a storm, followed by the death of the king, then another attack which was concluded by the Siamese forcing their way through a gate and capturing the city. Angkor Thom was recovered by the Cambodians in 1357, and in 1360, during further fighting, barrels of gunpowder aboard a Cambodian boat exploded. Four decades later, in 1394, the Siamese again invested Angkor Thom, and began the work of filling in the ditch before launching three bloody assaults, which the defenders repelled with stones. After seven months, the Siamese secured entry when someone treacherously opened a gate for them, so it is apparent that once again, the defences were themselves adequate to resist the equipment available to the besiegers.[1] By now, the Cambodian state was in disarray, its military capabilities weakened by Buddhist-inspired rebellions that called for the heavy taxation once used to construct palaces and fortresses to be diverted to temples.

There are also accounts of Kubilai Khan's efforts to impose his rule on the states of modern Vietnam, leading to major operations against the kingdom of Champa. In 1282, after failing to force submission, the Khan sent his commander Chagatou by sea, since Annam had refused passage by land. This host disembarked and found itself facing defences (called Mou Tcheng by the Yuan Shih) of a four-sided citadel, added to with palisades more than two leagues (20 *li*, ten kilometres) long with turrets, and three storey-high barracks for the defending troops (January 1283). The Chams deployed 100 pieces of artillery, described as stone-throwers, of the type *hui-hui p'ao*. Undaunted, the highly experienced Mongol general selected points of attack and launched so fierce an assault that he broke through and King Indravarman v (1265–85) abandoned the city and withdrew into the forests. Chagatou was unable to capture the king and eventually abandoned the invasion in 1284. A furious Kubilai Khan ordered new attacks, which were only ended when the rulers of Champa and Dai Viet prudently agreed to recognise the Khan as their overlord.[2]

The Chinese description of the Champa artillery as of the "Muslim" variety, part of the Mongol/Yuan state's armaments since 1274, makes a number of interpretations of the account possible. The large battery of trebuchets may have been constructed locally but the design imported from the north; or the technology

[1] A Leclerc, *Histoire du Cambodge depuis le 1er siècle de notre ère*, 1914, 201–13. The first part of this translation is reproduced as the source for H Quaritch Wales, *Ancient South-East Asian warfare*, 1952, 204–5.

[2] G Maspero, *Champa*, 176–85; Mus, Balistes, 340.

may have come independently to the kingdom through its trading relations with the Muslim world anyway. The translators suggest that the weapons must have had Muslim crews. Whatever the truth, unlike the impact of the weapon in Kubilai Khan's conquest of Sung China, his commander in Champa does not appear to have been much troubled by facing an enormous array of the same weapon deployed against him, and the 100 *hui-hui p'ao* failed to hold the city of Mou-Tcheng.

Whatever the solution to this question, this evidence suggests the military technology of this region was not among the most advanced. As for gunpowder weapons, there is evidence that firearms were becoming available by the end of the next century, when it is recorded that the then king of Champa died in a war with Annam from a musket shot. Vietnamese sources tell of further wars between the neighbouring realms and captures of cities by siege during the 1440s, and in 1471 the Annamese captured the city of Vijaya using bombards to blast a breach prior to storming the city, bringing the kingdom of Champa to an end.[1]

Japan

The defeat of Kubilai Khan's two efforts to invade Japan had been largely the result of the weather rather than Japanese resistance. The history of the kingdom during the next century does suggest that the defences would not long have resisted the Mongol power, although it would have been quite another matter to retain control of the country: it was a task that proved beyond the native dynasties. The very distorted and frequently fictionalised "war tales", highlighting the nobility, courage and skill of heroic samurai figures, offer sufficient information contained within some of the tales to reveal that just as anywhere else in the world at the time, fortifications and their defence and capture played a not insignificant role. Whereas the tales dealing with events of the twelfth century make little reference to such warfare, and where they do they suggest that the main tactic was nothing more sophisticated than taking an enemy by surprise by nocturnal attacks, and setting the overwhelmingly wooden buildings and palisade alight, by the fourteenth century things had moved on a little.[2] The Taiheiki (translated as "peace tale"),[3] which survives in many versions having been added to by subsequent poets, was based on an account begun in the 1330s and telling the story of the period of intense political struggle between 1318 and 1333. If the detail of individual events is inevitably fictionalised and if it is therefore well-nigh impossible to separate fact from invention at this level, the main events described all happened, so the source, although not history, is also not useless for our purpose. A new emperor, Go-Daigo, had ascended the throne, and oversaw a ferocious series of attempts to overthrow the seat of real power, the shogunate of the Kamakura, which initially failed but was ultimately successful.

The castle of Kasagi stood in a high location approached by a twisting path,

[1] Maspero, 231–8.
[2] P Varley, *Warriors of Japan as portrayed in the War Tales*, 1994, for a scholarly analysis.
[3] *The Taiheki*, ed./trans. H C McCullough, 1959, is the version used here.

which the defending imperial forces had strengthened by digging ditches and throwing up rock barriers. The opposing army overcame all these obstacles before arriving outside the castle's outer gate, where the defenders were arrayed, the wooden towers held by archers. The first assault was hurled back as the garrison shot arrows and hurled stones, but a surprise nocturnal attack by 50 men with grapnels to scale the wall got in and forced their way into one of the towers which they set alight, leading to the defenders fleeing in the dark.[1]

The victors moved on to attack another castle at Akasaka, which was "a stronghold of hasty devising", without a proper ditch, and with a simple wooden wall plastered over with mud. Tempted by its apparent weakness, the attackers rushed it from all sides at once, but the 200 defending archers did great slaughter among them. After other happenings, the army returned to the castle with a plan to scale the wall, but found this time there was a double wall, the outer rampart cleverly held in place only by ropes which the defenders cut at the right moment, leading it to fall and crush many of the would-be assailants. The culture of the tale is revealed in the comment about their shame at being held off by a few men "in a flatland castle". So a third attack followed, this time the men protecting themselves with their shields as they brought up grapnels. But as they crossed the ditch the defence hit them with boiling water spooned over the wall with ladles up to twenty feet (six metres) long. Thwarted again, they decided to starve the castle out.[2]

The historicity of the tale is questionable, as a description of a later attack on the same castle provides a physical description that is obviously of somewhere else. On this occasion, the fortress was strongly protected by a deep ditch and a powerful towered wall on the one side where it could be approached from the flat. The key this time was fire. The garrison had water supplies set up to extinguish fire arrows, but the attackers noticed this, dug down into the ground and uncovered a pipe leading from the higher ground, and blocked it. Their supply cut off, the defenders could not prevent the burning of two towers by arrows, and they quickly surrendered.[3] Their captors however had them all killed (a suspiciously precise number of 282 is given), the news of which led the defenders of nearby Yoshino castle, simultaneously under attack, to decide never to yield, but the place was taken by another secret attack by a select group from an unexpected point. The combined forces from these two captures moved on to attack the castle at Chihaya, another naturally well-protected site in very rough terrain, which was at first held by archery against direct frontal assaults. The besiegers settled down to a blockade, but the warriors quickly became bored and began to succumb to the temptations of an idle camp life. The commander ordered the building of a movable bridge, for which 500 carpenters were summoned from the capital. The result was some five metres across, and 65 long, and was moved forward by thousands of ropes worked by pulley. The device was successfully propelled across the ditch, and a host of warriors stormed across, but the defenders used a pump to project

[1] *Taiheki*, 70–79.

[2] *Taiheki*, 85–89.

[3] *Taiheki*, 171–4.

oil onto it, then hurled torches to set it alight. As those at the back continued to push forward while those at the front, in the flames, tried to retreat, the resulting carnage was fearful. At last, the attackers decided to undermine the defences with spade and mattock, and within three days had brought down a tower.[1] Events elsewhere rescued Chihaya.

This story, which spins tales of individual heroes and uses the language of the Japanese knightly order in a way comparable to the troubadours of western Europe, nonetheless talks of real sieges and real places, even if it is easy to suspect the detail, especially when one siege begins to sound suspiciously like another, or when a castle changes from being a temporary and vulnerable erection at the beginning of the tale to an impregnable stronghold later on. But some elements of siege warfare are common. Defenders relied on archery, and on hurling stones down on attackers at the foot of the wall. The archers took position on wall tops, but most effectively on towers, where they operated both from the tops, and from behind windows in the structure itself. Apart from the foundations, fortresses were wooden, and a contemporary illustration shows this clearly: defending bowmen crouch behind their wall, while the projecting rectangular tower has its panelled parapet thickly peppered with embedded hostile arrows.[2] The attackers relied on escalade or climbing hooks (grapnels), or capture by surprise or, as often, trickery of the defence. Sometimes they would call upon the method of sapping, or even less often, use the skills of common craft workers to manufacture special equipment like the bridge used so ineffectively at Chihaya. The only alternative was blockade. The importance of water supply was reflected interestingly in the Taiheiki, but mainly to demonstrate the wisdom of certain commanders.

There was a remarkable absence of sophistication in Japanese siege warfare. This was not due to any shortage of construction skills available in the country. Nor was it down to any absence of access to knowledge. The Taiheiki refers more than once to Chinese sages of ancient times whose writings on warfare were part of noble education. The victorious leader ordered a new palace to reinforce the image of the new regime, which was to be based on a Chinese model dating from the Ch'in dynasty, and it was of immense size, being 4280 by 2308 yards (3950 by 2130 metres) with 12 gates.[3] The specific and detailed instructions of the Chinese regarding fortifications and their defence would therefore have been easily available in Japan. However, the particular dominant culture of the whole period meant that the more effective and productive methods of attack and defence were consciously ignored. Perhaps the lack of engagement with forces that had a different philosophy meant that the warrior class did not have to change its style of waging war. A Japanese emperor himself wrote to his Ming contemporary in 1381, explain-

[1] *Taiheki*, 174–90.

[2] *Taiheki*, facing 63.

[3] *Taiheki*, 205 (speech from T'ai Kung, general in the Chou dynasty), 347–50, 363–4 (palace).

ing that his country was small and "its castles and moats are less than sixty", but confirming that their military knowledge included the classic work of Sun Tzu.[1]

Timur's empire after Timur

Eventually, after a murderous struggle among Timur's sons and grandsons, new regimes took shape in the various segments of his vast but personal empire. The most significant of the successors was Shāhrukh, who ruled for 42 years (1405–47). His wars included fighting to take major cities in Mesopotamia, of which no detail survives,[2] but the Timurid empire became separate and much weaker entities in Transoxiana, in Khurasan and in Iran. Syria returned securely to the rule of Egypt, while the Ottomans recovered their position in Anatolia, and eventually resumed their advance into Europe.

[1] Ryusaku Tsunoda, L Carrington Goodrich, *Japan in the Chinese dynastic histories*, 1951, 110–11 (from the Ming Shih).

[2] Medzpeh, ed. Neve, 95, 110–27, ed . Bedrosian, 70–1.

5 The early fifteenth century: changing times

THERE ARE TIMES IN HISTORY when it is generally accepted that the changes that took place represent such significant developments as to amount to the end of one era and the beginning of another. The fifteenth century has commonly been accepted as such a period. While this was certainly not so true for the great nations of the east as it was for the states of the west, the consequences of the economic, social, cultural and political revolutions in Europe would quickly be felt across the whole world. We need only to list them, in no particular order: in the arts, the flowering of the renaissance; the invention of printing, with far-reaching social consequences; the opening up of the Americas and the arrival of Europeans by sea in Africa and Asia. Underlying it all, the growing economic weight of capitalism was undermining previously dominant feudal relations, and would over time begin to look to convert economic strength into political power. In this process, the example of powerful city states in Italy would be emulated in Germany and the Low Countries, where newly self-confident cities in which merchants were dominant would consolidate their effective independence, although usually not without a struggle. At the ideological level, the challenge to the monopoly of the Catholic church in most of Europe by the Protestant reformation of the sixteenth century was prefigured on a nation-wide scale by the Hussite movement based in Bohemia a century before. The complete extinction of the already defunct East Roman empire represented only a symbolic turning point. But greater long term significance attached to the repeated victories of the non-noble Hussites of Bohemia over the armies mobilised by the Holy Roman Empire, under princely and noble leadership.

By 1453 it was evident that older forms of defence had been transcended by the new form of attack. The second half of the interminable Anglo-French struggle would see, not once but twice in Normandy, how quickly whole provinces where the defence was based on old walls could be overcome. While Henry V's recapture of his ancestors' homeland after Agincourt was remarkable for its speed, the reconquest by the French crown thirty years later was even faster. Both examples in fact will confirm that other factors than the relationship between instruments of attack and defence continued to play their part in determining the outcome. Although the role of cannon in these campaigns has long been acknowledged, it is also necessary to place them in the context of other contemporary sieges, where despite the presence of cannon, the defence repeatedly prevailed. We will explore how the defence began to make the material improvements needed to resist the devastating impact of the new weapons on stone walls. Over the century the reversal of the previous shift to higher walls, back to lower and more compact works, and indeed the renewed recognition of the defensive value of earthworks, set the trend for the future. Slowly, the development of the bastion trace that became universal during the sixteenth century began in the fifteenth, and again the Hussite wars provide some very early precursors. Alongside this process, the distinction between the kind of fortification that a state could afford, and one that a nobleman could aspire to, reached ever greater levels. This did not mean, however, that the tower house of the settler landowner in Ireland, for example, did not continue to have a defensive function as well as a residential and symbolic purpose.

Developments were taking place simultaneously across a wide geographical area. In order to maintain some chronological coherence, without which comparisons become difficult to make, this chapter will start by looking at some significant conflicts at the beginning of the century, in those parts of the world where Timur's presence had not been felt.

The Baltic region

By 1400 the theocratic state of the Teutonic Knights was already a peculiar anachronism, its official *raison d'être* gone since the conversion of the Lithuanian dynasty, but it still managed to provide, by virtue of rather tortured logic, an opportunity for German and other western knights to fulfil crusading vows, and it continued to receive sufficient in the way of contributions and recruits to maintain its rule, based on a network of forts and convent-castles across Prussia and Livonia. During the last decades of the previous century, all parties in the area had acquired the most modern technology along with sufficient expert practitioners to use it with effect. All were now also states with substantial resources and were able to maintain substantial armies for long enough to undertake the sieges that remained the basis of all warfare.

The Teutonic Knights undertook several naval expeditions to the island of Gotland, a place of interest equally to them and to the Danish kingdom, but also a resort for those making their living by preying on merchants trading across the Baltic. In 1398, the Order's commander of the castle of Schwetz (Święcie) conducted a major expedition to the island, 5000 men from Prussia including 50 Knights, to capture the chief town of Wisby. The island was not very heavily fortified, there being altogether three castles, and three tower houses in the town itself. A force of 200 was left behind after the capture. It was necessary to send more troops, this time with supplies comprising two small cannon and one large, one ton of gunpowder and 8,400 arrows. The Order's troops based themselves in the old towers. The Danes and Swedes sent a fleet to secure the island for themselves, this one carrying both *bliden* and cannon, and in January 1404 they began a regular siege of Wisby. The Order responded by sending a second relief force, which also carried gunpowder weapons. Landing in March, they successfully raised the siege, and in turn laid siege to part of the Danish force that had withdrawn into one of the castles. After a week of bombardment and combined land and sea blockade, the defenders, who had also had artillery, surrendered. The besieging force had comprised some 700 men, with 49 wagons and a large siege gun. The Order's involvement with the island ended in 1408 when it was sold back to Denmark.[1]

In the same year as the Gotland expedition, a new castle (*Hus*) was built at Godeswerder, in the region of Memel (Klaipėda), and it was erected in just four weeks, presumably therefore constructed largely of timber. The chronicle records it was provided with a good garrison,[2] indicating its clearly military purpose. Other castles were erected at the same time. Four years later the Lithuanians burnt the town of Memel, and their Grand Duke Vytautas (1392–1430), whose cousin Jogaila (Jagiello) was now king of Poland, led an army across the Memel to attack Godeswerder, equipped with many guns, and the place capitulated after just three days. The following year, while the Order organised a raid on the traditional route across the Prussian-Lithuanian "wilderness", Vytautas seized Dünaburg (Daugavpils) along with many prisoners, and then captured the small fort of Georgenburg (Jurbakas, on the Lithuanian frontier south of Memel) and destroyed it.[3] Four years later, the Order's chronicler noted with pride the manufacture at their headquarters at Marienburg of a gun weighing one hundred *stuckin*, "a size the like of which had not been seen in any German land".[4] The words confirm that this was the time during which new artillery of much greater size, and effectiveness, was beginning to be constructed, and suggest that 1408 was the year of its arrival in the lands of the Order. Access to more powerful siege weaponry, however, was not to save the Knights from their most crushing defeat ever.

Tannenburg and the siege of Marienburg

Going to war against their undoubtedly fellow Catholic neighbours of Poland, they faced a combined Polish-Lithuanian army at Tannenburg (Grunwald) in Prussia and went down to a decisive defeat on 15 July 1410, their grand master Ulrich von Jungingen and many knights being killed. The war had begun with some successes, as the Knights had captured castles at Dobrin, Berberen (Bobrownicki) and Sloterie in August and September 1409, using "big guns" in the process, possibly including that which had caused such marvelling at Marienburg the year before,[5] but the destruction of their army meant that instead of consolidating success, the very existence of the Teutonic state was in question, at least in Prussia. King Jagiello followed up his success with a march against the Knight's headquarters, but was thwarted. The reason for this is traditionally ascribed to the heroic defence led by Heinrich von Plauen, the commander of Schwetz (Święcie),

[1] Annales Thorunensis, Johan von Posilge and Detmars von Lübeck Chronik, *Scriptores Rerum Prussicarum*, III, all cover the events, 217–18, 273. For a detailed analysis, F Benninghoven, 'Die Gotlandfeldzüge des Deutschen Ordens 1398–1408', *Zeitschrift für Ostforschung*, 13, 1964, 421–61.

[2] Johan von Posilge, 220.

[3] Johan von Posilge, 235, 257–8, 264–6.

[4] Johan von Posilge, 292.

[5] Annales Thorunensis, Johan von Posilge, 302.

who would become the next grand master as a result. No doubt his leadership was decisive, but he was not entirely without resources. He found the fortress undefended and without weapons (it was said), but managed to summon up both weapons and men from neighbouring towns and fortresses, including 400 sailors who travelled up from Danzig (Gdansk), and in the end he could deploy some 4000 fighting men after just a few days. Jagiello arrived ten days after the battle, and remained there for eight weeks. He concentrated his attack against the outer defence, the *Vorburg*, deploying both guns and *bliden*. The Teutonic garrisons at other key fortresses in the region, Thorn (Toruń) and Elbing (Elbląg), sent in supplies of food and gunpowder. The news of the Knights' defeat had carried very quickly, and urgent help had been promised. Letters from King Sigismund had reached Marienburg, promising relief, and this no doubt cheered the defenders further. On 19 September, hearing of the approach of relieving forces, Jagiello broke up the siege, and the Order survived.[1]

Several factors stand out as explanations for this, aside from the decisive leadership of Heinrich von Plauen. The Order's domain in Prussia was controlled through a dense network of administrative districts with strong castles, and clearly while the garrisons had been thinned to provide the troops that had been crushed at Tannenburg, sufficient numbers remained in the brick-built and powerfully-walled compounds to prevent the country being immediately overrun. Next to this, was the strength of Marienburg itself. While the massive rectangular inner castle, surrounded by strong high walls and now restored, attracts most attention, it was the outer defences that held off the Poles in 1410. Repeatedly expanded, developed and strengthened during the previous century, the defensive complex measured some 800 by 250 metres, and, standing on the river Nogai, was surrounded by water-filled moats. Beyond the High Castle stood the Middle and Lower castles, each with their own galleried and machicolated walls, towers, and gatehouses. The brick walls of the outer castle had been constructed during the later fourteenth century. To the south of the castle complex stood the walled town. The combined effect of the tier upon tier of modern defences, held by a large garrison under a determined commander aware that the very survival of his state might depend on the outcome of the siege, was sufficient to hold off the victorious Poles. It seems that no assault was made on the castle, perhaps because it was so hard to get near enough to cross the moats or make a breach, and no reference is made to any attempt to mine, perhaps because of the wet ditches. But it is interesting that it was after 1410, when the Order was again secure in its headquarters, that yet more defences were constructed beyond the existing ones on the eastern side (that not protected by the river), specifically to improve the protection against cannon following the experience of the unsuccessful attack by Jagiello's artillery.[2]

The war with Poland continued for many years, and left the Order exhausted

[1] Annales Thorunensis, Johan von Posilge, 318–20.

[2] Marienburg (Malbork) has been illustrated in many books. For clear plans see Kaufmann, Kaufmann and Jurga, *The Medieval Fortress*, 2001, 241–2, and B Guerquin, *Zamek w Malborku*, 1960, with a summary in English, 91–3. A modern description: A Emery, 'Malbork Castle', *Castle Studies Group Journal*, 21, 2007–8, 138–56.

and diminished by the loss of the land around the river Vistula down to the coast at Danzig (Gdansk). Raids, devastation and sieges marked its progress. A previously unknown siege has been described based on a letter. In August 1422, Polish and Lithuanian forces laid siege to Lautenburg. They tried to drain the water from the moat, but were thwarted because of the fierce defence mounted by the garrison from behind the defences of the inner ditch. At that time the defenders were encouraged by the news that the besiegers had only one cannon, but they were worried that they might try to burn the town by using fire-arrows. They had learnt that the Poles were bringing up more guns on wagons, and the letter asked for an attempt to intercept this convoy. The presence of a battery was apparently considered decisive. In the end, the siege was broken off. Gollub (Golub-Dobryzn), a quadrangular castle with corner towers between Thorn and Rheden (Radzyń), was taken after bombardment from 14 guns, but Schönberg (Szymbark), held by a garrison of 100, held out against attack, with the siege being abandoned in September and the artillery sent back to its arsenal. It was in that month, in fact, that peace was negotiated.[1]

Russia

The principalities of Russia continued under the overlordship of the Golden Horde, a tributary relationship enforced by military expeditions from time to time. The Polish-Lithuanian state to the west also represented a powerful and sometimes dangerous neighbour, with Smolensk the subject of repeated sieges. None of these threatening neighbours prevented there being strife between the principalities. In 1398, Novgorod challenged the authority of the Muscovite Grand Prince Basil (Vasili) I (1389–1425), and weaponry was deployed as they besieged Orlets on the river Dvina for four weeks before it yielded to them.[2] In 1400, the northern city strengthened its defences again with the foundation of the stone wall of the citadel. Previously, the traditional earth and timber ramparts had sufficed. At the same time, Vytautas waged war against his Russian neighbours, failing in battle at the river Vorskla despite having large amounts of gunpowder weapons, while the next year he again attacked Smolensk with artillery, but withdrew after four weeks. This struggle continued over the next few years. Vytautas attacked Smolensk again, but briefly, in 1402, and made a more substantial effort in 1404, when he stayed at the siege for seven weeks, but despite again attacking with guns was unsuccessful. In the end it would be treachery that would admit him to the city. The Russians evidently relied on their fortifications – the prince of Tver built a fortress at Opoki, and Basil did so at Pereiaslavl.[3]

The Khan of the Golden Horde organised a sudden attack on Moscow itself in

[1] S Ekhdahl, 'Der Krieg zwischen dem Deutschen Ordens und Polen-Litauen 1422', *Zeitschrift für Ostforschung*, 13, 1964, 619–25. Winnig, *Der Deutsche Ritterorden und seine Burgen*, 1956, illustrates Gollub.

[2] *Chronicle of Novgorod*, 171; *Nikonian Chronicle*, IV, 113.

[3] *Chronicle of Novgorod*, 173–6; *Nikonian Chronicle*, IV, 117, 136–41.

1409. At Kostroma, Basil prepared for resistance by burning the suburbs to clear the ground in front of the walls, and successfully held off the Horde with guns and arrows. The Tartars devastated large parts of the princedom including Pereiaslavl where the new defences apparently were ineffective or perhaps incomplete, but it was noted that when the Khan ordered the prince of Tver to supply him with artillery, the latter refused. The war ended with a peace in 1410.[1]

France, the British Isles and the Low Countries at the Turn of the Century

The conflict between England and France continued at a more modest level. There was political upheaval in England. Richard II was replaced in a revolt by Henry of Lancaster in 1399, ushering in continual domestic unrest, in which the Scots and French did their best to interfere, and a remarkable revolt by Owain Glyn Dŵr in Wales nearly removed English rule from the whole country before it finally collapsed. Meanwhile English rule in Ireland, already ineffective in much of the country, and now unsupported, retreated further. In France, meanwhile, the germs of large scale future conflict were growing between wings of the ruling family, with the growing territorial power of the dukes of Burgundy in the low countries and the lands between France and the empire moving towards the establishment of an effectively independent state.

In the north of England, the old border fortress of Wark on Tweed had been seized by the Scots in 1399 while its custodian was helping resolve the affair of the English crown. During the course of the insurrections faced by the new dynasty, much of the opposition would arise from the powerful northern family of the Percies: but in the fighting, sieges did not feature, the powerful castles of the north, such as Alnwick, surrendering rapidly when fortunes turned against their party elsewhere.[2] The question here was apparently either that often the fortresses were not garrisoned or provisioned for war, or that if relief was not to be expected, there was no point trying to hold. There was an exception in 1403 when the young Henry Percy ("Hotspur") followed up victories against the Scots in the field with an attack on border defences, gathering forces to take on Cocklaws tower, a small tower house in Teviotdale. While archers kept down the heads of the defenders, other besiegers brought up engines including *trabuculis et mangonellis* to batter the wall. This may refer to an outer wall, but no remains exist. But the Scots held on, reaching an agreement to surrender if not relieved in six weeks (30 May to 1 August, actually two months). The siege was raised, as Hotspur headed off to meet his death at the battle of Shrewsbury. But events surrounding it are interesting. The Scottish council could not agree whether to risk the kingdom for "such a little tower", and ultimately decided to seize and destroy an English-held tower.

[1] *Nikonian Chronicle*, IV, 170–1.

[2] Walsingham, 242; Anon, *Annales Ricardi Secundi et Heinrici Quarti*, ed. Riley, 1866, 414.

The array of engines seems rather large, especially in light of the decision to accept a truce, and does not include guns.[1]

In other parts, fortifications continued to be defended and to be the object of attack. In 1405, the count of St. Pol tried to take the castle of Marck, part of the outer defences of Calais, with a large French force, but was driven off with heavy losses after three days of attack by a sortie of the garrison of Calais. Two years before that, Breton troops had landed at Plymouth and burnt the town, having broken in by night, and the next year, Dartmouth, had suffered the same treatment, although here the raiding force had been heavily defeated by local countrymen. The king issued a licence to the town of Plymouth to erect a wall of stone and lime, and to construct fortalices and towers for the protection of the harbour, but this was done after the burning.[2] The Tudor traveller Leland gave a strong hint that new work involved a strong quadrangular castle with round towers, the remaining structure being in the form of an outwork (deduced from its continuing name of barbican).[3] This is also suggested by the Spanish count of Buelna, who accompanied the French fleet on one of its expeditions, and who noted that there was a small fort at Plymouth, but indicated that the attackers were driven off by the fire of bombards from the town.[4]

On the French side of the frontier, there were a number of incidents recorded during these years worthy of notice. In 1405, and in response to the preparations of French fleets at Sluis in Flanders, an English force was sent to attack the castle, but was driven off and its commander, the earl of Pembroke killed along with 60 of his men by a vigorous defence conducted with cannon.[5] Meanwhile, the struggle had been continuing in English-ruled Gascony. In 1404, while an English force tried in vain to trick its way into La Rochelle, the French attacked a strong hillside castle at Corbetin, which the English and Gascon garrison used as a base for extracting tribute from a wide area around. The attack, led by the Constable of France himself, involved all kinds of siege equipment, including stone throwers and rams. After 12 weeks' attack, the outer wall of the castle had been destroyed, provisions were running low and there was no sign of relief so the castle was surrendered.[6] In 1405, in Saintonge, the main English fortress was at Mortagne, where they sheltered in a seaside castle with a double wall and high towers. Lord De Pont led an army to capture it, and hired carpenters to construct his siege engines. The stone-throwers were effective, hurling enormous stones that shook the walls and towers to their foundations, and the bombardment was so continu-

[1] Bower, *Scotichronicon*, ed. Watt, 8, 49–57.

[2] Walsingham, 271; *Calendar of Patent Rolls, Henry IV*, II, 346.

[3] *The Itinerary of John Leland*, I, 214: "The mouth of the gulph wherein the shippes of Plymouth lyith is wallid on eche side and chained over in tyme of necessitie. On the south west side ... is a blok house: and on a rokky hille hard by it is a stronge castel quadrate having a eche corner a great rounde tower. It seemeth to be no very old peace of worke" (written around 1535–45).

[4] Gutierrez Diaz de Gomez, *The Unconquered Knight. A chronicle of the deeds of Don Pero Nino, Count of Buelna*, ed./trans, J Evans, new ed. 2004, 118.

[5] *La Chronique d'Enguerran Monstrelet*, ed. Douet-D'Arcq, 1867, I, 107.

[6] *Chronique du Religieux du Saint Denys*, III, 181, 201–5.

ous that there was no time for the defenders to repair the damage. After seven weeks of this, the garrison fled by ship.[1] This account suggests that the trebuchet had not entirely yielded pride of place to cannon. In the same year, the duke of Orléans laid siege to Blaye with engines used on both sides. Orléans then besieged Bourg, attacking its gates and walls with his artillery, causing much damage, but not enough to capture the town, and the siege was given up after three months. Meanwhile, the duke of Burgundy had made another attempt to recover Calais. His preparations included the cutting down of trees to construct two large *bastilles* to convey to the siege (the local terrain being barren, as previously noted), as well as other equipment including *bricolles* and scaling ladders. He was already provided with great quantities of wagons, and *canons, bombardes,* and *artillerie* but was recalled by the king of France.[2]

In 1411, the two emerging parties in France, the Armagnacs (Orléanists) and the house of Burgundy were in conflict. The count of St. Pol was sent to lay siege to the Burgundian-owned castle of Coucy, noted by the chronicler as being of miraculous construction, but now it was the powerful enceinte with its massive towers rather than the palatial interior that had to serve against cannon that had not existed when it was built. Mines were dug against the Porte de Laon and other of the great towers on the wall. Liègeois miners were hired to provide the requisite expertise, but one mine collapsed while being inspected, causing much loss of life to the besiegers. When the mines were fired, the gatehouse collapsed but most of the wall remained standing. Eventually, after three months, the garrison negotiated.[3] Meanwhile, on the way to seize Étampes, the duke of Burgundy, John the Fearless, himself had to deal with the castle of La Bretonnière, the garrison of which fled when the stone throwers were deployed. At Étampes itself, the town prudently surrendered but the castle did not, confident in its strength because its location precluded mining. Initial assaults having failed, the duke called up stone throwers and bombards from Paris and their massive stones created a breach. The garrison withdrew to the citadel, the mid-twelfth century donjon, but this could be mined, and it surrendered after a corner tower had been thrown down by this method.[4] In 1412, an army led by the king, Charles VI (1380–1422), attacked Dun-le-Roi in Berry, held by 400 men. Initially, the siege engines had little effect on the walls, so a large gun called *Griete* was brought up. It took twenty men to move it, and fired immense balls. Its detonation could be heard four miles away. On the first day of use, it destroyed a tower. The second day, it fired twelve stones and demolished lengths of the wall. As the besieging force was demanding to be allowed to storm the town, it surrendered to avoid a sack. The attack had taken nine days. Moving on to Bourges, both attackers and defenders disposed of engines, stone throwers, crossbows, and bombards. External events including an outbreak of

1 *Chronique du Religieux*, III, 275–9.

2 *Monstrelet*, I, 132–6.

3 *Monstrelet*, II, 214–16; Juvenal des Ursins, *Histoire de Charles VI*, ed. Michaud, 1836, 474. Summary of Coucy: Mesqui, *Châteaux forts*, 135–8.

4 *Monstrelet*, II, 223; *Chronique du Religieux*, IV, 571–77; Ursins, 472.

widespread illness in the siege camp encouraged the rival parties to reach a truce, however.[1]

New types of artillery

The great gun, apparently Flemish from its name (but not the later *Dulle Griet* on display in Ghent), is another example of the impact that new and more powerful artillery was now having, albeit that such weapons were still rare. Since the late fourteenth century, larger guns had been constructed in the French kingdom, and by the Burgundians, and they were made either of iron or bronze. One described in detail weighed 384 pounds (174 kgs) and fired a 60 pound (27 kg) shot from a barrel with a diameter of 0.3 metres. Another dating from 1377 fired a shot weighing 200 pounds (91 kgs). Increasingly, the shot would be made of metal rather than stone, enabling a more powerful impact and a smaller calibre. However, the period between the 1370s and the 1430s was the high point for the manufacture of ever larger weapons, some of iron, some of bronze, weighing from 3,500 up to an astonishing 8,200 kgs (a 1411 construction at Braunschweig). The weight of their shot varied between a modest (!) 70 kgs (a 1404 gun built in Austria), through a more common size of about 150 kgs, up to an almost unbelievable 550 kgs for the Braunschweig monster. The love affair with sheer size, however much it flattered the ego of the princes and cities that commissioned the weapons, incurred a number of drawbacks.

The guns still had to rest on the ground, or on some kind of structure to absorb the recoil, thus limiting their mobility considerably.[2] Transport became a serious logistical problem. Their growing impact, while still sufficiently rare to attract the notice of chroniclers, was also indicated by the siege of Autel in 1413, described below, and was similarly indicated by the writings of the contemporary Christine de Pisan on warfare, composed around 1408/9 at the French court. She wrote that a besieger needed to have great guns firing 500, 400 and 200 pound (225, 180 and 90 kg) shot and plenty of small guns firing lead pellets.[3] Much of what she wrote was based on Vegetius, but anything about gunpowder artillery can of course only have been based on current practice. The need for a range of sizes expressed in her writings probably reflected an understanding by experienced commanders of what worked best in practice. The inventories of the duke of Orléans' arsenals at Blois reveal 18 small and four large *canons* in 1418. By 1421 seven copper guns had been added. In 1431, the castle had three large *canons perriers* each set in a different

[1] *Monstrelet*, II, 269–70, 282; *Chronique du Religieux*, IV, 651–7; Ursins, 476–8.

[2] Partington, *History of Greek Fire and Gunpowder*, 113–15; Beyaert, *Quelques problèmes . . .*, 76; Clephan, *Ordnance*, 85–6.

[3] Christine de Pisan, *The Book of Fayttes of Armes*, translation by Caxton, ed. Byles, 1932, 154. There is a modern edition edited by S and C C Willard, 1999, as *The Book of Deeds of Arms and of Chivalry*.

tower.[1] The variety of weapons in use in the field is confirmed by the example of the failed Castilian siege of Setenil in Granada in 1407, an account of which by a participant talks of a varied train of artillery containing both old and modern types of weapon, but makes particular reference to the presence of both one very large bombard, and a small one. The latter could be drawn by a single pair of oxen, but the large one required twenty, and the problem of moving them required the Castilians to send back troops to help the rearguard withdraw the guns from the abandoned siege, a task which took a whole day's work.[2]

The Netherlands in conflict

In the northern Netherlands, the county of Holland and the neighbouring prince-bishopric of Utrecht, and in the east, in the lands of the bishop of Liège, this was a time of intense struggles, in which the neighbouring rulers (Brabant to the west, Gelderland and Cleves to the east), were also engaged. The warring parties included the nobility, sections of whom in Holland were in a party usually opposed to the count, the developing towns beginning to assert themselves, and in Liège the interminable conflict between the citizens of the city and its bishop.

In 1387, Bishop Florence of Utrecht with the aid of the citizens besieged the lord of Montfoort in his castle, deploying "a great mangonel" and many bombards, with which they shook and damaged the walls of the castle, but failed to bring about surrender, the struggle being eventually mediated after 17 weeks.[3] The original castle was obviously strongly built and sited. Six years later, the bishop's neighbour, Albert count of Holland, who was also, as well as a Bavarian prince, ruler of Zeeland, Friesland, Hainault and tracts of the Rhineland, fell out with his son William. He assembled the large resources available to him quickly, seized the castles of William's noble supporters, and laid siege to the town of Gorinchem, which had been taken over by the rebels, and the neighbouring castle of Holthen. A settlement was mediated after the castle had been bombarded with trebuchets and bombards, and the defences were slighted. In 1397, with reconciliation achieved, William was sent by his father to deal with his disgraced court treasurer, who had taken refuge in Gelderland leaving 90 supporters in the strong castle of Loevestein. The castle was quite modern, much of it dating to the mid fourteenth century. It comprised at this time a rectangular three-storey main building with projecting rectangular towers at each end, enclosing a small walled courtyard on three sides, on the north bank of the river Maas with (probably) a wall containing a gate on the fourth side, protected by wet moats and facing a separate outer

[1] Dupré, 'Inventaire de l'artillerie du château de Blois en 1418, 1421 et 1423', *Revue des Sociétés Savantes des Départements*, ser. 4, v, 1867, 311, 316–17.

[2] Gutierrez Diaz de Gomez, 196–202. The siege is also interesting for reporting the decision of the Castilians to erect a siege tower.

[3] F Sweertius, *Rerum Belgicarum Annales, Chronici et Historici de bellis, urbibus, situ et moribus Gestis*, 1620, 305: a compilation of earlier chronicles, this being Gerbrand of Leiden.

ward. The Hollanders concentrated heavy fire on the outer defences, destroying the main tower (*major turris*). This was one of two circular corner towers, the surviving one of which is more than eight metres across with walls two metres thick. The town of Dordrecht sent its own bombards to join the attack, and this bombardment caused a fire that seriously damaged the outwork, which was then stormed. The defenders retreated to the "upper (inner) castle" but the attacker sustained such an intense attack with arrows that no one could stand at the windows, so after just one day the castle capitulated.[1] The complete disappearance of the rest of the outer ward makes it impossible to estimate the effectiveness of the besiegers' weaponry.

Not all attacks were so successful. In 1401, for example, the new bishop of Utrecht sent troops to take possession of the northern city of Groningen, which he had decided was historically part of his episcopate. The citizens did not agree, and prepared to defend their medieval walls with the help of additional ditches and outworks, and with the assistance from the outside of the Frisians, who were proving reluctant subjects of Albert as well, forced the Utrechters to lift the siege after three weeks, although it had been supported by siege engines. Count Albert meanwhile was having trouble with the lord of Arkel. In 1403, the count assembled sufficient force to lay siege to both the town of Gorinchem, and the castle of Arkel, simultaneously. Gorinchem,a town established by the Arkel family, was strongly walled, particularly on the Holland side, and had ten "thick high towers". The castle was adjacent.[2]

During a twelve week siege the great tower of the castle was demolished by the bombardment, but again there was a mediated settlement. But the same conflict erupted again with Albert's successor William VI in 1405, when the count joined with the bishop of Utrecht to besiege three of the lord's castles, building forts to blockade them while erecting siege engines against a selected target, and succeeding in bringing about surrender. The struggle resumed again in 1407, when Arkel's men captured Gorinchem by night-time escalade of a tower, from which they opened a gate. From here, the victorious rebel tried to recover his castle, held by the count's men, and he was able to deploy his own bombards, ballistas and trebuchets. There was now war between Holland and Gelderland, the latter accepting the inducement of an offer of Gorinchem. Peace was made in 1412.[3] In all these attacks and defences of castles and towns, it is notable how frequently bombardment by artillery from a encircling siege camp was the only significant means of attack. The flat and waterlogged country did not lend itself to attempts at mining, but the resistance put up to attack seems often to have been lengthy, and sometimes successful.

The troubles faced by Counts Albert and William of Holland were mirrored by the factional conflict in Liège, complicated by struggles with neighbours. In 1398,

[1] Sweertius, 313. On Loevestein, *De Monumenten van geschiedenis en kunst in de provincie Gelderland: I, Nijmegen, part 1: Bommelerwaard*, 130–47; F van Gulick, *Nederlandsche Kastelen en Landhuizen*, 1960, 138–45.

[2] C H Peters, *Nederlandsche Stedenbouw*, 1909, I, 50, 160–64.

[3] Sweertius, 320, 322–3, 326–7, 330–40.

there was war between Brabant and Liège on the one hand, and the count of Gelderland on the other, in a particularly richly recorded siege of Roermond, another fortified settlement on the river Maas. Roermond was yet another place that had taken advantage of its location both for commercial opportunity and defensive possibilities. Standing on the confluence of the little river Roer and the Swalm stream on the east bank of the mighty Maas, the circuit had been extended during the fourteenth century, a few stretches of the defences remaining to give some clue as to what then stood. The wall may only have been completed in 1388, and the work included the diversion of the course of the river itself to bring it up to the town walls. Other parts only had a dry ditch. Round towers of two storeys stood at the corners (the "Rat tower" being four metres high), with half-round towers on the intervening stretches. The assault was carried out "as much by bombards as by stone throwers", and the place's walls and towers were severely damaged during a three week long bombardment: one source says that a six metre length of rampart was demolished. It appears however that the bishop (John of Bavaria) made a private peace with the duke of Gelderland, leading to an abandonment of the attack.[1] The town's site made it very difficult to cut it off without access to a fleet, and the waterworks no doubt narrowed the options for a land assault.

Civil strife within the bishopric broke out again in 1406. The citizens rose up and drove the bishop and his supporters (the noble party) from the city, and they set themselves up in their castles and in the ancient city of Maastricht. The popular party in the bishopric, calling on the people of Liège and Huy for its manpower, attempted to capture the castles held by the bishop's supporters, and the first trial of strength took place at Bouillon, the strong but old castle to the south. The town fell, but the castle resisted for two months while the Liègeois set up stone throwers and engines that shook the walls. It was hunger however that led to the garrison surrendering with their lives.[2] Once again, the militias and craftsmen of the city had shown themselves able to compete with the nobility, at least in siege warfare.

Sieges of Maastricht 1407–8

Meanwhile, the bishop gathered support from neighbouring princes, keen to unite against a popular revolt. The Liègeois decided in 1407 to capture Maastricht, which the bishop had left well provisioned and strongly held. The city was situated on the west bank of the Maas, originally set up to guard a Roman crossing of the river, and still performing that function. The strong rampart constructed from 1229 had already been outgrown by the 1290s, when a much larger area was enclosed with a strong stone wall protected by a wide, wet moat fed by the Maas and a small stream that joined it south of the city. Across the river, the suburb of Wijk

[1] Jean d'Outremeuse, 232–4; Dynter, III, 638–41; *Chronique de Jean de Bavière*, 148; *Chronique Latine de Jean de Stavelot*, 101–2. On the town: I Keuler, Roermond: de geschiedenis zijner vestingwerken, monumenten en voorname bouwen, in *Gedenkboek ter gelegenheid van het 700-jarig bestaan van Roermond*, 1932, 360–3.

[2] *Jean de Bavière*, 168–9; Jean de Stavelot, 118.

was also strongly defended, with powerful walls and towers including one known to have had walls four metres thick. During the fourteenth century, in modernisation of the defences, round towers projecting from the circuit, and up to fifteen metres in diameter, had been added. Machicolations were a regular feature. It was, all in all, a big challenge, especially as the Liègeois siege did not begin until 24 November. It was not long (7 January 1408) before the harsh cold weather forced it to be lifted, although in the meantime damage had been inflicted by bombardment by bombards (apparently an almost countless number) and stone throwers on the city walls and on the houses within. The attack resumed in May and lasted sixteen weeks. Initial hopes of a quick capture soon disappeared, as the bishop's supporters remained loyal and launched various sorties covered by archers and gunners. So a blockade was set up all round, and ditches dug to protect the besiegers and stop access or exit. They also attempted to erect their own timber bridge across the Maas. In June, the defence was encouraged when bishop John returned to the city, but a powerful bombardment was being sustained. On the night of 4 July alone, 100 heavy stones were rained in on the city, demolishing houses and churches. The tide now began to turn. The bishop's brother, William of Bavaria, marched into the land and devastated rebel towns. The Liègeois did not yet however give up their siege, constructing new wooden, wheeled stone throwers with which they carried on smashing down the city's buildings. The admission of defeat came on 21 September, after 113 days. The duke of Burgundy then intervened to crush the rebels.[1]

Duke Anthony of Brabant, second son of Philip the Bold of Burgundy, was also occupied against his own subjects and neighbours. In 1413, he marched to secure control of the duchy of Luxembourg, which was resisting him, deploying cannons and bombards against several castles. At Autel, this artillery was used to demolish the walls, towers and gates. The chronicle records that the city of Mons sent along a bombard called "black Margaret", and others were sent by Trier and Utrecht along with "men very expert . . . in such sciences". As his campaign continued, castle after castle fell before his onslaught, although only one was recorded as being undermined. Other castles then made their peace with the duke. The lord of the castle of Brandeburg continued to resist, and the duke erected bulwarks in front of the gates, each held by 50 men at arms and crossbowmen.[2]

Owain Glyn Dŵr

Meanwhile, England's Henry IV faced a rebellion in Wales led by the Welsh gentleman Owain Glyn Dŵr. This arose from a quarrel between Owain and the local English potentate Lord Grey of Ruthin in 1400, and quickly

[1] *Jean de Bavière*, 168–9, 177–8, 185–203; Dynter, III, 672–3; Jean de Stavelot, 108–19. On Maastricht's defences: Peters, *Nederlandsche Stedenbouw*, I, 23, 52–3, 60–2, 68 (with illustrations including of defences now disappeared.)

[2] Dynter, III, 716, 861–2.

turned into a full scale national uprising.[1] In the ensuing fighting, castles (including many that had not been required for military service for a very long time) were to play a significant role, and the immense castles built by Edward I in north Wales and his colleagues in the south proved their worth. Frustratingly there is little firsthand evidence of the attack and defence, most of the contemporary descriptions being themselves based on second-hand reporting. Owain's guerrilla army did not leave written accounts of its military operations at all.

Owain's 1400 rising with a small band of companions quickly expanded following his first victory. By 1402, the Welsh were extending their operations from the north, where they had the traditional inaccessible bases in Snowdonia, to other parts of the principality, the towns of Cardiff and Abergavenny being burnt, and the following year Owain achieved his greatest success yet in capturing the mighty fortresses of Harlech and Aberystwyth. In 1405, the tide turned, one of Owain's forces was crushed in battle, and Glamorgan, the Gower and Cardigan recovered for the English crown. In 1407, there was a showpiece siege of Aberystwyth leading to its surrender the next year, when Harlech also yielded, and Owain simply disappeared from sight, to become the stuff of legend.[2] There had been many other events along the way involving the fortresses, to which we will turn.

Guerrilla armies do not generally carry siege trains with them. Conwy, a castle that if occupied would surely have resisted any contemporary attacker, was seized by surprise, as attackers in disguise slew two warders on Good Friday (1 April) 1401, and secured the gates, allowing entry to 40 of their companions. Such an insult to the crown was not to be sustained, and very quickly the forces of the country were assembled under the titular command of the young prince of Wales (the future Henry V) to blockade the fortress. Having no way to exploit its seaside communications and not having brought any supplies with them, the rebels capitulated at the end of May. The king responded to Henry's request for 120 men at arms, 300 archers and for *bastiles* to carry out the siege by pointing out that the castle had been lost by the negligence of the prince's constable.[3] The Welsh setback here was only temporary, and more English-held castles and towns became targets including many in the former Welsh march, none of which had needed to sustain any defensive duties since the defeat of the last Llywelyn. Even if the walls still stood, the problem was to find sufficient archers and men at arms, and the money to pay them, to make them safe, and to purchase sufficient supplies for their victuals. For as long as Owain continued to operate as a guerilla, on internal lines of communication, he could select his targets among many.

The English forces during 1402 therefore tried to contain the situation by maintaining their castles, and the same castles would soon enough serve as secure bases from which to organise to crush the uprising. Both Harlech and Aberystwyth were resupplied, at considerable expense, but both fell into Welsh hands

[1] Best among many accounts, R R Davies, *The revolt of Owain Glyn Dŵr*, 1995.

[2] Such is the bald outline provided in a 15th century Welsh Annals reproduced in J E Lloyd, *Owen Glendower*, 1931, 150–4.

[3] *Chronicle of Adam of Usk*, ed. Thompson, 1904, 96; Henry IV Royal and Historical Letters, *RS*, 1860, I, 702.

after very long blockades, in 1404. In some of the captures, there was possibly an element of betrayal: it is quite likely that at least some of the defenders were themselves Welsh. There is certainly no suggestion that they were stormed. Where Owain did try direct assaults, he was usually unsuccessful. The town and castle of Caernarfon, perhaps the most potent symbol of foreign rule, were the subject of attack. A letter to Henry IV reports that the governor had managed, in turn, to inform them, via a woman because no man had dared to try to make it through, that the Welsh, with help from the French, were organising an attack with very long ladders, engines and "sows", and that he had only 28 fit men to defend the great circuit. Remarkably, he did.[1]

To the south, Owain led a powerful Welsh army along the river Tywi in 1403, laying siege to the castle of Dynevor, and ransacking Llandovery before marching on Brecon. The castle of Dryslwyn was occupied, and it seems that none of the fortified locations listed had any means of resistance. The powerfully-situated crag-top castle of Carreg Cennen was defended, but fell. Certainly, the number of defenders was very small. Subsequent building accounts indicated that severe damage was inflicted to buildings and to the defending walls.[2] Royal orders directed troops from various parts of England to assemble rapidly and embark for transport to the relief of garrisons in Carmarthen, Cardiff and elsewhere.[3] In 1404, news of the imminent intervention of the French in Wales caused further alarm, and French troops were certainly with Owain when he succeeded in occupying the town of Carmarthen, albeit briefly, as it seems he could not take the castle. Other castles also had to resist what seem to have been blockades rather than close attacks. Llansteffan, on its headland, was held by 30 soldiers paid out of its governor's resources.[4] Coety (Coity) was submitted to regular blockade as well in 1404, occupying as it did, along with nearby Ogmore castle, an important route into Glamorgan, but was supplied by sea up-river via Ogmore and held out until eventually an English army raised the siege. The high walls of its inner ward and deep ditch, although of ancient (Norman) origin, were apparently powerful enough to deter assault, and this is just one of the many sieges of this conflict in which it does not appear that the attackers disposed of any sophisticated siege equipment. The outer ward, however, was much more weakly protected and there is some suggestion that there were breaches in this wall that were repaired after the crisis was over, and of entrenchments nearby that may have sheltered the besiegers.[5] The first significant defeat of the insurgents took place in 1405, when the Welsh tried to attack the ancient castle at Usk, which was among many frontier

[1] Ellis, *Original Letters illustrative of English History*, 2nd ser., I, 1827, 33–4, 35.

[2] Henry IV Letters, I, 138–40, 170; J M Lewis, *Carreg Cennen Castle* (guide book), 1990, 12–13.

[3] *Calendar of Patent Rolls, Henry IV (II)*, 1401–5, 439–40.

[4] *Calendar of Patent Rolls, Henry IV (II)*, 339, 439; Anon, *Annales Ricardi Secundi et Heinrici Quarti*, 415; Walsingham, 272.

[5] *Calendar of Patent Rolls, Henry IV (II)*, 475; R A Griffiths, Owain Glyndŵr and the siege of Coity Castle 1404–8, *Morgannwg*, XLV, 2001, 10–22.

castles "put in some condition for defence", but instead found inside a significant English force that sallied out and defeated them, causing heavy casualties.[1]

English records reveal much detail about contemporary ideas. During the planning for more than one of the generally unsuccessful full scale interventions, there are orders for the gathering of the king's artillery and other weapons from arsenals such as the Tower of London, to the borders (both west and north). Carpenters were commissioned, apparently with a particular task of providing transport for the guns and the accompanying tuns of gunpowder. The records also reveal the size of the forces now being paid in each of the castles held either by the king or prince directly, or else still in the possession of a nobleman, but subsidised by the royal treasury. While the English had still been mainly on the defensive in 1402–3, garrisons in the north Wales castles varied between just 16 at Flint and 40 at Harlech to 100 at Caernarfon, 150 at Denbigh and 165 at Beaumaris. The numbers suggest the role that these garrisons had to play in suppressing rebellion and asserting English control in the surrounding countryside, rather than simply holding the castle. In the south, during the same period, garrisons were being paid at Cardigan, Carmarthen, Haverford(west), Brecon, Hay, Kidwelly, Neath, Newport, Usk, and some of these were again of large numbers.[2]

Unfortunately, little detail survives of the long sieges of Aberystwyth and Harlech, the surrender of which marked the effective end of the revolt. The king commissioned Lord Berkeley to cut down and provide timber from the royal forests around Bristol for the engines, bastilles and so on required for the capture of Aberystwyth on 1 September 1407, and to hire twenty carpenters, of whom one was to be "a good and curious surveyor and scientific designer" of engines. It is known that Aberystwyth's defenders promised to yield if not relieved, but the attackers were (from the English point of view, fraudulently) prevented from taking it on this occasion, and had to resume the siege and continue into the following year.[3] Although they obviously had a significant battery, it is not known for sure what damage was caused to the massive defences of Harlech, where it has been suggested that the destruction of much of the outer curtain on the east and south of the concentric castle was due to this attack. Too little remains of Aberystwyth to attempt such an estimate. Harlech also revealed a pile of stone balls,[4] that may date from this event. The expenses of the rebellion continued to rise with repair work to the castles that had been engaged, and payments to garrisons maintained while the settlement was enforced.

[1] *Adam of Usk*, 152. Although the chronicler had been abroad during this time, it is likely that he was well informed of such events local to his home.

[2] *Calendar of Patent Rolls, Henry IV (II)*, 285; *III*, 60 (guns), 6 (commissions); Ellis, 13–14; *Proceedings and Ordinances of the Privy Council of England*, II (1410–22), 64–9 (payments to garrisons).

[3] *Calender of Patent Rolls, Henry IV (III)*, 361–2; Walsingham, 277.

[4] Annals of Owen Glyn Dŵr, in Lloyd, 153–4; A Taylor, *Harlech Castle* (guide book), revised ed. 1988, 8–9.

Ireland

There was little in reserve to assist English rule in Ireland, and the annals recorded non-stop raiding and feuding right through the fourteenth century, with even the coastal towns with their walls and strong houses and tower houses not secure: Limerick had been burnt in 1369, while in 1396 the "whole town" of Sligo met the same fate. The Irish rulers had acquired a taste for the castles originally imported by the settlers, from whom indeed they were anyway by now often indistinguishable. The O'Conors of Connacht based themselves in the substantial late thirteenth century castle of Roscommon, with another family residing in nearby Ballintober castle. In 1409, Roscommon was provisioned to enable it to resist the attention of raiders from Connacht. In 1413, a party of raiding gallowglasses broke into the "bawn" of the castle, but the constable was killed by an arrow shot from the castle. In 1418 or 1419, in order to attack the "great castle" there, a "small castle" was thrown up by the O'Kellys inside 15 days to stand against it. The O'Conors led a "large army" in an attempt to destroy the new fort, but were successfully resisted, so they abandoned the effort, and instead left their own stone castle well supplied.[1] Generally though, the fighting ended in the capture and burning, or sometimes the successful defence, of the strongholds, which often took the form of a tower with an attached courtyard (bawn). Against a significant armed force, such as that led by Earl Mortimer in 1381, there was no effective resistance, and on that occasion Mortimer had recovered the vital and substantial castle at Athlone, but such events were now very rare, and without such an army there was no force available except that of the local Irish (and Anglo-Irish) rulers. This was itself limited and dispersed, hence the nature of the events recorded in the monastic chronicles.

It is therefore not surprising that the landscape of Ireland became dotted with many hundreds of tower houses. This simple description, however, conceals a great deal of uncertainty and, therefore, scholarly speculation. Towers were built in Ireland from the fourteenth to the seventeenth century, and frequently documentary evidence is lacking altogether to allow any particular tower to be identified (and dated) with a written reference. Frequently too, all that remains is a featureless stone rectangle without outbuildings, impossible to date precisely. Features that might help with dating, such as window designs or machicolations or, more rarely, gun loops, could as easily have been inserted later as have been part of the original design. Records confirm the widespread establishment of *fortalices* from the fourteenth century, with the government encouraging landowners to build them both for their own, and for communal and government protection. They are explicitly stated to be there to guard the marches in the middle of the century, although it cannot be proven that they were tower houses.[2] This process

[1] *Annales Connacht. The annals of Connacht 1224–1544*, ed. A Martin Freeman, 1944, 339, 405, 439–41; Four Masters, *Annals of the Kingdom of Ireland*, ed. O'Donovan, 1848, IV, 747, 799, 837.

[2] T Barry, *The archaeology of Medieval Ireland*, 1987, 181–7, gives the references. But see

was further promoted with the 1429 grant of the sum of £10 to those prepared to erect a tower house in the government-controlled area of the "Pale" around Dublin, a grant later extended to other areas. While it is clear this time that towers were intended, and common design and size gives clues as to the assignment of some towers to the response to this grant, the argument around the date of any individual construction can still be circular! Elsewhere, the more prosperous followers of native Irish dynasties were themselves building towers, and in both cases the connection with the more simply domestic form known as the "hall house", closely akin to the manor house becoming common in England and Scotland at the time remains a topic of debate.

There is sufficient to indicate that buildings erected from around 1400 did contain defensive features, which may well reflect the changing political situation as much as it may reflect the social trend of land owners to seek greater privacy. In contrast to developments elsewhere, guns were still very rare in Ireland, as they were not manufactured in the island, and the scale of the warfare seems not to have encouraged such limited expeditions as there were from England to introduce cannons yet, something that was not to become common until the time of the Tudors.[1] The near-total disappearance of the bawns makes it impossible to know whether gun loops were inserted in their walls. The occasional detail provided in the chronicles does not suggest a very sophisticated level of technology being deployed. The fighting was not often between professional soldiers. Alongside the occasional reference to arrows, even more basic means of defence were deployed: in 1454 Donnell O'Donnell, having been released from being chained in the tower of Castle Inch (Donegal) when his rival Rury O'Donnell broke in by burning down the door, went to the roof, threw down a stone that killed Rury, and thus won leadership of his clan.[2]

The towers themselves were protected from the roof tops, with their crenellations and, sometimes, projecting bartisans. Doors were defended by the iron yett, familiar also from Scotland, and by either external box-shaped machicolations above, or by a simple hole in the floor above the lobby behind the door. At this date, the chronicles that tell of the numerous attacks on and defences of the "castles" (*caislen*) of small-scale landowners simply do not describe what actually happened. It is therefore another assumption that the measures described were intended to be used for defence, although that is surely a logical conclusion. But the truth remains that concrete evidence for exactly what was being constructed at this date, and what it was for, does not exist.[3] The arguments will continue.

also A Jordan, 'Date, chronology and evidence of the county Wexford tower house', *Journal of the Wexford Historical Society*, 13, 1990/1, 41–2.

1 T McNeill, *Castles in Ireland: feudal power in a Gaelic world*, 1997, 226–8.

2 *Four Masters*, 989.

3 As well as Barry, Jordan and McNeill cited in p. 222 n. 2 and n. 1 above, see D Sweetman, *Irish Castles and Fortified Houses*, 1995, 39, and the same, 'The hall house in Ireland', *Archaeology Ireland*, 12, 1998, 13–16; McNeill, 'The origin of tower houses', *Archaeology Ireland*, 19, 1992; C J Donnelly, 'Frowning ruins: the tower houses of medieval Ireland', *History Ireland*, 4, 1996, 11–16. A few of the even more humble moated sites, present in very large

Italy

Florence and Pisa

There is a large gap between the resources available to rival leaders in county Donegal and their contemporaries in northern Italy. By 1400, a clearer pecking order was emerging among the cities, as strong leaders secured power in some former republics, while other cities retained their republican or communal form of government. Smaller cities were falling into the power of larger, as the former land of city states gave way to a land of territorial states ruled from a major central power base. None, however, was strong enough to defeat all its neighbours, and indeed there was a constant tension between seeking to expand one's own territory and to deny the same to a neighbour. The result was incessant warfare and constantly shifting patterns of alliances.

In Tuscany, Florence, glorying in its republican traditions and artistic flourishing, finally ended the independence of its neighbour Pisa, although not without a protracted struggle that reveals much of the siege techniques of the age. It is no surprise that this highly literate world should have produced a detailed written record of the events. The time when Pisa had been a rival to the greatest cities of the peninsula was long gone, the silting up of the harbour having been a serious blow, but the Black Death had brought devastation so severe that recovery was impossible. Stubbornly the city clung to its glorious past, but by 1400 was a fruit waiting to be plucked. In 1405 this was done, the city's dominant families literally selling Pisa to the Florentines. During the coming conflict, a key to Florence's success was their diplomatic skill in ensuring that their own rivals were prevented from intervening to interrupt this extension to its power and resources, although what began as a simple coup turned dramatically into a prolonged and expensive war.

The Florentines acted promptly to secure their new acquisition, despatching a force to take control of Pisa's citadel (30 August 1405) and, they thought, thereby secure their power. Chroniclers recorded that the weakness of the citadel was known and understood, although there may be an element of hindsight in their criticism of the Florentine commander, Lorenzo Raffacani, who entered it with 500 men. The walls of the city itself were fairly strong, dating back in origin to a long period of new construction beginning in 1155 but not completed until 1286, when the suburb south of the river Arno, Chinzica, was finally enclosed. During the fourteenth century, this circuit had been strengthened with outworks in front of the gates and with new towers and bastions following the city's fall by betrayal into the hands of the emperor in 1328. Standing up to 40 feet (13 metres) high, and provided with clear space both in front and behind by edict, this represented a powerful obstacle, additionally protected by the river Auser on the north. Fortified bridges crossed the Arno which separated the city from its southern suburb.

numbers, may have included a defensive function as well, according to Barry, The defensive nature of Irish moated sites, in J R Kenyon, K O'Conor (eds.), *The Medieval Castle in England and Wales*, 2003, 182–92.

One of these bridges joined the old citadel, constructed onto the town's wall in the south-west corner.[1]

Not only was the citadel apparently ill-provided for and in a dilapidated condition, but it possessed a fundamental design fault, whereby a major tower (called St. Agnes) on its circuit also connected directly with the town, forming an angle with another wall which made the whole structure impossible to isolate and hold against an attack from that direction, and gave the citizens an easy approach. This was precisely the direction the Pisan citizens attacked from, and apparently had access to ample supplies of weaponry including siege equipment specifically including bombards, *mangani* and trebuchets. First they penned in the Florentines with a ditch and palisade from the river Arno to the St. Agnes tower, then began their attack with bombards and stone-throwers. The Pisans were able to force the garrison out of the tower by weight of bombardment, then used bombards to blast down the inner door and erupt into the citadel. Racing around the galleries that topped the rampart, they fought their way into the towers on the circuit and the Florentines fled or surrendered (6 September 1405).[2]

Dismayed, the Florentines rejected Pisan pleas to be left free and organised for full scale war. Commanders were appointed, the city's militia mobilised, and additional professional troops hired. Bertoldo Orsini was ordered to march into Pisa's territory to conquer the countryside with its *castella*, and to secure the city itself. He was given authority to provide garrisons for forts and castles as he judged appropriate, and to construct new fortifications at his discretion (5 October 1405).[3] He first marched right up to the city, and was met with a barrage of fire from the city's cannons (28 October). Orsini therefore decided to lay siege to the castle at Vicopisano, just ten miles (16 kms) upstream from Pisa, confident that if this outlying fortress fell, the city would not resist. In this decision he was proved badly wrong. The castle stood on a hill beside the Arno, with another river providing further protection, and had been the subject of frequent attack by the people of Lucca throughout the previous century. As a result, the Pisans had indeed seen it as an outwork of their city, and had built a new and modern castle on the hill in the 1330s, although nothing now remains of the walls.[4]

It was then held by a determined and well-armed and provisioned garrison, obviously aware of how much their city relied on their resistance. The initial attack was made by substantial forces, also well provided with weaponry, although the form of the assault seems to have been rather archaic by modern standards. Figuring prominently in the arsenal were a wooden siege tower of great height, necessitated by the castle's lofty site, another of lesser size, and very long ladders made

[1] E Tolaini, *Le citta nella historia d'Italia: Pisa*, 1992, 51–5; J Ross, N Erichsen, *The story of Pisa*, 1909, 13–14.

[2] Matthei Palmerii, De captivate Pisarum liber, *RIS* XIX (II), 1904, 10–11; Nafri di Ser Piero Cronaca, in Corazzini (ed.), *L'assedio di Pisa*, 1885, 8–10; Gino di Neri Capponi, Commentari dell'acquisto ovrere presa di Pisa, *RIS*, XVIII, 1731, 1127–31; Piero di Giovanni Minerbetti Cronica, *RIS*, II, 1790, 531–2.

[3] Instruction of the Ten to the Captain General, in Corazzini (ed.), 100–104.

[4] Palmerii, 13–14; Nofri di Ser Piero, 29–31; Capponi, 1131; Ross and Erichsen, 402–7 (description of Vicopisano).

of fir. No fewer than 40 master carpenters worked on the larger tower. Wooden cats were also constructed in numbers to provide shelter for the attackers. This work was done with remarkable speed, and the order for assault was given on 24 November. A bombardment had been carried out in preparation although there were not many guns deployed (as yet). An eye witness described the preparations in detail down to the names of the officers in charge of each element of the attack. Teams carried the fascines used to fill the ditch, others were involved in setting up trestles to cover the excavations through which it was hoped to drain water away. Protective fire was provided by bombards, being brought up close to the wall, and by crossbowmen mounted on top of the wooden tower, which was moved forward by teams of oxen. Boats full of soldiers attacked simultaneously from the Arno. The two assault structures were manoeuvred forward under the skilled direction of their constructors, but the attack failed. Frustratingly, the eye witness account has two pages missing at precisely this moment, and the other versions, written at second hand, give no detail. Following the failure of the assault, the Florentines turned to a close blockade, accompanied by continuous bombardment. More and more bombards were deployed, and more old-fashioned stone throwers were constructed by engineers from the north, these *bricollas* being sufficiently new to Matthew Palmerius to justify him writing a detailed description. Throwing stones of "immense weight", they were apparently more destructive than the bombards, although these included weapons of considerable size too. Orsini was relieved of his command, having been found to be without sufficient experience in matters of war.[1]

The Florentines were now aware that Pisa was not going to abandon its independence easily. They did not, however, attempt a regular siege of the city itself. No doubt they recognised that the population was sufficiently determined and well armed to put up a strong resistance from behind strong defences, the long circuit of walls would require an immense besieging force, and there had to be cheaper ways of using Florence's superiority in all aspects to bring Pisa to its knees. The conquest of the city's territory, denying it food and resources from its hinterland, was being carried out by Florentine detachments carrying artillery with them. In general, most of the small castles surrendered, although sometimes it was necessary to set up ready for a siege before the defenders agreed to an honourable capitulation. Some needed to be taken by escalade, on which occasions the small number of occupants were generally unable to offer serious resistance.[2] But the main decision on how to bring about the capture, and the one that proved ultimately decisive, was the order to block the river Arno itself between Pisa and the sea, and thereby to prevent any food breaking the otherwise rather distant blockade. The means to this end involved the construction of what the sources call bastions, forts built of earth and timber, well protected by ditches, ramparts and palisades and held by crossbowmen and cannon, opposite each other on the banks of the river. Between the two was built a bridge, itself provided with fortifications,

[1] Palmerii, 14–16; Capponi, 1132; Minerbetti, 540; Nofri di Ser Piero, 31–44 (for the detailed day by day version, breaking off in mid-sentence).

[2] Instruction to Rinaldo Gianfigliozzi, in Corrazzine (ed.), 113; Minerbetti, 544–5.

to both block the river entirely, and to enable contact between the two forts. Six thousand peasants from the Florentine countryside were enlisted for the work, as were 200 carpenters and other craftsmen from Florence, Pistoia and other places, as the work was carried out in the early months of 1406, a massive feat of engineering. With the river blocked, and Florentine forces denying the Pisans access to their own countryside, the outlook was now bleak for the beleaguered city. Their very numbers, a source of strength against direct attack, now became a source of weakness as access to food supplies was cut off, and "useless mouths" (such as monks and other religious) were expelled. The elements offered a glimmer of hope. In May 1406, the Arno was in flood, and the Pisans seized the moment to prepare wooden beams to float down against the bridge. It worked, the bridge being smashed. A large Pisan force immediately marched out along one bank, determined to capture one of the forts and break the blockade. But the Florentines saw this, and sent reinforcements over by boat, sufficient to repel the desperate Pisan assault and drive it back in rout.[1] The fort was now provided with a strong garrison of 300 mercenary soldiers, and where the bridge had been an iron chain was slung across the river.

Meanwhile, as the siege had progressed, attempts made by Pisan ships to convey food supplies up the river had been regularly thwarted by the Florentine fleet. Florentine commanders decided to attempt a surprise capture of the city by escalade (June 1406). Attacking on a Sunday morning along the wall from the gate of Scampace to that of S. Marco, they planned to get inside to open the gate for their compatriots, more than 5,000 strong, outside. They were discovered, the city was awoken by the blast of trumpets, the citizens rushed to defend the walls by the light of candles and lamps and the Florentine attackers were all killed or captured.[2]

The end was however in sight. On 16 July, Vicopisano at last surrendered, after a siege and close blockade of epic proportions. Accounts from both sides were agreed on the heroic nature of the defence. The besiegers had used bombards large and small, and traditional stone throwing weapons, and had reduced the castle's buildings to rubble, and also smashed great breaches in the walls. The surviving defenders huddled in underground caverns and cellars to escape the incessant, day and night, bombardment, and suffered appalling casualties. For all that, it was the exhaustion of food supplies that finally produced an honourable capitulation. Of the 200 men who had formed the defence, only 90 survived, and the fact that it was hunger rather than the immense arsenal of artillery that had done the job was noted by the obviously deeply impressed chroniclers.[3]

The city itself was now helpless, and surrender followed in October 1406, with famine rampant everywhere within the walls. As the Florentines occupied the prostrate city, their first move was to inspect the defences and take immediate

[1] Palmerii, 19–21; Anonimo Pisano Cronichetta, in G Odoardo (ed.), *L'assedio di Pisa*, 61–2; Nofri di Ser Piero, 53–5; Capponi, 1134–5; Minerbetti, 552–3.

[2] Palmerii, 21–2; Anonimo Pisano, 64; Capponi, 1135.

[3] Nofri di Ser Piero, 56–7; Anonimo Pisano, 66–7; and Palmerii, 24–8, give the greatest detail.

steps to bring in engineers and masons to undertake a massive programme of new work, including summoning back Domenico di Matteo from Venice. We know, of course, how well provided the victorious city was with craftsmen, artists and builders. But the account given here also serves to remind us that these masters were as much at home building fortresses as erecting cathedrals and grand palaces, as medieval masons had always been in every country. Now they were assigned to designing and building "citadels and towers", pulling down anything that stood in the way including churches and monasteries. The weakest stretches of the defences, facing to the west (and unprotected by the rivers) were the first areas to be worked on, along with the old citadel whose defence and recapture had begun the whole bloody saga more than a year before. A large garrison comprising 800 lances of mounted troops and 1500 hired infantry were stationed there, too, while the work of modernisation was carried out. The walls of Vicopisano were also reconstructed. Work at Pisa continued into the 1430s, and resumed in the 1440s when an entirely new citadel was built in the south-eastern corner connected to a new fortified bridge. From what remains, it seems to have been a quadrangular fortress with massive projecting circular towers connecting a powerful curtain with a sloping outer face. The famous architect Brunelleschi was in charge of this work.[1]

The Florentine republic had used its wealth and resources to mobilise large military forces including apparently unlimited numbers of engineers, masons and carpenters to manufacture great quantities of siege equipment and to operate large amounts of artillery both old and new, including small and large cannon able to demolish both walls and houses. The relatively modern castle at Vicopisano had nonetheless successfully repelled every effort to capture it by storm with a garrison of just 200 soldiers for nine months. The city of Pisa had a powerful circuit of walls and towers, and courageous defenders, including hired professional soldiers whose role was highlighted by writers on the Pisan side. It was the simple earthworks astride the river Arno, and the roving bands of soldiers who denied the Pisans access to their immediate hinterland, that had brought the defenders to their knees, both at Vicopisano and Pisa itself. Nonetheless, the conquerors now devoted even more money and crafts to the providing of the subject city with the most modern walls and towers, alongside, in the short term, a strengthened citadel for the Florentines to ensure their control, and in the longer term a brand new fortress in another corner. No doubt, the new works powerfully symbolised the new order and the hopelessness of challenging it. But they also provided Florentine garrisons with modern defences facing both inward, and outward.

[1] Minerbetti, 561–2; Anonimo Pisano, 75–6; Palmerii, 34; Tolaini, 72–4. Vasari, *The Lives of the painters, sculptors and architects*, trans. Gaunt, 1963, I, 291.

Plate 17. Rumeli Hisar (Turkey). The artillery fortress built by Mehmed the Conqueror in 1452 to isolate Constantinople mounted guns at sea level that controlled the Bosporus. Photo © Sarah Speight.

Plate 18. Spiš Castle (Slovakia). The enormous outer ward was added in the fifteenth century as a base for an army. The older castle rears up behind. Despite their late date, the outer ward defences are not designed for heavy guns.

PLATE 19. Threave Castle (Scotland). The outer wall and towers designed to mount guns at more than one level were added in 1447–50.

Plate 20. Dartmouth Castle (England). Built in 1481 to mount guns from the towers and adjoining batteries to command the entrance to the harbour.

Plate 21. Nauplio (Nafplio) (Greece). The Ottoman Turks built the artillery fort ("Bourtzi") on an island in the harbour. The main tower had heavy guns on the roof and the outlying bastions also carried guns.

Moles et turris sancti Nicolai dirupta:et pugna mari et terra.

PLATE 22. Siege of Rhodes, 1480, showing the failed Ottoman attack on St Nicholas Tower. Note the handguns firing from inside the tower. From Nicolas Caoursin, *Obsidionis Rhodiae urbis descriptio*, 1496, with permission of the Museum and Library of the Order of St John.

PLATE 23. Auxonne (France). Louis XI's citadel built with massive towers to control the newly-seized duchy of Burgundy after 1477 once had embrasures for large guns.

PLATE 24. Fougères (France). The powerful castle on the Brittany/Normandy border was strengthened with large gun towers such as this late in the fifteenth century. Note also the machicolations.

PLATE 25. Prague Castle (Czech Republic). The vast Mihulka (Powder) Tower was added by Benedikt Ried at the end of the fifteenth century. It contains artillery positions at all levels, and commands both the ditch and the land beyond the castle.

Plate 26. Salses (France). One of the first true artillery fortresses, built by the Spanish from 1497. Seen here, the main gate from the entrance barbican. The talus was added after a failed French siege in 1503, in the process blocking the moat-level gun embrasures.

Plate 27. Chinchon Castle (Spain). A late fifteenth century castle, with upper levels demolished, nonetheless shows the intention behind the design.

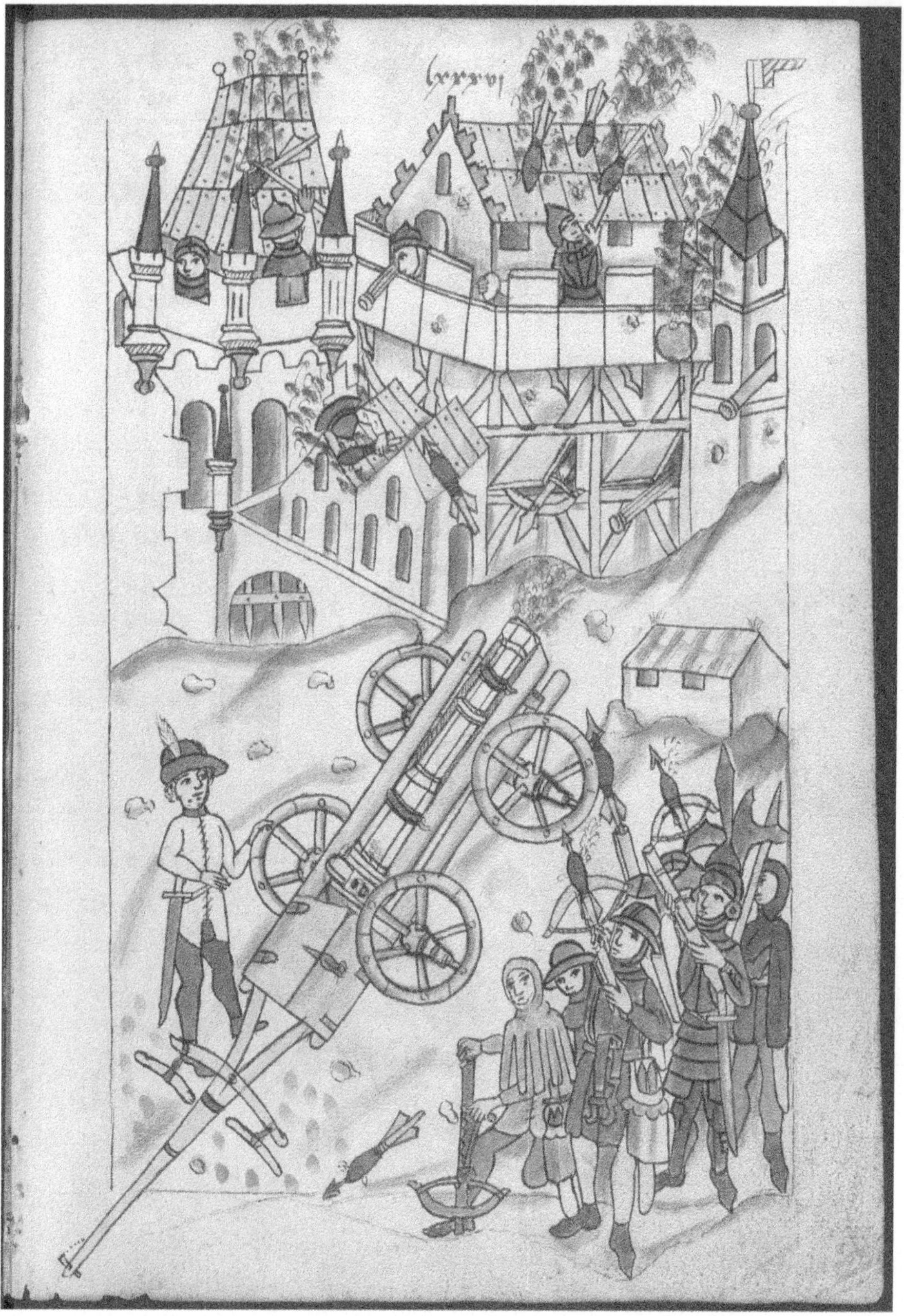

PLATE 28. From a MS German "Firework book" of the fifteenth century, showing a gun on a wheeled carriage and crossbows firing explosive arrows that are causing fires in the city, with defenders seen trying to remove them. The defence has crossbows and guns of its own. By permission of the Royal Armouries Museum.

PLATE 29. This late fifteenth century illustration shows both handguns and heavy artillery in a siege with defenders also using the hand gun. Flemish translation of *Life of Alexander the Great* by Quintus Curtius Rufus, British Library Burney MS 169, fol. 127^{r}.

PLATE 30. A fifteenth century *veuglaire* (fowler) at Castelnaud (France). A standard medium-sized artillery piece here provided with a reconstructed mounting that is a good reflection of how it would have looked.

PLATE 31. The fifteenth century bombard *Dulle Griet* on display at Ghent (Belgium). See pp. 275, 292.

PLATE 32. The fifteenth century bombard *Mons Meg* on display at Edinburgh Castle (Scotland).

Venice, Aquileia and Dalmatia

Florence's alliance with Venice enabled it to carry through the conquest of Pisa uninterrupted, and it was the growing power of the Adriatic republic dominated north-west Italy and Dalmatia. Since its close-run victory in the war of Chioggia, Venice had substantially extended its power by land while continuing to develop its possessions in the Aegean and mainland Greece. Many of the cities of the Lombard plain that had not already fallen under the influence of Milan were now under the rule of Venice. Aggressive diplomacy had achieved most of the conquests. In the north east, however, the kingdom of Hungary had for long now been a rival in Dalmatia, and the combination of Hungary with the empire in the person of King Sigismund[1] represented another challenge. The Dukes of Austria were an immediate neighbour and rival. Inbetween stood the extensive lands ruled by the patriarch of Aquileia. The last time Aquileia featured in military history was when it was besieged by Attila. Now the territory was about to be absorbed by its equally ancient maritime neighbour.[2]

In 1409, the childless King Ladislas of Naples sold Dalmatia to the Venetian republic for 100,000 ducats and thus launched a long period of conflict with the Hungarians. The Senate sent a force to secure Zara including 300 mounted crossbowmen and 300 foot soldiers, along with a fleet of galleys and ships diverted from other destinations. It seems that the defences had been neglected, because it is noted that they set about digging a substantial ditch more than twenty metres wide, and five deep.[3] The Hungarians sent large forces into the area and into Friuli and Istria, where several towns raised the banner of revolt against the troops the republic had despatched into the area to occupy its fortresses. Work at Zara continued into 1410, the defences being strengthened from sea to sea, a castle being constructed to protect the port and a citadel being constructed on the landward side. Meanwhile, Venetian forces laid siege to the town of Sebenico (Sibenik). The Hungarian defenders resisted strongly and it seems that the defences were strong enough to prevent a direct attack. Certainly, the siege was continuing in 1411, when they erected forts to maintain the blockade, and placed galleys against the harbour and in support of the two siege forts, to maintain a naval blockade as well.[4] Sebenico finally capitulated after a long blockade on 8 November 1412.[5]

In 1412, Sigismund engineered the election of his candidate, Louis Duke of Teck, as Patriarch, and had his ally the powerful count of Gorizia invest him with the patriarch's lands and castles. But the alternative candidate for the job, the local

[1] Sigismund of the House of Luxembourg was King of Germany from 1410, king of Bohemia from 1419, and of Hungary from 1387. He died in 1437.

[2] P Leicht, *Breve Storia del Friuli*, 3rd ed. 1952, is a well-respected short introduction to the history of the region. See my *History of the early medieval siege*, 10–11.

[3] Marino Sanuto, Vitae Ducum Venetorum, *RIS*, XXII, 1732, 846. Sanuto is a slightly later compilation based on contemporary accounts, and naturally biased in favour of his Venetian home city.

[4] Sanuto, 846, 855–6; A Bellono, Vitae Patriarcharum Aquilejensium, *RIS*, XVI, 1730, 64–5.

[5] Sanuto, 872.

aristocrat from the equally powerful Savorgnan family, retained control of many castles. In 1413, Sigismund led a powerful force to attack the Savorgnan castle of Ariis. The site, which probably dated back to Roman times, situated next to a river, had been constructed or repaired regularly. The Hungarian host gave up the attack after 25 days.[1] Soon, a five year truce was agreed. Sigismund was soon heavily involved in the major church council of Constance (1415) that would at last bring to an end the "great schism" in the Catholic church.

On the expiry of the truce, fighting however resumed. The outcome was determined by the result of several sieges. The chief cities were Cividale and Udine, Aquileia itself having become largely deserted. In Spring 1418, the fortified town of Sacile opened its gates to Venice, and as a result the whole territory west of the river Tagliamento fell into Venetian hands except for the castle of Pinta, which had to be besieged, and it fell and was razed in September. The patriarch withdrew to Hungarian territory, returning with a force of 4000 men to lay siege to Cividale, held by Venetian troops provided with crossbowmen and bombards. Meanwhile, the Venetians and their local supporters had attacked Udine, the centre of resistance to the republic. There was a prolonged defence. The ancient city had a powerful set of walls, the circuit having been expanded to completely enclose the original castle and citadel on its hill, the walls stretching for 7,117 metres with an extension of another 3,400. The castle itself at this time was probably comprised of a palace and other buildings beside a powerful tower inside a set of double, towered, walls. If defended and supplied, it represented a powerful and difficult objective, and the population had already thrown out occupying Venetian troops. The latter's supporters were trying as stated above to recover Cividale. This equally ancient city had its defences restored during the fourteenth century. It was low-lying, adjacent to the river Natasone, and contemporary illustrations show that river feeding a wide wet moat. Powerful outworks protected the gates and the illustrations show immense towers (generally rectangular) and bastions placed on a tall, crenellated circuit wall: again, a formidable proposition for an attacker. On this occasion, it was too strong for the Hungarians, who gave up after just fifteen days and withdrew, citing the cold weather as a reason. Their bombards and stone throwers had clearly not made sufficient an impression on the walls to persuade them to remain outside in mid-December.

In 1420 Venetian forces occupied the castle of Feltre, after demolishing its walls with bombards and storming it, and in June the city of Udine was betrayed to them, the patriarch having been unable to obtain help to prevent the loss of his remaining city, and the Venetians having begun to press a close attack with artillery.[2] The republic's success was consolidated when the Dalmatian town of Trogir was taken in the same year after a stretch forty paces long of its walls and towers was demolished by artillery and the place then carried by storm despite fierce resistance by its Hungarian defenders.[3]

[1] T Miotti, *Castelli del Friuli*, II, 47–50.

[2] Sanuto, 931; Bellono, 65–6; Vitae Additamenta Patriarcharum Aquilejensium (same volume), 85–6; Miotti, II, 357–83 (Udine), III, 125–31 (Cividale).

[3] Sanuto, 931.

Sigismund now needed to switch his attention to Bohemia to deal with the Hussites, so he had little choice but to acquiesce in the suppression of the temporal power of the patriarch, and the consolidation of a large tract of land into the republic. The Venetians themselves were taking no risks, with substantial building work done on the walls of Cividale to improve the protection against cannons. They also entered in close relations with the semi-independent aristocrats whose counties now stood on the republic's frontiers. In January 1424, for example, they negotiated with an embassy from the count of Segna who wanted several castles returned to him that had been occupied by Venetian forces in the preceding war "on the grounds of expediency and military necessity". The senate would not meet these demands, but did agree to provide the count with quite a substantial battery of weapons from the city including a hundred ballistas, and a like number of "crochos", translatable (presumably) as "pots" or small guns, along with saltpetre and sulphur.[1]

Venice and Milan, the siege of Brescia 1438

In 1437, Venice found itself facing a serious opposition led by the Visconti duke of Milan, who brought over to their side Venice's previous ally the marquis of Mantua. Central to the conflict became the defence of the city of Brescia, regarded as the key to the republic's mainland empire. In July 1438, the alliance's commander Nicolo Piccinino laid siege to Brescia, having occupied the surrounding countryside including its castles with little opposition, although the castle of Montichiao had to be reduced by bombards, surrendering to overwhelming force in three days.[2] The period of manoeuvre before the siege began had given the Brescians time to lay in substantial supplies of food and munitions. Venetian captain-general Gattamelata had supervised these operations, while the city's commander, Francesco Balsaro, organised an upgrading of the defences. The walled city was of ancient origin, and in 1438 represented an irregular quadrangle. At the highest point, inside the city, stood a "new" citadel, while there was an "old" castle on the enceinte. The stretch of wall most exposed to attack ran from a corner tower called Mombello via two gates to another tower called Ravaratto, according to an old plan. The garrison repaired the walls and cleared the ditch. Additional defences were created in the form of temporary bulwarks in front of many of the gates, and by the placing of spikes in the ditch, at least some of which were water-filled. The population was organised into "centuries" commanded by centurions and constables, assigned to specific sections of the city, to serve alongside the professional soldiers who included crossbowmen and hand gunners.[3] An active defence was undertaken, with garrisons holding neighbouring castles or launching attacks on

[1] F Cusin, 'Documenti per la storia del confine orientale d'Italia nel secoli XIV e XV', *Archeografo Triestino*, 3rd ser., XXI, 1936, 30–2.

[2] Cristoforo da Soldo, Cronaca, *RIS* XXI pt III, ed. Brizzalara, 1938, 7–10. Da Soldo was directly involved from the Venetian side, serving in Brescia as a participant.

[3] Soldo, 10; E Manelmi, *Commentarium de quibusdam gentis gestis in bello gallico ill. v. Francisci Ballsari, praefecti praesidii Brixiae . . .*, 1728, 8–15; G Eroli, *Erasmo Gattamelata da*

those held by the enemy, while during the course of the siege the republic organised troops to operate in the background, and ships operating out of the castle at Sirmione to combat Milanese vessels on lake Garda.

It was the autumn before Piccinino began a serious attack on Brescia. He ordered the building of bastions against selected points of the walls on which he could mount his artillery, although this had to be done against the threat of sorties by the defenders. A bastion was completed against the Mombello tower, and a battery of no fewer than 15 bombards, began hurling stones as large as 300 pounds (136 kgs). The defenders also had their artillery deployed. Two bombards stood on the Mombello, others were placed on other threatened spots, as well as a number of stone throwers (bricollas). The whole population had to be mobilised, including women and children, and the religious, to repair the damage being done by the bombardment. Early in November, after a large scale sortie by the garrison and citizens, Piccinino concentrated his attack on the rampart from Mombello to Ravaratto, with batteries of large bombards concentrating their fire non-stop, day and night, on these points. The effect was drastic, whole stretches of masonry being flattened, forcing the citizens to plug the breaches with earth and timber, a hazardous task as it was carried out under the continuous bombardment. The defence certainly was not passive, either. Artillery was mounted by the city's engineers so as to fire down on the besiegers' bastion against Mombello and render it ineffective. With powder running low, the defenders had to manufacture the ingredients themselves.

The attackers turned to their miners, digging under the St. Andrea tower, by the Ravaratto, but the excavation failed to bring it down. On 30 November, having drained the ditch around Mombello, the attackers tried again here but were held off in a desperate day-long struggle. Conscious of the new threat of mines, the Brescians constructed an inner rampart behind their ruined walls, and provided it with hand guns and other light firearms (*arbotane*). On 12 December, a new set of mines were fired, but the wall did not fall into the ditch. Nonetheless, a full scale assault was launched the next day at three different points, and the assault was sustained for two days. Our chronicler, Da Soldo, was the officer in charge of the defence at Ravaratto. The enemy had placed a barrel of gunpowder overnight and blew it up to create a breach in the patchwork rampart of earth and timber fascines, but the defenders held on, hurling down extemporised lime-filled bombs and firing their hand guns while women carried up the supplies. On 15 December, the assault having failed to break through, the besiegers dismounted and removed their bombards and the next day withdrew.[1] The defence was recognised as a heroic episode by the Venetians. The defenders had also had to cope with the presence of plague, and the commander's careful arrangements for the burial of the dead were also praised.

In summer 1439, there was a critical struggle at Verona, first attacked in June

Narni. Suoi monumenti e sua famiglia, 1786, 386–90. This book reproduces the oldest plan of Brescia's medieval walls, allegedly as they were in 1438, 393.

[1] Da Soldo, 17–27; Manelmi, 18–34.

by Piccinino and the marquis but relieved by the Venetians. In the autumn, however, they got into the city and occupied it, although the Castelvecchio, the great fortress built by the Scaligers with its independent access to the other side of the river Adige, and two other towers on the ramparts remained in the hands of troops loyal to Venice. This factor enabled the latter, accompanied by troops from Brescia, to make a nocturnal entry (19 November) and, assisted by a popular uprising, to rout the Milanese and retake the city. In an interesting detail, Da Soldo described how his Brescians seized and disabled a Milanese bombard by what would later be called spiking it.[1] Throughout the year 1440, the Venetians' army led by Gattamelata conquered numerous castles in the region, a common feature being that shortly after the bombards were mounted, the place either surrendered, or was stormed within a couple of days. The *rocca* of Marcharia was captured on the second day, Villafranca in four days, the castle of Peschiera was unusual in resisting bombardment for nearly three weeks.[2]

The Marches

A region that formally owed the pontiff allegiance was known as the Marches. By the beginning of the fifteenth century political power was disputed between rival territorial lords and local civic communes, each of them able according to their wealth to enlist the military service of a *condottiere* who would provide and lead his own army. The region around Montefeltro owed allegiance to the duke of Urbino, that around Fermo to its bishop, Piceno to the king of Naples, while the city of Ancona was traditionally loyal to the papacy. A detailed study suggests a complicated picture of the development of fortifications in the province, varying from communal refuges and watchtowers to powerful urban defences, from rural hilltop castles modernised to be defended bases for noble power, to urban citadels. The cities tended to buy the services of experienced Lombard engineers to reconstruct their fortifications, but during the first half of the fifteenth century only some 15 of the numerous fortresses of the region might be described as being of modern design. The everywhere dominant condottieri wanted strong bases from which to exercise political control, and preferred not to have to tie up their professional troops in sieges. Rather, the rural defences that came into being were intended as refuges for the local population during the periodic ravaging of the countryside that rival armies would engage in. Routinely, they would comprise of rebuilt older towers. Such was the small refuge castle of Truschia taken over by the Commune of San Severino in 1438 with the appointment of a castellan to look after it, but left to decay following a peace treaty.

Even the major fortresses were of doubtful military use. Alongside the expense of modernising the walls, was the continuous cost of providing them with arms. The important city of Macerata was still reliant on its thirteenth century defences until the end of the century, but was trying to keep up by equipping itself with

[1] Da Soldo, 32, 38–40.

[2] Da Soldo, 46–50; Manelmi, 62.

artillery. In 1437, its inventory showed that it had 15 small bombards along with two larger, and 29 crossbows in good condition: not exactly a large provision for its citizen militia. Such militias were sometimes called on to resist attack from professional soldiers. In 1444, Francesco Sforza besieged the small town and castle of Monsampietrangeli, which was holding out for its ruler, the pope. After several months of attack, the place fell, but not until most of the houses had been flattened, the whole wall demolished, and the surrounding vineyards cut down. The report stated that Sforza used bombards, engines and *trabucchi*. If so, this is a very late date for such a weapon still to be in action. Since the information comes from a letter from a cardinal pleading with the pope for resources to rebuild the town, it is possible there is some exaggeration.[1]

The Hussites and their neighbours 1420–1430s

The Hussites were the Czech followers of religious reformer John Hus, burnt as a heretic in 1415 by the assembled bishops at the Council of Constance. Over the following decades, they defeated the strength of the Holy Roman empire time and again. The armed forces gathered by the "heretics" had the advantage of fighting for something they believed in, whether this was purely religious, or also included an element of Czech national consciousness opposed to the influence of a ruling power represented by foreigners (Germans).[2] Historians have paid much attention to the tactical skill shown by the Hussite commanders and their innovative use of gunpowder weapons on the battlefields, but for the most part the struggle in Bohemia was no different from that elsewhere in that it revolved around the attack and defence of fortified towns or castles. The resulting sieges were documented by eye witnesses and there are many contemporary records. The Hussites demonstrated sophisticated techniques in siege warfare both in attack and defence.

The new united papacy of Martin v having called for an end to schism in 1417, the task of crushing heresy fell naturally to King Sigismund, especially after he additionally succeeded his brother as king of Bohemia. He was supported by the German-speaking towns and sections of the Czech nobility, but much of the rest of the country was Hussite, or sympathetic to them. As with many revolutionary movements, the Hussites were themselves prone to divisions, and their most effective leader, Jan Žižka, left Prague to establish at Tábor his own wing of the movement. The Táborites became one of the most radical, and militarily effective,

[1] P Jansen, Rocche, Châteaux et villes fortifiées: leur rôle dans les guerres des Condotte italiennes à la fin du Moyen Âge d'après l'exemple des Marches, in M Cambet, R Herin (eds.), *Château et Guerre. Actes des Rencontres d'Archéologie et d'Histoire en Périgord 25–7 septembre 1998*,113–35; R Piciaroni, 'La ricostruzione di un castello sanseverinate alla fine del sec. xv: Truschia', *Studi Maceratesi*, 24, 1991, 522–30; G Benaducci, *Della Signoria di Francesco Sforza nella Marca, e peculiarmente in Tolentino*, 1892, Documents, pp. lix–lxi, letter dated 17 April 1444 from Cardinal Capranica to the Pope.

[2] A good introduction is F Heymann, The Crusades against the Hussites, in Setton, *History of the Crusades*,iii, 1975.

of the Czech armies. Sigismund declared a crusade in March 1420. His position appeared favourable, with royalist troops already occupying the two hill top castles of Prague. The royal complex on Hradcany hill, thoroughly modernised over the years, controlled access from the west, while Vyšehrad, stood on a steep hill high above the Vltava on the other bank, outside the old town, dominated the south and east. Both were really self-contained towns in their own right. These were under blockade by the armed forces of Prague, which was itself well walled. The much smaller Old Town had its own wall and ditch, but the New Town, constructed outside as a planned settlement during the mid thirteenth century, also had a full set of walls completed by around 1250, with numerous towers, so the rebels of the city were themselves adequately shielded. Reconstructions suggest a triple ring: the towered inner circuit fronted by a low outer wall, which (either now or later) was looped to allow for small guns, and then beyond this another ditch, a rampart and palisade.[1]

Žižka brought a substantial army to Prague's aid. Vyšehrad was isolated by a deep ditch running around it to the river, which also served to protect the city from its garrison. The besiegers established a substantial ditched and ramparted base camp, with separate siege works behind the surrounding ditch at appropriate positions, while blockhouses were erected on the island in the river Moldau, linked to the shore with floating fortifications. On their way to Prague, the Táborites had themselves deployed siege arsenals (including bombards and "ballistas") to take several castles. At Benešov, Žižka captured the town but was thwarted by the fortified monastery defended by 400 men organised by the local baronage. Now, batteries of guns were emplaced including in churches facing the Vysehrad fortress (in one of which the place where the wall was knocked down can still be identified),[2] and serious damages was inflicted on defences and defenders. By June, the substantial royalist garrisons in Prague were running short of supplies. Sigismund marched on the city with a large German and Hungarian army, arriving on 12 June. At first he was able to achieve little, despite the size of his army and his siege arsenal, while the forces of Prague found themselves between the Germans and Hungarians outside, and the royal garrisons inside. There was fierce fighting around the temporary fortifications erected by Žižka's men to protect their position, in particular two forts on mount Vitkora Hora, in which the imperial troops were worsted. Eventually, Sigismund managed to send in fresh supplies to the castle. The crusading force blockaded the city for several weeks, but were eventually compelled to withdraw as the money and supplies ran out.[3] The Hussite sieges of the castles were then resumed, while other castles and towns

[1] V Humel, Research in Prague: an historical and archaeological view of the development of Prague from the 9th to the 14th century, in Alcock and Austin, *From the Baltic to the Black Sea,* especially 275–8 (with plan); reconstructions in Staňková, Štursa, Voděra, *Pražska architektura,* 1991, 32, 51, 55.

[2] T Durdík, Zum derzeitigen Stand der Studien zu Belagerungswerken vor Böhmischen Burgen, in Wagener and Laß, 309, (sketch of Hradcany siege works), 324 (church).

[3] Laurentius Březová, Kronika Hustitska, *Fontes Rerum Bohemicarum,* v, 362, 371, 377–81, 400; Chronicon Universitatis Pragensis, *Fontes Rerum Austriacarum, Scriptores,* II(1), 39–40, 43–5; Chronicon veteris Collegiati Pragensis, same volume, 80–1.

in the region were gradually brought under rebel control. In November, an army sent by Sigismund to raise the siege of Vyšehrad was catastrophically defeated by a Hussite army.

1421 would see the capture by the Bohemians of the castles of Prague, as they capitulated: the royalist/catholic forces were chastened by their defeats and unwilling to stand against the Hussites in the field. With these two powerful fortresses in their hands, the Czechs could be confident of the strength of their major city. Their forces had continued to expand the territory under their power, with castles and towns falling to their attack throughout the year. Žižka attacked Pilsen (Plzeň) on 14 February. The town was defended by many local noblemen with their retinues and many priests: the two sides of the struggle directly personified in this siege. The Táborites captured the suburbs but battered in vain against the inner walls for four weeks, despite having bombards that caused serious damage. In the end, a truce was made. Žižka went on to take the castle of Chomutov in three days.

Meanwhile, Chrudim fell to the Praguers, and Juromierz was stormed. The Prague army encamped by the ditches and dug their own entrenchments. With bombards and three large stone throwers they maintained a daily bombardment with stones destroying roofs and shattering the defences until the defenders gave in after 25 days. In April, the castle at Broda was attacked by a Hussite force, encircled, the ditch filled in and the walls scaled all in a few hours. Prague castle gave in at last in June. Then in July, the Prague troops attacked the castle at Bílina, allegedly impregnable. With two great siege engines and many bombards from the city, this reputation was quickly shattered. In a rare setback, in August, the Prague forces laid siege to the town of Most, but were defeated by a relief force led by Frederick Margrave of Meissen. The besiegers left behind 400 dead, and abandoned their two stone throwing machines and many big bombards. Frederick and the margrave of Brandenburg tried to follow up their success with an attack on Zaczen, but were driven off by the defenders with heavy losses.[1] The report provided for one of the contributing German cities, Nürnberg, stated that the city had provided 115 mounted men with provisions for eight weeks and a bombard weighing ten hundredweight, with supplies of powder and shot, and people to operate it including a "master gunner". In the same campaign, a participant's letter to the city of Ulm described the capture of the castle of Maschau. Arriving on 2 September, the German forces awaited the arrival of their artillery, which was then emplaced. At this, their enemy surrendered, some being massacred and others imprisoned.[2]

In May 1422, the Hussites attacked the royal castle of Karlštejn with enormous bombards and five great stone throwers, killing many with their bombardment. Investigation has identified the various battery positions on neighbouring heights, located to be able to hit the castle from four sides. It took a full scale crusading army to prevent the castle falling, while another invasion from the north recovered

[1] Březová, 466–72, 476–8, 505–7; Chronicon veteris Collegiati Pragensis, 83–5; Kronika Bartoska z Drahovic (Bartoš), *Fontes Rerum Bohemicarum*, v, 590–1.

[2] *Chroniken der Frankischen Städte: Nürnberg*, II, 1864, 33–5, 37–8; the same, I, 370–1.

Chomutov.[1] The symbolic importance of this castle of Charles IV was considerable, alongside its practical value as a base and arsenal.

Hussite fortifications and siege works

Bohemia had been largely peaceful for a long time when the Hussite revolution began, and most castles had evolved as comfortable dwellings. The rapidity with which so many fell to attack in 1421/2 should not therefore be a surprise. New defences were erected under Žižka's command at Tábor that indicated an understanding of modern needs. Building onto the thirteenth century construction he found, Žižka had two lines of defences created, an inner defence with half-round, open-backed towers, and a shield wall with a half-timbered parapet set up for guns mounted in triangular projections. Three polygonal bastions were added every 250 metres to an already very innovative defence work completed by 1433. It is these that were the most radical in design. Two of them, the Souckenica and Reznicka towers, were pentagonal, and stood at the end of projecting "necks" to offer greater flanking. The third, called the Žižka tower (see Plate 14), itself projected 23 metres. Each tower had a ground floor with large ports for heavy guns, and was three storeys high in all. A round tower (that outside the Bechynska gate) was rebuilt at the castle of Tábor, Burg Hradiště, in 1430, which is the oldest known gun tower in central Europe (although predated by a generation in western Europe), provided with ground floor gun ports and upper floor loops for hand guns (see Plate 15). During the 1420s, Žižka ordered the construction of a castle at Kalich which also had projecting provision for artillery, this time a bastion connected to the castle by a fortified corridor.

Other contemporary defence-building in Bohemia revealed a range of different styles, but all of them were determined by the need to mount guns in such a way as to cause maximum protection to the castle or town, and the maximum damage to an enemy trying to get close to the walls. The need to provide outer defences (termed "shield walls" in central Europe) that would both mount guns, and offer effective protection for older stone walls behind them, was also being thought through in these conflicts. The precise shape of the defence work best suited to mounting, and resisting, cannon was still under development while the gun itself was still evolving, so the claim of the Hussites to be among the innovative forerunners of the standard defensive work of the early modern and subsequent ages is very strong — certainly their work on bastions and artillery towers was to be further developed in Bohemia in the 1460s and 1470s by King George Poděbrady (see Chapter 7).[2] Their innovations on the battlefield, in particular the mobile

[1] Chronicon veteris Collegiati Pragensis, 86; Bartosi, 592. Karlstein siege works: Durdik, in Wagener and Laß, 311 (with sketch).

[2] T Durdík, Abriss der Entwicklung der böhmischen Artillerie-burgfortifikationen des 15 und Anfangs des 16 Jahrhunderts, *Castella Maris Baltici,* II, 1996, 36 ff, and the sources (mainly in Czech) cited there; J Muk, 'Táborsky Hrad Hradiste ve Středověku', *Husitsky Tábor,* 5, summary in German, 176; plan of castle, 169; D Libal, 'Burgen und Städte als tschechische Erbgut', *Castrum Bene,* 6, 1999. In English, see T Durdík, The city walls of Tábor

"wagon fort", reflected the fact that their strength lay in infantry, the dramatic successes they won led their opponents to try to copy them. It similarly cannot have escaped the notice of the crusader commanders that Hussite defences were more effective than they were accustomed to when it came to using gunpowder weapons to prevent capture. As elsewhere, actual survivals of guns from this time are very rare, but a Hussite gun dated to 1430 was found in the castle of Křivoklát, complete with timber mount. It is a small weapon, with a calibre of 38 mm, and 1480 mm long. The reconstructed mounting indicates though that while standing on the ground, its elevation could be adjusted making it into an effective weapon for siege warfare.[1]

Hussite commanders also applied their skills to organising sieges to take full advantage of the weaponry they possessed. Recent studies by Czech scholars of the archaeology of siege works have permitted an evaluation of a topic usually prevented by the disappearance of these inevitably transient constructions. The size and location of the besieged castle would determine whether there would be a continuous, or separated siege work. Generally, there was a headquarters area, often heavily defended from all sides. Ditches were backed by ramparts, sometimes provided with their own bastions, and sometimes it seems that the wooden core of these was prefabricated. The attack was based in sites prepared as batteries, whether for guns or for trebuchets. Some were square shaped, some round, some polygonal. The actual locations of these batteries has been identified at Bechyně, Konopiště, Zborený Kostelec and Karlstein. At Lichnice castle, during the Hussite siege of 1428, a regular-planned five bastion siege fort was built, mounting guns and stone throwers. At Grabstein (Grabštejn) castle, a siege fort was erected on a spur 250 metres from the castle, and overlooking it by 5 metres, in which heavy guns could be sited, and the work was defensible all round. Where, as here, outside interference was anticipated, the besieger would erect such protection. Blockhouses provided another form of covered siege work. Hussite works were among the most advanced in adapting traditional approaches to the needs of the new artillery for both offensive and defensive purposes.[2]

The fourth and subsequent crusades

When the external pressure was lifted after the failed efforts of 1422, the Hussites themselves fell into civil war. Žižka died in 1424, but the new leader, Prokop, was able to take the continuing battle into the territories of neighbouring princes. It was 1427 before Sigismund was ready to declare a fourth crusade, during which

— a breakthrough in the development of central European fortification, in G Perbellini (ed.), The town walls in the Middle Ages, *Europa Nostra Bulletin*, 53, 2000, 71–5.

[1] T Durdík, M Pertl, Tarasnice z Hradu Křikovlátů, in M Richter (ed.), *Urbes Medii Aevi*, 1984, German summary, 154–8, illustrations 150–1.

[2] T Durdík, in Wagener and Laß, 405–24; J Frolík, 'Nálezy z obléhání hradu Lichnice v Letech 1428–29', *Castellologica Bohemica*, 8, 2002 (German summary 408); J Kypta, J Richterová, 'Opevněné postavení oblétalu hradu Grabštejno', *Castellologica Bohemica*, 9, 2004 (German summary, 290).

time the Hussites had already successfully marched into both Saxony and Austria, taking the castle of Usti (Aussig) with guns and mining despite a relief attempt by the Saxons. The 1427 invasion was planned to be overwhelming and decisive, with four separate armies due to march into Bohemia. The royal court documented the immense arsenal they had summoned. Each elector was expected to bring master gunners and all they needed for their artillery, wagons, powder and shot. Further, the ecclesiastical electors had to supply other demands. The lands of Trier were expected to bring four chamber-guns (*kammerbuchsen*), four *darrabuchsen*, 200 fire arrows, and sufficient stone, shot and master gunners for the lot. Others were ordered to provide similar weaponry, although some additionally were called on for both hand guns (*hantbuchsen*) and "big stone(-firing)" guns, albeit usually just one each of the latter. The force provided by Nürnberg included a substantial train of wagons and horses, 24 guns carried on four wagons, and one large gun, six smaller, very many handguns, along with 400 cannon balls and four tons of gunpowder.[1]

In the end, two armies materialised. One laid siege to the town of Stříbro, attacking with bombards, but it was routed by the Hussite army. Prokop followed up his victory by attacking the town of Tachov, using guns and mining. The town fell after one week, the castle a few days later.[2] Prokop brought back to Prague the spoils in triumph, including a large bombard previously lost. The immense effort of the empire had been a colossal failure all round.

In 1428 the Hussites carried the war to their neighbours. Marching into Silesia, they besieged Nysa (Neisse), capturing the suburbs but proving unable to take the castle as a result of a strong defence. Sigismund sent orders to the region to equip itself to resist attack, calling on the Silesians to arm themselves with protected wagons conveying hand guns, in imitation of their tormentors.[3] Routinely the Hussites were recorded attacking castles with arrays of siege engines, bombards and other fire arms. Two years later, the German towns of Bayreuth, Auerbach and Plauen fell to them, leading the citizens of Nürnberg to spend resources on digging their ditch and improving their defences, and repairing the stonework.[4] Yet another crusade was planned in 1431, and the Hussites defeated the three separated armies in detail. The cities of the empire alone had been called upon for 103 bombards between them.[5] The defeated army of Frederick Margrave of Brandenburg abandoned no fewer than 3000 wagons and 130 guns.[6] It was not long before a negotiated settlement was initiated, and "moderate" forces on the Hussite side fell into armed conflict with the more radical wing of the movement, the defeat of which in 1434 enabled a conclusion to the war to be reached. There

[1] *Die Chroniken der frankischen Städte: Nürnberg*, II, 47–8; E Windecke, *Denkwürdigkeiten des Zeitalters Kaisers Sigismunds*, 1893, 222–3, 226–7: an eyewitness at the royal court.

[2] Chronicon veteris Collegiati Pragensis, 88–9; Bartoš, 594–7.

[3] Windecke, 309–10; Chronicon veteris Collegiati Pragensis, 90.

[4] *Chroniken . . . Nürnberg*, I, 376, 444–46.

[5] This listed in Die Metzer Chronik des Jacques Dex, *Quellen zur Lotharingischen Geschichte*, IV, 388–93.

[6] Bartoš, 604; Chronicon veteris Collegiati Pragensis, 92.

was still fighting with the unreconciled radicals. In 1438, Sigismund's successor, the Habsburg Albert II (1437–39) tried to take Tábor itself with an army of Saxons and Bavarians, but the defenders were strong and motivated, and engaged in daily skirmishing with the besiegers, who gave up after three months. Many casualties had resulted from the discharges of siege weapons. The Hungarian chronicler stressed the strength of the fortress.[1]

Hanseatic towns and northern princes

Compelled to look after their own interests by the absence of effective royal authority, such as that which promoted urban growth in England or France, the German merchants of the north had linked together in the Hanseatic League since the thirteenth century. Their domination of many trades, but in particular the bulk trade in grain, brought them wealth and guaranteed the adhesion of numerous towns and cities. These places sometimes required protection from the attention of neighbouring princes and noblemen, hence they were walled. But by the end of the fourteenth century Hanseatic towns, with Lübeck recognised as the most senior member, were independent players on the political field. Accounts from the rich record of town chronicles give details of some of the siege warfare of this region during this time.

When in 1378, in a rare visit to the north of the empire, Charles IV joined with the duke of Saxony and Duke Albrecht of Lüneburg to attack the castles of Dannenburg and Luchow, the emperor brought with him 600 men and two trebuchets (*bliden*), and the city of Lübeck sent along a contingent which included "much good equipment which frightened those inside". It was noted that there were no *bussen* there. The castles were quickly taken. This was a time when the interests of the duke and town of Lüneburg were the same, but did not remain so. In 1396, the duke built a strong castle in the town, complete with walls and ditches, and then handed it over to the town council. The same year, a joint army from Lüneburg, Lübeck and Hamburg, led by the Burgermeister (mayor) of Hamburg, stormed the castle of Horburg, used by its Lord to block the river access to the town, using *bliden unde bussen*. Clearly then, the northern towns were commonly using gunpowder weapons, alongside more traditional equipment.[2]

The death without adult heirs of the Duke of Schauenburg, who ruled Schleswig, enticed Queen Margaret I of Denmark (ruled 1387–96) to occupy Schleswig. Her death (1412) led to feud and civil war, in which the city of Flensburg was heavily involved: it was taken in 1412 when the town wall was stormed at a point where it was particularly low and weak. In 1416, the dukes of Schleswig and Holstein (both called Henry) assaulted the Danish castle of Glambeke, spe-

[1] Bartoš, 622; Chronicon veteris Collegiati Pragensis, 98; Thuroczy, *Chronicle of the Hungarians*, 101.

[2] Herman Korner Chronik, *Die Chroniek der niedersachsischen Städte: Lüneburg*, 1931, 128, 131; Detmar Chronik, same series, *Lübeck*, ed. Koppmann, Bruns, 1884, I, 560; Rufus Chronik, same series, 1902, III, 4–5. Nothing remains of the medieval fortifications of Lüneburg.

cifically strengthened by the Danes, and assailed it for two months with guns and other siege equipment including cats, in freezing weather. It surrendered after two months. In 1426, Margaret's son Erik VII (1396–1439) besieged the city of Schleswig by land and sea. The city had good defences, with wide ditches and strong walls and outlying bulwarks, well provided with gunpowder artillery, and the duke of Schleswig was assisted in defending his towns with men and materials from Hanseatic Hamburg. But the Danes pressed a strong attack on the Schleswiger strongpoint on the hill of Hesteburg, supplying themselves with both men and engines by sea. Fighting elsewhere in the region between the bishop of Bremen and the duke of Lüneburg prevented the latter from assisting against the Danes, but some 200 Holsteiners managed to seize (once again) the castle of Glambeke, using guns and taking it by escalade. The garrison, knowing that King Erik was tied down at Schleswig, fled, and the attackers found the place full of victuals and a large bombard, which they carried by hand (so perhaps it was not so big as all that?) to help protect the town of Glottorp, also under Danish attack.

The following year, with full backing from the Hanseatic towns, the duke of Holstein assaulted Flensburg, which had outlying forts only recently constructed to modern standards to control the approaches to the town, with a citadel-like fortress called Duburg to control the town itself. The early-fifteenth century reconstruction of the walls involved their heightening, and the provision of a ten-metre square outer defence for the octagonal tower originally built in 1300 as the governor's residence. These defences had resisted all attack — indeed, Duke Henry of Schleswig died assaulting Duburg in 1427. But four years later (1431) the town finally fell by the rather more basic method of a surprise attack, once the gate had been blocked open with a cart with a broken wheel, and combined land and sea forces of the Hanse and the duke starved the citadel into yielding after more than five months.[1] The attack was carried out by guns and cats, many of them supplied by the Hanse towns, along with victuals and 800 troops. A Danish relief attempt by sea was held off. Inside, the garrison was kept in by digging ditches and erecting bulwarks. The new forts were then razed — they do not appear, by contrast with those constructed by the Hussites, to show any break from traditional approaches.[2]

The Balkans and the Turks 1420–1440s

Sigismund's authority entailed many responsibilities in geographically far-flung regions stretching from the North Sea to the Balkans, where Hungary faced not only a rivalry with Venice but also a resurgent Ottoman dynasty,

[1] Rufus Chronik, 89–90, 239–50, 410–7; *Flensburg: 700 Jahre Stadt Flensburg 1284–1984*, 1984, 24–7; O Pelc, *Im Schutz von Mauern und Toren. Die Befestigung der Schleswig-holsteinischen Städte in Mittelalter und Neuzeit*, 2003, 34–8 (Flensburg); H F Schult, 'Die als Steinbau errichtete Turmburg in Flensburg', *Offa*, 53, 1996, 317–24 (summary).

[2] Die Ratschronik, *Die Chroniken der niedersachsischen Städte: Lübeck*, IV, 1910, 21; Lüneburger Chronik (continuation), same series, *Lüneburg*, 1931, 139–40.

whose defeat at the hands of Timur had brought only a few years' respite to their northern and western neighbours. The other beneficiaries in Constantinople also took such steps to protect themselves as they were able to resource. In 1413, Emperor Manuel II Palaeologus (1391–1425) sought to strengthen the defences of the imperial lands in the Morea with a reconstruction of the defences across the isthmus of Corinth, the Hexamilion ("six mile") fortification involving a wall of 3,800 yards (3,500 metres) topped with 153 towers: a massive work. Defences were also improved at Thessaloniki.[1]

Siege of Constantinople 1422

In 1422 Murad II (1421–51) decided to attempt the conquest of the famous city. His forces arrived on 8 June, when an army of 10,000 laid waste the countryside then constructed a siege camp just a bow shot from the triple walls. They seem to have been well organised for the attack. They built great wooden shelters and siege equipment to protect themselves from the arrows of the garrison and the stones fired from its bombards, and planned their own onslaught at points where they calculated the ditches were shallow and the walls most decayed from age. Against these points they arrayed "giant" bombards firing great stones. However, the defences were better than they anticipated, and despite landing many hits they failed to make a breach. Meanwhile, wooden siege towers running on iron wheels were drawn up to the edge of the ditch hauled by oxen, while sappers tried to dig into the ditch. On 22 August, there was a general assault, the Turks advancing with ladders and tools of all kinds to help them get across the ditch, but they were held there, then put to flight, and the victorious defenders were able to take possession of the siege engines and destroy them by fire. Disheartened, the attackers gave up a couple of days later, having lost (it was said) 1000 men, as against just 30 killed and 100 wounded among the defence. It is hard to verify the figures, as the Ottoman accounts skate quickly over the reverse, leaving both the city, and the account of the siege, to the Byzantines.[2] Although these versions suggest a deadly serious attempt to secure the city, was an army of 10,000, large as this was by medieval standards, really enough to attack such enormous defences, and to do so without naval assistance against the sea walls as well? Nonetheless, the lessons of this failure were not lost on the Ottomans.

Sigismund's defences

Ottoman aggression was far from tamed, as the following year the Hexamilion was stormed by Turkish forces, and raids into the Balkan principalities were regular occurrences. Since Nicopolis, King Sigismund had also good reason to pay attention to his Danubian frontier when business elsewhere allowed. Generally, his approach had been to cultivate relations with the rulers of Serbia, Bosnia and

[1] G Sphrantzes, *Fall of the Byzantine empire*, trans. Philippides, 1980, 22. The writer lived 1401–77 and was an imperial servant from 1418.

[2] J Cananus, De bello Constantinopolitano 1422, *PG*, 156, cols. 63–82; Sphrantzes, 27.

Wallachia, supporting them as what might now be termed buffer states against the Ottomans. This method sometimes failed, as when in 1427, the town of Golubac which stood on a cliff above the river was sold by its captain to the Ottomans, forcing Sigismund to try to recover it. However, the Ottoman forces came against him deploying "engines of war of great size", and the king had to abandon his attack.[1]

The position of the small Balkan principalities was precarious. George Branković succeeded as Despot of Serbia in 1427 when much of the south of his country was under Ottoman occupation, leading him to establish a new capital by founding a fortified settlement at Smederevo, prudently done with the Sultan's permission. This was built very fast, and excavations suggest an initially small fortress forming a triangle between the rivers Morava and Danube, completed by 1430 but at the cost of having to impress thousands of his subjects into virtual slave labour. By 1439, a larger outer enclosure had been completed.[2] Sigismund also restored royal influence, and defences, in Hungary, recovering large numbers of alienated royal castles, and also encouraged the modernisation of the fortifications of Danubian towns. The castle at Györ, for example, was adapted for guns, and the double circuit of walls at Sopron (completed during the fourteenth century) was further modernised early in the fifteenth. A co-ordinated effort was made to organise defences. An Italian, Pipo Scolari, was in charge of organising the construction of new fortifications between 1410 and 1426, and 14 new castles were erected along the Danube frontier between 1419 and 1429, including Orsova, Szoveny, Temesvar and Golubac.[3]

Sigismund had also conducted negotiations with the Teutonic Knights to bring them from the Baltic to the Balkans. With the aid of the emperor's mediation their war with Poland had been concluded and the scheme revived, this time in the form of having the Order set up a military colony along the Danubian frontier, now exposed by the collapse of Serbia. The (very reluctant) Knights under the command of Nicholas von Redwitz were to take over command of the whole zone between Belgrade and Severin, the expedition arriving in Hungary in 1429. The partial survival of meticulous records in the Order's Königsberg archives enabled the reconstruction of what was involved. Garrisons and associated non-military personnel for 16 fortresses and strong houses ranged in size from just twenty (with

[1] Thuroczy, 71.

[2] The description of Smederevo by Deroko (in Serbo-Croat) is summarised in J Fine, A tale of three fortresses, in T Miller, J Nesbitt (eds.), *Peace and War in Byzantium*, 1995, 182–3.

[3] Engel, *The realm of St Stephen*, 216–18, 236–8; L Gerevich, The rise of Hungarian towns along the Danube, in Gerevich, ed., *Towns in Medieval Hungary*, 1990, 35–40; I Holl, The development and topography of Sopron, same volume, 90–100; I Petrovics, The role of towns in the defence system of medieval Hungary, in P Contamine, O Guyotjeannin, eds., *La guerre, la violence et les gens au Moyen Âge*, 1996, 1, 263–71; K Kozak, 'Adatok a Györi püspökvar történetéhez', *Arrabona*, 3, 1961, German summary 55–6; Fugedi, Medieval Hungarian castles in existence at the start of the Ottoman advance, in Bak, Karoly, eds., *From Hunyadi to Racoczi*, 1982, 61–2; same author, *Castle and Society in Medieval Hungary*, 1986, 125–33.

four guns) to 400 (with 56 guns) assigned to the fortress of St. Laszlo (Ladislas), built in a rush by Sigismund on the opposite bank to Golubac and its Turkish occupants. All the garrisons were made up of a balance of soldiers, including a proportion of crossbowmen and (or) gunners, and workers to maintain the structure. But the cost for just one year amounted to 314,822 florins, an immense sum that was raised (almost) by assigning the Knights specified sources of royal income.[1] The second problem with this particular example of contracting-out was that it did not work. In 1432, the truce with the Sultan having ended, a Turkish attack routed the knights and the local Hungarian forces and captured three of the castles. Sigismund had to take the frontier back into his own control.

Ottoman advance continues

Turkish attacks continued, often taking the form of raids designed to collect booty and prisoners (as a reward and encouragement for the soldiers), sometimes with more permanent conquest intended. In 1430, they secured most of Albania, and Thessaloniki, Sultan Murad ordering the preparation both of guns and of stone throwers that were loaded onto ships and galleys and transported there. The city had been held by the Venetians since 1423, and there is a surviving eyewitness Byzantine account. The citizens and the Italians hurriedly extended the outworks, while the most fit among the people were selected to watch and defend the ramparts. Murad arrived soon after and had his men construct an enclosing fortification, while transporting his artillery and other equipment on camels and carts. The stone throwers were soon battering the defences, their stones landing "like thunder" and driving away the guards. At the same time, the Ottomans tried to dig a tunnel under the outer wall to allow them to get close to the inner. On the fourth day, it was stormed, the sultan having declared that his troops were free to plunder it. Archers forced the defenders to keep their heads down while ladders were placed, the harbour (Trigonion or chain-) tower was itself captured the same way. While the Venetians fled by ship, Turks who had got in opened the city's gates and Thessaloniki was plundered.[2] Work was then carried out to strengthen the city's ancient defences. However, it seems that the Trigonian Tower, where the Ottomans had made their entry, as it survives now, and the similar White Tower, are sixteenth century gun towers.[3]

1 E Joachim, 'König Sigismund und das Deutsche Ritterorden in Ungarn 1429–32', *Mitteilungen des Instituts für Österreichische Geschichtsforschung*, 38, 1912, 87–112.

2 Asik Pasha, trans. Kreutel, 164; Giese, *Die altosmanischen anonymen Chroniken*, 88; Sphrantzes, 45; Joannis Anagnostae, Narratio de extremo Thessalonicensi excidio, *PG*, 156, cols. 595–607.

3 Papahadjis, *Monuments of Thessaloniki*, 9; see Braun, Faucherre, Spieser, 'La Tour Blanche et la Tour Trigonion de Thessalonique', *Byzantinische Forschungen*, XI, 1987, 269–70. But dendro dating indicates that whatever was constructed after 1430 was rebuilt around 1535: M Kiel, 'A note on the exact date of construction of the White Tower of Thessaloniki', *Balkan Studies*, 14, 1973, 352–7.

First Ottoman siege of Belgrade

Belgrade was the key to the Danube frontier, and the Turkish sultan tried to take the city in 1440, having captured Smederevo after a two month siege in 1439. A visitor a few years before left a description of the defences as being very powerful, with five sections of which two stood on the river and three on the heights, with double walls and towers and apparently a substantial provision of artillery. He reported seeing in the citadel four large bombards, one of which impressed him by its great size, with a diameter at the business end of 42" (45 cms). Aware of the strength of the target, the Ottoman army was well provided with siege equipment including all kinds of artillery. The stone throwers and cannon were effective against the high walls and towers of the fortress, demolishing large parts of the defensive curtain, but the city was well protected by ditches, the damage was repaired night by night, and the defenders conducted a resolute defence, making many sorties to disrupt the Turkish attack. The decisive event was an attempt at mining that went horrendously wrong. The Turks cleverly began a tunnel from behind high ground out of sight of the city, using thousands of men to speed the construction. Someone apparently shot an arrow into Belgrade with a message revealing the location of the tunnel. This enabled the defenders to organise a countermine which was skilfully filled with gunpowder. When the Turkish tunnellers were heard approaching, the massive bomb was set off at the right moment in the confined space, causing a devastating explosion killing all the attackers, and leading Murad to abandon the siege after three months.[1]

The sultan had better fortune on the Corinthian isthmus. The fortification at the Hexamilion had a ditch filled with sea water in front of a wall with five bastions, which the Turks had to fill in under the cover of a bombardment by cannon. Breaches having been made in the wall by the bombardment, free plunder was offered to the Turkish troops who immediately stormed the place.[2] This was just one of a number of captures and conquests the Ottomans secured during this period. The new king of Hungary and Poland, Ladislas III/I (Poland 1434–44, Hungary from 1440) organised a new crusade, but it was destroyed in battle at Varna in 1444. A further Christian defeat at Kossovo in 1448 set the scene for the next phase of Ottoman conquest.

The Mamluk sultanate in Cyprus and Syria

The Mamluk sultans ruled Egypt, Syria and Arabia and periodically extended their control still further. Their acquaintance with modern artillery was first noticed at Alexandria in 1365. It may have been employed in the

[1] Thuroczy, 102–4, 121–4; Asik Pasha, 173–4; Giese, 89. The Ottoman accounts are very basic. The defences in 1432–33 were described by a Burgundian who wrote his account on his return home: G R Kline, (ed.), *The 'Voyage d'outremer' by Bertrandon de la Broquière*, 1988, 135–6.

[2] Asik Pasha, 178–9.

successful attack on Cyprus in 1425 and 1426, although the language is ambiguous.[1] It is however more clearly described in Taghri Birdi's descriptions of warfare in Syria, where the sultans had frequently to deal with Turcoman rulers and their own rebels against the Cairo-based élite of slave soldiers. In 1429, an expedition against the Turcoman prince Qara Yuluk recaptured Edessa (ar-Ruha) by siege. The town fell immediately to assault but the citadel resisted for two days, during which the Mamluk army set up "cannon and guns and began to breach the walls" whereupon the defenders yielded. It is not certain in what the condition the defences of this once major city were. In 1433, another expedition besieged the ancient fortress of Amida. Qara Yuluk had ordered the banks of the river Tigris broken but despite the flooded terrain the Mamluk forces pressed forward to the edge of the moat. Filling the obstacles with timber and stones, they attempted a storm but were repelled. So they turned to the construction of their own timber fortress against the city, and used cannon to hurl naphtha containers inside. The account also refers to ballistas. The siege ended when the Turcoman negotiated for a treaty after 35 days.[2]

These accounts confirm that the Mamluk empire was as familiar with and as effective in using artillery as any of their contemporaries, and indeed was possibly more advanced than their neighbours, especially to the east. Although it is not evident how this artillery was manufactured (the difficulty of obtaining suitable materials being a challenging obstacle) or its use organised – it is perhaps unlikely that the Mamluks themselves, formidable mounted fighting men, dirtied their hands using guns – they must nonetheless have had suitably experienced and trained engineers and artillerists to cast, move, mount and operate what seems to have been an integral element within their siege operations.

Eastern and central European siegecraft

The contemporary accounts cited here reveal the rapidly increasing role of guns in warfare, and a large variety of different sizes of weapon serving different purposes. Hand guns would begin to change battlefield tactics to the advantage of foot soldiers as the century progressed, although not yet decisively. In sieges, large bombards were now commonplace, with the power to flatten stone walls, while smaller guns were regularly deployed by, in particular, defenders, using them against both the personnel and the less substantial protective works of the besieger. The Hussites, among their other innovations, also pioneered a kind of fortification designed both to use, and to resist, the power of gunpowder. But the old methods did not immediately disappear. Several accounts

[1] La Monte, Zaida (eds.), 'Bedr Ed-Din Al-'Aini's account of the conquest of Cyprus 1424–1426', *Annuaire de l'Institut de Philologie et d'histoire Orientales et Slaves*, VII, 1939–44, 241–60; Makheiras, *Recital concerning the Sweet Land of Cyprus entitled "Chronicle"*, ed./trans. Dawkins, 1932, paras. 671–93; Abu'l-Mahasin ibn Taghri Birdi, History of Egypt, trans. W Popper, *University of California Publications on Semitic Philology*, 18, 1958, 36–8.

[2] Taghri Birdi, History of Egypt, trans. Piper, 18, 65–66 (ar-Ruha), 99–17 (Amida).

have mentioned stone throwers alongside the guns, and they would certainly not have been deployed merely for reasons of nostalgia. Several sieges witnessed siege towers in use, a form of attack that might have been thought particularly vulnerable to a defender's heavy guns. The final stages of attack also remained unchanged. A breach still had to be stormed, and to reach it a ditch had to be filled. If anything, the greater power of gunpowder weapons against older walls was in fact promoting even more rapid advances in counter-measures by the defence, with earthwork and timber outworks in advance of the stonework very common, and establishing an approach that would hold good for centuries to come as a protection against artillery.

The effect of Hussite innovation

Although Czech scholars in particular have repeatedly highlighted the innovatory aspects of fortress design during the years of the Hussite wars, with particular attention paid to the defences of Tábor, it remains to establish where, and when, there was diffusion of these new ideas. While conditions of warfare and revolution prevailed, there was little leisure to indulge in new building within Bohemia itself. It would be the second half of the century before widespread adoption of the main features prefigured at Tábor took place.[1] Where adaptations were made, they took the form of additions to existing structures, rather than re-modelling of the entire design. Such adaptations were to be found in castles held by Hussite leaders. Examples were the earthwork bastion at Kalich, that brought a concentration of guns to bear on the only approach, and the stone-enclosed bastion at Sion: both were freestanding, not part of an overall system. Starý Bernštejn saw the addition of a round tower to provide flanking. At Choustnik and Zvíkov, owned by Catholic lords, rectangular bastions for artillery were added to the outer defences. Elsewhere, where new castles were established, location outside of the artillery range of neighbouring heights seems to have been a consideration in the planning.[2]

The war had a dramatic impact on those neighbouring states that had been embroiled in the fighting. The regular involvement of Nürnberg has been cited. But the works carried out to improve the protection of the city were not particularly forward-looking, being initially stone cladding of the inner wall of the ditch. Only later, in the 1450s, was the decision made to provide effective defence against

[1] See, for example, T Durdík, Batterietürme der böhmischen Burgen, in Geró Laszlo, *Nyolevarlötödek születécnapjára, Tanulmányok. Múvészettötének-Múemlékvédelen,* VI, 1994, 237–8.

[2] T Durdík, Böhmische Burgen der Hussitenzeit, *Interdisziplinäre Beitrage zur Siedlungsarchäologie. Gedenkschrift für Walter Janssen,* eds. P Ettel, R Friederich, W Schier, 2002, 47–52; T Durdík, 'Česká hradní architekbura doby husitské (Böhmische Burgenarchitecktur zur Zeit der Hussiten)', *Casopis společnosti přatel starožitnosté,* 99, 3/1991, German summary 158; plans of Sion, 160, Kalich, 161, Zvíkov and Choustnik, 164. See also the older works by D Menclová, such as 'Vliv hustitkych válek na pozdně gotickou fortifikiční', *Uměni,* IX, 1961, German summary, 471. Menclová's work still provides many of the plans used in later studies of Bohemian fortresses.

artillery by creating an outer defensive circuit of a rampart and ditch fifteen metres advance of the old walls.[1] Other cities also took limited steps to improve their defences against the new threats, with Dinkelsbühl doubling its ditches on its most vulnerable side by 1435. At Rothenburg ob der Tauber, where the defences remain standing, improvements to the existing walls to incorporate guns had been concluded by around 1400, but the Hussite threat led to new works from 1427. These comprised a stone outer wall around the ditch, 12 low projecting towers on the Zwinger on the more vulnerable sides, new outer ditches and ramparts to protect the existing gate towers, and additional drawbridges.[2]

More evident during the Hussite conflict and in the wars in northern Italy, was the indefensibility of older types of fortress against modern attack. Yet it continued to be the case that resolute and numerous garrisons, regardless of the age of their walls, could still mount a resistance that could prevent capture. Such well defended fortresses might also, especially if provided with guns, succeed in their defence, although all the examples have suggested that private castles were now very unlikely to resist for long, while larger defended entities like towns and cities, such as Brescia, were far more likely to hold out long enough to discourage the attacker. Similar lessons were meanwhile being learnt in western Europe, as the Anglo-French wars resumed.

Siege warfare in the time of Henry v of England

For the unhappy French kingdom, already torn by internal conflict taking the form of vicious civil war involving substantial armies, the arrival on the scene of the militaristic new English king Henry v (1413–22), blooded as a youth in the suppression of the revolt of Owain Glyn Dwr, promised nothing but further trouble. For the historian, the second half of the "hundred years war" offers a rich array of contemporary accounts and surviving records. The propaganda element is easily identifiable behind the texts, although by balancing those from both sides it is usually possible to arrive at a consistent version of the facts.

Even while the Lancastrian was planning to resurrect his claim to the throne of France, fighting had continued among his future foes, witnessing the French siege and capture of Burgundian Compiègne followed by the bombardment, storming and sacking of Soissons using a mixture of guns (including a giant named *Bourgeois*) in 1414.[3]

[1] H Hofmann, *Die Nürnburger Stadtmauer*, 1967, 28; D Libal, *Castles of Britain and Europe*,1992, 100.

[2] K Heller, *Rothenburg ob der Tauber in Wehr und Waffen*, 1920, 32–4. But see also D Libal, *Castles*,139, for the later dating. The most reliable estimate of the dating of the different phases of fortification is in K Borchardt, E Tittman, 'Mauren, Tore, Türme. Zeugnisse zur Geschichte von Rothenburg ob der Tauber', *Jahrbuch des Vereins Alt-Rothenburg*, 1, 2005.

[3] Juvenal des Ursins, 498–9; *Journal d'un bourgeois de Paris*, 53; *Monstrelet*, III, 6–7; *Chronique de Jean le Fèvre, seigneur de St. Rémy*, ed. Morand, 1876, I, 158–65. This later account is largely based on Monstrelet but contains additional detail.

1415: Harfleur

Henry v's siege of Harfleur at the mouth of the Seine in 1415 was described in great detail by the English side (in particular the anonymous royal chaplain who compiled the *Gesta Henrici Quinti*) and much less so by the French, but it is accepted that the length of the siege and the losses suffered by Henry in his capture of the port determined that his much-vaunted expeditionary force would have to seek a way back to friendly territory at Calais, and in the end would pull off an improbable but critical victory at Agincourt. This laid the foundations for the subsequent conquest of Normandy, achieved by a campaign of many successful sieges in a very short time.

This was all in an uncertain future as the invading army encircled Harfleur, Henry arriving there on 17 August. The town stood on the bank of the Seine, and was further protected by another river, the Lezarde, which filled the ditch on one side, and entered the town through a water gate and culverts. The other side was dry, but had a double ditch. The town itself, polygonal in shape, was quite small, having a perimeter of 2,900 metres, but had been fortified well since it became royal around 1300, with high walls and 22 towers, and with walls about two metres thick. Its three gates, each placed between a pair of towers, were additionally shielded by stone barbicans with their own towers, and were provided with drawbridges and machicolations. These defences were apparently specifically provided with apertures for cannon. The garrison appears to have been of about 400 men commanded by the Sire d'Estouteville, not a large number for the circuit, but sufficient for as long as the moat kept the enemy at a distance.[1]

The king took command of one side, the duke of Clarence of the other. Clarence on his march intercepted a convoy coming from Rouen conveying cannons, barrels of powder and arrows and bolts.[2] The author of the *Gesta* refered often to the advice given to princes by Aegidius Romanus (Giles of Rome) on siege warfare as the basis for the actions taken by Henry and Clarence in this siege, including the fortification of their own camps against possible relief attempts, confirmation both that the education of princes included such studies, and that they were put to practical use in the field.

Having had his summons to surrender rejected, Henry entrenched his force and placed his cannons and machines against the walls to bombard the defences. Large planks were used to protect the guns, with hinged sections that were raised to fire, then lowered again, while fascines were mounted atop the trenches to protect the gunners. Concentrating the attack on the main barbicans, the English force gradually drove its trenches closer to the walls, their cannon effectively demolishing the wall, towers and outworks despite the fire from the defence — the *Gesta* stated that the attackers particularly targeted the towers and works from which the French were firing their artillery — while guns mounted on higher

[1] *Gesta Heinrici Quinti*, ed/trans. Taylor, Roskell, 1975, 28; E Dumont, A Léger, *Histoire de la ville de Harfleur*, 1868, 61–80, with a plan reconstructing the defences, a study of then surviving remains and of the town archives.

[2] *Gesta*, 34.

ground caused great destruction to the buildings inside the town itself. By night, the garrison sought to plug the gaps in their ruined walls with timber, fascines, barrels filled with earth and dung, and stones. They assembled a supply of jars of sulphur and quicklime on the walls ready to hurl in the event of an assault. Their guns and other engines (*ballistae*) were relocated in new sites amongst the ruins when the original works became indefensible. On Clarence's side, an attack by mine was organised, but the chronicler noted that because of the terrain they had no choice but to begin within sight of the enemy, contrary to the advice of Giles of Rome, thus losing any element of surprise and enabling them to countermine. Clarence's own entrenchment served both as a protection against relief, his force being isolated by geography from the king's, but also against sorties. He also reached the ditch and converted the defence into a base for close-range attack on the ramparts. Clarence's men protected their ditch against possible attack by driving tree trunks and stakes into the embankment.

As the attack progressed preparations were made for a final assault, with the construction of timber belfries and ladders as tall as the walls. Great bundles of wood ten feet (3 metres) long were readied on both sides to fill the ditch, but awareness of the incendiary devices prepared by the defenders caused this method to be abandoned. A sustained attack with arrows bolts and stones from the engines (where the solitary reference to *fundis* hints at the presence of more old-fashioned artillery than the guns) drove the French from the walls and enabled a successful storm of the barbican through the breaches created by the guns. The attackers set fire to the defences, starting it with fire arrows then sustaining it by application of "inflammable powders", enabling them to force an entry into the outwork while driving the defenders back into the town proper. The next day, as preparations were made to launch an assault at the places where the guns had smashed the defences and made the attack practical, the governor surrendered (22 September 1415). None of the main works of defence had fallen, but their state made the final conquest inevitable.[1]

The siege had lasted five weeks, and had involved the whole of the English invasion army in taking one small town. The struggle had cost Henry the services of several thousand killed, wounded or sick from his original army of some 12,000 fighting men. Also notable from this account was the failure of the mine. Although the French sources refer to the town being mined, our eyewitness made no reference to the mines having any impact on the outcome. The large presence of siege artillery in the form of guns was sufficiently noteworthy to attract the attention of all the chroniclers, and its value in creating breaches was apparent. However, the presence of substantial outworks shielding the older stone walls, and the courage and determination of the garrison, meant that weeks rather than days

[1] *Gesta*, 38–48; *Monstrelet*, III, 83–4; St. Rémy, I, 225; Hardyng, *The chronicle from the firste begynnyng of England*, chapter CCXIII. This verse account talks of the burning of the defences with "shot of wildfire"; Walsingham, 307–8; *Chronique du religieux*, V, 531–43; John de Wavrin, *Collection of the Chronicles of the ancient histories of Great Britain, now called England*, trans. Hardy and Hardy, 1887, 185–6; *Les cronicques de Normandie*, ed. Hellot, 1881, 16–17.

had been required to bring the attack to the point where a final assault could be envisaged, and the brave defence of the French troops honourably concluded after a short period of grace after which surrender would follow if relief did not arrive. Further, in order to secure the much-damaged port as a base of his own, Henry had to leave a substantial force to hold (and set about repairing) Harfleur. These precautions were certainly necessary. Despite Agincourt, a French force and fleet organised by the Marshal of France tried to recover Harfleur in 1416. They did not attempt an assault, suggesting that the repairs had been quickly executed, but intended to deny supplies to the garrison by preventing foraging and by blockading by sea. The English were anxious enough for the safety of the one permanent gain of their 1415 campaign to send the duke of Bedford with a fleet which was victorious, and brought fresh supplies for the beleaguered defenders of Harfleur.[1]

From Agincourt to Meaux 1417–1422: Caen, Falaise, Cherbourg

The catastrophic defeat of the French forces in battle in 1415 gave their enemy a reputation for invincibility sufficient to deter Charles VI's men from risking open battle for a long time. This also left Henry much more room to manoeuvre as he next set about the conquest of Normandy. His new expedition, said to have been 16,000 strong with sufficient numbers of carpenters and workers to construct siege equipment, and engines of war, landed in the province in 1417, and the defenders of Touque fled without a blow as he deployed the artillery, although it was apparently a "strong town with a noble castle". The captain of the garrison paid for this cowardice with his head.[2]

From here, he marched on Caen, the chief town of southern Normandy. It was considerably stronger than the city that had confronted Edward III. Following the siege of 1346, a substantial stone wall had been erected around the town, and the adjacent abbeys (St Stephen and Holy Trinity, and the church of the Holy Sepulchre) had been provided with their own defences in the form of walls and ditches, and equipped with guns. It was well provisioned and had an adequate garrison, perhaps a total of 7,000, although the majority was citizen militia.[3]

The shape of the city and its prosperity, which had given rise to extensive suburbs, represented however military weaknesses. The unwalled Île des Près, between the rivers Odon and Orne, protruded into the heart of the city, separating the walled old town and the walled Île St Jean to the south. As news came of the English landing, the people of Caen successfully resisted the arguments of the soldiers that the suburbs and the abbeys should be razed. The duke of Clarence seized the abbey that dominated the town, anyway. Despite the decision of the council of war, steps had been taken to prevent the English using the site, but a monk placed the survival of his monastery's buildings over that of the town

[1] Walsingham, *Ypogdima Neustriae*, ed. Riley, 1876, 477; Hardyng, chapter CCXVII; Elmham, *Vita*, 92; Juvenal de Ursins, 333.

[2] Elmham, *Vita & Gesta Heinrici Quinti, anglorum regis*, ed. Hearne 1727, 78–82, for a very detailed account; T Basin, *Histoire de Charles VII*, ed. Samaran, 1933, I, 49.

[3] L Puiseux, *Siège et Prise de Caen par les Anglais en 1417*, 1858, 18–26.

by going to Clarence and seeking his intervention, warning that the French had undermined the church and the surrounding walls ready to bring about their rapid demolition. Having taken the abbey by escalade, then raced around the city to the fortified religious sites on the other side too, the English had a strong and superbly-sited position from which their guns (and stone throwers, according to one of the French accounts) could shoot into the town, so it is unlikely that the people of Caen had any reason to be grateful for the monk's sensibilities. The king established his headquarters at St. Stephen, from the tower of which he could see straight into Caen. Thus established, and a heavy bombardment having been carried out against the whole circuit, but especially against the old town from St. Stephen and the Île des Près, and on the opposite side against the Île St. Jean, Henry and Clarence organised a co-ordinated assault on the town walls. The defenders held off the English on the east, but Clarence's troops broke in on the opposite side and took the French in the rear. They then stormed the fortified gate that connected Île St. Jean to the old town, and poured into the narrow streets. There was a fearful massacre of anyone found carrying arms, the survivors retreating into the castle. The town secured, they then had to take the ancient castle of the dukes of the Normandy, just as Edward III had done in 1346. Now the king's guns were much more powerful and effective than they had been on the previous occasion, and they quickly rendered the castle untenable and induced its surrender after some three weeks.

Henry then besieged Falaise, starting on 1 December. It was not normal practice to maintain military operations in the middle of the winter, the conditions not being conducive to remaining in the field, and finding supplies being a critical concern. Henry however succeeded in maintaining his army in the field despite the season. The town surrendered by 2 January "in terror of the guns", and the old castle standing above it resisted for longer, facing a dreadful bombardment from the guns hurling stone shot against it, and also being mined. Capitulation followed on 16 February.[1] The terms of surrender included the provision, one that was to be common to the surrender agreements reached with the Norman towns, that their "artillery, . . . powder, cannons, crossbows . . . (etc)." should be handed over intact.[2] The surrender of Falaise induced the surrender of Bayeux.

Meanwhile, the duke of Gloucester was besieging Cherbourg, bringing it to surrender after a blockade of ten months, according to one source its capitulation encouraged by bribery. There is little first-hand evidence for this prolonged siege. The town was in a strong seaside location, well-walled and with an adequate garrison, which had prepared for the coming siege by laying in provisions and destroying the suburbs. Acting as a citadel and focal point of the defence, the castle had 16 towers, and powerful outer works had been erected. It was not a difficult choice for Gloucester to decide that, although he would order close siege works laboriously constructed all round, the fortification of his own encampment, the

[1] Walsingham, *Ypogdima*, 479–81,and *Historia*, 18–26.

[2] *Rotuli Normanniae in turri Londonensi asservati*, ed. Hardy, 1853, 314.

breaking down of bridges leading to the town, and the setting up of batteries to fire on the town, his best chance lay in forcing surrender through blockade.[1]

Rouen

Henry returned to the attack in 1418 with the intention of finishing the conquest. Several towns capitulated without resistance as Henry organised the isolation and siege of the capital town, Rouen, taking Pont de l'Arche with artillery to secure the crossing of the Seine. French accounts offer various explanation of the surrender of this important fortified town, including both treachery, and the failure of the French and Burgundian courts to offer any help.[2] The capture was critical to the success of the coming onslaught on Rouen itself. That city refused a summons to surrender, firing guns at the envoys, and was then encircled on the landward side, with the English camp protected by its own defences against sallies, with a rampart topped with pointed stakes, while iron chains were stretched across the river Seine. The outlying fort of St. Catherine, lying on higher ground east of the city, was captured, apparently falling for lack of supplies. An eyewitness described Rouen as strongly walled and towered, and further protected by a deep and wide ditch, the bottom of which was turned into a greater obstacle by the digging of "pitfalls" a spear's length in depth every two feet (0.6 metres). The ground between the ditch and the rampart was further impeded with caltrops. Each of Rouen's many towers was provided with three guns, and between the towers were emplaced large cannon at a lower level. Additionally, the defending artillery included stone throwing engines *(trebgets)* mounted on towers.[3]

The noose tightened around Rouen (Plate 8) as the duke of Gloucester arrived from Cherbourg, a bridge was thrown across the river, and English ships arrived. A force of 1,500 was ordered to set up defensible positions in the forest of Lyons to deter any attempt at relief. At some point the English must have decided that to storm the city would a costly and uncertain exercise, and had instead resolved on a blockade. This was not without its risks, and the besiegers seem to have taken heavy losses from the French fire with guns, trebuchets and springalds. However, the large number of people sheltering within the walls exhausted the food supplies as the siege stretched into December, and the defenders ejected thousands of the poor. Henry would not let them through his lines, and they were left to dwell helplessly in the ditch in the winter weather, many dying there. In a gesture of charity much praised by English chroniclers, Henry supplied them with food at Christmas, but not otherwise. Within the city, every cat, dog, horse, rat and mouse had been eaten. Negotiations then began leading to the city's surrender, though not before a thousand people were said to have died of the famine. The city opened

[1] St Rémy, I, 320; Elmham, 147–62 offers a detailed description of the English siege works; *Monstrelet*, III, 242–3, describes it as the "strongest in Normandy".

[2] Salch, *Atlas des Villes*, 312–13 (Pont de l'Arche); Juvenal des Ursins, 542; *Monstrelet*, III, 277; St. Rémy, 342; *Journal d'un bourgeois de Paris*, 105; *Chronique du religieux*, VI, 259.

[3] From John Page, *The siege of Rouen*, 1876 edition, 16, a verse account that must be that of a witness; St Rémy, 345.

its gates after negotiating a period of eight days to seek relief, and Henry entered in early January 1419. As the English poet Page observed, "for hunger brekythe the stone walle". The completion of the conquest of the province was now only a matter of time, as towns and castles fell with little resistance to the invader. Château Gaillard, Richard I's mighty fortress, resisted bravely for six months, then fell somewhat bizarrely not because of English assault or the lack of food, but because of a shortage of ropes with which to draw up their water supply.[1]

The failure of the French king to come to the aid of his Norman subjects was because he had problems to face closer to home, where the duke of Burgundy was playing a devious diplomatic game designed to promote his own interests. The next stage would be brought to a climax by the shocking assassination of Burgundian duke John the Fearless in 1419 at Montereau, an action that determined that his successor Philip the Good (1419–67) would throw in his lot with the English. Meanwhile, during 1417, having taken further steps to put Paris itself in a state of better defence against domestic foes including laying in quantities of bombards, the French king's troops had been working to secure strong points around Paris itself. Étampes was mined and taken, Montlhèry with its gigantic donjon in a strongly-walled and towered enceinte built by Philip Augustus, and despite a garrison of one hundred, was forced to yield, and another castle at Mercoussis taken when the French diverted the water defences on which its small garrison depended. In 1418, Charles tried in vain to capture Senlis despite spending 200,000 francs (it is claimed) on the costs of the siege, much of it going on artillery. The duke of Burgundy besieged the Dauphin's town of Crépy in the Laonnais in 1419, and took it after two weeks after arraying his engines against the walls and gates.[2] The same year, the Dauphin's men besieged Burgundian Parthenay, a strong fortress with double walls and ditches, but a population divided in its loyalty, a non-military consideration that proved decisive, leading to its rapid surrender.[3]

With Normandy in his hands, Henry V could turn his attention to the crown of France itself. But he was not unaware of the need to ensure that Normandy was effectively organised to support his efforts. An official was appointed Purveyor of Provisions for the garrisons of the castles in northern Normandy and another for the south. Clerks of the Ordnance were appointed to look after the weaponry. Caen was made a depot for gun stones, some of which were quarried locally and others imported from England. Repeatedly, carpenters were recruited to construct engines, and to make carts for their transport. Stone cutters were recruited to make the ammunition. Government records indicated a regular series of com-

[1] *Monstrelet*, III, 277 f, 333–38; St Rémy, I, 341–53; Walsingham, *Ypogdima*, 485–8; *Journal d'un bourgeois*, 105; *Chronique du religieux*, VI, 259, 349; John Page, 8–20; *Journal de Clement de Fauquebergue, greffier du Parlement de Paris*, ed. Tuetey, Lacaille, 1903, I, 186, 212–14, 230–2, 252–3, records the reactions of the government of Paris to the news from Normandy, the steps taken to secure the city, and the continuing civil war with troops of the Dauphin impeding the provisioning of the capital. D Seward, *The Hundred Years War*, 1996, 158, for a sketch plan of the siege of Rouen.

[2] *Monstrelet*, III, 376; *Journal d'un bourgeous*, 86; *Chronique du religieux*, VI, 85, 179–83.

[3] Juvenal des Ursins, 547; Salch, *Atlas des villes*, 420–1, describes a massive thirteenth century fortification (Parthenay).

missions for craftsmen and labourers to go to France, in particular stone cutters, carpenters and smiths. One order calls for men "for the making of the king's cannon", along with supplies of copper, brass, tin, iron, lead, timber and sea coal for the construction both of the guns, and of the gun powder. Ships were commissioned to transport the artillery. Hundreds of men were recruited for these tasks and substantial industrial activity was being encouraged by royal demands. The maintenance of the castles was placed under direct royal control.[1] Major work was undertaken to modernise the defences of many places over the first years after the conquest. A study of the accounts of the town of Lisieux, for example, showed that the town's defences, which had been an incomplete palisade in 1417, were brought to completion by 1423, then strengthened with stone gates and towers and a more substantial ditch.[2] In 1424, the council of Normandy instructed the governor of Verneuil to reinforce the defences of the town with bulwarks (*bolevers*) in front of the gates of both town and castle.[3]

Montereau and Melun

The year 1420 saw the treaty of Troyes whereby Henry became heir to the French throne. It also saw more significant siege activity against supporters of the thereby disinherited dauphin, including all those who did not wish their country to be ruled by the English. The future Charles VII, aware of the imminent danger, established large garrisons in many towns and castles north and east of Paris, such as Melun, Montereau, Montargis, Meaux and Compiègne.[4] A Burgundian army stormed the castle of Alibaudières near Troyes using several large bombards to create a breach, the defenders capitulating after resisting a five-hour long assault the previous day. In June, the English joined the Burgundians east of Paris to seize the town of Montereau-faut-Yonne, which had four or five hundred defenders, French and Scottish. The English pressed the attack from one quarter, driving trenches close to the wall, and their artillery made breaches. For the Burgundians the town had a terrible symbolic importance, and they were hot in pursuit of revenge for the murder of their duke. A week after starting the siege, the town was stormed by escalade (23 June). The engines were then deployed against the castle, with devastating effect on its walls and buildings, and surrender followed on 4 July.[5] The joint army of Charles VI, Henry V and the duke of Burgundy, apparently very large, then moved on to lay siege to Melun, held by some 600 or 800 men. The besiegers set up powerful batteries of artillery with which to bombard the walls, causing considerable damage. The size and effect of the shot hurled

[1] R A Newhall, *The English conquest of Normandy 1416–24*, 1924, 247, 253–4, 263–7, and the sources cited there; *Calendar of Patent Rolls, Henry V*, II, for example at 148, 175, 200 (the guns), 329, 384, 396, 387 (tents).

[2] R Jones, Les fortifications municipales de Lisieux dans les chroniques et dans les comptes, in Contamine, Guyotjeannin, eds., 235–40.

[3] *Letters and papers illustrative of the wars of the English in France*, II, 54.

[4] Wavrin, 289.

[5] Elmham, 270–4; *Monstrelet*, III, 404.

by one large bombard called "London", constructed by and transported over to France at the expense of that city, was sufficient to impress one chronicler. At the same time, mines were begun from several different points. However, the defence was able by night to repair with earth and timber the damage done by day to the walls, using filled hogsheads as fascines, and the English attempt at mining was successfully interrupted by a counter-mine. All the accounts speak of the junction becoming the location for chivalric duels between knights of both sides, but clearly this mine was not going to lead to the capture of Melun. However, the siege was pressed, with the trenches being driven right up to the ditch. Burgundian troops successfully stormed an outwork on the further bank of the Seine, and one account tells of a rash attempt at a storm by escalade carried out to fulfil the ardour of the Burgundians and the visiting duke of Bavaria, against the advice of Henry v, which was a costly failure. Certainly it was hunger in the town that finally brought surrender, although the walls were seriously damaged, after some 18 weeks.[1]

Meaux

The joint offensive continued across northern France in 1421, the chronicles recording the capture of numbers of castles, some surrendering quickly, others resisting and, in the case of Rougemont (in Beaune) suffering a grim fate as Henry v had all the defenders drowned in the river Yonne.[2] Having set up camp near the river Marne, Henry had his workmen prepare more siege engines so that he could begin an attack on the city of Meaux, situated in a loop of that river. The ancient city's Gallo-roman walls had been rebuilt with semi-circular towers, and stretched for around 900 metres.[3] Separated from the town was a smaller and similarly strongly protected area called "the market", reached across a stone bridge and with river defences on three sides, and a canal on the fourth. The siege began on 6 October, the English putting bridges of boats across the river to ensure communications. Weather conditions were bad and illness was rampant. Meanwhile, the garrison of one thousand men faced the normal bombardment by cannons, bombards and engines which battered the walls and gates. An English source refers to the besiegers also having *trepgettes*.[4] Surviving chambers from seven bombards apparently used at the siege demonstrate the variety of different sized weapons employed, and that the guns were of considerable size.[5] Soon, recognising that the battering had made the ramparts untenable, the defenders withdrew into the market, and the besiegers, witnessing them doing this, stormed in behind them to occupy the town. The English occupied a small island in the river facing this

[1] *Monstrelet*, III, 409–14, IV, 12; St Rémy, II, 15–19; *Chronique de religieux*, VI, 443–9; *Chronicle of John Strecche*, ed. Taylor 1932, 49 (for the gun); Juvenal des Ursins, 559–61; Elmham, 276–88.

[2] *Monstrelet*, IV, 70.

[3] Salch, *Atlas des villes*, 415.

[4] *The Brut or chronicles of England*, ed. Brie, 1906–8, part II, 428.

[5] L Larchey, *Origines de l'artillerie Française: planches autographiées d'après les monuments du XIVe et XVe siècles*, 1863, plate 9. The chambers range from 0.9 to 1.3 metres in length.

last citadel, and placed big guns there with which destroyed the houses, while the besiegers captured the mills so prevented the townsfolk from grinding their corn. The siege took on the form of a classical operation. The attackers constructed various great wooden devices, including a form of siege tower (belfry) that was moved up to the end of the stone bridge reaching across the river, and from here overlooked the defences opposite. This machine also covered the attackers while they placed wooden ladders across the gap knocked in the bridge. Even more spectacular was the floating tower constructed on a floor of timber beams laid across two large ships, the top of which overlooked the tower facing it. This "monster" machine was about to be ordered into use as a means of crossing over to the defences when the city gave up and surrendered at mercy (May 1422), after a siege of seven months.[1] The most significant victim of the siege was Henry v, who contracted the illness in the swampy conditions that would later that year prove fatal.

English conquests, and the impact of Joan of (Jeanne d') Arc 1422–30

The war continued in France, with the dauphin based at Bourges and the English commanders, with the duke of Bedford operating as regent from Paris, struggling to extend in practice their notional rule over most of the kingdom, but facing resistance from many walled towns and castles, including in the north. Here, the sheer number of fortifications made it a vast challenge to establish garrisons everywhere, the English even reverting to forms of military service in exchange for land. There were sieges of St. Valéry on the Somme, taken after three weeks' artillery bombardment (June 1423), of Pont du Meulan, and Montaiguillon (Burgundy), while the French failed to recapture Cravant.[2] In 1424, Bedford recaptured castles in Normandy such as Gaillon and Ivry.

The Earl of Salisbury laid siege to Le Mans. His cannon blew breaches in the old walls, largely based still on the Gallo-roman circuit, and up to four metres thick. This did them no good, as a large hole was created near the bishop's palace, and the town corporation, aware that relief was not coming, negotiated a surrender (17 August) in which, in addition to yielding to English occupation, they also undertook to pay the expenses of the siege, set at 1,500 *escus*, a term which seems to have been common in negotiated capitulations and reflected the immense cost of the war. St. Suzanne then resisted attack with artillery for ten days, until the walls, which were much more recent, were smashed. Here, the garrison had to surrender as Salisbury's prisoners. Mayenne la Juhais next suffered a similar bombardment, and was mined as well before being subject to assaults and surrendering on terms. Finally, the earl blockaded La Ferté Bernard for four months, with the same result.[3]

[1] *Monstrelet*, IV, 71, 83–93; St Rémy, 45–50; Elmham, 319–27; Juvenal des Ursins, 561–3; Wavrin, 357–8, 369–74; *The Brut*, 428.

[2] *Monstrelet*, IV, 102, 136, 154, 158; Mesqui, *Châteaux forts*, 199–200 (description of Ivry).

[3] Cousinot de Montreuil, *Chronique dite de la Pucelle*, ed. Vallet de Viriville, 199–202; Salch, *Atlas des Villes . . .*, 191–2 (Le Mans), 357 (Ste Suzanne).

Resistance at Mont St. Michel

The isolated fortress of Mont St. Michel, however, never fell to the English. On its offshore island with limited land access, the ancient monastery was protected with rather less religious towers and ramparts of the most modern kind, and had been provided with a new cistern (September 1418) and a garrison, whose relations with the monastic community and the local people must sometimes have called for careful management as a blockade that would last no fewer than 28 years was established. An order of the governor, preventing his men taking women, or English prisoners, into the abbey precincts, is a reminder of the inevitable tensions. The governor, in ordering the strengthening of the defences by the construction of a new round tower (1426), and a postern gate, had to be careful to confirm that the work did not reduce the monastery's property rights.[1]

The detailed records maintained by the French side enable a reconstruction of events that casts much light on the conduct of war in these times. The country remained fundamentally loyal, meaning that when there were no foreign troops around, towns returned to their French allegiance. It was not until 1423 that the occupiers were moved to try to do something about the stronghold off the coast, when troops came in February to built a fort at Tumbelaine to try to control the ravages of the garrison. Its garrison (at 23 July 1424) comprised 30 men at arms and 90 archers. This proving not equal to the task a second fort was built at Ardevon in September 1424. The defenders of this fort included crossbowmen and the relative sophistication of the work was shown by the fact that its construction included the provision of a drawbridge. The garrison (November 1424) consisted of 20 men at arms and 60 archers. English ships tried in vain to attack the Mount in 1424 and 1425.[2] The blockade never prevented communications with the defenders. Records survive of payments to messengers for repeated journeys carrying instructions from the royal court to the governor of the Mount.[3] Yet another fort, at Genet, was constructed in 1428, with a garrison of 120 men, as with the others the cost being borne by the locality. In fact, it was not until 1434 that a serious attack was attempted by land, when Lord Scales organised an assault with troops from all the neighbouring English garrisons under cover of the fire of a large battery of artillery but entirely in vain, the English suffering heavy losses and the defenders, remarkably, none killed at all. Rebuffed, the blockaders resorted to constructing yet another fort,[4] and the partial blockade continued year after year.

Elsewhere, French attempts at reversing the gradual progress of conquest had so far proved fruitless, and had involved further crushing defeats by smaller English armies in the process (the battles of Cravant and Verneuil). Even their attacks on fortresses were unsuccessful. In 1427, for example, the Constable of France assaulted the castle of St. James de Beuvron, from which the English were ravaging Brittany, but his troops were routed by an English sally from a postern gate, and

1 *Chronique du Mont St Michel*, ed. Luce, 1879, I, 22, 87, 221, 250–1.
2 *Mont St Michel*, I, 24–7, 125, 154, 160, 165–6.
3 *Mont St Michel*, I, 260, 282.
4 *Mont St Michel*, I, 35.

fled abandoning their artillery and supplies. In 1428, the inhabitants of Le Mans schemed to secretly readmit French troops to the city. The small English garrison retreated into the ancient tower d'Obrindelle, and held out while sending for urgent aid to Alençon. From there, the fearsome John Talbot (earl of Shrewsbury) raced to the city, caught the French asleep and without a watch, resulting in their routing and the recovery of Le Mans by the English. Talbot then took Pontorson after a lengthy blockade, and Laval by escalade. Salisbury meanwhile was busy reducing other towns and castles, heading towards the Loire where Mehun capitulated. We know that his expedition of some 2,500 men included guns, stones, and carts to convey them, four master gunners being paid 20d per day, and ten miners at 9d per day.[1] His troops then approached the major city of Orléans, but on this occasion were driven off by French forces marching out of the city. Undismayed, Salisbury went on to take Beaugency on the right bank of the Loire, where his guns bombarded the castle and the stone bridge. Although the main interest of the castle lies in its immense four-storey donjon 36 metres high, the defences in 1428 were a towered enceinte, standing near the town end of the bridge. Jargeau surrendered after a few days (5 October), then on 12 October the English arrived again before Orléans, this time determined to achieve a capture.[2]

Siege of Orléans

With a force of perhaps 7,000 men, Salisbury was doubtless optimistic. The accounts show payments to four master gunners, who were in charge of a varied battery of ordnance, and forty miners under two master miners.[3] In fact, the different surviving records show that Salisbury had cast his net wide, commissioning equipment and supplies from various sources,[4] therefore it is hard to establish how many guns and gunners were serving in his army at any one time. The prize was worth the effort: the progress of the siege of Orléans would be watched all over Europe, the outcome being widely seen as decisive.[5]

The citizens took their movables back behind the walls and burnt their suburbs themselves, while the English began by attacking the French outworks at the far end of the bridge across the Loire (the fort of Les Tournelles), which put up a fierce resistance. The besiegers established a protected earthwork against it in which to mount their guns, and also prepared mines under the rampart. The artillery bombardment had such effect that no defender dared stand on the rampart. They organised ladders to make an assault, which was carried out on 21 October. The defence was vigorous, ladders being thrown back into the ditch and

1 Stevenson, *Letters and Papers*, 410 (muster of the expedition, June 1428).

2 *Chronique dite de la Pucelle*, 210, 221–5. Salch, *Atlas des Châteaux forts*, 447, 450 (Beaugency).

3 L Jarry, *Le compte de l'armée anglaise au siège d'Orléans 1428–9*, 1892, 28–9, 118–19, 143–4.

4 Stevenson, *Letters and Papers*, I, 408–10, records another indenture to John Parker of Cheshunt for guns and equipment.

5 For an impression of the views of well-informed observers, *Chronique d'Antonio Morosini*, ed. G Lefèvre, 1898–1902, III, 9–17.

all kinds of missiles including quicklime and hot ash was hurled upon them such that the attack was abandoned after heavy losses. However, overnight the French abandoned the defence and withdrew, breaking down arches of the bridge behind them to prevent easy crossing, and the besiegers turned it into a siege fort against the city.[1]

The French sent assistance to the city, which was well provided with guns. One of these was responsible for the shot that killed the earl of Salisbury while he was inspecting Les Tournelles on 24 October. On their side, the English army too received reinforcements. In December they resolved that they could not take Orléans by storm, but would have to blockade it into submission. To bring this about they constructed a series of siege forts all round, on both banks of the Loire and on an island in the river, there being thirteen in total. However, and crucially, the withdrawal of the Burgundian contingent the following April meant that there were insufficient numbers to complete the enclosure, and the forts to the south and east in particular were very isolated. Perhaps the English were relying on the fact that the defence was largely passive, so as supplies were consumed, the capture of the city looked likely to follow. The growing risk was certainly notified to the dauphin, since the gaps between the English redoubts to the east and north east meant that there could be fairly easy communication with the city from the outside. New arrivals included the remarkable Joan of Arc who arrived on the scene with the blessing, and the army, of the dauphin, determined to raise the siege and drive the English out of her country. Troops, supplies and guns were brought into the city by the Marshal de Rais, the English being kept behind their defences by a large scale diversion, and Joan herself arrived there on 4 May 1429, easily making it through the line of English redoubts.[2] She had no intention of continuing the passive defence.

Accompanied by sceptical French aristocrats and troops and artillery from all over the region, Joan organised the defenders, now definitely superior in number, to tackle the redoubts on the left bank, but starting with the isolated English fort of St. Loup on the very next day, 5 May 1429. She led the attack while other troops prevented Talbot from sending aid from neighbouring forts, and overcame the 100–120 defenders. Other forts were subjected to the same treatment, including Les Tournelles and Les Augustins. The Maid, although wounded, organised an attack from two sides with scaling ladders, while a wooden bridge secretly constructed was used to cross the gap in the Loire bridge and launch an attack from that side as well. The forts were conquered (7 May). The next day, the English abandoned the siege and all their equipment in an undignified flight, following which Joan led the recovery of many castles and towns in the region, much of Champagne returning to the allegiance of the French royal house.[3] Jargeau was

[1] *Chronique dite de la Pucelle*, 227–9; Salmon, ed., 'Chronique du siège d'Orléans', *Bibliothêque de l'école de Chartres*, 2nd ser., III, 3–5.

[2] *Chronique dite de la Pucelle*, 230–5; 'Chronique du siège d'Orléans', 4–6.

[3] *Chronique dite de la Pucelle*, 251–66; 'Chronique du siège d'Orléans', 7–10; *The Brut*, 435; Basin, 117–21, 135–9; P de Cagny, *Chroniques*, ed. Moranville, 1902, 140–6; Morosini, III, 185; *Journal de Clement de Faucquembergue*, II, 313.

recovered by a force led by Joan and the count of Alençon. The garrison under the earl of Suffolk himself rejected a summons to surrender, which proved a fatal mistake, because the French set up bombards and cannons on 11 June, and assaulted the very next day, killing all but 50 of the substantial garrison. If this was intended as an example, it worked: the defenders of Beaugency capitulated after a couple of days' attack, on 16 June.[1] The victory at Orléans was a turning point, the French were encouraged into vigorous action and the weakness of the overstretched English position was exposed. Charles VII was crowned king at Rheims in consolidation of his claim to the succession.

Bedford had no intention of abandoning Henry V's achievements easily, and while the Burgundian alliance remained in force his position remained sustainable. When Joan attempted to recover Paris later in 1429, she was repelled. The citizens laid in supplies of artillery, emplaced them on the ramparts, and cleared out the ditches (August). As a consequence, when Joan's army approached the city in September, apparently bringing carts with materials to fill the ditch between the St. Denis and St. Honore gates, there was heavy artillery fire and they were driven off with losses after four hours skirmishing, and Paris remained loyal to Henry VI. It was recorded inside the city that the French had been reliant on a popular uprising in their favour but were disappointed, the militia standing firm at their allotted positions on the ramparts.[2] The Maid's extraordinary career came to an abrupt end the next year, when she was captured defending Compiègne against an Anglo-Burgundian attack: her plan had been to launch a counter attack supported by the fire of cannons and crossbows from the walls, but the failure of her own side to carry out the plan led to her being cut off in the outwork in front of the gate and captured.[3] Her subsequent fate is well known. One of the wonders of her brief career remains how she acquired such knowledge of tactics including of siege warfare, and inevitably the possible role of the professional soldiers around her must be considered.

Sieges of the 1430s: Compiègne onwards

The Anglo-Burgundians could no longer defeat Charles VII, but nor could the latter drive his enemies out. The siege of Compiègne was actually a significant event in its own right, and in a way symbolised the stalemate that was being reached. There is a detailed description of the operation from a Burgundian chronicler, Georges Chastellain, who was very close to the action. The duke himself led the attack. Having changed hands more than once already, it was in 1430 strongly

[1] Cagny, 150–1, 153. Cagny was in the service of Alençon so his account of these events is particularly valuable.

[2] *Chronique dite de la Pucelle*, 290–2; *Journal d'un bourgeois*, 239–45; Cagny, 166–8; Morosini, III, 210, expressing the expectation that the city would fall; *Journal de Clement de Faucquembergue*, II, 322–3. The same source later records the diversion of moneys to pay for repairs to the ditch at the gate St Honore (entry of 3 July 1430, II, 351,) suggesting that more damage was done to the defences than was indicated.

[3] Lefebvre Saint Organ, *Compiègne*, 242–9; *Monstrelet*, IV, 409; Georges Chastellain, *Chronique*, ed. Kervyn van Lettenhove, 1863–6, II, 36–9, 46–50.

protected, and its bridge across the river Oise was defended by a powerful redoubt. The garrison also possessed large quantities of artillery, and this was sited so as to be able to cause severe damage to any besieger getting too close, the hill top site being very helpful to the defence, and the suburbs having been razed in the siege of 1414 to deny shelter to an enemy.[1] The Burgundians tried to take the redoubt, erecting their own bulwark as close as they dared, and driving trenches towards it, reaching to within a bow shot of the bridge itself. They were obliged to construct deep and well-protected works to offer some shelter from the defenders' guns, one of which was described as a culverin without equal which was in the skilled hands of a cordelier from Valenciennes named Noirouflé, who was said to have been responsible by himself for killing 300 of the besiegers with his lethal weapon. The defence also mounted frequent sorties, and when the besiegers began a mine, it was observed and countered.

The siege continued, and was reinforced by 2,000 English troops led by the earls of Arundel and Huntingdon, with whose help the day and night attack was maintained for another two months before, at last, the redoubt could be assaulted. It was by now entirely indefensible, and many Frenchmen were killed trying to escape back into the town. The duke himself left the siege, and the count de Ligny was left in charge. A council of war recognised that because of the geography and their insufficient numbers, they had never been able completely to surround Compiègne. It was therefore agreed to erect a new siege fort opposite the main gate, which was done, and entrusted to a force of 300 men. This proved effective in completing the blockade, and by September food was running low. The French were not willing to let Compiègne fall without an effort however, and the count of Vendôme was sent with 4,000 troops and plentiful supplies to see what he could do to break the ring. Showing tactical superiority over the Anglo-Burgundian force, Vendôme managed to draw his opponents away from the town, then slip in behind them to send through the provisions, while linking with a major sortie to storm and capture the siege fort, using scaling ladders and guns, succeeding at the third attempt. The outmanoeuvred, worsted and demoralised besiegers gave up the siege the next day.[2] The French forces went on to attack Clermont, using the bombards and cannons abandoned by the Anglo-Burgundians at Compiègne, but were driven off by the earl of Huntingdon, who recovered the lost artillery, this being noted as some kind of small compensation for the setback. Some local lords in the region then decided to throw in their lot with Charles, and banded together to make an attempt on the town of Corbie. This small town had few trained fighting men in it, but was bravely defended by its inhabitants led by their abbé, men, women and religious alike standing firm on the town wall and hurling missiles at the French as they tried to cross the ditch, to such good effect that the attackers retreated, suffering many casualties but chiefly, one suspects, their pride.[3]

The impasse continued. In 1431 Bedford tried to consolidate his control of the Île de France by attacking Lagny. The earl of Arundel set up a great bombard to

[1] Chastellain, II, 31–2, 52.

[2] Chastellain, II, 53 (Noirouflé), 55–6, 65–8, 87, 98–105.

[3] Chastellain, II, 120–22 (Clermont), 200–1 (Corbie).

fire on the arch of the drawbridge, and smashed it with the first shot. He then stormed the outwork in front of the gate. Arraying batteries of guns (and possibly stone throwers too) he did much damage to the walls, but the garrison of 800 resisted, and a French army arrived and caused the lifting of the siege.[1] In August, the duke of Alençon tried to recapture his castle of Ste. Suzanne in Maine, and was outside for more than two weeks, but on this occasion was compelled to abandon his attack because of a lack of "enough stones ready for his bombards and cannons".[2]

1433 saw sieges of Passy and Avalon, and Belleville, forced to surrender after a one month siege and bombardment by engines. In 1435, in Normandy, there was a significant peasant rising against English rule, but the French were unable to take advantage of it beyond forcing the evacuation of the English fort at Ardevon (part of the blockade of Mont St. Michel).[3] St. Denis, held by 600 men, was attacked by the Anglo-Burgundians whose engines broke down the walls and gates. An assault was ordered, but the attempted escalade was repelled after two hours of struggle. However, the defenders saw they had no prospect of relief, and surrendered.[4] Then, the duke of Burgundy concluded the treaty of Arras with Charles VII, and immediately caused a drastic shift in the balance of power. Important towns such as Dieppe, Melun, Houdan, Corbeil, and Harfleur, returned to French control at once. Dieppe, Meulan and Pontoise were recovered by ruses, the latter by nocturnal escalade with sentries absent. Perhaps having learnt that a direct assault was not a good way to capture Paris, the French were now adopting a blockading approach, seizing surrounding towns to prevent supplies reaching the very populous city. Melun closed the river Seine, Lagny the Marne. Vincennes castle was taken, again by escalade with ropes and ladders. By April 1436 the English in Paris were in desperate condition, and the citizens were in secret communication with the French. On 8 April, the citizens rose in revolt and seized the St. Jacques gate, to give admission to regular troops, and the English garrison retreated into the bastille, whence they capitulated on 13 April. The unhappy Parisians faced a different blockade by now hostile English garrisons at Meaux, Creil, Montereau, Mantes and Vernon.[5]

The strain placed on the English by their attempted occupation of so much hostile territory was illustrated by a muster of their garrisons in France in 1434, before the Franco-Burgundian rapprochement made it even harder to hold on. Spread among no fewer than 40 fortresses in Normandy and northern France were 347 lances of men at arms, and 2,415 archers, signifying that none of the garrisons was very large, and would be both vulnerable to attack, and not in a position to provide troops for a field army.[6] Surviving records also illustrate the extent

[1] *Monstrelet*, IV, 441, V, 35; Basin, I, 169–71.

[2] Cagny, 180–1.

[3] Cagny, 192; Basin, I, 205–19.

[4] *Monstrelet*, V, 70, 100, 184–7; Cagny, 194–212.

[5] Basin, I, 227–33; Cagny, 212–16.

[6] Stevenson, *Letters and Papers* ,II, 540–6. The towns and castles listed were: Honfleur, Avranches, Bayeux, Pont de l'Arche, Coutances, Tumbelaine, Cherbourg, Caen, Évreux,

of the arsenals available to the commanders: there are inventories from 1428 and 1435 from the bastille in Paris, and from 1435 for the arsenal in the citadel of Rouen. In the bastille St. Antoine in Paris were only fairly modest supplies. 17 hand guns were there, two of copper, the rest of iron, but none had their chambers. There were also two small "fowlers" (*veuglaires*) on wooden mounts, five large crossbows and four large horn bows, and some thousands of bolts and bullets. On the terrace stood two further timber-mounted fowlers, while five more, of various sizes, stood in the courtyard.[1] By 1435, with the city under serious threat from its rightful owners, the arsenal was not a great deal more impressive. Alongside three chambered cannons were two made of copper and four "culverins" of iron, one large cannon with two copper chambers, another "big cannon or bombard of iron", six feet (two metres) long, and a number (unspecified) of springalds. Possibly the same five small guns still occupied the courtyard.[2] In the same year, a record survives of the arsenal stored at Rouen, the centre of English rule in Normandy and now the main base in northern France. Here were to be found a whole battery of guns, the heaviest identified as a bombard named "Bedford", and nine "fowlers", two of them specified as being of copper and respectively eight and seven feet long. The largest culverin fired an eleven inch (280 mm) calibre shot, the next largest firing a shot of seven inches (178 mm). The culverins were also of various sizes, there being six large ones and 37 small. Of the older form of artillery there was a small representation in the form of two bricolles, but ballistas remained very popular, the inventory listing no fewer than 303 weapons. Also present were the ingredients for gunpowder and tools and vessels for the use of the workshop in preparing everything for use.[3] The list illustrates the accumulation of military hardware during a long war, and the great variety of sizes and calibre, which must have created major problems of maintenance and supply.

Scottish involvement

The renewed interest of the Scottish kingdom in taking advantage of their old enemy's preoccupations took military form, but was marked by a singular lack of success over this period. The Scottish evidence nonetheless indicates that they were operating with the most modern siege equipment available. In 1430, King James I (1406–37) brought immense bombards over from Flanders. In that same year he was paying one Nicholas for the metalwork associated with bombards and other engines.[4] In August 1436, the king besieged the frontier castle of Roxburgh with the major part of the kingdom's total stock of artillery, but the attack was in

Dreux, Caudebec, Verneuil, Carentan, Vire, Rouen (town and castle), Pont de Rone, Vernon, Pontoise, Mantes, Meulan, Possy, St Germain en Laye, Château Gaillard, Dieppe, Alençon, Fresnay le Vicomte, Essay, Argentan, Domfront, Pont Odon, Gisors, St Lô, Renville, Touques, Lisieux, Danremarche (near Dreux), Harfleur, Conches and Arques.

[1] 'Inventaire de la bastille de l'an 1428', *RA*, 12, 1885, 332–43.

[2] 'Inventaire de la bastille', 346–8.

[3] Stevenson, II, 566–73.

[4] Fordun, continuation, 1287; *Exchequer Rolls of Scotland*, IV, 1880, 677.

vain, with the Scottish nobility feuding, and the great quantity of guns, wagons and supplies was abandoned, although the sustained fifteen-day bombardment with artillery and arrows had seemingly caused many losses among the garrison.[1] The king was still paying German engineers for their role in the siege in 1438,[2] suggesting that the country did not yet have access to sufficient native skills to construct or operate the machinery it was using, albeit so far with little effect.

Attack and counterattack in France 1436–40

In 1436, the duke of Burgundy attempted to seize Calais from the English. His large force included engines "beyond counting" (Basin), and he occupied the outlying forts without difficulty. His Flemish troops were camped half a mile from the walls, and found that they were too far off to do any damage, so they moved closer, erecting a redoubt in which to place their artillery. For fifteen days they sustained an attack with bombards. Meanwhile the English defenders, who would be reinforced with a large contingent from England, organised a major sortie against this redoubt, which they attacked by escalade, finding the Flemings asleep, and massacred them. The siege was abandoned ignominiously.[3]

Having lost Paris and their continental allies, the English government now sent out no fewer than four armies, including the force raised to defend Northumberland against the Scots. One went to Normandy, where the duke of York besieged the town and abbey of Fécamp, a "soft" target with few provisions or defenders which had quickly to surrender after three weeks (September 1436). Lord d'Estouteville was sent to recover it in October with an army of local levies and 700 men at arms. According to one source, he succeeded by entering the town by breaking down an iron grille that blocked the passage of a stream through the wall.[4] By the end of 1436, despite numerous efforts in many places, the map looked much the same as it had at the start.

Although 1437 began with a surprise recapture by the English of Pontoise (11 February), the troops wearing white cloaks to disguise their snow-covered escalade,[5] the next year however would witness some more solid gains by the French. Starting in April 1437 with the successful assault of Malesherbes, the French constable Arthur de Richemont went on to take Château-Landon by storm on the fourth day of the siege, the garrison troops being killed as they had refused to surrender. The town of Nemours followed, surrendering after several days of bombardment by French artillery.[6] Richemont then went to Paris to gather more gunners, before attacking Bray-sur-Seine. The chief target was

1 *Liber Pluscardensis*, ed. Skene, 1877, I, 380; Bower, vol. 8, 297, 378.

2 *Exchequer Rolls of Scotland*, v, 32.

3 Basin, I, 241–5; Cagny, 227.

4 Cagny, 226–7; Basin, I, 249.

5 Cagny, 229.

6 *Chronique d'Arthur de Richemont, Connetable de France*, by G Gruel, ed. A le Vavasseur, 1890, 132–3; Chartier, *Chronique de Charles VII*, ed. Viriville, 1858, I, 236; *Monstrelet*, v, 294; Cagny, 238–9, says Château-Landon took 14 days.

Montereau-faut-Yonne. Here, he constructed a small bastille on an island in front of the castle, and worked overnight to dig ditches and erect screens to protect his troops, because the garrison (of about 400) was well provided with guns. The next day, he ordered the preparation of the approaches and the digging of mines, and even ordered the diversion of the river Yonne. Incendiary devices hurled into the town set houses ablaze. The river continued to thwart an assault, and a boat brought up sank under the weight of the armour placed around it to protect it from enemy fire. Eventually, bombards flattened part of the wall, and the place was taken by storm after some six weeks, with, according to one account, the king in person wading across the water to lead the assault, an action that seems a little out of character! The artillery was then moved and relocated against the castle. The defenders offered to surrender.[1] The main event of 1439 was the siege of Meaux, the vital stronghold east of Paris. The constable began on 20 July, setting out the siege camp, and constructing a bastille on the island. His artillery was organised by a certain Jean Bureau, who would become the master organiser of the French artillery. By contrast with the slow pace of the siege conducted by Henry v, De Richemont's was a rapid affair, apparently fearing that the English would march to the town's rescue. The assault was ordered after 20 days, and succeeded.[2] The distant blockade of Paris was finally broken.

The pattern continued. In 1440. Somerset tried to recover Harfleur, attacking by land and sea. He deployed bombards, cannons and several "big flying engines", possibly stone throwers but more likely cannon. The garrison of 400 was short of food and hard pressed. The French were now able to mobilise relief forces, but this one was held off by the English archers and the town surrendered, having lasted out for four months.[3]

Pontoise and Creil, 1441–43.

The French in 1441 attacked the town of Creil. The town, on the south bank of the river Oise, had a castle occupying an island in the river with elaborate defences protecting the bridge that crossed the river through the castle into the town. The latter was a small rectangular site with six gates. Remains suggest that the wall was more than six metres high behind a water-filled ditch four metres deep, but was only 1.75 metres thick. Rightly judging that taking the town was the key to the outcome, the French invested Creil on 19 May, and a few days later set up their "great engines" which caused much damage to the defences on the eastern side. An initial assault was repelled, but the garrison chose to consider that honour had been served, and they surrendered and marched out after twelve days.[4]

The attack on Pontoise in July 1441 proved much tougher and involved large French forces, which were established in fortified camps either side of the Oise.

[1] Richemont, 135–7; *Monstrelet*, v, 294; Cagny, 239–41.

[2] Richemont, 146–7; *Monstrelet*, v, 387–8; Basin, 261; Chartier, I, 250–2.

[3] *Monstrelet*, v, 418–24; *Cronicques de Normandie*, 87–8.

[4] *Monstrelet*, VI, 6; Chartier, II, 16; Boursier, *Histoire de la ville et châtellenie de Creil (Oise)*, 1883, 18–23, 137–9, including plans of the town and castle.

The English made strenuous efforts to bring relief, Talbot twice achieving this, and when the duke of York led a major army up and used a bridge of boats to approach where the French camp was weakest, Charles withdrew, but left his son Louis occupying a strongly fortified position around the abbey of St. Martin. York found the countryside bereft of supplies and had to withdraw too, allowing the French to resume the siege. Facing the same shortages, Charles decided to bring matters to a head with an assault, although the garrison was large (1,200 men) and well gunned. The attack was launched under a fierce protective fire from artillery after the guns had created breaches in the walls. Gunfire also had the effect of driving the garrison to abandon one of their outworks, a redoubt built at the far end of the bridge over the river that was then used by the besiegers as a base for part of their attack. The ditch was crossed in fierce fighting, the defenders responding with stones, darts, incendiaries and even, according to one source, boiling oil. The French gunners were however decisive in forcing the English off the wall and permitting the attackers to place their ladders and gain entry, and the city was taken by force on 19 September, the defenders being slaughtered.[1]

In 1443, Talbot led a force to attack Dieppe. The English built a strong fort on the hill opposite the town's weakest point, just a quarter of a league away, but left inside only a weak garrison. This was attacked by the Dauphin and the count of Dunois who forced their way in despite the fire from the defenders' bombards and arrows, and wiped out the garrison (August 1443). The fort was illustrated in a miniature which, if even remotely accurate, suggests a deep ditch and a wooden palisade with flanking in its shape, provided with a large number of small gun loops. Even if not actually illustrative of the fort at Dieppe, it is presumably representative of such forts in general.[2]

War in Gascony 1442–3

In 1442, King Charles sent a large force to the south, led by Gaston count of Foix, (whose rather later biography gives some fascinating details), and including Jean Bureau, with artillery comprising 30 or 40 big culverins, 12 big serpentines, not counting the really heavy guns, the bombards, *courtaulx* and *canons peryes*, with all the accompanying supplies and ammunition.[3] Their first target was Dax, where a strong redoubt had been erected in front of the main gate. The French guns were set up and their bombardment was effective in demolishing the walls of the town and redoubt, so that after several days an attack could be launched. The guns called *courtaulx* hurled incendiaries into the town, causing fires. The redoubt

[1] *Monstrelet*, VI, 6–22; Basin, I, 263–75; Chartier, II, 20–5; Richemont, 166–71, all give detailed accounts. The reference translated as "boiling oil" is in Basin, who describes "ferrentes aquas vel quas invenire possent pinguedines incensas fundere ac spangere", suggesting animal fat of some kind.

[2] Basin, I, 285–9; *Cronicques de Normandie*, 95. The miniature is in Larchey, *Origines de l'artillerie française, planches:* plate 91.

[3] *Histoire de Gaston IV, comte de Foix*, par Guillaume Leseur, ed. Courteault, 1893, I, 5–7. The editor points out that the chronology of this text is wildly inaccurate.

fell after fierce fighting, as the day ended, the French attacking using ladders wide enough to allow two men abreast, cats and other equipment, and the next day the town itself yielded (July). The old castle with its donjon and small walled courtyard seems to have played no independent role of the siege.[1] The English hastened to reinforce their defences, and Bordeaux, the capital, was further fortified with bulwarks and supplied with "gunnes, engynes" while the ditches were scoured and cleared out. Loops and *canonnières* were inserted into the defences. The city, which was a centre of metal working, had been able to construct its own supplies of guns since the previous century, according to the town records, including a "grand canon" built in 1421. The weaponry available in 1442 included the whole range of throwing machines, *bridas, coulevrines* and *canons*. Gunpowder was stored at each of the four main gates, and stone shot was quarried locally.[2] Meanwhile the French were free to attack more than one place at a time, and while besieging Dax were also reported to be besieging Bayonne.[3] Once having gathered sufficient forces, the English organised a counter attack, and managed to recover Dax in August. Here, they stormed the town castle, capturing in the process "al the adversaries grete gunnes".[4]

The prime target for the French was the fortress of La Réole, on the river Garonne. They began their attack on 4 October, and stormed the town four days later. The castle however continued to resist and held on for six weeks before being forced to capitulate, having suffered heavy bombardment.[5] St. Sever, which had been lost and regained, was attacked again by Gaston de Foix, with artillery commanded by Bureau. The trenches were prepared overnight with the sound of the digging concealed by trumpet, and large culverins and serpentines were mounted where they could bombard the gate, following which the suburb was stormed. The heavy guns were now set up and soon created breaches, but the defenders repaired them. Mines were dug, and the garrison countermined. Eventually, after some five weeks the town was running out of both food and gunpowder, and surrendered.[6]

At this point negotiations began for a truce between the old enemies, and in 1444 it might have appeared that the hundred years war was finally over. It would be the folly of the English to resume the war a few years on.

[1] *Monstrelet*, VI, 54–5; G Williams, ed., *The official correspondence of Thomas Bekynton, secretary to King Henry VI*, 1872, II, 187, 196; *Gaston IV*, I, 12–16; Gardelles, *Les châteaux du moyen âge dans la France de sud-ouest*, plate XV (Dax).

[2] Y Renouard, *Bordeaux sous les rois d'Angleterre*, 1965, 420–1, 509; Bekynton, II, 189.

[3] Bekynton, II, 196.

[4] Bekynton, II, 247.

[5] *Monstrelet*, VI, 56.

[6] *Gaston IV*, I, 32–8.

Siege warfare in the west by the mid-fifteenth century

One of the well-known effects of constant warfare, aside from its catastrophic consequences for those that are victims of it, is the encouragement it offers to technological progress. Over the early decades of the fifteenth century, in both west and east, gunpowder artillery developed such as to become quite soon the only weapon of choice for sieges. Although there have been a few references to the continued use of trebuchets, bricolles and blidas during the early decades of the century, by 1440 the only stone thrower is the gun. Time and time again, the accounts tell of these weapons creating breaches and leading either to surrender, or to successful assault. Ballistas and, occasionally, springalds, continue to feature as projectors of bolts, the former especially useful when mounted on stable platforms on towers, but the springald had largely disappeared. In their place, although not in place yet of crossbows or longbows, small scale gunpowder weapons firing metal balls became prominent in siege warfare, their range and penetration now sufficient to make them much more effective than before. The Hussites may have pioneered their use as field weapons, but every prince now looked to establish substantial arsenals of guns, large and small. The new scale of larger and more penetrative artillery coming into use at the beginning of the century was now itself being superseded, and the work of the Frenchman Jean Bureau (and his brother) in realising the potential of the gun would make the French army that ended the Hundred Years War the leading force in this area.

The nature of defence was changed fundamentally through the changing character of the attack. Particularly noticeable is the fact that fortifications created or strengthened during the fourteenth century, and therefore quite modern anyway, had to undergo considerable improvements to be modern by the standards of the mid-fifteenth century. Surviving records are revealing. At Harfleur, already strongly defended against Henry v's guns, the English occupiers did not simply repair the walls and towers, but the records show substantial new works too. In the harbour, a large round tower was created, now disappeared, but apparently larger than the enceinte towers and crenellated and machicolated. On the main circuit of walls, the Tour du Milot was provided with two arrow loops, and two specifically designed gun embrasures. The need to protect gates and stone walls with outworks was evident everywhere. At Harfleur, a surviving bulwark outside the Porte du Rouen on the south east of the town has the shape of an arc, projects 34.5 metres, and has four semi-circular turrets and its own drawbridge, altogether a formidable outer defence work. Whether the stonework was erected after 1415, or after the final conclusion of the wars, is not clear, and generally the outworks of this time were of earth and timber, with the advantage of rapid construction, and if properly put together, not necessarily an inferior defence.[1] Similar design

[1] B Le Cain, Les fortifications de Harfleur au début du xve siècle, in Baudry, ed., *Les fortifications dans les domaines Plantagenet*, 2000, 101–11.

features were found in the rebuilding done at the castle of St. Sauveur-le-Vicomte in the Cotentin peninsula, where the older defences were modified to incorporate embrasures for small guns.[1] The new works done to provide walls for Lisieux included work in 1431/2 that might have been gun platforms in front of one of the towers.[2] Records also show new work being done on the English possessions in Gascony, in particular *boulevards* provided with artillery at Bordeaux (1442) and at Langon and St. Macaire (1432–4).[3] Another miniature shows a clearly temporary wooden palisade built against the moat in front of the main gate of a fortress, provided with shot holes for small guns.[4]

Placing guns inside towers made a number of changes necessary. There had to be room for recoil, somewhere for the smoke to escape, and a way of sighting the weapon. By the final third of the century, the round apertures with sighting slits, built into specially-designed chambers with walls of great thickness and ventilation channels, would become familiar. By the middle of the century, they were being prefigured. There were circular gun loops in the circular donjon added by Charles VI to the castle of St. Malo.[5] It used to be thought that the Tour Boucle at Mont St Michel, which has the form of a pentagonal four-storey bastion, dated from the governorship of D'Estouteville (i.e. 1426–45), which would certainly have made it a candidate, along with Žižka's Tábor, for the earliest specifically designed artillery bastion. However, the construction has been reassigned to the 1480s, while the otherwise advanced design of the whole rampart with level access to the tower tops may well have been the result of necessity, because inside, the houses were built right against the wall.[6]

An example of a structure largely built from new was the massively-reconstructed castle of Bonaguil, which featured large stone platforms ten metres wide around the thirteenth century donjon, and flanking towers, and added massive machicolations to the tower tops to provide cover for close defence. There is good reason to think that the owner was erecting something of a grandiose modern fantasy, but the (dramatically overstated) provisions, whatever their actual intent, are still good evidence for what contemporary design was meant to be.[7] At this time, though, the most usual step was to place heavier guns in suitably prepared and defensible emplacements rather than inside or on top of existing towers,

[1] Mesqui, *Châteaux forts*, 335–6.

[2] Jones, *Les fortifications municipales de Lisieux*,244.

[3] Evidence cited by M Vale, New techniques and old ideals: the impact of artillery on war and chivalry at the end of the Hundred Years War, in Allmand, ed., *War, Literature and Politics in the late Middle Ages*, 1976, 61–2.

[4] Larchey, plate 91B from a MS of the Chronique de Hainaut, dated 1459.

[5] Mesqui, 334–5.

[6] G Bazin, *Le Mont Saint-Michel*, 1933, 211–33, 332–5; corrected by N Faucherre in M Bayle, P Bouet, J-P Brighelli, H Decaens, M Dosbart (et al.), *Le Mont Saint Michel, Histoire et imaginaire*, 1998, 145–8, following M Nortier, P-A Lablaude, *Millenaire Monastique de Mont Saint Michel*, v, 1993, 177–201. Bazin nonetheless contains magnificent illustrations of the Mount's fortifications.

[7] P Lauzun, *Le château de Bonaguil*, 1884, 9–38; Mesqui, 65. For a modern revision of the underlying intentions of the builder, M Coste, *History, description and interpretation of the castle of Bonaguil*, 2001.

while smaller guns could be fired from wall tops, including such useful sites as the abbey tower of St. Stephen at Caen in the siege of 1417. Another building created anew at this time with provision for guns was Raglan castle (Gwent), the parts of which that have been attributed to the 1430s–40s included combined gun and arrow loops within the immensely thick walls of the hexagonal great tower (unique in Britain) and others protecting the gatehouses (Plate 16). Quite what, if any, defensive purpose their aristocratic builder felt was needed is unclear, but the provision was certainly suitable for hand guns.[1] If this example was built by a rising regional landowner, the Cow Tower at Norwich was erected by the city, a free standing structure whose actual function is also unclear and the subject of debate, discussed in chapter 11. The roof offered space for larger weapons, with wide crenellations. It has certainly been interpreted as an artillery tower.[2] The curtain wall built around the vulnerable sides of the island-based tower house owned by the Black Douglas family at Threave in western Scotland in 1447 was an outer defence for an artillery age. The lengths of wall are provided with traditional arrow slits, but the round corner towers are gun towers, of three levels, with dumbbell and inverted keyhole embrasures (Plate 19).[3]

Handguns were capable of being used with much less substantial building work. Even if their role in defence has tended to be overlooked in favour of the heavier guns, there seems little doubt that such weapons were major features whenever a siege reached the point of an assault. Another miniature (this time, appropriately, from a fifteenth century manuscript of Froissart's chronicle) shows no fewer than three handguns in use resisting an assault by escalade on a fortress. One gunner is shown firing from behind a drawbridge, while two others have fired their guns through the round holes of the standard gun loop set in wall towers.[4] Technological improvements in the design of these weapons included the step of providing the gun with a wooden stock, making it much easier for the soldier to hold, aim and operate. The precise date at which this first appeared is not known, but certainly it was well before the middle of the fifteenth century (as discussed in Chapter 3) and became commonplace in manuscript illustrations during the second half. The penetrative power of handguns rendered them useful for both distant and close quarter fighting. Continued progress towards the arquebus of the sixteenth century would occur throughout the fifteenth, with an ever increasing role for such weapons both in siege warfare, and in battle conditions (Plate 29).

With the widescale use of guns in defence, older means of attack that would be easily vulnerable to such weapons seem to have simply fallen out of use. Mentions

[1] A J Taylor, *Raglan Castle* (guide book), 1950, 21–3, 38–40. A new edition by J R Kenyon (2003) offers an updated interpretation. Here the outer wall erected around the foot of the tower (about 1460) obscures the loops.

[2] A Saunders, 'The Cow Tower, Norwich: an East Anglian bastille?' *Medieval Archaeology*, XXIX, 1985, 109–18; B Ayers, R Smith, M Tillyard, 'The Cow Tower, Norwich: a detailed survey and partial re-interpretation', *Medieval Archaeology*, XXXII, 1988, 184–98. An alternative suggestion is that it was an unfinished temporary work dating from much later (same source, 199–202).

[3] C Tabraham, *Threave Castle* (guide book), 1994, 19–24.

[4] Larchey, plate 77.

of siege towers ("belfries") and movable shelters ("cats") are extremely rare, and of battering rams so rare that they are possibly anachronistic inventions of the chroniclers. On the other hand, ditches still had to be crossed and walls mounted, so there are frequent references to the preparation of faggots and bundles capable of being conveyed, under fire, to the edge of the ditch, and many assaults were carried out by scaling ladder. To get close enough to the defence to launch an attack called for extra protective measures and by mid century few are the sieges where there is no mention of the attackers advancing by digging approaches. The day of the trench had arrived with a vengeance. Nonetheless, the limited effective range of the artillery meant that any exchange of fire would continue to be at terribly close quarters. Such close action was repeatedly illustrated in the rich survivals of manuscript miniatures, tapestries and other forms of imagery dating from this century. A biblical scene preserved in Reims shows a fortress being attacked by archers and by gunners. The latter are firing a large bombard standing in an elevatable wooden frame behind a tilting wooden mantlet, while the hand gunners shelter behind portable wooden screens, held up with a prop, and of the same height as the men. Taking into account artistic licence and lack of interest in scale or perspective, the images certainly look practical and realistic.[1]

An older common feature of sieges was the siege fort. But the modern siege castle, although it shared the function of its predecessor in providing a besieger with a means of establishing a blockade when the length of the wall precluded a continuous encirclement with available forces, or to enable a commander to leave a small force to pen in a hostile garrison which they could not reduce by immediate action, also sometimes had a more active role too. Such forts, redoubts, bulwarks or bastilles might simply hold a garrison (as at Compiègne in 1430), or they might represent a strongly-held base for the attack itself, a secure emplacement for artillery, to protect it against sorties, and a starting point for the digging of approaches. In such a case they would need to be sited that much closer to the defences. A major skill was called for to get this right. In a fascinating sidelight on the operations at Orléans, the quasi-mythical account by Jean de Bueil explained how the English had to set up camp at such a distance to be safe from the defenders' artillery, but at the same time had to construct their fortifications at the right range to allow their bombards to be effective.[2]

Although there are references to teams of expert miners being deployed, they are becoming rather more rare in the west, suggesting that the power of guns in creating breaches made this form of attack more efficient than the slow, dangerous and, once begun, inflexible, attempt to undermine. It remained a form of attack that never entirely went out of use, and during the course of this century the first use of gunpowder to increase the destructive force of the mine was noticed. The siting and the strength of the defences of the fortress under attack would determine whether undermining could still be the most effective form of attack. Although references to miners in Burgundian pay are less frequent during the

[1] M Sartor, *Les tapisseries, toiles peintes et broderies de Reims,* 1912, 175. The story of Judith: siege of Bethelia.

[2] J de Bueil, *Le Jouvencel,* ed. Favre, Lecestre, 1887, I, 163–4.

fifteenth century, leading one scholar to suggest they were becoming less used, the Bureau brothers' reform of the French artillery specifically included specialists identified as sappers.[1] Salisbury's army at the siege of Orléans, as noted above, included a substantial corps of miners. But overall, as our account has shown and will show again in the story of the final expulsion of the English from Normandy and Gascony, it was unquestionably the skill and power of the gun rather than the ingenuity of the miner that had become decisive.

Poliorcetics

The poliorcetics of the age had to reflect current usage. Christan de Pisan added modern artillery to the precepts of Vegetius early in the fifteenth century, and both Vegetius and Vitruvius continued to be popular sources of learning despite the dramatic changes in technology since their days. Henry v and his commanders were educated on Giles (Aegidius) of Rome, so must have learnt about modern siegecraft from elsewhere. Philippo-Maria Visconti, duke of Milan, was educated in the use of military machinery.[2] Konrad Kyeser of Eichstadt's work, *Bellifortis*, was written around 1395–1405. This text, like Christan de Pisan's, is a mixture of the ancient and the new. Alongside illustrations of trebuchets and blidas are shown wheeled versions of the stone thrower, and battering rams, but also numerous bombards and cannons.[3] This work was a best seller of the age and continued to be copied for a long time to come. One of those influenced was Mariano Jacobus (Taccola) of Siena. His *De Machinis* a work of 1449 (originally written about 1430), described how the general needed to know about the use of bombards, alongside the skill of launching things using a mangonel, and constructing tunnels, a mixture of the old and the new learning. But Taccola added information explaining how a bombard could be rendered ineffective: getting humidity or water inside it, driving a plug of hard wood into the touch hole (an early form of spiking that we have encountered in use already), or mixing camphor with the gunpowder. His work also included descriptions of large winch-wound crossbows and other useful engineering accomplishments such as bridges of boats which were being used regularly by commanders, alongside other practical and frequently used devices such as different forms of ladder suitable for attacking a defended wall, items that rarely attract any detailed attention from chroniclers or witnesses.[4]

1 Brustem, *L'armée bourguignonne*, 16; Dubled, 586, and see below.

2 Decembrio, Vita Philippe Mariae Vicecomitis, ed. Candido, *RIS*,xx, part 1, 1925, 316–17.

3 M Berthelot, 'Histoire des machines de guerres et des arts mécaniques au moyen âge', *Annales de Chimie et de Physique*, 7th ser., 19, 1900, 289–420, reproduces the De Kyeser drawings. See also Gille, *Renaissance engineers*, 57–62, and Lindgren, Waldburg Wolfegg, Beminger (eds.), *Konrad Kyesers Bellifortis*, 2000, describing a later copy of the MS.

4 Mariano Taccola, *De Machinis. The engineering treatise of 1449*, ed. G Scaglia, 1971, 69–70; M Berthelot, 'Pour l'histoire des arts mécaniques et de l'artillerie vers la fin du moyen âge', *Annales de Chimie et de Physique*, 6th ser., 24, 1891, 466.

One of the difficulties in interpreting what these works were for is that they are liberally strewn with fantasies, anthropomorphic creatures and impossible machines, next to detailed drawings of pieces of equipment. Did their portrayal of the counterweight trebuchet mean that such weapons were still part of the armoury? And if they were when *Bellifortis* was written, does the fact they still appear in Taccola mean anything other than that the latter was copying the contents of the earlier writer, rather than reflecting current reality? The chronicle accounts suggest the latter. But it does mean that one must be very careful before considering the appearance of a machine in one of the text books as sufficient evidence that they were (or ever had been) in actual use. The appearance of a trebuchet in the anonymous text called the "Anonymous of the Hussite wars", now convincingly argued to date from no earlier than 1472 rather than the assumed date of 1430, certainly does not indicate that this weapon continued in use at such a date, but the drawing, although similar to that of Konrad Kyeser's, is not the same. So why has the anonymous writer troubled to insert it at all?[1] For the same reason, perhaps, as he has presented other devices that were no longer in use, and indeed in some cases, never could have been constructed. This might also explain the insertion in a number of German manuscript military handbooks of 1400–1450 of a drawing of a torsion-powered mangonel, interestingly more or less the same illustration copied over.[2]

Gunpowder and guns

One work for which there is evidence of practical application was the anonymous so-called *Feuerwerkbuch* dated to around 1400. It reads as if it were a handbook by a master gunner, concentrating as it does on very practical questions of manufacturing gunpowder, beginning with the selection, preparation and purification of the ingredients, preparing and loading the gun, aiming and firing it, and preparing different kinds of missile and shot to be propelled. It contains advice on how to obtain the greatest range and the most powerful impact. So popular was it that it was actually put into print a century after it was written, and of the surviving manuscripts in German at least 20 date from before 1450.[3] There is an unresolved debate whether German was the original language, or whether one of the French language versions, the *Livre du secret de l'artillerie*,

1 B S Hall, *The technological illustrations of the so-called "anonymous of the Hussite wars", codex latinus Monacensis 197*, I, 1979.

2 R Leng (ed.), 'Anleitung schiesspulver zu bereiten, buchsen zu laden und zu beschiessen. Ein Kriegstechnische Bilderhandschrift im cgm 600 der Bayerischen Staatsbibliotheek, München', *Imagines Medii Aevi*, 5, 2000, 125.

3 The 1529 version is reprinted and also rendered into modern German in Hassenstein, *Das Feuerwerkbuch von 1420*, 1941. The actual date of writing was anywhere between 1414 and 1428, and the actual title is that it describes how to make "guns, powder, fireworks . . . to defend a town, castle or fortress besieged by the enemy." There is now an English translation, G Kramer (ed.), 'The Firework Book: Gunpowder in medieval Germany', *Journal of the Arms and Armour Society*, XVII, 2001.

might have been the progenitor of the rest.[1] This book was by no means the only such text to be written during this time. Leng has identified, alongside 58 versions of the German text, and 47 of *Bellifortis*, in German archives, other technical handbooks variously penned around 1411, 1415–18, somewhere from 1410–30, and 1425 in the Vienna, Munich and Nürnberg archives.[2] According to his analysis, these were written for the use of engineers (gun masters) themselves, not for the education of rulers. Another text not in this list, something which itself hints at the even larger number originally written, has now been published. Johan Bengedans' *Kriegkunst und Kanonen*, found in Copenhagen, was written by a German engineer who sold his skills to the king of Denmark and the Teutonic Knights in the years before 1450. The manuscript covers much familiar ground such as how to make gunpowder effectively, as well as detailed instructions on using quadrants to work out how to hit an enemy tower with a gunshot, and, particularly interesting, much attention was paid to designing incendiary and explosive arrows to be fired from ordinary bows and guns (Plate 28).[3]

Having access to the range of artillery was now an essential requirement for any campaign. But the larger guns essential to demolish walls were vulnerable: heavy and immobile, they could readily fall into enemy hands if relief forces were operating. Although gun-making was a significant industry, lost weapons could not necessarily find instant replacements. De Bueil's mythical king had to abort a siege of a castle because he had lost most of his artillery,[4] and we have listed a number of occasions where a besieging army abandoned its weaponry.

There had been a steady increase in the size of the largest guns. Now, at the middle of the century, the apogee was to be achieved with the great guns built both by the Ottomans (as we shall see in the next chapter) and in the west. Generally, evidence of dating for the manufacture of particular guns is very rare, and surviving examples are thin on the ground, an obsolete weapon being far too tempting a source of expensive metal for reworking.[5] The bombard known as "Mons Meg" (Plate 32) has sister weapons of a similar type, and a study has suggested that they were all made at Mons by a craftsman such as Jehan Cambier who worked there in the middle of the century. Mons Meg retains a documentary history and is the only bombard to do so, and was built in 1449 and sent to Scotland. The giant weapon is made of iron staves welded together longitudinally, held by wrought iron hoops. It is 15 feet (4.6 metres) long with a calibre of 18 inches (0.46

1 On which see H Dubled, 'L'artillerie royale française à l'époque de Charles VII et au début du règne de Louis XI (1437–69). Les frères Bureau'. *Mémorial de l'Artillerie française*, 50, 1976(4), 563.

2 R Leng, Ars Belli. 'Deutsche tacktische und Kriegstechnische Bilderhandschriften und Traktate im 15. und 16. Jahrhundert', *Imagines Aevi*, 12 (1,2), 2002.

3 J Bengedans, *Krigskonst og Kanoner. Kriegskunst und Kanonen*, H Blosen, R A Olsen et al.(eds.), 2 vols, 2006. One of his letters to the High Master of the Teutonic Knights boasts that he can devise six types of fire arrow to shoot from crossbows and seven types of explosives to be fired from a gun (II, 81–3).

4 J de Bueil, II, 207.

5 A Burgundian record reports the construction of ten culverins from the recycled metal of six older weapons in 1431, cited by Brustem, *L'armée Bourguignonne*, 1953, 32.

metres). "Dulle Griet" (Plate 31), the Ghent gun we shall also see in use shortly, is even bigger with a length of over 16 feet (5.01 metres) and a bore of 25 inches (0.64 metres). The two guns differ in the design of the powder chamber at the rear of the weapon. Mons Meg weighed 15,366 pounds (6,969 kgs). Transporting such equipment was a substantial logistical operation. Transporting their ammunition, and loading and operating the gun itself, would also have been a slow task. Information about the movement of the duke of Burgundy's artillery is suggestive. In 1409, a bombard weighing 4,000 kgs needed special carts reinforced with iron, 32 oxen, 31 horses and 25 drovers, and travelled at five kms per day. In 1436, the duke's gun park needed 300 carts to move it.[1] Roads had to be remade and bridges rebuilt to cope with the weights, so the burden on the people recruited locally and the effect on local economies, must also have been considerable. Where it could be done, artillery was transported by water. But the immense load still had to be off-loaded and moved to where it was needed, and the poor state of medieval roads turned each such operation into an organisational challenge of critical importance for the success of the ensuing siege. It is a remarkable testimony to the ability and resourcefulness of medieval commanders that the task was, usually, accomplished without by itself determining the outcome.

Fortunately for all concerned, many people understood that simple size was not everything, and smaller guns would still be able to deliver a heavy blow. Large weapons continued of course to be made. The two poorly preserved bombards at Mont St. Michel (the "michelettes") are said to date from the English siege of the 1420s (although without actual proof). They are not as big as the great Mons weapons but still stand at the top end of the scale. One is 217 cms long (seven feet) with a bore of 41 cms, while the other is 365 cms (almost twelve feet) long.[2] They are in keeping with the type of larger weapon noted in the inventories of the arsenals. But the great majority of weapons listed or described were much smaller. Larchey's study of the evidence for French guns showed that the great majority of weapons were of middle or small size. The inventory of King Ferdinand of Naples had 13 large bombards among a total of 133 guns, 22 of the balance being small bombards and no fewer than 65 culverins.[3] There continued to be much experimentation, too, during this time. Illustrations show many attempts at multiple weapons, for example several cannon mounted side by side, and other, perhaps fantasy, multiple weapons in the text books. The ribaudequins as multiple mountings, however, were attested in chronicles and inventories.[4] The big variety of guns deployed at many of the sieges (despite the logistical difficulties it must have caused) seems generally (from the accounts given here) to have worked the intended results. This, the history of their use, is the practical evidence, then,

[1] Gaier, *Art et Organisation militaries dans la principauté de Liège et dans le comté de Looz au moyen âge*, 1968, 96–8, citing evidence in Garnier, *L'artillerie des Ducs de Bourgogne*.

[2] R Smith, R Brown, *Bombards. Mons Meg and her sisters*, 1989, analyses and describes these and other bombards.

[3] Larchey, iv.

[4] Larchey, plate 57 is a very good illustration of a three-gun ribaudequin on an elevatable wooden platform.

for the improved effectiveness of siege guns during the first half of the fifteenth century. It was clear to commanders that the artillery at their disposal would have different functions, ranging from demolishing walls and towers to picking off individual enemy soldiers or, from the side of the defence, preventing the attacker from building siege works close enough to do harm.

On the other hand, among the obvious disadvantages arising from having such a variety of weapons was the absence of standard calibres, meaning that each gun needed its own ammunition and could not share. Medieval master gunners would manufacture the shot suitable for their particular gun on the spot, following the tradition of the operators of stone throwers dating back over centuries. Gradually the technical weaknesses were eliminated or at least reduced, each change being related to others and combining to lead to a gigantic leap forward in performance and reliability. The considerable problems in manufacturing gunpowder were gradually resolved by experimentation. Early gunpowder had many impurities, leading it to absorb water too easily, while powder that had been corned (grained) burned too fast. By the 1420s there were refinements to cope with both problems, echoed in the writings of Kyeser and in the *Feuerwerkbuch*, and through the rest of the fifteenth century improvements were maintained.[1] Properly grained gunpowder, commonplace by mid-century, permitted a reduction in the amount required to achieve the same propulsion, and more consistent effect. This was also better fitted to guns made of bronze. During the first half of the century there had been continuous experimentation in gun manufacture. Wrought iron guns were relatively inefficient, cast iron guns were stronger, and now became commonplace as methods of manufacture improved, but, paradoxically, were also more prone to blowing up. As a result, even if powder could be made to have greater ballistic power, the risk of destroying the gun was so great that it had to be made weaker. Copper was the best material to resist this danger, but was so expensive that hardly any weapons were constructed from it, although a culverin recovered from the Loire at Orléans and claimed to date from the time of the siege was of this metal.[2] A list of the largest known guns made between about 1394 and 1430 identifies bronze as the favourite material over iron by a margin of three to one.[3] Guns of bronze alloys became the standard material for artillery until late in the nineteenth century. The boost to metal industries was considerable as the demand was relentless, and countries (like France) that did not yet have native industries had to import the various metals from their neighbours, including countries (like England) with which they were more or less permanently at war.

Stone shot had obvious problems associated with the effect of an uneven surface, possible friability, and so on, and large numbers of shot cutters had to be employed at sieges to turn quarried stone into missiles. Methods of compensating

[1] B S Hall, The corning of gunpowder and the development of firearms in the Renaissance, in B Buchanan, ed., *Gunpowder: the history of an international technology*, 1996, 87–110, describes the science in relation to the historical evidence.

[2] M Desnoyes, 'Les armes du siège de 1428', *Mémoires de la Société archéologique et historique de l'Orléanais*, 18, 1884, 449–52.

[3] Table in Hassenstein's edition of the *Feuerwerkbuch*, 145.

for the weaknesses included coating the ball in metal, generally iron because copper was too expensive, and eventually stone shot would be replaced by metal altogether for English and French guns, allowing for a reduction in the mass of the projectile while simultaneously actually increasing its effectiveness. Taken together, these improvements meant that lighter and more mobile guns, less prone to blowing up and to the wearing-out from use that might quickly render them useless, were able to fire more powerful missiles over greater distances. Indeed, it is very hard to determine what range these guns had. It is generally accepted by students of the subject that the normal maximum was between 1000 and 2000 metres, although of course there is a significant difference between a maximum range and an optimum range. Even the lower of these figures represents a massive increase over the range available to the trebuchet, and clearly shows why the better guns of the fifteenth century very quickly replaced the old stone throwers.

Wheeled carriages were introduced for lighter weapons, appearing in illustrations of the first half of the century, meaning along with greater mobility and ease of aiming that the gun's own structure could absorb the recoil (Plate 30). Without the wheeled carriage, smaller guns could still be mounted on adjustable frames enabling their aim to be altered.[1] As a design improvement, lugs (trunnions) appeared on the guns, having been invented for Philip the Good by François de Surienne, but not before the second half of the century, although this improvement was one of many that have also been attributed to the period in which the Bureau brothers oversaw the artillery of the king of France, in this case in an earlier, more experimental form. For a long time, heavy guns in particular would rely on fixed mounts, and a fifteenth century manuscript illustration (of Aretin's *Première guerre punique*) shows side by side a large and a small cannon, both set in fixed wooden frames with trunnions allowing elevation, next to soldiers armed with handguns in wooden stocks (suggesting a date later in the century) and crossbows.[2]

Developments in organisation: France . . .

Probably the single most significant advance was in organisation, where both the French king and the Burgundian duke took the vital step of establishing a distinct artillery park. The French king's guns were under the direction of Pierre Bessineau from 1420–40, then came under the control of the brothers Jean and Gaspard Bureau. A great deal has been attributed to these two, little of which can actually be proved, but the fact remains that during the period in which they held office there

[1] An illustration from MS Cotton Nero Eii, part 2, f.1 (of about 1415), printed in P Porter, *Medieval warfare in manuscripts*, 2000, 37, shows a small gun mounted on quite a complicated wooden frame standing in front of the gate of a town. There is also what seems to be just a fence standing directly in front of the gate, perhaps indicating a temporary outwork.

[2] R Smith, Artillery in the hundred years war, in Curry, Hughes, *Arms, Armies and Fortifications*, 157–9; Beyaert, *Quelques Problèmes*, 73–7; J F Fino, 'L'artillerie en France à la fin du Moyen Âge', *Gladius*, XII, 1974, 13–26; Dubled, 575–8, 616 (illustration).

were changes in the artillery that amounted to a revolutionary transformation. Both were, as was common with medieval officials, people of many talents. The elder brother, Jean (1387/93–1463) was at various times a general, a diplomat, and treasurer of the kingdom. The younger, Gaspard (1393/1400–1469) had a more technical role alongside periods as a commander in the field, and documentary records survive telling of his responsibility for organising the artillery. He was deputy to Bessineau before taking over from him. The role of both in transforming the French artillery was attested by contemporary chroniclers, although the evidence simply does not permit judgements about who first identified the various technological changes that took place over the decades in the middle of the fifteenth century. Nonetheless, those in charge must have had a role, so it is not unreasonable to allow the brothers considerable credit for this.

There is little doubt about their responsibility for improving the organisation. In the mid-1440s, Charles VII created in effect a standing army for the first time. The long-term significance of this step, of course, was dramatic, and similar steps would be taken by the Burgundians. Although the royal ordinances did not talk of the artillery, it seems very likely that the Bureaus' changes to this arm accompanied the overall transformation. They established both a royal artillery park, and a force of men to service it. These, the *franc archers de l'artillerie*, had specified responsibilities for the transport and protection of the guns, for the construction of artillery emplacements, and for the preparation of approaches, saps and so on. Additional experts, such as carpenters, would as before be recruited as needed for a particular campaign, but now there was a standing core of experts available to service the royal artillery.[1]

. . . and Burgundy

Significant evidence illustrating how the dukes of Burgundy developed their own artillery arm during the middle of the century was collected from the archives of the duchy by the historian Jean Garnier. The Burgundians had an official called *L'artilleur du Duc*, who was in charge of manufacture and maintenance of the guns and gunpowder. Other officials called variously controllers, masters and receivers of the artillery appear in the records for the following decades, François de Surienne being *Maître* in 1440 as part of his varied career. The records show some of the lower-ranking positions. The crew being paid to operate a bombard comprised: a knight or "expert gentleman" in command, although one suspects that the *canonier* (with his valet) might have had the real role of expert, a carpenter (with his own valet, too), presumably in charge of constructing the gun's platform, and six workmen. Other records confirm that it was the *canonier* who was the expert, attracting good pay, and offering his services wherever he was wanted, regardless of nationality.[2]

From the middle of the century, there was a considerable increase in the

[1] Dubled, 582–6.

[2] Brustem, 6–11.

quantity of artillery held in the duchy, with a substantial proportion of it being smaller weapons. There were 206 *veuglaires* manufactured in 1445, stored in the ducal arsenal at Dijon, while *canons*, a term signifying at this time smaller weapons, were constructed at Binche in 1446 and stored in the arsenal there. Other arsenals were sited in Tournai, Metz, Bruges, Lille, Brussels, The Hague, Malines (Mechelen), Nivelles, Aire, Luxembourg, Namur and Huy, meaning that the duke was able to call upon a train of artillery close to any part of his scattered domains. The duke was also able to mobilise guns owned by his subjects. Sometimes these belonged to the aristocracy, Garnier reporting that Jean of Luxembourg lost his own guns at the siege of Compiègne in 1430, while the duke borrowed guns from local nobles in 1443. Towns had their own arsenals. The inventory of Bruges from the first half of the century counted 113 copper, iron and bronze serpentines, one of them a real monster at 17 feet (5.2 metres) long but most of them small, six ribaudequins, and 21 large *veuglaires*, one of them more than five metres long weighing 2,583 kgs. The inventory of the city of Mons listed 40 serpentines, most of them on wheeled carriages, and some firing iron balls weighing ten pounds (4.5 kgs), confirming their relatively small size, 84 *veuglaires*, some wheeled, 11 *canons*, 136 culverins, and as hand weapons, 284 hackbuts, indicating that this list is probably of a later date in the century.[1]

. . . and elsewhere

Everywhere, in fact, steps were being taken to ensure that powers established some structure and organisation for their artillery provision, whether it was German cities or the Venetian republic. Some of the manuscripts cited earlier seem to have belonged to the *Büchsenmeister* appointed to organise the guns of cities such as Nürnberg, Landshut and Worms. In Venice, although there had been a factory for manufacturing artillery since the late fourteenth century, it was 1440 when a supervisor of gunpowder supplies was appointed, and by 1451 payments were recorded to Ferlino (previously employed by Savoy and Milan before his capture by Venice) to construct a large bombard weighing 6000 lbs (2720 kgs) and firing a 400 lbs (182 kgs) shot up to two miles. From now on, there was a continuous line of chief gunners and gun makers in the pay of the Senate.[2]

[1] Brustem, 13–14, 35–6. See also N Michael, G Embleton, *Armies of Medieval Burgundy 1364–1477*, 1983, 16–21.

[2] Leng, Ars Belli, 214–15; M Mallett, J R Hale, *The Military organisation of a Renaissance state: Venice c. 1400 to 1617*, 1984, 81–3.

6 The late fifteenth century, 1: France, Britain, Central Europe and the Balkans

THE MIDDLE OF THE FIFTEENTH CENTURY suggests a moment when the power of the offensive seemed unusually dominant. In the year that Constantinople fell, the armies of France with their newly powerful and effectively organised artillery completed the expulsion of the English with remarkable speed. Yet even now, there was evidence that the equilibrium between attack and defence was again about to be restored, in the failed Ottoman attack on Belgrade, or the successful defence of Neuss nineteen years later. We have examined the significant progress in the quality, quantity and in particular in the organisation of artillery achieved by the middle of the century. In response there was a continuing adaptation of defences to cope with the threat of the gun, both in the nature of the defences, and the skills now called for in commanders of garrisons. Likewise, attack and defence of towns and fortresses continued to be increasingly expensive in resources. The cost of the large numbers of people needed to sustain an army in the field, and their equipment and supplies, had long been a component in the consolidation of central government power, and warfare was a prime mover in the establishment of bureaucratic apparatuses. Successful states were now richer and better (or at least, more) governed states, and fewer and fewer were the lesser landowners able to challenge a prince militarily: although both landowners and cities were among those who took the drastic step of rebellion during this time.

The Metz campaign of 1444

THE IMPERIAL CITY OF METZ found itself in conflict with René, duke of Lorraine, claiming to be its overlord, who was also duke of Anjou and a cousin of Charles VII of France. The latter having now concluded a truce with the English, found himself at liberty to assist René, as well as having to find immediate employment for the hordes of unemployed soldiers in his kingdom. But Metz was no easy target. Its successful resistance to the attack of 1324–5 was described earlier. The defences of 1444 are better known, and were substantial, although little now survives. No fewer than 43 towers projected from the ramparts, which were fronted around two thirds of the city by the rivers Moselle and Seille. Each tower was named after the craft or trade whose organisation took responsibility for its upkeep. The walls had been extended during the thirteenth century to enclose the growing suburbs, creating a total length of some six kilometres. Bridges led to massively fortified gates, in some cases with a bent entrance, with barbicans which themselves crossed the river. The most recent tower was the *Tour Camoufle*, erected in 1437 to reinforce a corner of the rampart. Three storeys high and cylindrical in form, it was named after the mason who built it. The walls had been set up to protect guns, with embrasures both for large weapons and handguns. The citizens determined to resist attack, and to equip the city with "all kinds of artillery": bombards, serpentines, *courtaulx*, hackbuts, culverins[1] and great quantities of lead. Captains were appointed to command each battery, and supplies were provided for the substantial garrison of soldiers. Repairs to the defences were ordered. Each parish was then given responsibility for clearing out the ditch in its sector, and, in a move of considerable tactical sense, small numbers of soldiers were sent out to hold the strong places in the surrounding countryside, especially those close to the walls. These steps were taken during August, and the first French forces, led by the count of Richemont, arrived in the area on 10 September.[2] The city's concern to upgrade its defences was continuous, and during October orders were issued to construct two earthwork bulwarks as additional defences in front of the two of the gates, erected by master mason Jean de Commercy.[3]

The French accounts tell a similar story. The French forces caused great destruction to the surrounding countryside, reducing large numbers (perhaps up to 30) of the local castles and strong houses. The Metz chronicles support this version, but explain that some were held by as few as the 15 soldiers who held the castle at Ancey from 10–17 September, or the three men who forced the French to storm the castle at St Blaise before surrendering on 16 September. The fortification at Veucy, to the south, was held by 22 men, while just a dozen Metz soldiers, provided with guns and supported by villagers, held a "strong house" at Ennery.[4] The French were forced to bring up their own artillery in carts, find a suitable emplacement and begin bombardment, although it took only three or four shots from a serpentine to end the resistance here (23 September). On 7 October, two large bombards had to be set up against the house of Waldrinowe, but here the defenders had their own bombards which were well prepared, and their shooting at the French was effective. It was 19 October before Veucy was taken by storm, backed up by the onslaught of seven large bombards.

Meanwhile, with the French forces reducing these tiny outposts, the rest of the garrison of Metz was organising daily sorties from the city, capturing many French troops and killing others, sometimes accompanied by workers to demolish places

[1] For an attempt to summarise what was meant by the different names of gun types, see the glossary.

[2] Chronique dite de Praillon, ed. De Saulcy, Huguenin, *Relation de la siège de Metz en 1444 par Charles VII et René d'Anjou*, 1835, 222–3; Philippe de Vigneulles, *Chronique*, II, 279–84; on the walls: A Boinet, *Le vieux Metz*, 1923, 81–4; M Grosdidier de Matons, *Metz*, 1957, 23–30; J Schneider, *La ville de Metz aux XIIIe et XIVe siècles*, 1950, 32–5.

[3] Vigneulles, II, 290.

[4] *Praillon*, 227–9; D'Escouchy, 28; Basin, 13–17.

that might be used against the city, and in one sally 3500 workmen were escorted by 800 men at arms to cut down trees. Regularly, convoys of hundreds of carts were sent out under strong escort to collect as many supplies from the countryside as possible, and Basin claimed that the city had two years' supplies.[1] The garrison also carried out attacks to recover or destroy the castles and strong houses already captured by the French. Goin was taken on 23 September by a group of just six, who seized the booty then burnt down the house. On 1 October, a party of 200 men captured the strong house at Longeville, along with ten French men at arms and their horses and equipment. A much larger group of French troops, including 38 men at arms, was captured when the strong house of Magny was taken by storm on 12 October.[2]

The king could see no prospect of bringing Metz to surrender through continued prosecution of his attack. Perhaps he had expected that the flooding of the countryside with his hordes would bring the city to its knees. Indeed, the wild indiscipline shown by much of his own force suggested that a formal siege of the powerful and resolute fortress, with its substantial supplies and well motivated garrison, that would necessarily have to begin in mid winter, was not a good choice. There was very little effort to establish any kind of direct attack on Metz itself. On 22 September, 1500 French troops had marched up to the Île des Mors, north-west of the city ramparts, and set up a battery of bombards which fired a total of four or five shots into the city. They then withdrew. That was the total of it. When another party of 600 marched close by the ramparts in October, they were skilfully targeted by three Metz bombards located at Woudrinowe.[3] The devastation of the territory was another serious problem for his own forces, too. It was therefore no surprise when a treaty was concluded on 28 February, with the city paying Charles VII a large sum and recognising René's lordship. The events, although lacking in dramatic grand moments, were of some significance. The importance of holding outworks to prevent a would-be besieger getting close had apparently been well understood by the captains of Metz, and the ability of even small defended outposts to cause delay and disruption had been well demonstrated. To Charles VII, the main lesson seems to have been that a large force of ill-disciplined soldiers mainly concerned to gather loot and plunder was of little military value, and chroniclers date the beginning of his reform of his army to his experience in front of Metz.

To the Metz governors, the attack seems to have caused more alarm than the sources might indicate, since almost at once they ordered the reconstruction of the fortifications of the *Porte des Alemans,* to replace the temporary bulwark erected for the siege (Plate 13). The new work comprised a pair of powerful cylindrical towers in front of the existing towered gateway, with the effect of creating a much more effective barbican embracing the islet and bridge by which access to the city was obtained from the north east, across the river Seille. The foundations, 18 feet (5.5 metres) thick, were laid in eleven days by the mason Henri de Ranconval.

[1] *Praillon,* 233–43, 248; Basin, II, 16–17.

[2] Vigneulles, II, 287–9.

[3] Vigneulles, II, 286, 292.

Later illustrations of this still surviving defence (despite "restorations") show it liberally supplied with gun loops.[1]

The recovery of France completed

Normandy

During the truce of the mid 1440s between England and France, the latter began a programme of organisational, military, and financial reforms to consolidate their victories. The former began the descent towards the domestic chaos characterised as the Wars of the Roses, a struggle for political power that would soon turn into bloody civil war and the capture and recapture of the throne between rival claimants around whom factions of ambitious and ruthless rivals would gather. Before the political class had broken apart, the duke of Somerset led the government of the feeble Henry VI. Did he believe that overseas victories would be a valuable outlet for the ambitions of English nobles and gentry, which had now spent since time immemorial seeking wealth and fame at the expense of the country's near neighbour? If so, he made a fairly fundamental miscalculation if he believed that the weakened English regime in Normandy and Gascony would long resist a sustained French attack. Certainly, the attitude he adopted to the breach of the truce with Charles VII appears suicidal in retrospect.

The French demanded control of Le Mans in return for extending the truce, but the English garrison refused to leave. The French army under the count of Dunois set up siege works around the town, and the defenders agreed to leave (16 March 1448).[2] The following year the fragile truce collapsed. The adventurer François Surienne, called L'Aragonnais, who had previously served the duke of Burgundy, organised a party of 600 English soldiers with whom he seized the powerful castle of Fougères by a daring night-time escalade, sacked the wealthy town, and held on to the heavily defended castle, from behind the walls of which they could ravage and plunder across the land of Brittany (23/4 March 1449). Somerset refused to restore it to the French.[3] The French government organised at once for war, with the intention of recovering Normandy first. In preparation for this campaign, they carried out their own surprise attack, at Pont de l'Arche, on 15 May. Operating according to a carefully prepared plan, 400 troops gathered quietly in nearby woods while a carter, who had built up a relationship with the men watching the gate, now succeeded in blocking open the drawbridge of the bastion in front of the gate with his cart and overpowering the sentry, enabling the troops to rush in. The garrison of 120 English troops was overwhelmed.[4] The capture of this town immediately posed a threat to Rouen.

[1] Vigneulles, II, 303–4; Salch, *Atlas des Villes*, 86–7, and the sources listed at p. 282 n. 2 above.

[2] Basin, II, 61.

[3] Basin, II, 75; D'Escouchy. 154–5; *Les cronicques de Normandie*, 99; R Blondel, *De reduction Normanniae*, ed. Stevenson, 1963, 5.

[4] D'Escouchy, 163–5; Basin, II, 81–3; Blondel, 24; *Les cronicques de Normandie*, 104.

Other English-held castles and walled towns quickly fell to French attack. First to go was the ancient fortress of Conches with its twelfth century donjon, which was as incapable as the curtain of resisting fifteenth century artillery but fell to a ruse similar to that used at Pont de l'Arche (10 July), while the assistance of the local population also helped bring about the capture of Vernueil. A surprise attack carried the town, a miller having shown the French where and how to drain the wet ditch, and the small English garrison fled to the castle (the *Tour Grise* which was Philip Augustus's round donjon, with its separate defences). Blondel commented that this was so well constructed with compacted cement holding the masonry that siege engines could not break it up, but relief being impossible, the defenders capitulated (23–8 August).[1] One account says the garrison was 120 strong, another that it was only 30, but the number at least was sufficient to hold a tower had the will and the supplies been present. Other locations were physically weaker, such as Pontaudemer, defended by a palisade. Here a French royal force led by the count of Dunois joined local forces, and hurled in incendiaries (called Greek Fire by Blondel), which set fire to the timber and in the ensuing chaos the town was stormed and some 400 defenders were captured. The French accounts say that the damage was actually done by the old-fashioned method of fire-arrows from bows, their force here having only a few small bombards. Gournay castle capitulated, but Harcourt had to be attacked, and the artillery was deployed: bows, crossbows and culverins. The guns hurled shot against the ramparts and, battering the lower walls of the castle, breached them leading the garrison of 150 to capitulate (15 September), apparently "fearing greatly the effect of the cannons" (Chartier). Harcourt had not been modernised since the thirteenth century, so the defenders' fears were well-founded. From here, a detachment equipped with cannon was sent to the castle of Roche-Guyon on the Seine, the 50 defenders of which handed over the site in return for freedom to depart, but left behind their "cannons and culverins". The captain was also married to a local lady, which may have influenced his decision! Meanwhile, the town of Lisieux, without defenders, was peacefully handed over through the intervention of its bishop.[2]

With beaten English forces taking refuge in Rouen or in other fortresses further inland, the duke of Somerset took no steps to challenge the French in the field, which allowed them to continue to organise simultaneous attacks on several isolated garrisons at once. The support of the local people for the invaders made the English position so much more vulnerable. Argentan was given up to the count of Dunois as the garrison took refuge in the castle, but like so many others this had not undergone rebuilding to enable it to resist modern attack. A single shot from a big French bombard knocked a hole in the curtain wall big enough to let a cart through: the defenders retreated at once into the donjon from which they negotiated to surrender. A French force marching up the Seine reached Château Gaillard, and put it under siege. Commentators noted that it was (still)

[1] Basin, II, 84–90; Blondel, 31–2, 54–55; Berry, *Le recouvrement de Normandie*, ed. Stevenson, 1863, 258–65; D'Escouchy, 189–90.

[2] Blondel, 69, 86–7; D'Escouchy, 191–4; Basin, II, 95–99; Berry, 269–75; *Les cronicques de Normandie*, 110, 115; on Harcourt: Mesqui, 457.

reputedly impregnable, and could only fall to hunger. By comparison with many of the other Normandy fortresses, Château Gaillard lasted longer than most, but still it surrendered after six weeks, its 120 defenders marching out on terms. The fall of this fortress made feasible an attempt to recapture Rouen. The citizens were keen to open their gates but the English did not agree. Some Rouennais tried to let the French in by seizing a stretch of rampart and holding it for escalade from the outside, but Talbot, drove them out (16 October). Nonetheless the combination of external attack and internal revolt forced the English back into defensible strongpoints: the fort around St Catherines on the perimeter, the ducal/royal palace, and the gates. Thus separated and cut off, now their fate was sealed, as powerful artillery batteries were set up against them on all sides. They were also short of provisions for the large numbers of English who had gathered in the city. Trenches were dug around the palace, and cannon and bombards were emplaced. Somerset, inside, was compelled to seek a truce in order to negotiate surrender, which was brought about on 3 November, when King Charles VII finally took possession of the chief city of his Norman duchy again. The English surrender also involved giving up the neighbouring towns and castles of Tancarville, Arques, Caudebec, Montivilliers and Lillebonne.[1]

Showing both real determination and the benefits of an effective organisation of his troops, artillery and supplies, Charles did not rest on his extraordinary achievements, but kept up the pressure despite the arrival of winter weather. His forces now laid siege to Harfleur and Honfleur in the frost and snow, the bombards being set up and battering great holes in the walls of the towns. Harfleur, which had resisted Henry V so bravely, had been modernised after that experience, gave up on 1 January 1450. As before, mines played an important part in weakening the walls, crumbling them into the ditch, while the repeated blows of the artillery contributed to the shaking of the defence, and it might be concluded from this that it was only the strength of the garrison (some 1500 men) that enabled the town to hold out at all. The king himself supervised the emplacement of 16 immense bombards protected by deep trenches, along with a "great quantity" of other guns, although the acknowledgement of the presence of Gaspard Bureau in charge of the guns, and his brother Jehan, "fort subtil et ingenieulx" (Chartier) in charge of the mines might give a more accurate picture of the leadership here. From this success the French moved on to Honfleur on 18 January, whose 300 defenders resisted till 18 February in the face of artillery, approach works and mines which shook the walls severely and persuaded them to agree to give in if not relieved within the specified time.[2]

Earlier, the fortress of Fougères, the surprise attack on which had prompted the collapse of the truce, was also recovered by the French, led by de Richemont, duke of Brittany. The duke's army which had taken St Lô, Coutances and Valognes with very little resistance, set about besieging Fougères. The commander organised

[1] Chartier, II, 118, 132–4, 154 ff; Blondel, 119–40; Berry, 281–9, 301–6; *Les cronicques de Normandie*, 120–1, 125–35, 140; D'Escouchy, 210–26; Basin, II, 119–29.

[2] Blondel, 152–55; Berry, 324–7; Chartier, II, 176–89; Basin, II, 131–33; *Les cronicques de Normandie*, 141–2.

the establishment of proper siege works, including bastilles against one side, approaches and batteries of artillery to fire upon the strong fortifications, where the masonry was preceded by outworks erected by the professional soldiers inside, led by their captain François Surienne. After two weeks, the Bretons stormed one of the outworks. It took a month overall for the trenches to get close enough to the defences and for the bombardment to clear the way for an assault, at which point (5 November) the defenders, believing honour to have been served, surrendered. In a recorded statement by L'Aragonnais dated 15 March 1450, he asserted that he had requested additional supplies of arrows, powder, and culverins to help hold the place. When besieged by the duke of Brittany, he stated that he needed 1000 men to hold Fougères, but only had 400. He had held out for five weeks awaiting relief, but was forced to surrender "pour la forte baterie des bombardes et autre artillerie".[1]

The English now held onto only the Cherbourg peninsula and to towns around Caen, Bayeux and Falaise. Somerset assembled a large army and sent it to Cherbourg to attempt to recover losses, or at least prevent further defeats. Thomas Kyriel collected a bombard and a "flying engine" from Cherbourg and laid siege to Valognes, causing it much damage by bombardment and leading to its surrender within three weeks. Meanwhile, however, the French army had assembled to resist the new invasion, and Clermont and Richemont crushed the English at the battle of Formigny (15 April 1450). There was again no force in the field to impede the sieges by which the conquest of the province would now be completed. Avranches held out for three weeks before yielding to Richemont's mines and cannons, which knocked down stretches of wall. Bayeux, meanwhile, resisted only 15 days. During this time the French dug trenches, prepared mines and emplaced guns, and their bombardment reduced the town wall, never strong, to rubble. It was only a swift surrender that saved the place from the sack that the French soldiers were keen to inflict on it. The major castle of St Sauveur-le-Vicomte gave up in June after just eight days, despite having a garrison of 200. The Constable of France had organised approaches but could not bombard since all the guns available were then in use at Caen.[2]

The capture of Caen would be the key to resolving the campaign, and significant French resources were gathered for the effort (5 June). Once again, the abbey Des Hommes (St Stephen) would be a major starting point for the attack on one side, the other abbey (Des Dames) the headquarters for the attack on the opposite side, although the French forces, being in sufficient strength, were able to attack from all quarters at once, and the river Orne was bridged to permit communication. Cannon including 24 large bombards were deployed all round the city, some so large "that a man could sit in the mouth". The outwork in front of the Bayeux gate was stormed on the first day, while the count of Dunois led an attack on the outwork of Vauceules and took it after a long struggle on the second day of trying.

[1] Blondel, 142; *Chronique d'Arthur de Richemont*, 201; Berry, 321; D'Escouchy, 201–5; Stevenson, *Letters and papers illustrative of the wars of the English in France*, I, 289–94.

[2] Chartier, II, 203–13; Blondel, 161, 207–10; Berry, 339–45; *Les cronicques de Normandie*, 148–51; D'Escouchy, 275–92.

Meanwhile, the mines were started from the abbeys, and the wall and tower opposite St Stephen were brought crashing down. Basin noted how one of the big guns facing the church of the Trinity demolished a wall tower with a single shot. The creation of a breach led the English to negotiate, even though they were numerous, and well equipped with all kinds of weaponry. One chronicler noted that the king could have refused the request, since the town was his for the taking, but was perhaps wisely guided by the calculation that the separate castle of Caen could yet have posed a serious challenge. After about one month, the city yielded (25 June), and Somerset led out 4000 men but left behind all his large artillery.[1]

Cherbourg was the last powerful fortress held by the English in Normandy. It was also a powerful fortification, although certainly not (as the French chronicler Blondel fondly believed) the work of Julius Caesar! It was however built of very hard masonry, well protected with wet ditches, and also provided with outworks suited to the artillery age. The French set up a powerful battery of their best artillery, under the direction of Gaspard Bureau once more, using nine or ten of the largest bombards and they placed four bombards on the beach, which were covered by the sea when the tide was in, then were brought back into use when it receded, to the understandable astonishment of the English who cannot have expected an attack from this quarter. A corner tower was seriously damaged by this bombardment. As the siege unfolded, the much less modern defences of Falaise were under attack from an army led by the count of Dunois, with a battery of three bombards and one cannon directed by Jehan Bureau, which quickly breached the walls and induced surrender (20 July). Domfront then yielded after two weeks (2 August) to the same combination. The state of morale of the garrisons is hard to know, but cannot have been high, and even the marvellously strong Cherbourg negotiated a surrender after little over a month (12 August). Basin wrote that at Cherbourg, yet another innovation was recorded when the French discovered a new means of mounting their largest bombards. This may well have been a reference to some kind of fixture that enabled the immense weapons to be elevated. The effect, anyway, seems to have been at least partly responsible for the rapid capitulation of the fortress. During the same time, the castle at Valognes was again recovered by the French.[2] Thus was the duchy of Normandy completely recovered by the king of France in, as the chroniclers proudly noted, just one year and six days. The French chroniclers naturally hailed the achievement. Some of them noted that it was due to their possession of the largest arsenal of "grosses bombardes, de gros canons, *veuglaires*, de serpentines, de crapaudaux, de ribaudequins et de coulevrines" ever assembled, and to skilful and effective miners, reliable experts paid by the day, as with the other siege equipment builders, under the skilful leadership of the Bureau brothers (Berry). Others also drew attention to the good discipline and effective command of the reorganised French army. To these factors we might add that poor leadership of the English, their probably low

[1] Basin, II, 144–49; D'Escouchy, 308–11; *Les cronicques de Normandie*, 148–56; Blondel, 215–23; Berry, 350–53; Chartier, II, 211–21.

[2] Chartier, II, 231–7; *Richemont*, 210–25; Blondel, 232–36; Berry, 359–73; Basin, II, 151–55; *Les cronicques de Normandie*, 159–68; D'Escouchy, 312–17.

morale and the failure to attempt to put armies in the field (except once) left the French with a free hand to isolate and capture the fortresses.

Gascony: conflict resumed

Now, with the whole north cleared (except Calais), attention was turned to Aquitaine, the counts of Clermont and Dunois being sent there. However, the French could not rule out that the English would try to counterattack in Normandy, so they left a substantial proportion of their forces in the north, and entrusted the command in the south to local leaders including Gaston count of Foix, whose old-age recollections form one of the liveliest sources for the events. Gascony, a subject of the English crown for centuries, proved to be much more resistant then Normandy.[1] The initial attack however proved fairly easy, the path cleared with prudent (and subsidised) defections by local nobles. This was territory that had been subject to conflict at its borders already while the main effort took place further north. Gaston had captured the castles of Mauléon and Guiche in 1449, deploying light and heavy guns, despite defence using the same weaponry including a reference to "pierres de fer", perhaps meaning metal-clad shot.[2] In 1451 the French took the crucial towns of Bourg and Blaye,[3] the latter by heavy bombardment followed by an assault. Bordeaux itself, then Bayonne quickly surrendered, while Dax fell to attack by guns and (decisively), undermining of the walls.[4]

Return of the English, and final conquest (1452–3)

Capture proved easier than retention, and for various reasons including the effective lobbying of Bordelais exiles in London, a large English force (5,000 men led by Talbot) was readmitted through one of the gates of Bordeaux in October 1452 by a faction of the citizens, and much of the province revolted against their new rulers, whose troops were now pinned down in fortresses such as Bourg and Blaye. The province would have to be conquered again.

Sieges were once again the main events, and it was a siege that brought about the best remembered event of the war, the death of Talbot in a fight in which entrenched artillery for the first time played a decisive role in overwhelming a frontal attack, at Castillon. The French army took Gensac without resistance and then besieged Castillon on the river Gironde. The Bureau brothers were in charge of the works. They had created a strongly entrenched siege camp with a deep ditch, backed by tree trunk-reinforced earth ramparts on which were mounted numerous culverins and serpentines. The battle took place on 17 July 1453, and the town

[1] M G A Vale, 'The last years of English Gascony 1451–3', *Transactions of the Royal Historical Society*, ser. 5, 19, 1969, esp. 129–35, for an excellent short account of these events.

[2] *Histoire de Gaston IV de Foix par Guillaume Leseur*, ed. Courteault, 1893, 47–52, 65, 69–73, 94. Gardelles, *Les châteaux de moyen âge dans la France du sud-ouest*, 143–4 (Guiche).

[3] Basin, II, 163–5; D'Escouchy, 332.

[4] *Histoire de Gaston IV*, 113–21.

surrendered next day. Despite the death of their commander and the crushing of his army, the Bordelais continued to resist. Dunois took Cadillac on 19 September but only after a full-scale bombardment. He had his heavy bombards deployed in a single battery, the sight of which led the English to abandon and burn their own outwork. The guns were re-deployed against the gate tower, while fascines were readied to fill the ditch. The defenders decided to withdraw to the castle, whence they agreed to surrender, along with the defenders of Benauges (a quite substantial castle that had undergone some modernisation) and La Réole. The victors moved on to Blancquefort which offered little resistance (October). Charles now ordered all his forces to concentrate at Bordeaux itself. This time the Anglo-Gascon inhabitants determined to resist, but the force deployed against them was great, and the ability of Bureau to carry out his threat to flatten the town with his artillery was probably not in doubt. Additional modernising defensive works had been carried out a dozen years before, but the enceinte was still effectively that of the beginning of the fourteenth century, constructed at a time before guns existed, and they were facing what was now the largest and best organised artillery train in Europe, under the personal leadership of the most effective artillery commander there had yet been. Prospects of relief from England were also poor. In addition, King Charles wisely wanted to have a grateful city rather than a resentful and shattered one, so was willing to offer generous terms of surrender which were accepted in October, and he balanced his decision not to punish the rebellious Gascons too severely by ordering the construction of two citadels in the city to command its future loyalty.[1]

The Low Countries and the Burgundian realm

The detailed recollections of the Burgundian military adventurer Olivier de la Marche are among several accounts that detail contemporary siege warfare. The examples were much smaller scale conflicts, usually involving the house of Burgundy, still yearning to create a new "middle kingdom" for itself. Although the chroniclers present an image of the dukes as modern exemplars of an anachronistic chivalry, the sieges they describe were actually brutal and sordid examples of advanced fifteenth century warfare, in which the role of the heroic knight was far less important than the humdrum contribution of engineers, artillerymen, pioneers and other soldiers of lower class origins.

[1] D'Escouchy, 65–71; Basin, II, 191–201; Gardelles, 106. Blanquefort had always been an important defensive site because of its location on a route into Bordeaux, and although it did not resist long this time, the king ordered its strengthening when he entrusted it to a new owner, in order to guard against a potential English attempt to recover the province. Antoine de Chabannes began work on installing massive new artillery defences that would, in the coming decade, include two enormous circular artillery towers, on the circuit which enclosed the thirteenth century six-towered inner residence, rebuilt around 1300, as well as five turrets which, along with the curtain wall, included numerous embrasures prepared specifically for guns: Mesqui, 63–4, for a summary, plan and illustrations; Salch, *Atlas des Châteaux forts*, 328–30, for aerial view.

Recovery of Luxembourg 1443

The nobles, however, were always in command. In 1443, a force was sent to secure the duchy and city of Luxembourg (held by a German garrison commanded by the count of Click), in which was serving the illustrious knight Jacques de Lalaing, whose recorded deeds add to de la Marche's more detailed version. Luxembourg was protected by thirteenth century walls and towers, while the castle stood on a promontory, hard to take but easy to cut off.[1] The attackers, carrying ladders and other siege equipment, gained entry to the town by a daring escalade by a small detachment three hours before dawn, selecting a point on the wall where there was no gallery, from which they descended to open a postern and allow the others in. The defenders took refuge in the citadel, which the Burgundians isolated by occupying the abbey behind the castle while erecting a high screen made of barrels filled with earth and stone on the square in front, to keep themselves safe from the German troops' culverins. After three weeks the count of Click led a sortie against the abbey from which he escaped, leaving the castle, now low on provisions, to negotiate surrender.[2]

Another Burgundian force attacked the castle of Villy, a strong tower whose occupants were ravaging the country. The Burgundians established themselves in the castle bailey, blocking all the exits. Philippe de Vaudrey, described as the duke's master of artillery, then arrived with bombards and serpentines. Olivier de la Marche, then a young page, described how he was sent along to watch and learn. A close siege was sustained. The castle's owner, baron de Commercy, organised a sortie in fact designed to enable the defenders to escape. Some did, and the rest surrendered when the duke agreed to spare their lives (October), but the castle was razed to the ground.[3] By the end of the year, the task of conquest was complete except for Thionville.

The rebellion of Ghent

In 1451, Ghent rose in revolt against the regime, and not for the first time in its history found itself at war with its overlord. The Burgundian chroniclers' accounts drip with class hatred, but they are in close agreement on the events.

During early 1452, the Ghenters strove to secure the approaches to the city by capturing and fortifying surrounding castles at Gavre, Poucques and Schendelbeke. Forces were sent to help hold the towns of Alost (250 men) and Oudenarde (300 archers, 60 lances of regular soldiers). On 4 April, several thousand men from

[1] Although the defences to enclose the area below the city (the Grund) and the Plateau du Rham (the high ground on the eastern side) dated from the late fourteenth century and comprised a high, towered curtain and the Dinsel Gate (Plate 12). This latter was seriously damaged in the Burgundian attack, with its floors being burnt. For an accessible account, National Historic Monuments Commission, *Wenzel Itinerary*, 1995, 68–72.

[2] Le livre des faits du bon chevalier Messire Jacques de Lalaing, in Kervyn van Lettenhove, *Oeuvres de Georges Chastellain*, 1863–6, VIII, 35–8; *Les mémoires de Messire Olivier de la Marche*, I, 387–92.

[3] *Les mémoires de Messire Olivier de la Marche*, I, 369–71.

Ghent equipped with artillery marched on Oudenarde, forcing the commander, Simon de Lalaing, to retreat behind the walls. The Ghenters threw a bridge across the river Scheldt and crossed to attack the city from the side of Tournai, but found the defence well prepared with artillery and crossbows. So they built a second bridge to attack from the other side, while also protecting themselves with ditches and palisades. A sortie by the defenders set fire to the outer suburbs to deny cover to the besiegers. They resorted to heavy artillery bombardment (13 April), and they tried to foment dissent in the besieged town between the inhabitants and the garrison, but failed. Lalaing's troops helped keep the besiegers at a distance with cannons and serpentines. Some of the Ghent guns were set up to fire incendiary shots over the walls with a view to starting fires, but Lalaing posted watchers in the church towers to spot where the balls fell, and had supplies of water established at every corner from which the townswomen were directed to douse the flames. In due course, a relief force of 3000 men from Picardy led by the count d'Estampes came near. First it had to fight its way across the bridge at Espierres, where the Ghenters had fortified the church. The Burgundian force (in which Jacques de Lalaing was serving) was well provided with culverins, *veuglaires* and crossbowmen, under cover of whose fire they stormed across. They then met the footmen of Ghent in open battle and routed them, forcing them to raise the siege (24 April).[1]

Among the abandoned weaponry was the immense bombard *Dulle Griet*, here on its one and only military use. Despite its size, 500 cms long with a bore of 64 cms, and weighing more than 16 tons, and firing a shot that would have weighed around 340 kgs, it had made no impact at all on the siege. It seems that the defenders were sufficiently experienced in siege warfare to take measures of active defence that prevented their opponents getting their guns – which would certainly have blasted holes in the city's not very modern walls – into the positions they needed to be in to create breaches. *Dulle Griet* still bears the coat of arms of the house of Burgundy, not those of the city in which it still rests.[2]

Oudenarde saved, the Burgundian duke now took the offensive. Troops were sent to all the neighbouring cities, and bridges of boats were set up by his engineers across the network of Flemish waterways. Bulwarks were built at each from timber and earth to protect them from counterattack. In May, the Ghenters occupied the village of Nevele and erected a bulwark outside it on the Courtrai side, to strengthen their defence of trenches and wooden barriers, all provided with guns. The count d'Estampes having been informed, marched from Oudenarde at great speed to interrupt the work, and stormed the defence, eventually securing the village after its possession had swung back and forth twice.[3] The event, not significant in itself, is illustrative of the strategies of the two sides as the one sought to create a large defensible zone around Ghent in order to prevent a close siege, and the other strove to prevent them.

[1] D'Escouchy, 386–95; Chastellain, II, 221–48; *Olivier de la Marche*, II, 66–82.

[2] Schmidtchen, 'Riesengeschütze des 15 Jahrhunderts', *Tecknikgeschichte*, 44, 1977, 165–6; R Smith, R Brown, *Bombards*, 23–4.

[3] Chastellain, II, 268–75; *Olivier De la Marche*, II, 86–7; d'Escouchy, 400–5, 416–20.

In June, the duke ordered up a force from Holland which found a strong Ghent force occupying the village of Axel, with earthwork defences and artillery. Jacques de Lalaing was involved in operations that included the storming of these defences, and the capture of several other outlying defences, some of them newly built bulwarks, but two of them fortified churches. During the operation he had to defend as well, as a Ghent force launched an attack on the Burgundian troops holding the town of Hulst with a large train of artillery and other siege equipment, but was defeated by a sally of the defending force before it could begin a siege.[1]

There were also lesser conflicts in which attack or defence of fortifications featured, and Olivier de la Marche recorded some of those. A force led by the count d'Estampes attacked and destroyed several bulwarks at villages which the Ghenters had fortified, including Morbecque, and the abbey of Bordelois. This was described as having just one entrance, heavily protected by (presumably) light and hand guns. Olivier records that it was also extremely hot, so they decided not to attack. This suggests that it was now quite possible for small, local defence forces to create effective defences against professional troops provided they had guns, and the attackers, who were a mobile raiding force, did not have a siege train.[2]

June 1453 saw attacks on the three outposts seized early on by the rebels. Two large bombards were set up that quickly blew holes in the walls of the small fort at Schendelbeke. Assaults were then mounted seeking to take the place by escalade. The 60 Ghenters remaining alive inside surrendered at the duke's mercy after five days, but they were all hanged. On 2 July, the army moved against Poucques and repeated the treatment, with the same outcome. The attackers were able to seize the outer court, as the defenders had hurriedly constructed a bulwark to cover the bridge across to the inner. The Burgundian artillery was deployed against a weak-looking stretch of wall between two of the towers, and quickly caused it damage. A large bombard was added to the battery, protected by screens and mantlets, and Jacques de Lalaing, looking out from this protective screen, was killed by a shot from a *veuglaire* mounted in a tower. After nine days of attack, the end came when a long stretch of wall came crashing down. Gavre proved slightly harder to take, the 60–80 defenders holding out for two and a half weeks before giving in to the overwhelming fire power of the Burgundians. They met the same fate as their fellow citizens (23 July), after the defeat of a force sent from the city, despite deploying numbers of ribaudequins mounted on carts. In this siege, according to the later recollections of Olivier de la Marche, the Burgundians used mortars ("mortiers"),[3] which would be an early indication of the use of this kind of gun.

The citizens of Ghent now faced the probability of a direct attack from the duke's large army, which deployed in all its splendour before the city on 31 July. Duke Philip the Good (1419–67) now however lived up to his nickname and was generous in the terms on which he accepted submission. At no point do the records suggest that the rebels had any significant number of professional soldiers at their disposal, so the victory of the duke was not surprising, but the widespread

1 Chastellain, II, 316–19.

2 *Olivier de la Marche*, 111–12.

3 D'Escouchy, 81–9; *Olivier de la Marche*, 134–43; Chastellain, 357–67.

use of gunpowder weapons on both sides is again evidence that such weapons, ranging in size from siege artillery to hand guns, was available in large quantities everywhere in the Burgundian realm. Older weapons such as crossbows and longbows also featured regularly. The glorious deeds of the noble knights however, have to be looked for quite carefully among the actual fighting done by their social inferiors.

In 1456, the appointment of the duke's illegitimate son David as bishop of Utrecht provoked resistance in that city, which raised further troops from the duchy of Guelders. They decided to put their faith in fortifications, reported Chastellain, and erected earth and timber bulwarks on the great rivers that laced the terrain where the city stood. Duke Philip assembled a substantial army, including 5000 archers and 1000 men at arms, along with a big train of workers, and a large battery of artillery. Chastellain reported that this included cannons, *veuglaires* and serpentines and also included "new engines" that made a noise like "the heavens breaking" because, with one touch of the match, they hurled ten or twelve stones together. Although the defenders had placed their bulwarks to control traffic along the rivers, they nonetheless saw that their situation was hopeless and hastened to submit before the duke's formidable arsenal had to be deployed (30 July to 5 August). Further east, however, Deventer, lying mainly on an island in the river IJssel (Yssel), and surrounded by friendly country, decided to hold out against the new bishop. The duke arrived to besiege it in August, but the hostile country made supply difficult and hazardous, and heavy rain made the land boggy. Initially, the defenders fought outside their walls, but were driven back by the superior force of the duke, who then proceeded to assail a bulwark erected as an outwork, which stood sufficiently far in advance of the town to enable the people to come and go at will. Setting up their batteries, the Burgundians concentrated on this work, and after four days of bombardment, the holders abandoned it. The Burgundians also had to undertake the difficult operation of getting across to the island in the middle of the river, and still found themselves there a month later. Negotiations then began. Thus a small east Netherlands town, taking advantage of a strong water-based location, held out against a very powerful ruler for more than a month, and knew well enough to strengthen its old defences against artillery.[1] The accounts of these "internal" conflicts are no less informative than the histories of the international struggles in demonstrating the relative strengths and weaknesses of attack and defence, and, almost incidentally, recording the use of new forms of weaponry that had caught the attention of the writer, normally indicating that this was the first time that the author had witnessed such equipment in use. These descriptions from the 1450s have shown both mortars and multiple-barrelled large guns. The former must have been short range guns, designed deliberately to lob large shot up and over walls. Examples have been noted from illustrations of a slightly later date. Multi-barrelled guns in the form of the ribaudequin had been in use for a long time by this date, so the novelty described

[1] Chastellain, III, 126–31, 156–72; Wavrin, *Recueil des cronicques et anchiennes istoires*, ed. Hardy, 1891, v, 370–1.

here must have been in the size, and illustrations of a few such monsters have survived.[1]

The British Isles

In England, the Lancastrian regime was challenged, overthrown, restored and eventually replaced, while the Scottish kingdom saw a remarkably steady (and rare) father-to-son succession of the Stewarts, the dynasty that had succeeded the house of Bruce.

Scotland under James II (1437–60)

The existence of noble families strong enough to challenge the authority of their crowned rulers had been a feature of the political landscape of the later middle ages for some time, but no family exercised as much virtually independent authority as the Douglases of southern Scotland. In the 1440s their castle of Threave had been reinforced with a screen wall and corner towers intended for the use of artillery. The humbling of the Douglas "empire" by the young King James II in the 1450s after he came of age is not recounted by any contemporary chronicle, but this lack is fortunately compensated for by other records. A letter to Charles VII dated 8 July 1455 reported James's actions as he wanted them to be seen internationally but the detail rings true.

James began by attacking the Douglas stronghold at Abercorn (Lothian), which may have been a stone-built courtyard castle with corner towers. The castle was not prepared to resist serious attack. If his father's use of modern siege weapons had been largely unsuccessful, young James seems to have been better served. His guns, having been arrayed against the castle, caused serious harm to the curtain wall, leading to stretches collapsing. The king then prepared an assault by escalade. His men gained entry meeting little resistance, and the demoralised occupants yielded. The castle was demolished and the leaders among the defenders hanged. The royal forces scattered the assembling Douglas forces, then attacked Inveravon, a courtyard castle with corner towers standing on a promontory above the mouth of the river Avon a few miles further west. This does not seem to have been very strongly built, the wall of the surviving circular tower being 1.4 metres thick, and certainly it did not resist long. Payments were subsequently recorded for the purchase of numerous tools and instruments for the destruction of the sites.[2]

The king turned his attention to the west, and marched on Threave, around which his force still lay when he wrote to the French king, stating that he hoped

[1] See the plates in A Essenwein, ed., *Quellen zur Geschichte der Feuerwaffen*, reprint 1977, especially plates XL–XLIII from a manuscript dated to 1460–70.

[2] Burnett, ed., *Exchequer Rolls of Scotland*, VI, 12, 92; J Pinkerton, *The history of Scotland from the accession of the House of Stuart to that of Mary*, 1797 edition, I, 486 (copy of James II's letter); M Salter, *The castles of Lothian and the Borders*, 1994, 22, 63, for brief summaries of the remains of Abercorn and Inveravon.

for an early surrender of the castle, once again abandoned by its lord. The recorded payments show money spent on the tent for the royal camp, James continuing to oversee the sieges himself. Other payments were made for the cutting of the stone balls for the bombards, and a special reimbursement was made to the earl of Orkney for carriage of the "great bombard" from the arsenal in the east. This, however, suffered a mishap when a wheel broke on the way, payments being made too for the subsequent repairs to bring the journey to completion. Bills were still being settled as late as 1458, when John Carleil of Kirkcudbright was paid £11 for his role in the transport of the *artilarie* and bombards, while the year before there was a payment for the construction of seven stones for the use of the "great bombard". It is not clear whether the guns were used, or perhaps the threat was sufficient, but fall it did.[1]

Despite King James's interest and expertise, in the same year his attempt to recover Berwick was a failure. Although the whole nobility of England was engaged in civil war, the Scottish attack on the crucial border fortress was a fiasco, and he was compelled to retreat from the fortress by the stout resistance of the garrison and local levies, abandoning "all his ordenaunce and vitayle" brought along the coast.[2] Records suggest, in fact, that the government had been unable to afford the moneys needed to maintain the fortress in good condition.[3]

The king's keen interest in modern weapons brought about his untimely death five years later, when the gun he was standing next to when his army laid siege (successfully) to Roxburgh castle blew up, an iron fragment flying off and killing James. The incident serves to remind that as late as 1460, guns were still vulnerable to exploding, whether because they had not been properly made and tested in the first place, or as they failed through wear and tear from too much use.[4] As for the castle, the last major repair work had been undertaken over a period of nine years during the 1380s, at considerable expense, involving the erection of new and heightened walls and towers, but subsequent surveys had indicated that the works were not complete and that it was not defensible. Throughout this time the castle had withstood Scottish sieges successfully in 1417 and 1436, as noticed already. It is therefore difficult to work out how strong Roxburgh was in 1460 when it fell, at last, out of the hands of its English occupiers.[5]

1 Pinkerton, I, 487; *Exchequer Rolls*, VI, 119, 161, 199–202, 209, 294, 456; C Tabraham, *Threave Castle*, (guide book), 1994, 19–25.

2 *Short English Chronicle*, ed. Gairdner, 45.

3 Colvin, *King's Works*, II, 570–1.

4 Pinkerton, I, 244; *Brief Latin Chronicle*, ed. Gairdner, 1880, 168.

5 Colvin, *King's Works*, II, 818–21, details the works at Roxburgh. The commissions to the earl of Salisbury and the sheriffs of the northern counties to raise the levies of Yorkshire, Nottinghamshire, Derbyshire, Cumberland, Westmoreland, Northumberland and Lincolnshire to go the rescue of Roxburgh and the town of Berwick were clearly too late for the former, but indicate the importance attached by the crown to the vital frontier fortresses: *Calendar of Patent Rolls, Henry VI*, VI, 589, 612.

The Wars of the Roses

Unlike the fighting in France or the Balkans, little of the struggles known as the wars of York and Lancaster revolved around sieges, and pitched battles instead played a decisive role. There has been some debate and various theories advanced to explain why there should have been such a dramatic change to the normal form of medieval warfare. Some of the evidence is not very helpful: for example, just because large sums of money had been spent on a castle does not mean that this had anything at all to do with any intention of modifying it militarily. But there were still operations that revolved around the capture or defence of a fortification.[1]

The civil war was a struggle between powerful nobles and their respective, armed, clienteles. The cost of war in France had left little available to ensure the maintenance of castles in England, and although the Glyn Dŵr rebellion had led to immediate spending on sites in the region, this had not been sustained, and only on the Scottish border (as at Roxburgh) was there an effort to maintain the fortresses. There was much less reason to spend royal resources on the large number of now aged castles spread across the country, except when they were required as residences, prisons or administrative centres. Analysis of royal building works shows that by 1450 most royal castles were in disrepair.[2] One source incidentally demonstrates that the castles at Windsor, Wallingford, Pevensey and Pontefract were in use as prisons, while in London the Tower had long exercised that function.[3] In contrast, great nobles spent lavishly on their castles, especially in the north of England (such as Alnwick, Warkworth, Sheriff Hutton, Middleham, Raby, Brancepeth), or the earl of Warwick (the famous "king maker")'s work at Warwick itself. Seeking as they were to establish their position as leaders of society, conspicuous expenditure on grand buildings was then, as now, a common way of showing off. On its own, it tells little of what, if any, thoughts the nobility had about the military potential of their great homes. Although the development of more advanced artillery did mean that for the most part, existing castles in England would have been largely useless for defence, a few castles did feature militarily.

There is little evidence that towns had assumed a defensive function, as many had in Europe. There is a risk both of oversimplifying any explanation, and of taking advantage of hindsight. As always, people will take refuge behind even ancient defences when that is the best remaining option, and will try to bring them up to standard by erecting temporary works. And who in 1455 could have known who would wear the crown in 1461? Erecting new walls and towers was a slow and expensive business. It could not be achieved in response to immediate tactical requirements. The town of Coventry featured from time to time in the events of the

[1] See, for example, the detailed analysis of A Goodman, *The Wars of the Roses*, 1990 edition, 181–93; and the critique of J Petre, 'Historians and the role of castles in the Wars of the Roses', *Fort*, 12, 1984, 5–9.

[2] Colvin, *King's Works*, I–II, throughout.

[3] *John Benet's Chronicle*, ed. Hariss and Hariss, 1972, 210, 212, describing the imprisonment of great nobles in respective castles.

wars, but the walls and towers had been under construction on and off since 1250, with a particular building spurt dated tantalisingly to the mid-fifteenth century. However, it is inconceivable that the military events of the time can have served as anything other than a spur to raising the money required, given the slowness of the building work.[1]

Fortifications did feature in several of the fights, but usually in the side shows of the "private wars" that the breakdown of central government permitted to take place. The local attack and defence of Taunton castle in 1451 was such an example, of which no detail survives to permit a reconstruction. But the ancient work held out. In 1457, a French raiding force landed in Kent and seized the town of Sandwich. This was a substantial and well armed body provided with artillery, which suggests it intended to attempt a major target. Wavrin gave interesting detail of their moves. They landed two leagues from Sandwich at six in the morning on 28 August, and 1800 of them immediately marched on the town, but were impeded by bad roads. They reached the town and attacked the bulwark behind its water-filled ditch. The defences of this town still represent something of a mystery. The records speak of significant efforts having been made to provide defences during the previous century, the digging of a ditch and creation of a rampart, the provision of gates and, in 1451, a bulwark of two storeys provided with guns, clearly the work described by Wavrin. Excavations combined with documentary records suggest it may have been a stone tower with corner turrets, standing outside the Sandown Gate and dating from the late thirteenth century, and perhaps adjoining the town rampart. There seems to have been a fierce fight here before the French broke in, although it is not explained how they crossed the ditch. Anyway, the town now fell to assault launched both from the land side, and from French ships entering the harbour. The townsmen resisted street by street before being driven out, leaving the invader to plunder at will. However, the local countryside mobilised and joined the people of Sandwich to counter-attack and force the French back onto their ships after continuous fighting for ten hours which left the invaders exhausted, and suffering a large number of casualties themselves. Of the defences, apparently formidable but insufficient, stretches of earthwork rampart and two of the gates, remain.[2]

The years 1459–64

In these years of bloody fighting, the balance swung back and forth, the weaker party at each point having to flee for their lives, often leaving armed supporters behind whose resistance is recorded. It was such operations that tended to see the use of fortresses. In February 1460, for example, King Henry VI commis-

[1] Turner, *Town defences*,118–19, and the detailed references given there.

[2] Turner, 163–5; Gairdner, ed., A short English chronicle, in *Three Fifteenth Century Chronicles*,70–3; the same, *Brief Latin Chronicle*,166; Wavrin, *Recueil*, 385–88; *John Benet's Chronicle*, 218. I Stewart, 'Archaeological investigations at Sandwich Castle', *Archaeologia Cantiana*, CXX, 2000, 51–73, suggests reasonably that it may have resembled the so-called Ypres Tower at Rye, although its shape could not be confirmed.

sioned workers (wheelmakers, cartwrights and smiths) to provide transport for bombards and cannons to be conveyed from London to Wales, where the Yorkists were holding out, and taking advantage of the powerful walls still standing there. The earl of Pembroke, Jasper Tudor, was specifically commissioned to recover Denbigh castle, but he was unable to do so until after the death of the duke of York at the battle of Wakefield, and then only briefly, town and castle being historic and loyal possessions of the Mortimers. The castle was very strongly placed on a high hill, protected by a double circuit of walls and partly shielded by the walled town below. It was the latter that suffered most from the war, and was unusual in being given a royal grant by Edward IV towards the cost of repairing its walls.[1] Later that year, Henry issued an interesting commission to the Master of the King's Ordnance, Thomas Vaughan, requiring him to establish the whereabouts (and who was in possession) of the royal arsenal: bombards, cannons, serpentines, fowlers, culverins, powder, sulphur and saltpetre, lead, iron and tin, along with crossbows and longbows and pavises. He was then to transfer them to a safe place.[2]

In the same year, the king marched from London to the midlands. There, while based at Coventry, the Lancastrians decided to turn nearby Kenilworth castle into a major fortress. Their method was to order the transport on 40 carts of all the "guns and armaments" from the Tower of London to the castle, alongside unspecified other works to strengthen the defences. The castle, royal from 1399, had undergone much work making it into a more palatial residence. Perhaps it was its situation, surrounded by water, that encouraged the thought that it would be easy to transform it into a serious fortification by the provision of artillery, and perhaps the erection of temporary bulwarks to house the guns and protect the ancient stonework. The decision was overtaken by the Yorkist victory at the battle of Northampton.[3]

This victory left lord Scales in the Tower of London still provided with the guns he might otherwise have sent to the midlands. Against him, the Londoners supported the Yorkists and seem to have laid siege to the vast fortress behind its Edwardian curtain and strong water defences. Several chroniclers noted that the defenders' guns caused serious casualties among the citizens, although giving no detail of how they intended to capture the place. The end in fact came when Scales was caught and killed by the Londoners. The Tower then surrendered to the new government.[4]

Early in 1461, royal commissions indicate steps to mobilise local people to defend their localities against "evildoers". The commission to the Lincolnshire town of Stamford does not suggest that the very old walls of the place were expected to perform any military function, but the order to the bailiffs of Shrewsbury to

1 *Calendar of Patent Rolls Henry VI*, VI, 564–5, 604; Butler, *Denbigh Castle and Town walls* (guide book), 1990, 16–17.

2 *Calendar of Patent Rolls Henry VI*, VI, 612.

3 *John Benet's Chronicle*, 225; Colvin, *King's Works*, II, 683–5; Thompson, *Kenilworth Castle* (guide book), 1996.

4 *Brief Latin Chronicle*, 169; Wavrin, 295–6, 302–4; *John Benet's Chronicle*, 226–7.

restore their "walls, towers, dykes, gates, portcullises and bars" and hold them against any followers of the duke of Somerset indicates a different intention.[1] By March of that year, the situation had changed again, with Henry deposed and Edward IV installed after victory at Towton, and the latter ordered William Lee, a "joyner", to find carts, horses, oxen and men to transport "three cannon or great bombards" to "Thorpewaterfield" (Thorpe Waterville) in Northamptonshire, near Banbury, which was to be besieged. This was a large manor house, two wards with surrounding wet ditches, licenced to be crenellated by the bishop of Lichfield in 1301, and now occupied by a force of Lancastrian supporters who resisted for a few days. It must be assumed that they had taken up residence there because of the potential to organise it into a temporary fort, already assisted by moats, rather than because of any inherent strength of the buildings themselves, now represented by a barn. Remains suggest a well-built two-storey hall forming the main element of a substantial mansion while defensive works — if indeed they were meant as more than boundary walls — were limited to an outer circuit.[2] News of its fall had reached London by 4 April so resistance can only have been brief.[3]

The Lancastrian party, driven on by the determination of Queen Margaret, did not accept defeat but returned to land in northern England, where it took possession of the major fortresses, while yielding Berwick to the Scots in return for their support. So in 1462, Edward had to send substantial forces north. Castles such as Bamburgh, Alnwick and Dunstanburgh did not resist sieges but yielded, perhaps as a result of the by now rare measure of promising the defenders their lives, and the invasion seemed to have been defeated when a combined Scottish-French force returned to the attack in support of Henry VI, sending the local Yorkists once more into retreat. Edward was forced to set about the recapture of Bamburgh and Skipton, and the relief of Norham, castles, the latter invested by the French and Scots for 18 days before being relieved by the earl of Warwick. The only significant resistance to the large royal army was at Bamburgh, which was besieged on 25 June and was stormed, despite its off-shore location. A manuscript note recorded that it was held by Ralph Gray, and it was summoned because the king wanted to have the castle "hole, unbroken with ordennaunce". Having failed to obtain surrender by this method, the royal arsenal was then deployed against the walls, the "great gonnes" including those called "Newcastle", and "London", which it was noted were made of iron. Their effect was to send stones crashing down from the walls into the sea, while a brass gun called "Dysyon" shot through Gray's own chamber many times. Finally, the place was taken by assault by men at arms and archers. This confirms the chronicle account of William Gregory (mayor of London) who said that "great guns were laid against it and it was taken by force".[4]

[1] *Calendar of Patent Rolls Henry VI*, VI, 657 (17 and 20 January 1461).

[2] *Calendar of Patent Rolls Edward IV*, 28; Mackenzie, *Castles*, I, 340–1; Selby, *Associated Architectural Societies Reports*, VI, 1861, part 1, 64–7; Leland, *Itinerary*, I, 6 ("The ruines of the utter waull of Watervilles castelle").

[3] *Paston letters*, ed. Gairdner, 1904, III, 266–7.

[4] Warkworth, *Chronicle of the first thirteen years of the reign of King Edward the fourth*,

Meanwhile, the king ordered that all the fortresses in Wales be taken into his hands in 1461. A wise precaution, but one that did not entirely work, because the Lancastrian party took possession of the great fortress of Harlech. In June 1463 Edward IV commissioned William Herbert as constable, with what seems to have been the regular garrison of 24 soldiers. However, in repeating the commission in October, Herbert was given authority to accept the submission of the "rebels" then occupying it. It does not seem that any major effort was made to assault the mighty place, as it did not finally yield, by composition, until 1468, chroniclers again writing that it was too strong to take by assault. One contemporary records the name of Philip Vaughan, captain of Hay, who was the only person of rank to die in the siege of Harlech (killed by a gunshot).[1]

Caister 1469 and St Michael's Mount 1473

The fall of Harlech left the whole realm in the safe government of Edward IV, but this did not mean the end of fighting, and the very next year saw the siege of the castle of Caister in Norfolk, its fame having a lot more to do with the record of the events preserved in the letters of the Paston family than in any significance of the siege itself. But this little private war still reveals much about the nature of siege warfare by 1469. The Paston family was in dispute with the neighbouring magnate, the duke of Norfolk, as it had been as well with the earl of Suffolk. In this first dispute, the latter had attempted to occupy a manor at Hellesdon belonging to the Pastons in 1465, a party of 300 of their men approaching it with guns and other weapons that would have been sufficient to destroy the Paston party if they had attacked the place, but they were held off. However, they took the manor later and called on the local peasantry to pull down the walls and the lodge.[2] In November 1468, John Paston II informed his brother, John III, that he had hired four men to help him defend Caister: "they kan welle shote bothe gonnys and crossebowes . . . and devyse bolwekys". One was described as a good man, though a little given to drink! John Paston III was to hire other defenders locally. In May 1469, the latter was writing to the former, in London, to send word of someone skilled in repairing steel bows, since his little troop had managed to break three of them.[3] The siege was underway in September when the family matriarch, Margaret, wrote pleadingly to John II to tell him that his brother was in great danger at Caister, lacking supplies of food and gunpowder and arrows, and with the walls "sore brokyn with gonnes" by Norfolk's men. The duke had called out more troops

ed. Halliwell, 1839 (reprint 1990), 2–3, 37–8 (reprint of the MS account); *Brief Latin Chronicle*, 178–9; *Short English Chronicle*, 79; Wavrin, 432–3; *William Gregory's chronicle of London*, ed. Gairdner, 1876, 226–7.

[1] *Brief Latin Chronicle*, 231; *Calendar of Patent Rolls Edward IV*, 271, 352, 355; *Brief Latin Chronicle*,182; William Worcestre, *Itineraries*,ed. Harvey, 1969, 205; *William Gregory's Chronicle*,237; *Warkworth's Chronicle*, 3. It has been suggested that this siege was the event featured in the famous later song *Men of Harlech*.

[2] *Paston letters*, selection ed. Davis, 1956, 45–6, 56 (10 July and 27 October 1465).

[3] *Paston letters*,ed. Davis, 71.

for the following week, and had sent to (Kings) Lynn for more guns, so clearly intended to launch an assault. On September 15, John II suggested that his mother was exaggerating, and reported that a local truce had been concluded, but on September 25 John III wrote to report that he had surrendered Caister for lack of victuals, gunpowder and also, tellingly, "mens' hearts and certainty of rescue". Part of his letter was, interestingly, a plea to ensure that the four hired soldiers were duly rewarded for their services.[1]

The place at the centre of this fierce onslaught was a fortified manor originally built by John Fastolf, enriched by service in France, from 1432. It was of brick, a very rare building material at the time, and had a square inner ward and an outer court surrounded by water. There were corner towers, of which the original "great tower" remains, standing 90 feet (28 metres) high, and containing embrasures for guns, and machicolations. But the curtain walls themselves were thin and certainly not designed to resist a full scale attack. It was described by a contemporary visitor as having a great hall 59 by 28 feet, and certainly its domestic provision was its chief attraction, with large exterior windows confirming the secondary role of any defensive apparatus . It is suggested it may have been intended for coastal defence, but it was essentially a large and splendid house designed to demonstrate the taste and wealth of its owner, and provided with some elements that could also serve a defensive role, which was a sensible approach at such times of widespread banditry, not to mention political conflict. William Worcestre added a list of the names of the defenders, which reached a total of 27 men. Norfolk, having failed to obtain voluntary surrender, brought a force of no fewer than 3000 men provided with guns, culverins and bows.[2] We are not told how this immense force was used to force the capitulation of a garrison of fewer than 30, except to imply that the besiegers' weapons were located somewhere that enabled them to blast a breach in the walls, and that the numbers of attackers meant that it would have been quite simple to fill in the moat to prepare the assault. Norfolk, anyway, obtained possession of the disputed house, although he had to return it to the Pastons a few years later.

William Worcestre also provided the only helpful information when John, earl of Oxford, with a force of 80 landed on St Michael's Mount in Cornwall, disguised as pilgrims (30 September 1473). The defences of the island monastery were not of great strength, but the offshore location compensated for that, although it was, at its closest, just a "bowshot" from the mainland, and connected by a recently constructed causeway that was subject to the tides. Additionally, Oxford was able to secure large quantities of victuals from the surrounding countryside with which to stock the Mount. Edward IV mobilised a significant force of some thousands to blockade it, but they engaged in frequent truces with the garrison and even permitted it to re-provision itself, whereupon the king replaced their commander with a new sheriff (24 December). He engaged in a number of bloody skirmishes at low tide with the defenders, in which severe casualties were apparently suffered,

[1] *Paston letters*,77, 78, 81–2.

[2] William Worcestre, 189–91; O'Neill, *Castles and Cannon*, 30; E Smith, *The Story of Caister Castle*, nd. (local guide book).

and several ships were also hired to prevent the supply of the rock, which had its own protected harbour. But it was more than five months before Oxford was forced to capitulate after most of his men had been bribed to desert (15 February).[1]

No other siege operations took place during the reign of Edward IV, and the seizure of Bodiam castle in Sussex in 1483 by Richard III does not seem to have involved any resistance. It is therefore perhaps appropriate that as indicators of the stage that the arts of attack and defence of fortification had reached in England by the later fifteenth century we have the rapid capture with guns of a modern manor house in a private war at Caister, and the prolonged resistance at St Michael's Mount brought to an end by bribery. Each in their own way confirms once again that technology *could* play a very small part in deciding the outcomes of events. The crown did not use its access to the weaponry and engineers possessed by the state to crush Oxford's lengthy resistance on a little island, but relied on the ancient methods of blockade and, in the end, corruption, such as would have been recognised by theorists since ancient times. Meanwhile, a powerful nobleman turned to modern weaponry to force to surrender a tiny force occupying a house that he coveted in just a few days of fighting. These defenders had been able to hire some of the many experienced soldiers there were about, but not to defend the walls, towers and moats of Fastolf's castle against Norfolk's overwhelming local power.

The Holy Roman Empire

The Holy Roman Empire had seen the progressive reduction of central authority since the time of the Hohenstaufen dynasty, in favour of that of both territorial princes, and free cities. By the fifteenth century, the princes were many in number but very much localised in power, while the towns and cities were counted in immense numbers, and where they had the constitutional right, relied on their direct relationship with the emperor to deny the aspirations of their local princes and, everywhere, the thousands of knights and lesser nobility who lived in castles (10,000 of them already by 1300) with a handful of retainers. The conflicts were local and innumerable, between neighbours, between different factions within the cities, between cities, among the princes and between various and shifting leagues formed to counter the weakness of the individual entity. The emperors' strength lay only in their own hereditary lands. The holder of the office in 1440 was the young Frederick III (1440–93), latest of the Habsburgs. The death of Sigismund had brought the Luxembourg dynasty to an end, and his daughter's marriage to Albert of Habsburg had brought the latter the crowns of Bohemia and Hungary. However, Albert had died in 1439, and his heir Ladislas was born posthumously, leaving his cousin Frederick the only adult Habsburg. Frederick III had to spend almost the whole of the first decades of his reign in his own lands in order to secure his dynasty's hereditary rule. The continued claim

[1] *Warkworth's Chronicle*, 26–7; William Worcestre, 103, 111.

of the Habsburgs to the Empire was certainly not guaranteed. In the political uncertainty brought about by the birth of Ladislas, an infant originally entrusted to Frederick's care, a number of different and conflicting forces were in play. The Bohemians did not want a Habsburg ruler but agreed to accept Ladislas as king, but with a Bohemian regent (Poděbrady). The same obtained in Hungary, where Hunyadi was the regent. Neither solution was immediately agreed, and in the meantime the nobles of Austria, who included some great families like the Cilli, rebelled against Frederick (see below).

Meanwhile, the records of some of the cities, in this case Dortmund, Cologne, Basel and Nürnberg, offer illustrations of the widespread conflict. City governments everywhere were ensuring that chronicles were kept of their doings, permitting a fascinating picture of the concerns of these proud controllers of often very small independent statelets to be reconstructed. These activities included, frequently, preparations for local war.

Soest and Cologne

The small Westphalian city of Soest was in conflict (formally speaking, a "feud" (*Fehde*) which had its own rules of conduct) with the powerful elector and archbishop of Cologne (Köln), one of the great and ancient cities of the empire, its archbishop one of the senior princes. Archbishop Dietrich von Mörs could call upon many neighbours (with whom he had close relations) for military and financial assistance. Soest in turn, which was not a significant power, found help from the duke of Cleves (Kleve)-Mark, Adolph I, and his allies. The struggle raged for several years without any decisive events, raids being launched to burn the countryside and take prisoners for ransom, in the course of which a number of noble castles were captured. The young heir to Cleves, John, commanding the Soester forces, was heavily involved in such attacks, which stretched to an attack on the town of Werle, which suffered the loss of 140 buildings to fire caused by shot from the attackers' *tormentis*.[1] In October 1445, the archbishop's troops tried a direct assault, seeking to mount the walls of Soest by escalade by surprise at night. The defenders were however alert and ready, and the attackers fled, abandoning the many instruments they had ready with them for the burning of the town.[2]

In 1447, the archbishop raised a large army from neighbouring bishops (Munster and Hildesheim) and princes (Brunswick) and hired from Bohemia. The estimate of 13,000 men with 800 wagons is not impossible. This great force was to capture Soest in direct attack, following a destructive march there in which the town of Lippe and a number of other intervening walled places were subjected to attack. Several smaller places capitulated, but Lippe resisted. The besiegers set up camp in late July and began a heavy bombardment, focussed on one tower. The

[1] *Koelhoffsche Chronik 1499*, Die Chroniken der niederrheinischen Städte: Köln, III, 1877, 787–8, for the raid of 1446, for example; Ernest Joachim de Westphalen, Rerum inter Theodorum Moerseum Archiepiscopum Coloniensem et rempublicam Susatensem gestarum adversaria, *Monumenta inedita rerum Germanicarum*, 1743, III, 2225–9.

[2] Ernest Joachim, III, 2231.

Bohemians — described as Hussites in the sources — led the assault, filling in the ditch in order to mount an escalade, but were repelled by the townsfolk, losing 40 men. After a total of twelve days attack, the bishop saw that Lippe would not fall easily and decided to bypass it.[1]

Arriving at Soest, his army captured a bulwark erected outside the town and also the abbey of St Walberg, which stood close by the walls and gate. This had been held by troops of Cleves, who had been driven out in a violent assault led by the Bohemians. From here, they trained their cannon on the town and fired into it for three days and nights. The townsfolk had their own guns, and fired these back with good effect, although the Cologne forces had some 300 gunpowder weapons, including some large bombards. These were breaking down the walls and smashing houses, such that no one dared venture onto the street. Fortunately for Soest, John of Cleves had recently arrived with 400 men, who valiantly defended the town from the walls, ditch and bulwarks. On 18 July, the besiegers were ready to assault, and attacked from two sides, but they had made a fundamental miscalculation: their ladders were too short. So they failed, and lost one hundred men killed and wounded, the defenders losing eight men shot. The archbishop was himself struck by several arrows that hit his shield, a gift from the king of France. Sources note the heroic contribution of the town's women to the defence. The Bohemians, meanwhile, were apparently very angry at the mismanagement of the attack, cursed their ecclesiastical commander, and marched off home. A couple of days later (21 July) the chastened archbishop withdrew, leaving behind much equipment including apparently no fewer than 600 ladders (possibly an exaggeration by the hostile Dortmund chronicler).[2]

The defenders of Soest and the men from Cleves, including troops from Dortmund, were able to launch various counter-attacks, including attacks on enemy "strong houses". On 18 August, the owner of one such fired on his attackers with his own gun, causing wounds. The attackers therefore set up their own gun, and on the third shot knocked down the door in the gateway, giving them access to the fortified house.[3] The war ended inconclusively. The references to bulwarks confirms too that knowledge of what was essential to strengthen old defences to resist artillery was commonplace.

Central and Southern Germany: Rothenburg ob der Tauber . . .

Nürnberg shared with other cities in the centre and south of the empire an active political and military history throughout this time, something that was inevitably consequent upon its determination to maintain its rights and freedoms as an independent entity under the emperor. We will begin, though, by drawing upon the records of the allied city of Rothenburg ob der Tauber, whose fate was to be tied in

[1] Ernest Joachim, III, 2238–9.

[2] Chronik des Johann Kerkhorde von 1405–65, *Die Chroniken der westfalischen und niederrheinischen Städte*, 1887, 94–5; *Koelhoffsche Chronik 1499*, 788–89; Ernest Joachim, III, 2240;

[3] Chronik des Johann Kerkhorde, 105–6.

with Nürnberg's in the critical conflict of 1449 onwards. Both were much greater cities than Soest. Although they needed allies, they were capable of holding their own against local enemies.

Rothenburg had been a free city since 1172, and had undergone a substantial growth. Originating in a small settlement on high ground above the river Tauber, the place had quickly outgrown the circuit of walls begun in the twelfth and completed in the thirteenth centuries and during the fourteenth had developed a new circuit of powerful double walls and towers (of which there were 42). The immense cost of the circuit had been met by regular levies on the city's trade, especially wine, and imperial grants of this tax were regularly renewed. Although never actually subjected to attack during the Hussite wars, it was close enough to feel at risk, and had therefore undertaken further improvements during the first half of the fifteenth century, when it was decided to add a further twelve towers to the existing defences. The defences dating from the 1380s onwards contain very modern aspects: outer barbicans at the gatehouses, gun loops in the walls arranged for flanking, and small projections from the curtain taking the form of solid gun towers to protect the ditch. As the existing towers were not strong enough to mount heavy guns, arrangements were made to deploy them in the space between the inner and outer walls, to fire from emplacements behind the outer wall. While these defences were necessarily of a low height, the inner defences behind them were quite different. The *Faulturm* stood 38 metres high, and like some of the other high towers (such as the 28 metre *Galgenturm*) shared a function as a powerful symbol of the city with a practical role as a watchtower. These towers, several dating from the 1380s–1390s such as the *Klingentor*, were later provided with a *Vorburg*, creating a powerful defensive work with ditch, drawbridge, portcullis and gun loops as well as crenellated wall tops with fighting galleries.[1] The city records make numerous references to the defences, and the decisions that had to be made to reconcile the demands of ordinary life with those of sustaining security. In 1397, for example, an agreement was made with a nunnery that the windows which had been opened in the town walls, and through which the sisters might be seen, would be closed off, and no future openings would be allowed in either the inner or outer walls, explicitly to avoid harm to the town. Parts of the old town wall and ditch were however sold off to citizens for their own use in 1399.[2]

The concern of Rothenburg's leaders for defence and control of their space reached out beyond their city's walls. During the late fourteenth century, citizens carried out a planned and determined expansion of control over the surrounding territory, creating for Rothenburg a large hinterlands of villages and countryside,

[1] Rothenburg ob der Tauber retains much of its late medieval wall. M Vasold, *Geschichte der Stadt Rothenburg ob den Tauber*, 1999, 179–96, for a modern description, and for a revised dating see Borchardt, Tittman, Mauern-Tore-Türme, *Jahrbuch des Vereins Alt-Rothenburg*, 2005. For helpful illustrations see also K Heller, *Rothenburg ob den Tauber in Wehr und Waffen*, 1922. Examples of the means taken to finance the defences in Schnurrer, ed. *Urkundenbuch der Reichstadt Rothenburg*, 1999, I, 299, 327, 372. See Chapter 5 above for a brief analysis of the evolution of the defences themselves.

[2] Schnurrer, II, 1032, 1119.

totalling 400 square kilometres (roughly a quadrangle 20 by 20 kms). This territory was then itself provided with a means of defence, work beginning in 1432 on a *Landhege*. This took the form of an earthen rampart four metres high topped with a thick hedge, and with a wet moat either side two and half metres deep, itself provided with a rampart and hedge, the whole barrier being 18–22 metres wide. Construction of the whole immense rampart was not completed until 1480. Nine large square towers were built to overlook the places where roads crossed the blockage. It can hardly be that the primary purpose of the *Landhege* was military, as it was not designed for active defence, nor could the city have employed sufficient numbers of sentries to guard it all, but it would certainly have controlled routes of access very thoroughly. But nor was this all. Inside the ring, villages were turned into local defences by throwing up ramparts and digging ditches around churchyards or fortifying the church itself. No fewer than twenty such fortifications were created, with the intention of securing local people against raiders who might be Hussites, or robber barons. The use of stone-built churches with their towers, and of a yard already possessing a wall, to provide the basis of defence for villagers and their belongings, was scarcely original, but in the land owned by Rothenburg it was a planned and organised building operation.[1] There can be no doubt of the city's commitment to these immense works. The expenditure was continuous, but on top of that, people had to be found as well. In 1400 Rothenburg may have controlled 1,760 combatant citizens who owed military duty to the town, while the entire population was liable for service on a routine basis to provide watch from the walls and gates, a quite considerable burden that went along with the benefits of citizenship.[2]

Conflict with local landowners provides an example of the city flexing its muscles and showing why its citizens were ready to shoulder this burden. In 1441, an expedition 1500 strong (calling in aid from allied towns), and provided with 115 wagons, arms and armour and "good guns" marched on a castle near Ingolstadt. Its owner gave up, being massively outnumbered, and the prisoners included twenty knights, a master gunner and other more domestic servants. The castle was burnt. The expedition continued to attack Gibelstatt, home of another hostile landowner, which also fell with little struggle and was destroyed, the captured nobles being tried and five of them beheaded. Later in the year, similar expeditions were carried out . Five hundred men from Rothenburg went to join their allies in Schwäbish Halle with guns, powder and munitions, and together the townsmen besieged and capture a noble castle at Mangenfels, many of the defenders fleeing overnight.[3] These examples show that however well built, and however well-held, no single castle was able to resist the power of the region's cities. The next major military engagement of the city would be linked to the fortunes of their ally, Nürnberg.

[1] Vasold, 165–9 (map at 168).

[2] Vasold, 197.

[3] Chronicon Rottenburgense, in Raimundi, *Duelli Miscellanorum*, 1724, II, 204–5, 206, 208, 209–10.

. . . and Nürnberg

In 1444, that city had to organise an expedition against the lord of Waldenfels, whose followers were seizing merchants trading between Nürnberg and Leipzig. The expedition seized a castle that was held by just three people, getting in by the gate, and found inside it quantities of guns and other military equipment. They resolved to render the place indefensible by demolishing the tower but were hindered by the thickness of its walls. They went on to attack the castle of Lichtenburg, and disposed of their own artillery, including larger guns that were deployed first to fire into the outer court, then afterwards were placed in the church towers. The reporter told the city that they had to acquire stone shot from the town of Eger, and they were also short of powder, which they sent to Erfurt to acquire. He finished by asking the city to send them an extra 200 men, and also people to help build them shelters during their siege.[1]

A few years later, the city was in serious conflict with Albert Achilles of the Hohenzollern family, Burgrave of Nürnberg (and who would succeed his brother as Margrave of Brandenburg in 1470). The city stood under an old castle which was actually a double-ended fortress, the two parts originally properties respectively of the emperor and the Burgrave, and encircled it with its own city wall. Now, however, the castle was held by the citizens, the town defences had been maintained. Some of the defences were old, like the big pentagonal tower added to the castle in the thirteenth century, other parts were modern, like the outer curtain wall stretches of which had been completed during the wars against the Hussites. An instruction of October 1449 distributed duties between named citizens of serving as night watches, including on the bulwarks erected in front of the city gates, and allocated master gunners to these locations. Guards were named for each of the gates, and also for the wall towers. On the new Watertower, ten watchers were to serve in two shifts, either side of midnight. A number of towers were listed as being further protected by their own outworks. A large arsenal of different weapons had been accumulated, Nürnberg like other large German cities having been constructing guns for nearly a century.[2] The actual fighting did not involve any attempts to capture the strongly defended city, but was instead conducted by raiding, plunder and devastation.

. . . and both confront Albert Achilles 1449–51

Whereas Nürnberg was strong enough to thwart the prince's ambitions, other towns were forced to face even more severe attack, as Nürnberg's ally Rothenburg did. The enemy army, far superior to anything the city could mobilise, marched into their territory in 1449, setting up camp on the Franconian hills a few miles from Rothenburg, then set about seizing people and cattle and plundering the villages. The troops of the city fought back, storming the churchyard at Welhausen

[1] Die Chroniken der frankischen Städte: *Nürnberg*, II, 63–8.

[2] *Nürnberg*, II, 275–9, 283–90. For examples of arsenals see Würdinger, *Kriegsgeschichte von Bayern*, II, 342–56.

and then captured the "bulwark on the church yard" of Blaufelden. Albert himself arrived with more soldiers, and led an attack with cavalry, infantry and guns that stormed a village ("small town" of Bilshoven), brutally massacring the men and throwing young boys over the walls, leaving them to die. This action, for which there is only the account of the opponent as evidence, confirms the presence of defences around even the small settlements of the region. A hired lord arrived with cavalry from Nürnberg who were used to counterattack villages held by the foe, while a force of 600 Rothenburgers launched an attack on the churchyard of the village of Tannenbuhl, capturing it at the third attempt, killing 39 defenders. The force went on to take another village, storming over the ditches into the settlement, where the master gunner set up his guns to bombard the *Thorhause* (gate house), suggesting the presence of some fortified structures there.[1] The devastation of the city's territory must have been appalling. What is also suggested by the account is that the *Landhege*, already under construction for nearly two decades, seems to have played no part, and that the village defences were put to serious use, although each time they failed to hold out against a professional military attack. However, the city itself remained secure behind its walls, guns and armed citizenry.

The war with Albert Achilles having been concluded in 1451, in 1452 the towns of Augsburg, Memmingen and Kempten, with others, organised a war against one of the local nobles, von Rechberg, who had been in Nürnberg's pay during the previous struggle, but was now resuming the role of robber baron. The combined towns' forces besieged castle Ramstein for four days, taking it following a bombardment with guns. They went on to assail the castle of Ruckberg, employing the master gunner Conrad Merk whose weapons demolished the walls and buildings inside, leading to the fall of the castle after one month. Von Rechberg returned to the offensive and succeeded in looting and burning the towns of Sulz and Rottweil (1454). This war ended in 1457, with the towns having to pay the baron for his destroyed castles.[2]

Further to the south and west, the major imperial city of Basel was engaged in a fierce struggle with its Rhine-based neighbours at Rheinfelden in 1445. The conflict pursued a pattern similar to those already noted. The Baselers ravaged the surrounding possessions, while organising an attack on the place itself that involved a major army (5000 men with 200 wagons according to the bishop's chaplain), and two very large guns, calling for more than 40 horses to convey them, along with large numbers of smaller firearms that were used to direct a continuous bombardment into the castle. Two boats meanwhile were sent to launch an attack from the river, equipped with ladders. The resources available to a large free city, and its possession of the means to deploy them to effect, are clear from this description.[3]

In 1462, there was a large scale conflict between Albert, acting in the name of

[1] Chronicon Rottenburgense, 218–19, 220–26, 253.

[2] From sources assembled in Würdinger, I, 325.

[3] Die Chronik Erhards von Appenweiler 1439–71, *Basler Chroniken*, ed. Bernoulli, IV, 259–61.

the emperor, and the duke of Bavaria. On 4 February, the town of Monheim was attacked. It was a small place, held by just 200 men, but assisted by two master gunners despatched from Nürnberg, who caused considerable losses among the besiegers. But the town of Augsburg helped the attackers by sending four "middle"-sized guns, and these weapons were large enough to create a breach in the wall by bombardment, and thereby induce surrender on 2 March. The Bavarian attempted to besiege Augsburg, but was unable to make any impression. What was significant was the role the towns played (on both sides) as, in this case, providers both of artillery and of skilled gunners.[1]

Brandenburg

Meanwhile, the Hohenzollerns were establishing their mastery of their recently acquired territory (of electoral rank) in Brandenburg. Having brought the nobility into their orbit, they turned to reducing the turbulent towns to submission. The linked towns of Berlin and Colln on the river Spree had provided the strongest resistance, and in 1442 Elector Frederick II (1440–70) forced them to accede to his will. Of this, the most visible measure was the right to build a castle, the scope of which beside the river Spree, and across the town wall and ditch, was clearly laid out. This fortress was intended to serve as a base of the prince's authority and a symbol of the towns' subjection, and not for its defence against possible external aggression.[2]

In such a climate, the widespread presence of fortifications at every level is no surprise. The maintenance and upgrading of town walls, chiefly by the provision of additional outworks, has been noted several times. The references to blockhouses including those specifically intended to house guns indicates the development of different forms of sometimes temporary small fortifications to provide cover for gunpowder weapons. The vulnerability of castles, whose owners may often have been unable to afford any works, has also been noticed. The use of churches and monasteries as defensible refuges, and indeed as was shown at Soest as something more substantial than that, was to be found everywhere. In 1407, the margrave of Baden had replied to an imperial command not to turn the monastery of Maulbron into a fortress by stating that it was already and had been for a long time. Ostheim (in the lands owned by Rothenburg ob der Tauber) in 1436 had not only a fortified church, but the churchyard too was provided with a deep ditch and high walls. Near where there had been long-term raiding by the Hussites this was no surprise. Elsewhere, the absence of an effective dominant legitimate authority able to fulfil the duty to protect, naturally compelled people to establish places of refuge where they could.[3]

[1] Würdinger, II, 43, 53–8, for the sources for this account.

[2] G von Raumer, *Codex diplomaticus Brandenburgensis continuatus*, 1831 (reprinted 1976), I, 207–8 (1442), 209–10, 216 (1448).

[3] Würdinger, II, 415–17, and the sources listed there.

The Habsburg lands and Austria

From the beginning, Frederick III had been forced to deal with revolts, sending forces to take rebel castles with artillery, using the guns to force their capitulation and often then destroying the place afterwards, such as with Grub an der March in 1440, or with fortifications described as bulwarks, suggesting more temporary works.[1] In 1452, there was a full scale war as the emperor's enemies brought a great host of troops from all the duchies, Bohemia and Hungary. The fighting mainly took the form of sieges. The castle of Orth on the Danube was the subject of a ten-day attack with heavy guns. Orth, a hereditary imperial possession, was bravely held by a small garrison of imperial soldiers who defended the walls and ditches, actively seeking to disrupt Count Cilli's besieging forces. A fierce exchange of missiles, stone shot and arrows from each side, led to an attack with ladders and the capture and burning of the castle (15 August).[2] The combined rebel armies laid siege to the emperor and his household itself in their palace in Wiener Neustadt. The advancing forces began by storming the outlying defensive positions at the town's leper hospital, and in the mill close to the main gate, from which locations they set up their bombards to fire on the town. The siege camp was sited at only one thousand paces distance, and could clearly by seen from the town walls. This also placed it within range of the larger guns, however, and there was a ferocious exchange with many casualties and much damage resulting. The city of Vienna itself supplied two big bombards of such power that "walls of no matter what thickness or strength could resist them".[3] The church, viewing with horror the bloody struggle between the chief champion of Christendom and his subjects taking place a short distance along the river Danube from the empire of the Ottomans, intervened to restore peace, in which control of Ladislas was taken from the emperor.

By 1458, the death from plague of the young Ladislas had simplified dynastic issues (although in the short term leading to the regents of Bohemia and Hungary respectively seizing the thrones). In 1460, Frederick finally regained control of the castle of Orth, which was also being used as a base for depredations against the local country.[4] A year later, the Viennese had to send an expedition to destroy more troublesome bulwarks. In 1462, the chronicle noted the use of machines "commonly called mortars", providing further evidence that modern weaponry and siege techniques were in common use across the empire.[5]

[1] Ebendorfer, Chronica Austriae, ed. Lhotsky, *MGH SrG* NS XIII, 1967, 387–9.

[2] Ebendorfer, 416–17; Aeneae Silvii episcopi Senensis, *Historia rerum Friderici tertii imperatoris,* ed. Boecler, 1685, 107–12. This writer, future pope Pius II, was a close associate of the emperor.

[3] Ebendorfer, 417; Aeneae Silvii, 111–12.

[4] Ebendorfer, 449, 467–8. This chronicle records many other similar events.

[5] Ebendorfer, 512, 553–4.

Burgundy and the Empire: from Liège to Neuss and defeat in Lorraine

The aggressively expansionist dukes of Burgundy meanwhile were continually engaged in warfare against the French, or rebellious subjects, or those they wished to conquer. Louis XI of France (1461–83) faced his own problems too, and having lost (or at best drawn) a battle at Montlhèry (1465) also meddled in revolts against the Burgundians. In 1466, the people of Dinant rebelled again. Raising his army, the duke gathered at Namur in August and raided as far as Dinant, but the locals opposed him vigorously. On 18 August, the Burgundians laid siege to the town, which was protected by powerful walls and a wet moat. As with many towns, however, the presence of a strongly built monastery just outside the walls proved to be a military weakness as the besiegers occupied it and the count of Charolais set up a bombard there, which blew down the town gate, only to see the Dinanters repair it. From the opposite side, the attackers fired from the hilltops concentrating on a corner tower. There were guns deployed all round, and the bombardment was frightful, raising clouds of thick smoke and killing many of the people inside, who nonetheless refused a summons to surrender. The duke then ordered bridges thrown across the river Meuse in order to ensure easy communications. The assault was prepared, each soldier being provided with a bundle of straw and wood to fill in the ditch. The decision was not delayed even by an explosion of gunpowder stored in the camp, which killed 32 men, because news came of a relief column marching from Liège. The assault was about to take place when Dinant surrendered unconditionally, which did not save it from pillage, and the execution of the revolt's leaders. Liège then gave up too. That ancient warrior Olivier de la Marche noted the decisive role of the Burgundian artillery, commanded by Pierre de Hacquemberg.[1]

Beauvais

Duke Charles the Bold (1467–77) was a great deal more prone to military adventures than his predecessors, and overall, a great deal less successful. But no setback ever seemed to deter him from the next military operation. A large part of his reign was spent in conducting sieges, and most of his defeats arose from relief attempts.

In 1472, in renewed hostilities with Louis XI, Duke Charles besieged Beauvais. The events of the subsequent four weeks have been recorded in great detail, including an account by a resident of the town, which manages to fill in the personal stories often lost in the major chronicles. The place was well fortified but weakly held (without professional soldiers). The walls had been strengthened during the previous century, with new adaptations for artillery put in place. The enclosing wall had a talus, revetted in earth, and projecting towers with platforms for guns.

[1] *Olivier de la Marche*, 256–7; Wavrin, 475, 495–6, 514, 524–36.

The continuous ditch was filled by the waters of the rivers Therain and Andon. In places the ditch was more than 50 metres wide. The town gates had outworks, drawbridges and portcullises. Large sections of the circuit were also covered by marshy ground outside, making them almost unapproachable.[1]

At seven in the morning of 25 June, the Burgundian army reached the town and, its demand for surrender having been contemptuously rejected, its assaults against two of the gates and the rampart inbetween began at eight. The outwork protecting the suburb in front of the Porte du Limaçon, protected by 16 arquebusiers, was overrun, but the gate was closed and the Burgundians' initial cries of victory were muted as their attempts to raise ladders were thwarted by defenders — including the women of the town — hurling down on them from the wall top stones and anything else to hand. This attack lasted all day before it was abandoned. Meanwhile, at the Porte de Brêsles, where the attackers could find no cover from suburbs outside the wall, a similar tale unfolded. As night fell, reinforcements arrived, 200 lances of troops who had ridden all day from Noyon, a distance of 90 kms. While the defenders reinforced their defences with timber and earthwork, Duke Charles' men set up their camp and protected it with ditches. They also moved guns into the houses of the suburb facing the Limaçon gate. However, the defenders set their own guns to fire back, including in the large cylindrical *tour de Crou* which was located in the garden of the bishop's palace, and covered this gate. The defenders succeeded in setting fire to the Burgundian-occupied houses forcing their retreat. For the next two weeks, there was continuous bombardment from a large siege train containing guns of all sizes, in particular six large bombards, and great damage was caused to the town walls and the houses inside. On 9 July, there was an unsuccessful attempt to divert the waters of the river Therain, while efforts at mining were thwarted by the marshy ground: the garrison prepared two countermines, but these were not needed. Burgundian troops brought bundles of faggots to fill the ditch at the gates de Brêsles and Hôtel de Dieu, and this assault lasted three hours. While the defence stopped them at the ditch edge, women again kept up a continuous supply of stones and of pots and barrels filled with hot lime to throw down. The next day, the garrison launched a sortie at three in the morning, driving the enemy away from the ditch, and tipped two Burgundian cannon into it. The Duke finally gave up and marched off on 22 July. Throughout the siege, the town had been largely open to the south, and through this route had arrived continuous help in the shape of professional soldiers, arms including guns, and munitions and food, sent from Paris and the neighbouring area.[2]

The size of the town had prevented Charles from managing a complete encirclement, and the terrain had thwarted his efforts to use his superiority in

[1] J Ganiage, ed., *Histoire de Beauvais et du Beauvaisis*, 1987, plan at 113; M Renet, *Beauvais et le Beauvaisis dans les temps modernes: Époque de Louis XI et de Charles le Téméraire*, 1898, 125–8.

[2] Basin, *Histoire de Louis XI*, ed. Halphen, 1966, II, 120–38; Anon, *La Chronique de Lorraine*, ed. Marchal, 1859, 111–15; Anon, Discours du Siège de Beauvais par Charles duc de Bourgogne en l'an 1472, Beauvais, 1622, in P de Comines, *Mémoires*, ed. Godefroy, Langlet du Fresnoy, four vols, 1747, 205–15.

numbers, artillery and training to storm the walls. The determination of the whole population to resist, and the continuous arrival of reinforcements and supplies including artillery, gave them the expectation that they could hold out. Remarkably, despite the reported impact of the Burgundian bombardment, it seems that only 24 of the inhabitants were killed by this attack.

Gelderland

A year later Charles attacked the people of Guelders, who did not share their ruling house's willingness to hand over the duchy. This time, in rather different circumstances, the outcome was the opposite. The Burgundian assembled a large army including numerous guns, powder and gun stones. Marching up the river Meuse (Maas) via Maastricht, towns such as Sittard yielded immediately to overwhelming force. The late thirteenth century castle of Montfort, built on a low hill in otherwise flat and marshy ground, did not, and required the attacker to set up his artillery, but on the third day the defenders saw the hopelessness of their situation, and surrendered. Next was the town of Venlo, reached on 17 June 1473. Once again, the proud citizens decided to resist, but this induced four days and nights of non-stop bombardment by the Burgundian artillery after which surrender was agreed.

The citizens of the town of Nijmegen, strongly located between the great rivers of the Netherlands had used the days of grace allowed them before the enemy arrived to strengthen their defences. Setting everyone to work, the town had ordered the digging of a new, deeper and wider, town ditch, and the erection of new works to shield the gates and towers of the wall. Two suburbs were demolished, and the occupants of a friary were relocated. The old, and the very young were offered the chance to leave, and some did. The duke reached Nijmegen on 26 June, and deployed his army to besiege the town. He was joined by John duke of Cleves (Kleve) with his forces. The chroniclers claim that the Burgundian deployed no fewer than 500 bombards, serpentines and other weapons, which may be a pardonable exaggeration, but it was certainly true that a continuous bombardment was sustained, with devastating effect on the walls and towers. On 9 July, Burgundian troops stormed the New Gate, hoisting their standard above it. The defenders forced the assailants out after fierce fighting. The bombardment resumed, this time focussing on defences known to be weaker on the side of the river Waal. Amid the crumbling masonry, the townsfolk, with no hope of relief, began negotiations for surrender. The settlement of 20 July included a bill of 80,000 florins for the duke's expenses.[1] Again, old walls suitably reinforced against artillery and — crucially — defended by people with loyalty and courage, had forced the most

[1] Willem van Berchem, 'De Gelderse Kroniek', ed. J de Mooy, *Gelre*, 24, 1950, 130–35; 'Chronijk der Landen van Overmaas en der aangrenzende Gewesten, door eenen inwoner van Beek bij Maastricht', ed. J Habets, *Publications de la Société d'Histoire et d'Archéologie dans le duché de Limbourg*, VII, 1870, 38–42.

modern army in the west to organise a full-scale attack. But that attack had prevailed after a few weeks.

In Alsace, Lorraine and what is now Switzerland there was a patchwork of small territories that the duke coveted in order to bring geographical coherence to his realm, but in his way stood independent rulers such as the dukes of Lorraine and the Austrian (Habsburg) ruler of several counties in Alsace, and the rulers of a group of closely-allied cities, the chronicles of which provide much detailed evidence. What we now call the Swiss Confederation was known to its participants as the *Eidgenossen* (literally, oath-comrades), who were linked in mutual interest with the imperial city of Strasbourg, and all regarded themselves as part of the German emperor's dominion. Conflict with Burgundy had followed the duke's acquisition of the Austrian counties in Alsace, while the duke's governor, Peter von Hagenbach, had aggressively threatened with destruction the Swiss towns that resisted his autocratic demands. This behaviour led to his capture and execution, and war with his master. Basel, in December 1473, ordered its defences prepared, and organised citizens to each sector of the defence. Strasbourg issued regulations detailing everyone's duties in the event of attack. The castle of Ortenburg, a foundation of Rudolf of Habsburg, was one of those occupied by Hagenbach in 1470, and its condition was reported in a survey carried out for the duke's government in 1473. The place was a hill-top fortress with "very defensible" strong walls. There was a lower outer ward, which the surveyor, Mangin Contault, noted could not be used to threaten the higher castle, but was itself untenable in the face of the latter. He recommended that a garrison of ten to twelve men with provisions and artillery would suffice to hold it, but that overall, there was much work needing to be done owing to its neglect. Perhaps these repairs were not carried out, because the Strasbourgers came with a large force and two guns and recaptured it after Easter, 1474. It does not seem that the towering Bergfried dominating the site had any influence on the outcome.[1]

The castle and town of Héricourt, near Montbéliard, had also fallen into Burgundian hands in 1470. In 1474, an army made up of contingents from Fribourg, Solothurn and Lucerne were joined by the duke of Austria, and troops from Strasbourg and Basel. On 8 November, with two "great bombards" sent respectively from Strasbourg and Basel, they began the attack. The destruction was however limited: one witness noted how strong the defences were, while the attackers were unwilling to move their weapons too close. The weather was grim, and the Swiss urged that they would prefer to storm the walls rather than stay out in the cold. The dilemma was resolved by the news of an approaching Burgundian relief army. The besiegers met and routed them. Returning to Héricourt, they pressed the bombardment closer, now causing serious damage to the defences, and

[1] D Specklin, Les Collectanées, ed. R Reuss, *Fragments des Anciennes Chroniques d'Alsace*, II, 1890, 461–2; Johannis Knebel, Diarium, ed. W Vischer, H Boos, *Basler Chroniken*,1880, 41; L Stouff, ed., *La description de plusieurs forteresses et seigneuries de Charles le Téméraire en Alsace et dans la haute vallée du Rhin par Maître Mongin Contault, Maître des Comptes à Dijon* (1473), 1902, 46–8. On the castle, Mesqui, 271–3.

casualties among people behind them, and surrender by the 450-strong garrison followed on 17 November.[1]

Neuss

The long siege of Neuss by the duke of Burgundy in 1474–5 was a very different affair from the quick conquests of Dinant and Gelderland. In its outcome it resembled Beauvais, but in its epic duration was very different. The event itself is highly revealing of many aspects of the level of development then reached. It contains examples of siege techniques both ancient and modern within the same operation, of a highly modern army fought to a standstill by a resolute defence, itself driven to deploy the most up-to-date methods to resist. There are plentiful contemporary records which between them present a remarkable degree of agreement on the main facts. The most valuable are the eye-witness accounts of the Burgundian court historian Molinet, and the town secretary Christian Wierstraat's long verse chronicle.[2] Yet Neuss has become a subject of little interest to English-speaking writers in particular, an omission deserving of rectification.

The siege arose from a dispute between the small Rhineland city and the latest archbishop of Cologne, who insisted on his jurisdiction, while Neuss asserted it was a free city. The archbishop enlisted the aid of Charles the Bold, who in July 1474 began siege operations.

Neuss was of ancient origin, having been a Roman frontier fortress with a civilian settlement (Novaesium), reappearing in history when its fort was destroyed by the Northmen (863). In the twelfth century it was certainly walled, and by the fifteenth it was a prosperous place, with many stone-built houses.[3] North of Cologne, and close to Düsseldorf, Neuss stood in flattish terrain, but well protected by water. An arm of the Rhine, which then took a different course to today, ran by the eastern side, with a substantial island intervening between town and the main course of the river. A tributary, the Erft, joined the arm by the south of the town, and other streams ran in front of the western wall. Molinet eulogised the town's strength, although actually it was quite small and not originally especially strong. Its medieval walls had, however, undergone reinforcement, work which was speeded up as conflict developed. The inner defences seem to have comprised high, well-built stone towers, and the gates were substantial. The townsfolk, their garrison commanded by the Landgrave of Hesse, had constructed an outer line of defence. Bulwarks constructed of stone, bricks and earth stood in front of the older inner defences, linked by deep ditches, and further such works stood on the outer line, behind a ditch filled with water from the rivers. Each of the five main

[1] *Die Berner-Chronik des Diebold Schilling 1468–84*, ed. G Tolder, 1897, 176–85; Knebel, 122–40. The donjon alone remains: Salch, *Atlas des Châteaux*, 670.

[2] C Wierstraat, Die historici des belegs van Nuis, ed. Norrenberg, in *Chroniken der deutschen Städte*, xx, 1887. As there are several other editions of this chronicle, but all based on the original printed version, references here are given to the line rather than the page.

[3] P Stenmans, J Lange, N Bommels, etc., *Neuss in der Wandel der Zeit*, 1969, 66, 80 and many illustrations recording parts of medieval Neuss that no longer exist.

gates was protected by a bulwark "of the manner of a bastille", provided with plenty of weapons, chiefly gunpowder, with specific mention of culverins, arquebuses and serpentines. The defenders were experienced soldiers and the town was well provided with food and military supplies, although the state of the gunpowder stores was to be a repeated concern, as prodigious quantities were consumed. It was noted that Neuss contained two powder mills but still needed to import crucial elements like saltpetre.[1] The ability of the defenders to bring in fresh supplies on several occasions would prove critical in the end.

Charles surrounded the town with a multinational army drawn from all quarters of his realm. He set up his own quarters in an abbey not far from the main gate (the Obertor). Italian troops held the quarter on the Gelderland side, and they put up batteries of two big bombards, one smaller, and several other cannon. Against the Aachen side, where there was a particularly "marvellous" (Molinet) bulwark, was deployed a force including 200 English archers (Edward IV being an ally) and 200 Guelders culveriniers, and linked by "deep and spacious" trenches to the forces either side. On the Jülich side was the force commanded by the baron de Lannoy, while against the Cologne-side gate were deployed the men from the Low Countries with plentiful artillery. The bombardment began, and the defence responded, the attackers finding themselves vulnerable and the Italians in particular suffering considerable losses, the defenders setting the tone for the kind of defence they intended to have with sorties through the (northern) Rheintor in the early days capturing 200 of the besiegers, and through the south-western Zolltor killing many.[2] The besiegers also had to cope with sudden attacks launched on them from outside, as on 10 August when troops from Cologne made a successful raid, taking 70 prisoners.[3]

The duke decided to take the island in the river called Wait, that provided a channel of communication for the defenders towards Cologne, and from which troops armed with guns were causing loss and disruption. This proved to be a complex undertaking, involving the bridging both of the Erft and the arm of the Rhine. The initial attack was a fiasco, the troops who got across by boat being wiped out when the rope they had used to guide the rafts across the stream broke. A second attempt was made when more boats were obtained and 500 élite soldiers sent across. These captured the site, then set up their own bombard to fire on the town, and secured their position with a bridge of barrels. A counter-attack was organised from Cologne with a fire-ship against the bridge, leading Charles to block the channel for good with rubble, hay and anything to hand to create a dyke thirty feet (nine metres) wide.[4] The effect on the town of the loss was considerable, with the Burgundian guns able to inflict much destruction.

The duke now turned back to assaulting the town itself from his side of the

1 *Chroniques de Jean Molinet*, ed. Doutrepont, Jodogne, 1935, I, 29–32; Knebel, II, 102, 105. This diary reproduces the texts or contents of eye-witness letters sent from Cologne to Basle; Wierstraat, l. 140 f (erection of bulwarks where needed).

2 Molinet, I, 33–7; Wierstraat, ll. 177–80, 317–30.

3 Koelhoffsche Chronik, 835.

4 Molinet, I, 38–41; Wierstraat, ll. 346–450; *Olivier de la Marche*, 290–2; Knebel, 106.

river. The bombardment was beginning to cause serious damage to the town's older defences. The town secretary recorded the demolition by guns of the Niedertor (in the north-west), while two other gates were also pounded down, being (he said) not really very strong. But the citizens knew how to respond, their repairs consisting of stone and earth bulwarks, the stones providing the core while large earthworks on top provided defences able to resist gun shots. In effect, he wrote, the gates were replaced by bulwarks.[1] The duke was preparing to capture the main bulwark facing his own position, and had built a substantial bulwark of his own. The main component was large oaks, and the result was a semi-circular fortification of several storeys, thirty feet (nine metres) high, equipped for the use of gunners and crossbowmen. Within its shelter, the attack was prepared with ladders and mantlets, but disastrously the ladders proved too short by ten feet (three metres). The assault on 10 September lasted two hours, but the defenders responded by hurling down hot water, burning faggots and "boiling oil", and also killed many Burgundians with gunshots (300 died in the event), because, it was noted by Molinet, their gunports were at several levels, thus enabling them to cover their walls effectively. Apart from the "boiling oil", the account reads true of the horrific nature of close-quarter siege warfare at this time. Another contemporary account confirms the strength of the defending bombardment, but also suggested that an assault was not actually planned, and the losses of the Burgundians were quite light (eight dead, sixty wounded).[2] The defenders during this time were not quietly waiting on the Burgundians, but organised their own sorties, helped in this by the closeness of the siege works to the town. One attack, launched simultaneously from the Rheintor and the Niedertor, caused major losses to the besiegers, and destroyed much of the opposing works, and captured two tons of gunpowder.[3] Other sorties, the sixth and seventh of the siege, were launched on 14 October from the western Hamtor and the neighbouring Niedertor, and succeeded in capturing eight handguns and a serpentine from the besiegers' trench.[4]

By November, Neuss had been under attack by overwhelming numbers for many weeks, casualties among the garrison of about 2500 were high, and supplies in particular of powder were running low. Wierstraat reported that up to now, 30,000 arrows and 100 tons of gunpowder had been used, and only half a ton remained. The blockade was not truly complete, as a citizen got out to explain the town's plight and in an operation the success of which must have involved some quite complex organisation, five or six hundred men made it into Neuss from Cologne overnight through a gap in the lines, each carrying ten pound (4.5 kg) sacks of saltpetre, while the defenders organised sorties to cover the movement. The duke, furious, ordered the mobilisation of yet more troops from the Low Countries, and the digging of new trenches at the point where his enemies had

[1] Wierstraat, ll. 460–71, 475–500, 510, 525–47.

[2] The letter of Jehan Baugey to the Council of Dijon is reproduced in R Vaughan, *Charles the Bold*, new ed. 2002, 322–5.

[3] Molinet, I, 43–44 (the *huyle boulant* is at 44); Wierstraat, l. 749; Knebel, 117.

[4] Wierstraat, l. 957–90.

got in.[1] At a point where the ground permitted it, count Campebasse organised an attempt at mining under the key bulwark, but the excavation was betrayed to the defence who countermined successfully, leading to a dreadful underground battle in the darkness.[2] Professional miners from Liège provided the operatives, apparently on both sides.

The siege was attracting widespread attention across the empire, most of which had rallied in support of the town. Troops, supplies and money had been despatched in response to the emperor Frederick's appeals, the town of Cologne did not share its archbishop's hostility, and the surrounding countryside was therefore mainly enemy territory for the duke. At the same time, the duke's camp was obtaining the quality of a large town itself, with all the attributes of a comfortable court for the nobles, dignitaries, and visiting ambassadors, and all the facilities required by a now long-settled army of well over ten thousand. The defenders made sure by their regular sorties that the besiegers needed to remain alert. In turn, the townspeople were kept under constant pressure too during these first weeks of the siege, with the Burgundians maintaining their bombardment, but also creating numerous false alarms of possible assaults by sounding their trumpets, keeping the citizens awake for fear the attack would be real. There were frequent fires caused in the town by the bombardment, and the women of Neuss were organised to dowse them.[3]

Molinet recorded here that the duke had a Castilian in his camp, who was educated, and had read his Vegetius. He had identified a possible decisive weapon, called a "crane", and drew it for the duke, who was won over and ordered it to be built. It was an immense wooden tower, on four wheels, 20 foot (six metres) square, and forty feet (twelve metres) high, with even taller ladders and a drawbridge at the top. The ground was however entirely unsuitable for such a heavy device, which got stuck, while another machine prepared simultaneously in the adjacent quarter was destroyed when one of its wheels broke. The chronicler noted that such engines had become useless because they were vulnerable to gunfire,[4] although in this case it was the ground they stood on that had proved to be the critical problem. Perhaps if the Castilian had read Christine de Pisan as well as Vegetius, he might have reconsidered his proposal. The defenders of course were aware of the construction taking place under their eyes, but do not seem to have been much worried by it, or by other traditional designs of weapons tried by the Burgundians. That which caused the most concern was a weapon erected to propel fire arrows over the ramparts.[5] Winter was now closing in, making the prosecution of the siege more difficult and life in the siege trenches particularly unpleasant. The difficulty was compounded on 6 January 1475 when the sluices controlling the flow of water from the river were opened at a time of high water levels, forcing the Burgundians to abandon some of their siege works.

1 Wierstraat, ll. 1007–60; Molinet, I, 50–55; Knebel, 145; Koelhoffsche Chronik, 836.
2 Molinet, I, 46.
3 Wierstraat, ll. 840, 880–82.
4 Molinet, I, 44–5.
5 Wierstraat, ll. 818, 1165–70.

The Burgundian losses included 30 cannon and four bulwarks flooded and many hundred horses drowned.[1]

The new year also saw, at last, the empire responding, with Frederick III gathering an immense force drawn from numerous princes and cities on the Rhine to defend Neuss. The imperial army managed to gain control of several towns in the territory that the Burgundians had secured, such as Remagen, but they had to lay siege to the town of Linz, held by a Burgundian garrison. Olivier de la Marche was sent from the siege camp with a force to relieve the town (5 January), his men carrying supplies on their horses. In a well-planned operation, they crossed the Rhine by boat, then caught the besiegers (Saxons and men from Trier) by surprise, and stormed a bulwark after a shot from a culverin ignited and exploded the gunpowder stored there. The Burgundians then enjoyed themselves on the besiegers' food and their good Rhenish wine before withdrawing in good order. Linz surrendered anyway.[2] Meanwhile back at the siege, the Rheintor bulwark was finally taken by storm by Italian soldiers on 16 February, and this might have been a decisive if belated success in forcing the town's outer defences and opening the way for a final attack, but the defenders heroically held on, erecting a strong temporary bulwark immediately behind that which had fallen using earth and timber, and with the help of Liègeois miners prepared deep ditches. The approach of the emperor had a dramatic effect on morale, just as it also seems to have persuaded the duke to step up the attack in the hope of winning the town quickly, and one letter talks of incendiary bombs being hurled into Neuss. The defenders however, were as *animati* as they had been on the first day.[3] The forces of the relief army finally came into view on the opposite bank of the Rhine, where they set up camp and from where they were able to set up batteries to bombard the Burgundian siege works. Communication with the town was effected by the novel method of firing in hollow balls containing letters, but the range was considerable and there was surprise and joy when one, finally, fell safely in the market. No less usefully, fresh provisions were got into the town across the river.

The imperial army showed little sign of any intention to face the duke in open battle, so the siege continued. Further crises occurred. There was a fierce assault on one of the gates on 24 March, only just held off, and a prisoner treacherously opened a lock gate to drain the water from the outer ditch between the Obertor and the Zolltor. Renewing their damaged siege works, and restarting the slow work of mining, the Burgundians opened a new attack with a new bulwark against the St Quirins Gate. The defenders sallied out against this work on 8 April, took and destroyed it, drove the Italians from the mines they were preparing (Molinet says they fled), and then resisted three attempts to recapture the work. The stretch of wall here, where the ditch had been drained, was the focus of the fighting. The wall itself was at risk of collapsing and had to be reinforced with earthwork and

[1] Wierstraat, ll. 1406, 1445–74; Knebel, 173.

[2] Molinet, I, 66; *Olivier de la Marche*,294. The news reached Neuss, to much rejoicing, on 15 March: Wierstraat, l. 1815; Koelhoffsche Chronik, 838.

[3] Knebel, 166.

bundles of straw.[1] These attacks, carried out when the risk from imperial troops from the outside was growing, shows the determination of the Burgundian to achieve victory here. Ambassadors with him reported indeed that this last operation had been critical: the ramparts had been undermined and the assault was to have been launched when the bombards had done their work, but a countermining effort by the defenders had been successful so no result had been obtained. The besiegers were also aware of the persistent shortage of gunpowder in the town, and thought this would be decisive.[2]

The end was nigh. With envoys from France and the Pope working diplomatically, there was a sharp conflict in the field between elements of the two armies (21–3 May) followed by a truce. News from elsewhere decided Duke Charles to disengage, and the people of Neuss learnt of the raising of the siege officially on 28 May. The cost had been enormous, after eleven months of fierce fighting. 500 men remained of the original garrison of 1500, which had several times been reinforced during the siege. The losses of the Burgundian besiegers over the period have been placed as high as 12,000, or as many as were actively involved at any one time, and in the circumstances, with the likely losses from disease as well as injury, the figure is believable. In the town itself, two gates, seventeen towers and 300 houses had been destroyed, and the town had acquired a debt (150,000 gulden) that would take generations to pay, despite the support of the empire: the reconstruction of the defences began that same year.[3] Despite attempts to portray the outcome as a victory for the Burgundians by his many apologists, Charles had not succeeded in taking one small town despite using the entire resources of his immense inheritance.

Olivier de la Marche, than whom one would not find a more experienced warrior, concluded that Neuss would have fallen had it not been for three factors: the re-victualling of the town, the floods that swept away the duke's siege works, and the arrival of the emperor with 60,000 men.[4] These good points are a little unfair in omitting two additional considerations. First, the heroic determination of the citizens of Neuss and their military commander with his largely Hessian garrison to continue to resist all the odds was decisive. Similarly, their military skill in creating defence works that seem to have been perfectly suited to the weaponry of the day, both to resist bombardment and to allow the defence to use their own weapons to devastating effect, was instrumental in making the resistance effective. As for the defence works themselves, the fact of there being several lines was also militarily critical. It enabled a resolute defence to keep the besiegers sufficiently far away to prevent them blasting a breach, and it meant that a single attack being successful would not, necessarily, mean the fall of the town. Indeed, several times the Burgundians did carry the outer defences, but were unable to hold onto them. The resort to more ancient forms of weaponry by the attackers provides

1 Wierstraat, ll. 2015–20, 2035–74, 2237; Molinet, 74–5, 79–80; Knebel, 235.

2 Gingins de Sarra, *Depêches des Ambassadeurs Milanais sur les campagnes de Charles le Hardi 1474–77*, 1858, 1, 69, 106–7, 112, 136–7. The Milanese were allies of Burgundy.

3 Stenmans et al, 109–11; Molinet, 1, 102.

4 *Olivier de la Marche*,298.

a sidelight that caused some amusement to the defence, and caught the attention of the chroniclers, but had negligible impact on the action.

The end of Valois Burgundy: Lorraine and Switzerland 1475–77

Charles, undaunted by his failure before Neuss, faced continuing war with the towns and cities allied in the League of Constance.

In July 1475, the allied towns sent an army to capture the castle of Blâmont, the garrison of which was interdicting the trade and travel of the folk of Basel. The defenders, who included good gun masters, caused many losses to the besiegers. The men of Strasbourg erected their own bulwarks, and then, on 4 August, launched an assault that failed because the ladders were too short. After four hours of fighting in very hot weather, the attempt was abandoned. Reinforcements arrived, and the chronicle described the slow process of unloading the guns from wagons under fire from the defence. Once set up, however, their threat seems to have disheartened the Burgundians who surrendered what they had boasted of as the "strongest castle in Burgundy". The victors found inside plenty of powder, ammunition and supplies. They undermined the walls to destroy them (24 August). In fact, the castle dated back to the thirteenth century, the town wall to the fourteenth. The castle was triangular, with a circular donjon and towers 12 metres high, 7.6 in diameter. Well fitted with arrow loops, one must assume the defenders had established platforms on the enceinte, or used the rooftops, for their guns. Impressive though it was, it was not really that strong in the face of modern weaponry. The victorious allies pressed on, and over a period of two months, they captured nine Burgundian-held towns and castles, destroying the latter.[1]

Come the autumn of 1475, the duke of Burgundy attacked Nancy, the capital of the duchy of Lorraine, the first of three sieges of the city. Little now remains of old Nancy. The *Porte de la Craffe* is the only survivor of the medieval defences: this began life as a simple square gate tower in 1463, flanked with two circular towers. Slightly later engravings show it protected by water, with a circuit of stone walls with projecting round towers.[2] Now it was bombarded from several sides at once, while the Burgundians dug their trenches and fascines were readied to prepare for an assault. The heaviest battery was placed against the St Nicholas Gate, while the defence placed guns on all the high towers and returned fire. One of the wall towers, taller than the others, was a particular target for a big Burgundian *courtau*, which sent over large stone shot. In an interesting early example of spotting, the gun master Jacques returned fire from guns within the defences, guided in their aim by someone higher up the tower as to where they were landing. It worked, the *courtau* was hit, and dismounted. Concentrating their own guns in one place to provide covering fire, the garrison sortied successfully, burning the fascines and seizing two of the bombards and other smaller guns. A second sortie on 10 November, using the tactic of a false flight, took the Burgundians by surprise, killed

[1] Knebel, 276–81; *Die Berner Chronik*, 252–64; Mesqui, 61.

[2] P Marot, *Le vieux Nancy*, 1970, 15, 113, 117–18.

several of the duke's gunners, and captured more of the weapons. However, 17 days later, Duke René II of Lorraine (1473–1508) ordered the city to yield.[1]

The conflict now turned to the border regions between Burgundian and Swiss territories, and in February 1476, Charles attacked Grandson, held by a Bernese captain with 500 men. He first tried an immediate attack but was driven off with one hundred of his men killed, so he settled in for a siege, emplacing numerous bombards and other guns to batter the walls, then attempting another assault, with success. The defenders were all hanged. But the *Eidgenossen* quickly rallied their forces and scattered the hostile army in a dawn attack (2 March 1476), seizing an immense booty from their camp, and reoccupying the town.[2] Duke Charles was not short of contacts or resources, and having lost one army, he simply raised another, returning to the offensive in the summer. In May, he ordered a replacement artillery train from the city of Metz, comprising 43 guns with all their carts, including one monster weapon that needed 46 horses to move it. The target was Bern, but on the way he tackled the small town of Murten, held by 600 men, where he arrived on 9 June. He set up his own fortifications around the town, which he assailed from two sides, organising a continuous bombardment with serious effects on the ramparts: one source says the whole rampart was demolished on one side. An assault along this breach was however repelled after a two-hour battle, leaving many Burgundians dead in the ditch (18 June). A few days later, on 22 June, his army was again routed by a combined allied force based on Swiss infantry. All the artillery was distributed between the victors: Basel received two large bombards.[3]

Charles drew no lessons from his defeats, and raised another army. Meanwhile the followers of the ousted duke of Lorraine recaptured his capital. After a fairly desultory blockade, because the attackers were too few in number, their effort was stepped up in September when "the whole country" gathered to assist their duke. A close siege was now established, deliberately digging trenches as near as possible to the walls. Strasbourg sent a battery of bombards, that were emplaced close to the city wall and began to demolish it. Stones were shot of "marvellous size". The defenders were in trouble, having few stocks of provisions, while the besiegers were plentifully supplied. They tried a couple of sorties: one succeeded in finding two serpentines unattended and hurled them into the ditch. The second was much less successful: the attempt was routed, and the troops were pursued back toward the wall by German troops. The defenders inside the town feared that the besiegers would enter the city along with the fleeing Burgundians, so raised their drawbridge, leaving most of the Picard troops who had made the sortie outside to be killed or captured. The Burgundian governor surrendered on 6 October and Nancy welcomed back its legitimate ruler.[4] Within days, Duke Charles arrived

[1] Anon, La chronique de Lorraine, *Recueil des documents sur l'histoire de Lorraine*, ed. L Marchal, 1859, 175–81; Knebel, 313–21.

[2] Diebold Schilling, *Luzerner Bildchronik*, ed. R Durrer, P Hilber, 1932, 76–7; Knebel, II, 347–50.

[3] Knebel, III, 7–22; Schilling, 81 (illustration at 139); *Die Berner Chronik*,, 35–9.

[4] La Chronique de Lorraine, 228–35; Knebel, III, 40–64.

in the duchy, and the third siege of Nancy began when the Burgundian army deployed in the neighbouring towns on 28 October. Initially, he did not have heavy artillery, which had been stored at Metz, but once it had arrived he was in a position to begin a serious attack. The defenders were vigorous, launching sorties on the Burgundian camps, causing many casualties, but their supply position was parlous, having had little time to re-supply since the previous siege, and it being already winter by the time the serious fighting began. The conditions were no less hard for the besiegers, living in the open in December, and their own supplies often imperilled by being intercepted by René's garrisons. Despite contemporary accounts of a heavy bombardment being sustained, the main threat to the city was not an assault. Duke René had once again appealed to the League for aid, and messengers slipping out through Burgundian lines reported the desperate straits of the occupants, who were reduced to eating cats, dogs and rats.[1] Again, the allies responded by assembling a relief army. Not for the first time, they routed the Burgundian army, amid the snow and ice (5 January 1477); but this time, Duke Charles himself was killed. Having no direct heir, the waiting vultures descended on his realm, Louis XI swallowing up Burgundy itself, while the Netherlands possessions fell to Maximilian of Habsburg, future emperor.

The story of the end of Valois Burgundy again reveals the reality that although siege techniques had advanced to the point where guns were clearly powerful enough to demolish defences, it was unusual for this alone to be the reason for the fall of a well-defended town, although castles (even though associated with walled towns) were a different matter. The Burgundian artillery and that of the opposing towns and Swiss cantons was a potent force: without it, a siege would not be begun. But it confirms that it was far from being the deciding factor.

Ottoman conquests in Eastern Europe and the end of Byzantium

Throughout these conflicts ran a thread argued particularly strongly by the papacy: a new threat from Islam was now in the heart of Europe. The control of the Ottoman Turks was being consolidated across the Balkans and Greece, and the capture of Constantinople in 1453 marked not only the symbolic end of the Roman empire, but the arrival of another contender to replace it.

Despite this picture, however, aid was paltry and the empire of Byzantium was left to expire with a dignified but futile courage, and to find a new splendour and glory as Ottoman Istanbul. The sultans learnt from their errors, skilfully turning even defeats to their future advantage. The last failure to take the great city in 1422 had not been forgotten, and, after the crushing of the Christian crusade at Varna (1444), the capture of the Hexamilion (1446) and the further victory at Kossovo

[1] La Chronique de Lorraine, 245–67; Knebel, III, 69–99; Schilling, 85 (illustration at 145).

(1448) the Ottomans had pretty much a free hand in the Balkans. In March 1452, a new sultan succeeded. Mehmed II (1452–81) was to be a great builder as well as a conqueror. But nor did he delay the siege of Constantinople which can be said to have started when Ottoman troops set up camp in the straits of the Bosphorous and began to build castles there, and certainly his intentions were clear from the start as he set about recruiting the specialist workers – masons, carpenters, metalworkers, miners – in their thousands. Preparations on the new fortresses had begun during the previous winter with the assembling of timber and iron. A ferry was established across the Bosphorous. Work on the triangular-shaped fortress was carried out by staking out the ground plan, and working with very large stones. Tursun Beg was involved in the construction of Rumeli Hisar (Plate 17), which he described as having twenty portals open to the sea below the castle and close to the shore, with a cannon placed at each. Kritovoulos, the Albanian in Ottoman service, adds that the walls were designed to create the longest possible base for artillery to control the shore, and that the guns were pointing in different directions to achieve the maximum cover. The fortress walls and towers were of great strength, perhaps as much to cope with the effect of firing large guns from them as to resist bombardment themselves. Meanwhile, a similar castle was built on the opposite shore below Yeniz-Kale called Anadolu Hisar, each being completed in record time.[1] The straits were now blocked and the city's fate was, literally, sealed: one of the first shots fired from the new fortress sank a Venetian ship, the Rizzo, in November.

Ottoman plans and preparations

It was written on the Ottoman side that the Turks understood that the ancient walls would not resist modern artillery. The challenge was to assemble a sufficiently powerful train of weapons, and many accounts have been written of the awe-inspiring cannons deployed against the city. Several writers tell of the gun founder (a Hungarian) who had been serving in the city, but, his services unpaid, had left and offered them to Mehmed in 1452, who gave him the means to cast a massive bronze cannon. Completed in three months, the monster gun achieved a range of one mile (1.6 kms) in a test firing at Adrianople, hurling a metre-diameter stone weighing 400 kgs and making a crater six feet (two metres) deep.[2] This gun, only one of the Ottoman arsenal but the largest, had to have its own means of transport. Mounted on thirty linked-together wagons propelled by sixty oxen and with a corps of workmen each side to prevent tipping and in front

[1] Giese, *Altosmanischen anonymen Chroniken*, 99; Tursun Beg, *History of Mehmed the Conqueror*, trans. Inalcik, 1978, 33–4; Kritovolous, *History of Mehmed the Conqueror*, trans. Rous, 1954, 19–21; N Barbaro, *Diary of the Siege of Constantinople*, trans. Jones, 1969, 9–10; C Riccherio, in J Melville Jones, *The siege of Constantinople 1453: seven contemporary accounts*, 118. Both castles are discussed in more detail below.

[2] Doukas, *Decline and fall of Byzantium*, 200; Kritovolous, 43. For discussion of Ottoman access to the technology, K de Vries, Gunpowder weapons at the siege of Constantinople 1453, in Lev, *War and Society in the eastern Mediterranean*, 1997, 353–9; A Williams, Ottoman military technology, same volume, 363–66.

to clear the road, this genuine prophet of doom inched its way overland towards Constantinople during two months (February and March 1453), arriving five miles from the city where the rest of the besieging army was gathering. Well aware of the vulnerability of his cumbrous new weapon, the sultan had deployed his forces throughout the winter to prevent any attacks from the city, and to capture Christian-held forts on the Black Sea. All the evidence points to careful preparation, with the sultan and his officers working out in advance exactly where to place their batteries, based on an understanding of where they would have the greatest effect. Mehmed not only had the determination to succeed where every one of his predecessors had failed, he also had the means, the skills and the weapons. The last siege of Constantinople began on 5/6 April 1453, and would last just 55 days.

Constantinople's defences

The empire had long been desperately short of resources, although its last rulers seem to have understood their predicament to judge by their frequent efforts to enlist western aid, including efforts to end the schism of the churches. The double walls and ditches that had withstood Avars, Arabs, Bulgars and many others were still, fundamentally, the walls erected at the command of Theodosius. Records of works done to maintain the defence have been studied, some of it provided by chroniclers, some recovered from the inscriptions in the stonework. In the last century, there had been a major earthquake in 1344, following which the fallen stonework had been restored, and a parapet raised along the moat, recorded by Nicephorus Gregoras. In the most recent past, there had been extensive restoration by John VIII Palaeologus (1425–48), noted by inscriptions dated 1434, 1438, 1439, 1440, and 1444 on towers of the outer wall. The work seems to have been slow and subject to many interruptions, for which the most likely explanation is shortage of funds, although contemporaries sometimes attributed this problem itself to embezzlement. The outer line of the defence was to be the key battleground, rather than the soaring internal walls and towers, and it was to be the fate of the defence of the outer line that determined the outcome of the siege. One of the best sources for these events is the diary kept by the Venetian Nicolo Barbaro, who recorded the works ordered by the emperor (Constantine IX Palaeologus, 1448–53) as the Ottoman army's advance unfolded. He noted that the emperor concentrated on trying to reinforce the outworks of the imperial palace (the Blachernae) which stood at the northern end of the circuit and represented a weak point in the defence precisely because it had no bulwarks and the outer walls were not strong: men landed from Genoese galleys were immediately sent to dig new ditches at this point (14 March).[1]

The 700 well-equipped men who arrived in the galleys were a vital reinforcement, and their leader Giustiniani was appointed commander. His shortage of soldiers prevented him waging the kind of active defence beyond the ramparts

[1] Barbaro, 23–4; Ricchiero, 120, recorded that hopes rested on the outer walls because they were strong and in good repair; A van Millingen, *Byzantine Constantinople*,1899, 95–108.

that we have seen used to considerable effect in other sieges, but the Genoese had command of a mixed force of no more than around 7000 men to defend walls stretching three and a half miles (more than 6000 metres) along the land walls, without counting the sea walls and those on the harbour (another 11,000 metres between them) that also required protection against the Ottoman fleet. This meant that the Ottomans were free to array their siege camp without interference and set up their batteries without risk of disturbance during April, and one of the eye witnesses later observed that the city paid a terrible price for allowing their enemies to approach so close without trying to keep them further off.[1] Part of the battle was for control of the waters. Mehmed's success in moving a whole squadron of ships overland to refloat in the Golden Horn was critical in completing the isolation of Constantinople, and a remarkable achievement of organisation in its own right. A bridge constructed from one thousand wine casks completed communications with Ottoman forces on the Galata side, which also completed the encirclement of the whole city. It would also allow, in May, for one of the larger guns to be moved to the hill above Pera (Galata), from which it could be aimed to fire across the Genoese-occupied suburb and against the ships in the harbour, sinking one with its third shot and forcing others to move, thus also creating another area of concern for the stretched defenders.[2]

Attack by land, April–May 1453

By land, the sultan's great gun was deployed against the St Romanus gate, where advice was taken to prevent the gun drying and cracking between firings. The target of the first shot was about to be repeated when the Hungarian suggested moving it so as to fire at the same point but from a different angle, then repeated from a third point. The consequence of the triangulation was devastating, the whole stretch of wall being demolished as a result. Other reporters indicate that the Ottoman siege works were operating against the whole length of the land walls, the troops digging in behind barricades and what were described as "vaulted bunkers", and miners were also set to work at prearranged locations, although their role in the final outcome would be pre-empted by the success of the direct attack. Kritovolous gave credit for the observation of the weak points of the walls at the palace end to the sultan himself, and for the establishment of a battery of the heaviest guns against this stretch. Barbaro, from the inside, noted the four locations of the big guns that began their work on 11 April: three against the palace, three against the Pigi (Pegae) gate, two against the Charisius gate and the final four against the St Romanus, as already observed. The bombardment was sustained as the main form of attack for nearly a week (12–18 April), the really large guns only achieving a handful of shots each day, but the many (relatively) lighter weapons kept up a sustained battery firing shot of between 200 and 500 pounds (91–227 kgs). The effect of the heavy weapons was apparently truly destructive of

[1] Leonard of Chios, in Melville-Jones, 14–15.
[2] Barbaro, 45 (5 May).

the ancient masonry, Kritovolous talking of stretches of wall and lumps of towers being dislodged at one blow. There were two of these monsters, the largest we have already seen being built, its slightly smaller partner still firing shot weighing 800 pounds (363 kgs).

The defenders brought to bear what experience they had in the face of such a battery. Bales of wool were lowered on ropes in front of the wall, but this had little effect, so wooden palisades had to be thrown up backed with bales of brushwood, branches and reeds and screens of hides and skins, filled out and protected with earth. Unlike the crumbling masonry, the earthwork-based defences extemporised to fill in the breaches were more capable of absorbing the stone shot as it struck.[1] The movement of the largest Ottoman gun against St Romanus was also reported in Leonard of Chios' report to the pope, as well as its effectiveness in creating a great breach, and this source also reported an attempt at mining by gold and silver miners from Novo Brod. The defenders however detected this work, and countermined effectively, filling in the caves the diggers had created under the rampart, or driving out the attackers with smoke and water according to other accounts. The besiegers' intentions were apparently regularly passed to the defence by messages concealed in arrows and sent by some of the many non-Muslim troops who made up, already, the Ottoman army.[2]

Bombardment continued, but the Ottomans were now pressing home the attack from different quarters and probing the defences. On 7 May, an attack with a battering ram was driven off. On 12 May, there was an assault directed against the palace walls. On 14 May, the big gun that had caused so much distress to the fleet was moved again, this time against the Charisius gate, where it did little harm, and then against the St Romanus gate, held by 300 men with guns and crossbows, which it demolished, forcing day and night work to repair the breach. Two days later, another mine was discovered when the noise of men working underground was detected near the Caligaria gate (by the palace), and this was destroyed by a countermine. The inhabitants awoke on 18 May to the astonishing sight of a Turkish siege tower sitting against the ditch just ten paces from the wall. This was protected against the severe risk of destruction from guns by being filled with earth and covered with hides, and it had been moved up a trackway from the camp half a mile away, itself screened by camelskin-covered hurdles. This had, somehow, been put in position overnight, and now it was used as a base for hurling earth down to fill up the ditch in front. More mines were discovered and destroyed over the next week, generally in the palace area.

Final assault 28–9 May 1453

The besiegers' preparations culminated in a devastating artillery bombardment on 28 May. On the inside, the defenders made ready large numbers of timber mantlets with which to strengthen the ramparts, but, Barbaro noted, suffered from

1 Kritovolous, 36–40, 43–9; Tursun Beg, 34–5; Doukas, 211–19; Barbaro, 27–41.
2 Leonard of Chios, 14–18; Tedaldi, in Melville-Jones, 3–5; Riccherio, 118–20.

a rather ill-timed dispute about pay with the workers hired to move them. Long before dawn on 29 May, a first wave attacked with ladders but was repelled with heavy loss. The attacks were being directed at several different stretches of the defence simultaneously, in particular where the breaches were and the ditches were filled to the point at which they could easily be crossed. A second wave attacked the St Romanus gate, and was likewise held off. The sultan sent in the Janissaries, his élite corps, but first fired the great cannon again, with the effect of knocking down the repaired stretch of defences opposite. Although initially driven out of the outer defences, the Janissaries quickly rallied and returned to the attack, forcing their way in. Giustiniani was hit by an arrow, and the defence now crumbled as he withdrew to seek treatment but failed to appoint a deputy, and the Genoese troops were overrun. The emperor himself died in the mêlée as the Ottoman troops forced their way through an open postern into the city itself, the precise identity of which remains uncertain, but were thereby able to take the defenders in flank and rear in the space between the inner and outer walls, and rout them. With the sound of victory reaching the Ottoman ships, meanwhile, attacks by escalade were simultaneously launched from all around the city from the Ottoman fleet, everyone motivated by the prospect of fabulous loot. Turkish losses in the assault were enormous, and some 700 defenders were said to have died in the final battle. The city was sacked, and soon after the Genoese governor of Pera agreed terms of surrender with Mehmed.[1]

Some thoughts about the capture of Constantinople

Hindsight suggests the inevitability of the Ottoman victory, and the military odds clearly were, in all respects, overwhelming. But if the operation is taken instead in the context of other sieges of fortified cities at this period, as we have observed already in many places, and will see again shortly at Belgrade, it might instead be concluded that it was the absence of any kind of external support, and really only that factor, that made the fall of Constantinople the most likely result of Mehmed's attack. And even with that vital factor, there were also some tactical points that suggest the result need not, even then, have been absolutely predetermined.

There were hints that Mehmed was not far from having been defeated. The losses his army had suffered were huge, one of the negative consequences of planting his attack so close to the walls. Despite the massive damage inflicted by the Ottoman bombardment it was only at the fourth attempt, the second attack by the corps of Janissaries, that the breakthrough was made, and that itself was only decisive because of chance events. It is an impossible "what if? . . ." to ask what would have happened had the attack been held off, but it is not inconceivable that Mehmed might have been ready to give up at that point, as he would decide at Belgrade in 1456. The reason for suggesting that the result was not certain (without suggesting that in the long term it was other than inevitable) is that we have

[1] Doukas, 221–43; Tursun Beg, 34–7; Leonard of Chios, 30–41; Riccherio, 120–3; Barbaro, 52–66; Kritovolous, 52–93.

seen that cities, including many with fortifications that were not themselves modern, were resisting the most modern artillery attacks. Certain other factors needed to be present, such as a determined defence, which the Christian defenders of Byzantium certainly possessed, and adequate supplies, which the vast open spaces contained within the city walls certainly permitted the population to cultivate. Although the old inner walls had shown themselves very vulnerable to gunfire, it was the outer line of defences that were critical. These had been updated and reinforced, and they were the place where the actual fighting took place, with the ancient ditch in front nearly as effective a barrier to the Ottomans in 1453 as it had been to all previous attackers by land.

The defenders had also benefited by receiving "inside information" about Ottoman plans, and this might have helped Giustiniani to decide where to concentrate his limited forces. It might also have been critical in helping the defence to identify where the mining was taking place and to take the necessary counter-measures. Certainly, the Ottomans had carried it out on a large scale, and had it not been so successfully intercepted it might well have played a much more significant part. The contemporaries who observed that it was the failure to conduct an active defence that was critical were, probably, right too. Although the close siege that Mehmed organised from the start meant having to sustain continuous heavy losses, this was offset by being able to focus his attack wherever he chose and keep the defenders under continuous pressure. It meant that he could begin his mines quite close to the wall, and that extraordinary events like the vast apparatus brought up to the ditch overnight were possible, possibly a decisive factor in making it in turn possible to fill the ditch sufficiently to allow for an assault in the first place.

The weakness of the defence's artillery was, if anything, a more significant contributor to the conquest than the strength of the attackers'. Alongside the failure to keep the Ottoman forces at a distance and to impede their use of their own equipment and guns, the Byzantine guns also failed to destroy the Ottoman siege works and siege machines, suggesting that the guns they had were only hand weapons. The penury of the indigent Roman empire was responsible for this lack, although the Italians who formed the backbone of the defence had access to the most modern weaponry. A state that could neither afford to purchase the weapons nor had the resources to manufacture them itself was certainly facing a difficult future.[1]

[1] On the relative state of Byzantine and Ottoman military technology, see discussions by Bartusis, *The late Byzantine Army*, 334–41; De Vries, Gunpowder weapons, 343–59; Williams, Ottoman military technology, 363–66; Petrovic, Firearms in the Balkans on the eve of the Ottoman conquests, in Parry, ed., *Technology and Society in the Middle East*, 164–90; Agoston, 'Ottoman artillery and European military technology in the fifteenth and seventeenth centuries', *Acta Orientalia Academiae Scientarum Hungaricae*, XLVII, 1994, 16–29, 30–2.

Ottoman advances continue: 1454–56, Serbia and Belgrade

Mehmed spent little time consolidating his victory. Early steps were taken to repair the damage to the city's defences, with a great force of lime-slakers being mobilised to prepare the lime to repair the city walls, and to establish gun-making facilities there.[1] Future targets lay further afield: to the west, in the Balkans and Greece, and in the east, there remained Christian rulers on the north coast of Asia Minor.

The following year (1454/5) Ottoman forces attacked the fortified towns of Trepča and Novoberda (Novobrda, nr Pristina) in Serbia. Trepča's outer wall was immediately captured, and after a short siege the citadel fell. At Novoberda, the demolition of the walls by cannon took forty days, which was sufficient to ensure the victory. Both towns were locations of mines, the produce of which now went to the Ottomans.[2]

1456 brought evidence that the sultan could be resisted: Belgrade once again would prove to be the stumbling block. It was not that the sultan was unprepared. The siege this time would engage the most powerful guns the Turks could call upon, including many constructed anew to add to the existing battery, the Genoese merchant de Promontorio (who claims to have been a witness) stated that 28 great new bombards were built, and 100 new ships were sent up the river Morava.[3] But this time they would not call upon the skill of their miners in the time they had available, perhaps recalling 1440. Among new types of weapon deployed at Belgrade were mortars, according to witnesses. An Ottoman general was one of the more notable victims (reported Asik Pasha) of a shot from this early example.[4] Ottoman ships were also organised to attack from the rivers. From the start, the Serbian and Hungarian defenders undertook the kind of active defence that Constantinople had lacked, sending out forces that attacked and disrupted the Ottoman advance, causing damage to the siege train which, by the very nature of its size and weight, was incapable of being either concealed, or manoeuvred. The fortress on the Danube was able to rely on natural protection on two sides with the rivers Danube and Sava, and on the exposed side could count on a double line of walls, and a wet (or marshy) moat. Additionally, the regent Hunyadi was conscious that the holding of this place was absolutely critical to the continued safety (if not existence) of the kingdom, and undertook extensive mobilisation to resist the invasion.

The course of the siege offers both contrasts and similarities with that of Constantinople. The Ottoman forces, converging on Belgrade, were impeded by resistance before they reached the town, forcing them to ensure that their train of guns, which was being brought via Skopje, the larger pieces disassembled, by a host of foot soldiers, was under continuous protection. Additionally, the Danube fleet

[1] Agoston, 25–6; Asik Pasha, 198–99; Giese, 109–10; Tursun Beg, 37; Kritovolous, 93; Doukas, 241–3.

[2] Tursun Beg, 37–8; Kritovolous, 103; Asik Pasha, 205.

[3] F Babinger, 'Die Aufzeichnungen des Genuesen Iacopo de Promontario de Campis über den Osmanstaat um 1475', *Bayerische Akademie der Wissenschaften (Philosophisch-Historische Klasse)*, 1956, heft 8 (1957), 82–6.

[4] Asik Pasha, 206.

was resisted by a fleet created by the Hungarians, itself provided with armour to protect the ships against fire and shot, and the latter proved more effective than the Ottoman galleys, something which was critical in enabling contact with the besieged by the third significant difference, the substantial relief army that was assembling on the opposite bank of the river. Despite these impediments, the siege was eventually begun around 4/5 July with the Ottomans establishing their siege camp and emplacing their heavy guns where they could bear on the walls of Belgrade. Tursun Beg's account talks of having both cannon and catapults, but the latter seems unlikely, and may be a mistranslation (but see below). Thuroczy's description told how the bombardment was directed both against the walls and over them, creating an enormous noise and causing great clouds of sulphurous smoke. The old walls and towers of the city crumbled under the weight of shot crashing into them, and the ditch became filled in at various points. The Hungarian troops tried to compensate for the damage by erecting stockades and digging a new ditch inside it, but the Ottoman artillery proved effective also in breaking up the new palisades.

On the fifteenth day, the walls having been breached and the moats filled, the assault was launched (19 July). The ferocity of the attack and the numbers of troops involved were sufficient to storm the outer defences and the Ottomans quickly occupied the outer parts of the town. But their failure to immediately press on to attack the inner defences proved to be a critical mistake, as the reinforced defenders launched a powerful counter-attack while the Turkish forces were plundering the town, and drove them out again with heavy casualties (20 July). The sources indicate that they pursued them back out of the town and into their own siege works, where they sensibly set about disabling the siege artillery, and had brought with them iron nails with which they wrecked the guns by driving these into the firing apertures, a very early example of "spiking" cannons. This can only have been a pre-planned operation.

The Ottoman version put a different spin on the facts. They claim that the defenders were actually drawn out and then driven back with heavy losses, which seems an effort to put a brave face on a massive setback. They also blame a whole section of their army for not joining in the assault. In fact, the absence of discipline among the assailants, not unusual because it was the promise of plunder that was necessarily a significant motivator of the polyglot and multinational Ottoman armies, in contrast with the discipline of the defending troops, their overall defence led by the respected and experienced commander Hunyadi, and the defenders of the town itself motivated by the charismatic figure of 71-year-old friar Capistrano, was really the critical factor. The sultan restored order, but his losses had been very heavy and he abandoned the operation within a few more days, leaving behind much of his unmovable artillery. Letters sent from Belgrade brought the news of the victory across Europe where it was received with much relief, given the abject failure of the Christian rulers to come to the aid of the Byzantines.[1]

[1] Asik Pasha, 206–7; Giese, 148; Kritovolous, 111–15; Thuroczy, ed. Montello, 173–81; Gragg, *Commentaries of Pius II*, 84; Die Chronik Erhards von Appenwiler, *Basler Chroniken*, IV, 324–5, 391–3, with letters from Belgrade dated 21 and 23 July, 2 August;

The Morea and Balkans 1458–60

Further Ottomans expansion was not long delayed. Their empire was surrounded by individually weak old states, some of them independent, others the colonial possessions of maritime states such as Venice. Mehmed II continued his series of victories, most of them involving the siege and capture of substantial fortresses. In 1458, his forces marched through the Balkans and took Corinth in Greece after just a few days, shamefully according to the Byzantine historian of the event, whereupon most of the Morea capitulated except for the island fortress of Monemvasia and some castles held by Venetian garrisons. The Acrocorinth was a massive citadel and one version suggests that the sultan left a blockade there while his army subdued the peninsula, returning to the isthmus in July, the citadel then yielding at once. Meanwhile, a second Ottoman force led by Mahmoud Pasha was operating in Serbia. The Serbs refused to submit, and Tursun Beg notes that Mahmoud lacked siege weapons, which were presumably with the sultan. Nonetheless he decided to invade, reducing border forts with little resistance and reached the major town of Smederevo, which he attacked despite fire from the defenders' cannon and hand guns. The Ottoman described the defences as formidable, a trench and a double wall. However, they were held up before the inner defences and abandoned the siege. The fortress was indeed a significant place, of recent construction (1427–39), located at the junction of the rivers Danube and Morava, the natural triangle thus formed dictating the shape of the town and leaving a straightforward set of successive lines of defence on the "dry" side (see page 243 above). It had already changed hands twice, the Ottomans having taken possession of it after a long blockade in 1439 then ceded it back after the Hungarian-led crusade of 1443, and it was finally given up by the Serbs when the Turkish army returned the next year (June 1459).[1] Golubac, which surrendered after being surrounded by land and river, and Mitroviča, also fell. In 1460, the sultan sent his forces back into the Morea, where the citadel at Mistra, last refuge of the previous ruler, the Despot Dimitri, in the central Morea, also gave up without a struggle, marking the completion of the Ottoman conquest of mainland Greece.[2]

Ebendorfer, 432–5. In a detailed study of the various accounts, J Held reached the conclusion that the Ottoman forces cannot have exceeded 50,000 (including significant numbers of camp followers), in contrast with the extravagant claims of the time, and the defending forces of Hunyadi 16,000, many of whom had not arrived by the time the siege ended: The Defence of Nandofehervar (Belgrade) in 1456: a discussion of controversial issues, in Vardy, ed., *Society in Change, Studies in honour of Bela K Kiraly*,1983, 25–31. For a much less critical account using improbably large numbers but giving much fascinating first-hand detail, Nisbet Bain, 'The siege of Belgrade by Muhammed II, July 1–23 1456', *EHR*, VII, 1892, 235–51.

[1] Tursun Beg, 40–2, 44 (places the yielding of Smederevo in 1460); Giese, 149; Sphrantzes, 76; Doukas, 256; Fine, *A Tale of Three Fortresses* (description of Smederevo), 184–9 and A Deroko, *Medieval Castles on the Danube*,1964, 19–20. K Mihailovic, *Memoirs of a Janissary*, trans. B Stolz, 1975, 111–13, reports that the Ottomans cast four new guns at Corinth because they were unable to transport their artillery up to the fortress. Mihailovic's chronology is, however, jumbled.

[2] Tursun Beg, 42–4.

Black sea and Lesbos 1460–2

In the next couple of years, the sultan's forces completed the capture of remaining pockets of independence to the east. The Black Sea town of Amasra capitulated in 1460, then in 1461 a combined land and sea operation was launched against Trebizond, the last remaining Byzantine Greek statelet. The ancient city was blockaded by land and by sea, and batteries of artillery were emplaced. The ruler, the self-proclaimed emperor with the historic family name of Comnenus, surrendered after an operation lasting 28 days, Tursun Beg stating that no shot had been fired. The town of Sinope had already capitulated, by honourable surrender, to a similar attack.[1] Ottoman control of the Black Sea' southern shore was now secure. In September 1462, the sultan had to send an expedition into the Aegean to regain control of the island of Lesbos, where the governor Niccolo Gattilusio had rebelled, having previously submitted. The fleet arrived conveying with it a substantial arsenal of siege equipment. The defenders' only chance of survival lay in instant submission, but they refused, and the army encamped on the high ground above the town of Mitylene, and emplaced its artillery. According to the witness, Leonard of Chios, these included six large guns of which the biggest fired shot weighing 700 pounds (317 kgs). Bombardment began, and at the same time miners began to drive their deadly excavations towards the ramparts. After about a week, serious damage had been done to the buildings, and the walls of the city had been breached by the guns. With assault threatened the defence gave up, although surrender did not save their lives. Contemporaries describe the town as well prepared to resist, with adequate supplies, sufficient engines, and earthworks, presumably constructed to provide some protection for the old stone walls, and suggest that the garrison may have comprised a thousand men. But they also note that the Ottoman guns were effective in demolishing the targeted area of the outworks as well as the more vulnerable inner walls. What is left now, and by analogy with other fortification work of the local Gattelusi rulers, suggests that there had been considerable upgrading of the walls and towers during the previous decades.[2]

Another account of the Balkan fighting given by Jörg, a master gunner from Nürnberg, who had been sent to the duke of Bosnia to cast guns in 1456, but had been captured by the Turks who offered him employment. He continued to work for them until, in 1480, he found an opportunity to get away, became the pope's master gunner, then wrote down his experiences. His account confirmed such events as the attempt of the Venetians to re-establish themselves in the Morea, by building new defences on the Hexamilion, an effort abandoned on the approach of Ottoman forces, in 1462 (see below). Soon after, Jörg was to be found casting four new guns for Turkish forces facing the Hungarians in Albania, but coming off worst against the powerful resistance led by Skanderbeg in 1464. They re-

[1] Tursun Beg, 44–6; Kritovolous, 172–4; Giese, 149; Doukas, 257; Mihailovic, 119.

[2] Tursun Beg, 48–9; Kritovolous, 181–3; Doukas, 257; Leonardo Chiensis, De Lesbo a Turcis capta epistola Pii Papae II missa, in Hopf, *Chroniques Gréco-romanes*, 1873, 361–4. F Hasluck, Monuments of the Gattelusi, *Annual of the British School at Athens*, XV, XVI, 1908–10, describes the remains of the defences of Mitylene (XV, 261–5), Aenos and Chios.

turned the next year and constructed what Jörg calls "a fine town" in just 32 days. Jörg also accompanied the army that stormed Chalkis on Euboea (in 1470) after an attack of a month, using massive guns.[1]

Albania from Skanderbeg to final Ottoman conquest

Jörg's account included his work for the Ottomans as they fought in the western Balkans to subdue the Albanians. Ottoman sources for what happened here are few and lacking in detail, while, perhaps reflecting the tribal nature of Albanian society, first hand accounts from the other side are also rare. It is possible to reconstruct the main features, and recognise the extraordinary fact that it took not less than forty years for the Ottomans to suppress finally the resistance of this small country.[2]

Increasing Turkish settlement in the region from the 1420s sparked resistance from the local Albanian tribes, but this only became centralised when the scion of a leading family, George Castriot, who had become a Muslim with the name of Skanderbeg and had served in the sultan's army, revolted against the loss of family lands in 1443, reconverted to Christianity, and proceeded to demonstrate a remarkable level of skill both militarily and politically. Having taken the key town of Kroja by deception, Skanderbeg sent forces to secure the frontiers of the mountainous territory, but then had to deal with the fortified towns held by Ottoman forces. Some of these garrisons surrendered immediately taking advantage of generous terms (as at Petrella and Stellusa — perhaps identifiable as Matí, see below), sometimes they waited until the Albanians "planted engines". Svetigrad required a more formal attack, the garrison having decided to resist. The Albanians debated whether or not to attempt a storm, apparently having no artillery. In the end, Skanderbeg left a blockading force, and the town duly capitulated. The victory of the sultan at Varna that year allowed Mehmed to send troops to Albania, but on each and every occasion that his armies attempted to invade the country, Skanderbeg, expertly conducting guerilla warfare but giving battle when he judged it right or necessary, routed and drove out the invader.[3]

What is harder to reconstruct is the technical level of the Albanian fighters. Undoubtedly, they would have been familiar with artillery and with attacking or defending towns and castles, but that is not the same as being able to undertake large scale siege warfare. Most of the fortifications in the country were small, or very small — when the Albanians had to fight the Venetians in 1447, they besieged Danj, but the same chronicler later described it as a "small hamlet". Only Kroja and Berat were of substance. Dating from the ninth century, Kroja stood on a ridge

[1] Details from letters reproduced in D Malipiero, 'Annali Veneti dall'anno 1457 al 1500', ed. F Longo, A Sagrade, *Archivio Storico Italiano*, VII(1), 1843, 51–7. Jörg von Nürnberg, *Nachricht von der Turken*, 1500. The pages of this original printing are not numbered. On the value of the work see A Vasiliev, 'Jörg von Nürnberg', *Byzantion*, 10, 1935, 205–8.

[2] J Fine, *Late Medieval Balkans*, new ed. 1994, gives a modern summary political history.

[3] J de Lavardin, *The historie of George Castriot, surnamed Scanderbeg, King of Albanie*, 1590, 27–30.

600 metres high and could be approached only from one side. Boasting a Bergfried at one end, and substantial curtain walls with round towers, it would remain an important fortress until the eighteenth century. Skanderbeg is said to have ordered the repair of Albanian defences in anticipation of further Ottoman attacks. Two examples of building dating from this time illustrate the scale of the work.[1] At Matí, on a rocky summit 1225 metres long, there are ruins of a walled town (120 metres long) with a citadel 65 metres long, approached only by a narrow path. Remaining walls of mortared stone stood to 3.4 metres. A long history of occupation makes it hard to confirm when these walls were built, however.[2] More certain is the refuge or fort on Cap Rodon (Kepi i Rodonit), a promontory projecting into the Adriatic, protected by a cross wall 80 metres long. A stone hall stood inside, and a stone tower (barbican) was built lower down by the sea, connected to the hall with another stone wall. A lower outer wall 4.8 metres wide possesses gun loops, and is preceded by a deep ditch. Albanian documents are said to confirm its construction after the failed siege of Kroja in 1451. This work was destroyed by the Turks in 1467 and reoccupied by the Venetians in 1500, to whom the upper part of the defences have been attributed.[3]

In 1450, the Ottomans had again attacked Svetigrad, led by the sultan in person. Skanderbeg had ordered the defences reinforced with "certain bulwarks". Having had a summons rejected, the Ottomans emplaced two very large guns against the town's weakest side, and three days of bombardment created a breach, with the walls and its flankers (towers?) demolished. The ensuing assault had to be made up a very steep slope, and the defenders, who had spent the night filling the breach, were successful. An attempt at escalade was then made on the opposite side, but the ladders were hurled back. Three more days of bombardment followed, and three more attempts at assault, but it was noted that despite the wrecked walls, the location itself made the town impregnable. Once again the bombardment resumed, "balls of wildfire" were thrown in, mines were attempted. The defenders hurled heated stones and boiling water at a platform the Turks raised against the town's gate, and destroyed it. The end came through treachery: someone poisoned the only well, surrender was agreed. The siege lasted from May to July.[4]

In September, Skanderbeg tried to recover the place. His Albanian troops were mainly archers but he also had German and Italian arquebusiers and some small guns, described as being more to annoy the defenders than damage the ramparts. After establishing his lines behind a palisade at "arquebus shot" range from the walls, there followed three attempts at escalade, the second involving 10,000 men,

[1] M Sufflay, 'Städte und Burgen Albaniens, hauptsächlich während des Mittelalters', *Akademie der Wissenschaften, Wien. Philosophisch-Historiche Klasse,* 83(1), 1967, 19. For photographs, S Turnbull, 'The blunted arrowhead: the defensive role of the great medieval fortresses of Albania', *Medieval History Magazine,* 2/5, 2005, 46–51.

[2] S Anamali, 'La forteresse de Skanderbeg à Matí', *Studia Albanica,* 4th year, 1967, 95–101.

[3] Pirro Thomo, 'La forteresse de Skanderbeg au Cap de Rodon', *Studia Albanica,* 4th year, 1967, 103–7.

[4] Lavardin, 127 (Danj), 137–85, for a day by day account.

that failed. Meanwhile, the sultan was coming back. Skanderbeg therefore quickly raised the siege and raced to reinforce the garrison of Kroja, which he secured with a year's provisions and many firearms and crossbows. Over the winter, he ensured the defences were repaired, and added new "bastions", suggesting outworks in this context. The precautions were necessary. The Ottoman siege in 1451 involved heavy and light artillery, the former manufactured on the spot and the latter conveyed on the backs of camels. It seems that it took four days to set up camp, and 15 to manufacture the heavy guns, four of them being very large and firing shot weighing "six hundredweight". This array of great weapons was deployed on two sides. Bombardment began and in four days large sections of the wall had been demolished. As the sultan's troops prepared for assault, Skanderbeg attacked the Ottoman camp from the outside. The first assault took place and failed, and he made another attack, this time forcing the sultan to strengthen his own defences. Two further assaults were repelled with heavy loss, and a mine was dug, which the defenders located and destroyed by a sortie at night. When Sultan Murad died, the siege was ended, and the brave defenders could count themselves among the few who had survived a full-scale Ottoman onslaught. The new sultan, of course, had his mind on bigger targets. The Albanians, meanwhile, not aware that the next Ottoman target would be Constantinople, sought help from abroad. It is reported that Skanderbeg's alliance with Aragon-Naples included promises of guns, iron bars and assault engines, confirming the relative paucity of such equipment available to the Albanians.[1] Among the measures now taken by the Albanian leader was the erection of a fortification on the frontier to act as watchtower and refuge. Artisans were brought from Kroja and work was completed in six months. A few large guns were moved there from the same place (confirming their presence there in the first place), and the new garrison was trained in their use.[2]

Throughout the 1450s and 1460s, new Ottoman invasions took place, and Kroja successfully held off sieges in 1457, 1466 and 1467. Eventually, the Turks tried a change of strategy, and began to take the smaller, and coastal, towns, leaving behind garrisons when their main forces withdrew. The hero, Skanderbeg, died in 1468 and was followed by his son John Castriot. The Ottomans then went to war with Venice, leading to two large-scale sieges of Shkodër (Scutari). On the second occasion, Kroja fell at last, but only after a year's siege, as did other fortified towns in Albania (1477–78).

The town of Shkodër stood on the river Drina, above a lake and close to the sea. It had a roughly rectangular shape with walls two miles in length. It was occupied by the Venetians, but with a mainly Albanian population. The first siege (in 1474) had lasted three months, beginning in May. Using a battery of four large and 12 smaller guns that destroyed part of the rampart, the Ottomans then tried to storm the town but were repelled. When a second attempt also failed, this siege was abandoned. The second attempt was better prepared. Eventually, the Ottomans had their massive guns ready: two firing shot weighing 400 pounds, one 300

[1] Lavardin, 194–246, 280 (Aragon); R Kreutel, ed., *Vom Hirtenzelt zür Hohen Pforte: Derwish Ahmed, genannt Asik-Pasa-sohn*, 187, 264–66, 278 (Asik Pasha).

[2] Lavardin, 260.

pounds. Soon after, a monster firing 650 pounds was set up. The account of the siege tells of how many shots hit the walls each day: 35 on 5 July, for example, and this also gives a clue as to the slow rate of fire of these very large weapons. As the month went by, more and more heavy guns were made ready, emplaced and fired, including a battery on the hill above the town. The missiles included incendiaries designed to burn the houses. There is then recorded the use of mortars, firing shot of no less than 1200 pounds weight. The number of hits each day climbed rapidly — 102, 168, 180. The history intriguingly notes that the attack included "mortars, balistas and catapults", being used for incendiaries. The bombardment was not targeted on any particular point, but covered the whole wall. The assault was concentrated against the town's main gate. Although the attackers seized one outwork, they were driven out again, and thousands of Ottoman soldiers were casualties. A second and third attempt met the same fate. The sultan therefore left a force to maintain the siege while he marched off to capture the neighbouring fortresses, including Dristi (then called Drivast), a small hilltop settlement with a polygonal towered curtain, but not strong, which had been destroyed by the Mongols in 1242 and now faced a similar, and this time terminal, fate. Its walls were breached (although this took sixteen days), and it was stormed. Alessio (Lissus) was stronger, having been reinforced by the Venetians, including a modern rampart erected in 1451, but this fell easily. Having secured the whole country, and tightened his grip on Shkodër by using boats to seize the island on the Drina and set up bridges protected by towers at each end, the sultan left a large force to continue the siege. The siege ended when Venice made peace with the Ottomans including yielding up the bravely-held town in March 1479.[1] The population took the option available of leaving, and was counted at 450 men and 150 women. Ottoman rule in the country was not finally secured until a revolt by John Castriot was suppressed in 1488.

Conquests in the eastern Balkans 1460s–80s

Between the strong Christian kingdoms of Hungary and Poland (and the soon to be powerful realm of Moscow) were the two principalities that would later form Romania, Moldavia and Wallachia. The difficulty even of being certain of the course of events[2] has not however prevented historians and others from delving into the area, led on by the presence in Wallachia of prince Vlad Tepes (the Impaler). This bloodthirsty individual (even by the standards of the day) alternated

[1] Marinus Barletius, *De obsidione Scodrensi ad Serenissimum Leonardum Lauretanum Aristocratie Venetae Principem*. There are many editions of this text. This account is based on the very early 1504 printing, checked with the edition in the 1600 compilation by F Sansovino, *Historia universale dell'origine et imperio de Turchi*; for a different perspective, G Shpuza, 'La lutte pour la défense de Shkodër dans les années 1474 et 1476–79', *Studia Albanica*, 5th year, 1966, 181–9. Asik Pasha, 264–66; Sufflay, 25 (Drivast), 27–8 (Alessio), 28–30 (Shkodër).

[2] P Panaitescu, 'Les Chroniques slaves de Moldavie au xve siècle', *Romanoslavica*, 1, 1958, concluded that all Moldavian chronicles were derived from a single source, written at the court of Stephen of Moldavia (1457–1507).

between staying in power by paying tribute to the sultan, and posing as the heroic defender of Christendom against the Turk. His claim to the role of defender of the faith might be subject to a little less cynicism had half the Impaler's fights not been with his Christian neighbours. Rather more impressive was the work of his northern neighbour, Stephen the Great, who in a long reign modernised his small state and made Moldavia into a great regional force. While much of his success was due to skilful diplomacy, military force also played a part.

In 1465, Stephen took the two Black Sea cities of Cilia and Cetatea Alba. Cilia surrendered on the third day of bombardment, Cetatea Alba without a fight. The prince provided both with artillery and garrisons.[1] Meanwhile, the sultan had not taken his setback at the hands of the rebelling Wallachian lightly, and had marched an army into the province. Vlad's supporters had decided to rely on discretion and the leniency of the Ottoman, so the prince fled to Hungary. It was noted that the Turks approached the capital town of Tirgoviste, observing no troops on the walls, although someone fired off a cannon at them before disappearing. Of more interest was the open field occupied by thousands (allegedly) of impaled victims of Vlad.[2] But the story also confirms the possession by the Wallachians of modern weaponry.

Stephen meanwhile consolidated his regime, securing two castles in Transylvania from King Matthias of Hungary (1469), and in 1475 defeating both the Ottomans and Prince Radu (Vlad's brother ruling in Wallachia) before, three years later, undertaking new works to upgrade the fortifications of Cilia. In 1483, the chronicle records his foundation of another fortress at Smeredova, confirming his role as a founder and moderniser of defensible works.[3] Stephen's successes included restoring Vlad in Wallachia (1475), and occupying the fortress of Craciuna on his southern border to improve protection against raids. The drawback was that he had managed to alienate the sultan, and Bayezid, successor to Mehmed in 1481, was not having it. Mehmed had already seen to the re-submission of Wallachia and the death of Vlad. Now his son organised a major expedition which led to the capture of Cilia and Cetatea Alba. It is possible that the chief purpose was to secure towns which were known to have a significant commercial importance, controlling major trade routes across the Black Sea, and by the same operation, to reassert his domination over the "disloyal" tributary prince. The sultan decided that ships were the most efficient means of transport for his large train, including his siege weapons which included (according to Ottoman sources) both large numbers of guns of various calibres, and *manjaniq*.

The land army crossed the Danube on 26 June 1484 where it was joined by Wallachian tributaries. The siege of Cilia began on 8 July and it fell after just six

[1] G Urechi, *Chronique de Moldavie*, trans. E Picot, 1878, 95, 97–9; Prince Vlad's campaign of 1462 was described in a letter to Matthias Corvinus of Hungary after the Wallachian had taken possession of the town of Giurgiu from a Turkish garrison before launching a devastating and murderous raid into Ottoman Bulgaria: K Treptow (ed.), *Dracula: Essays on the life and times of Vlad Tepes*, 1991, 315–6, letter of 11 February 1462.

[2] Chalkondyles, book IX, reprinted in Treptow, 330; Mihailovic, 131.

[3] Urechi, 101, 135, 157, 163.

days. Five days later, the Ottomans marched on Cetatea Alba, arriving on 22 July, where they was joined by allied Crimean Tartars. Here, negotiations for surrender began on 5 August. Throughout, the forces of Stephen were quite unable to intervene. At both, the same procedure was followed. First the Ottomans set up their camp, making sure it was defensible. Then the artillery was emplaced and the bombardment begun of the walls from three sides. Meanwhile, trenches were constructed from the camp towards the ditch, probably adopting a zig-zag pattern to protect occupants against fire from the defenders. Arrived at the ditch, the next task for these troops was to empty it of any water then fill it in to enable an assault. At both towns, surrender was accelerated when the garrison commanders died in the bombardment, and the citizens decided to seek terms. The walls were severely damaged by the bombardment and clearly would not prevent certain sack and massacre. Bayezid had both places well provided with troops and artillery before withdrawing, a precaution that proved worthwhile when a Moldavian attempt to recover Cetatea Alba by secret escalade was discovered and repelled (perhaps in 1485).[1]

Mehmed II earned his title "the conqueror". The realm he had created stretched from the Hungarian border to the Peloponnese, from the Adriatic to the eastern limits of the Black Sea, and included much of Anatolia. The success of the government Mehmed created, and a succession of strong rulers beginning with Bayezid II (1481–1512) meant that the Ottomans would truly become the largest power in the middle east, and a still expanding threat to Christendom. Their success in absorbing the skills they lacked from people who had them, regardless of religion or ethnicity, helped them to compensate for an early lack of scientific knowledge or technological skills. By this time, they could call upon no less an array of siege know-how than any of their potential enemies, their mining was a potent threat, and their artillery, notwithstanding a now old-fashioned attachment to massive size, was proving lethally effective in destroying defences. The sultan's personal interest in gunpowder weapons no doubt played a considerable part in promoting the continued development of the arm in the Ottoman army. Jacopo de Promontorio's invaluable description of the household and government of the sultan around 1475 also confirmed the formal organisation of Turkish forces, and includes reference to a corps of "well paid master manufacturers of bombards".[2] These, along with the élite Janissaries, were often recruited from conquered lands. The non-stop construction of cannon, such as the report of a whole battery being constructed at Istanbul and Pera to strengthen defences against seaborne attack in 1473,[3] was a feature of Ottoman military organisation. The occasional references to slingshot catapults is puzzling. The repetition of the information from different

[1] Urechi, 169; Malipieri, 134–5; Forres, *Rustem Pacha,* 17–18. The campaign has been thoroughly analysed, using many untranslated Turkish sources, by N Beldiceanu, 'La campagne Ottomane de 1484', *Revue des Études Roumaines,* V/VI, 1960, 67–73; same author, 'La conquête des cités marchandes de Kilia et de Cetatea Alba par Bayezid II', *Südost Forschungen,* XXIII, 1964, 38–86.

[2] Babinger, 36.

[3] Malipiero, 86–7.

sources is too frequent to dismiss this as anachronism on behalf of the writer. If it is true that the Ottoman artillery included such weapons, it must be because they served a purpose that gunpowder weapons did not: perhaps for lobbing things over walls that could not be fired from a bombard. Whatever the truth, the wide geographical extent of the empire and its strong degree of centralised administration, meant that there was always somewhere, and someone, from which the necessary raw materials and technical skills could be obtained by a government determined to match the most advanced in their world.

What evidence, then, is there about Ottoman fortifications from this period? The short answer is: mixed. The speedy construction of the fortresses of Rumeli and Anadolu Hisar, to block the Bosphorous above Constantinople, saw the construction of effectively-designed artillery batteries that proved up to their task. The rest of the defences however comprised high walls and donjons (the largest of seven storeys) along with towers of varying shapes and sizes. The terrain naturally determined the overall shape, but defences dedicated to the deployment of guns were limited to the sea-facing outer ramparts. The inner walls, with a thickness of around two metres and no piercing for large guns, could have provided a base only for hand guns. With the fall of Constantinople, both fortresses lost their military purpose at once.[1] From the same period dates Yedikule Hisar (the "castle of seven towers"), built by the sultan in 1457–8. Located in the south east corner of Istanbul, it is an irregular pentagon abutting the ancient wall and enclosing the former stately entrance at the Golden Gate. Its design was unaffected by any previous structure. This fortress had some serious defences. The former Byzantine entrance was closed off and a new, much narrower gate created, while the inner gate towards the city was also a narrow towered arch closed by a portcullis. The wall walks were well shielded by tall parapets, and the walls were of considerable strength and thickness (12 metres high, five thick). These, and the projecting flanking turrets standing between the major angle towers and those of the ancient gate after which the place was named ensured that if needed, defenders could mount a proper resistance. The outer face of the castle was protected by the inherited triple wall and ditches against external attack: it seems far more likely that it was built as a base for a garrison to help ensure control of the recently conquered city, doubling up with becoming a prison.[2]

These three mid-century fortresses were brand new. Other Ottoman works fill out the shape of the picture. In Anatolia, Mehmed ordered the modernisation of Kayseri in 1465. The original Seljuk castle (a rectangle some 80 by 200 metres with 19 towers and two gates), was protected by a ditch beyond a low outer wall (largely plain and untowered), and the gates were each protected by an outer barbican with bent entrances and large-scale emplacements behind double loops in front of the gatehouse. These barbicans were the only apparent concession to modern artillery, although (again) defence from the tall wall tops by hand gun was

[1] A Gabriel, Châteaux Turcs du Bosphore, *Mémoires de l'Institut Français d'archéologie de Stamboul*, VI, 1943, 10–26, 18–73, for well-illustrated surveys of these two fortresses.

[2] Gabriel, 85–113.

possible. Otherwise, the ancient defences, dating right back to a rebuild by Justinian, seem to have been left substantially undisturbed.[1]

Other Ottoman fortifications in Anatolia have suffered serious destruction from many causes including earthquakes. At Tokat, an ancient castle on the town's central twin-peaked rocky hill was rebuilt by the Ottomans. Its location made it almost unapproachable, and the wall with semi-circular flanking towers was created from the only direction from which direct access was possible. By the same token, such isolation made it essential that the castle contained storage for sufficient supplies to resist a blockade, so the site was well provided with granaries and other storage facilities, a cistern, and a long staircase carved through the rock to provide sheltered access to the river below.[2] Mehmed had the fortress of Sivas (ancient Sebasteia) reworked, but this is now unrecoverable. At the time, the sultan's work was as modern as could be, taking into account materials and masters. The rapid expansion of his borders meant that places set up with a military function quickly lost them, leaving some to decay, while others took on a range of administrative roles. The attachment to powerful inner enceintes with flanking towers (including often one or more master towers described as donjons), despite the vulnerability of such walls to modern artillery, but protected by a concentric outer defence, indicate the multiple roles of such buildings: perhaps ceremonial or symbolic, regularly administrative, frequently a garrison, as much as having a defensive military purpose themselves. They had all been built by the time that further warfare with the Christian powers in the Mediterranean at the time of the death of the Conqueror led both sides to grasp the need for a different type of fortification altogether.

[1] A Gabriel, *Monuments Turcs d'Anatolie*, 1931–4, I, 18–30, with plans and illustrations. On Justinian's work, reducing an even larger circuit in size, Procopius, *Buildings*,333–5.

[2] Gabriel, *Anatolie*, 85–6; Ainsworth, *Travels and Researches in Asian Minor, Mesopotamia, Chaldea and Armenia*,1842, II, 17–19; Barth, *Reise von Trapezunt nach Skutari*,1860, 22–3. Each account suggests a different date for the later stages of the work.

7 The late fifteenth century, II: A time of transition?

At the other end of the Mediterranean from the growing Ottoman empire, the consolidation of the Christian-ruled states in Iberia reached a critical point during the fifteenth century, with Portugal becoming the first to turn to the creation of an overseas dominion, while the unification of the kingdoms of Castile and Aragon in the last third of the century laid the foundations for the emergence of what would become, in the following century, the dominant western European power. An initial stepping stone for the development of this potent force was the conquest by Ferdinand and Isabella of the last Muslim kingdom in Iberia. Granada succumbed to a sustained attack that reveals much about siege warfare at the end of the fifteenth century.

Portugal in North Africa

Siege of Ceuta

The Muslim states of the North African coast had become incapable of uniting or of finding sufficient resources to match those of their expanding northern neighbours. There had been several events since the beginning of the century to predict the outcome. Whereas the world chiefly knows the Portuguese prince Henry through his nickname of the "Navigator", his first claim to notice arose from the small kingdom's successful capture of Ceuta in 1415. The chronicles assign a significant part to King John (João) I (1385–1433)'s brood of young sons in the decision to attempt this conquest, from where it was easy for an effective navy to control the straits. It does seem that the presentation of this as being a contribution to the (otherwise dormant) anti-Islamic crusade may have been a genuine reason for the attempt. It was for all that an act of simple, unprovoked and undeclared aggression. John would rather have attacked Granada, but was hindered from doing so by the rather inconvenient truce then in place between that kingdom and Castile.

King John planned a meticulous operation, because it was no easy thing to convey an army by sea to capture a strongly-walled city in completely hostile territory. His spies surveyed Ceuta under cover of a diplomatic voyage to Sicily, reporting back that they thought it could be taken. They had studied the beaches and the currents, to enable a landing to be made safely, and they had also viewed the layout, height and thickness of the walls and towers "in order to know what artillery was needed."[1]

Over the next two years every carpenter in the kingdom was apparently employed as the fleets were built, while others set about construction of the siege weapons including numerous large bombards and cannons, and, as a spy noted, many scaling ladders.[2] The construction of such an armada caused anxiety in the region, and it is remarkable that the true target remained unaware, the king resorting to some remarkable deceits to confuse possible enemies. When the fleet sailed in August 1415, secrecy had been maintained. Seven days at sea brought it to the Mediterranean, where storms drove it close to the shores of Granada, which feared it was the intended target. Muslim troops rushed to arms, and cannon and crossbows were mounted on the walls and towers of Malaga while others rushed to the shores or took to boats to sally against the invaders. However, the fleet was too far offshore, and the cannon shots all fell short, while the boats were no match for the armament of the Portuguese galleys.[3]

The fleet was then able to reassemble near Ceuta around August 7, counting no fewer than 63 transports, 59 galleys and 120 other, smaller vessels. Although the ruler of Ceuta, Salem ben Salah, had by now had notice of the attack, the diversion of the fleet had led him to send away again the reinforcements he had summoned. Now, his troops manned the walls and towers of the city. Ceuta stood on a narrow headland, and comprised three parts: walled towns either side of the heights of Almina, where the citadel stood. The army landed on the beach below the heights, with Prince Henry and his brothers in the lead, while most of the fleet remained off the port itself, and it was this successful landing that seems to have been the key, since the governor had chosen to send the garrison out to oppose the landing. Defeating these troops, Henry and his brother Edward were able to fight their way in through the gates as they drove them back. The role of the princes was confirmed by a French adventurer who participated, although he wrote it down many years later. Another, contemporary, account, however tells of a massive artillery bombardment for many hours before the assault took place, which must have been delivered from the fleet. Having reached the town, some troops fought their way along the wall top to reach other gates and open them with axes. As the defenders fell back to the citadel, the attackers turned to plunder, but the Muslim forces were not in fact crushed, and returned to attack the now scattered Portuguese, and had it not been for the heroic role of the princes themselves in rallying the invaders, they might have been driven out. Instead, they held on, and forced the defenders back into their castle. At evening time, the Portuguese pon-

[1] Chronique du Roi João I, *Chroniques de Gomes Eannes de Azurara*, ed. Virginia de Castro e Almeida, 1933, 13–29; Mattheus de Pisano, *Livro da Guerra de Ceuta*, ed. Correa da Serra, 1790, I, 7–20. This later account (written around 1460) uses Azurara (properly known as Zurara), but also other sources. The author may have been the son of Christine de Pisan. A fine modern account is given in P Russell, *Prince Henry, the "Navigator"*, 2001, 29–53.

[2] Letter of Ruy Dias de Vega to Ferdinand I of Aragon, 23 April 1415, in *Monumenta Henricina*, II, 132 f.

[3] Chronique du Roi João I, 60; Mattheus de Pisano, 39–40.

dered their position. The citadel was a powerful fortification protected by thick double walls. In the end, guards were set to keep watch, but the garrison had lost heart overnight and fled. Ceuta was now Portuguese.[1] What is striking is how well the Portuguese were organised and led, and how ill-provided the Muslim defenders of Ceuta were to resist the attack, neither their undoubted courage, nor their substantial but not modern defences being a match for the invaders.

From Alcácer Ceguir to Tangier

Many attempts were made to recapture the city, or at least to keep up pressure on the defenders, who were commanded by Count Pedro de Menezes up to 1437. With the original population gone, no trade with the country outside, and a large garrison to maintain (3000 from the outset), it was a major drain on Portuguese resources. Few of the attacks amounted however to full scale sieges, the forces at the disposal of the local Muslim leaders not being organised or resourced for what would necessarily have been a very substantial operation against a garrison that could readily be reinforced by sea. Instead, the chronicles recorded numerous failed attacks by Muslim forces, usually driven off with heavy loss by the better armed (and armoured) Portuguese troops. On one occasion, it was noted that the Moors had brought up two very heavy bombards with which to devastate Ceuta. But the count organised his own artillery to fire on the enemy battery as it was being established, and his master gunner was good enough to score a direct hit, killing the Moorish gunner and three of his sergeants.[2]

The Portuguese also made a number of attempts to extend their realm, not always with success, an attack on Tangier in 1437 led by the sons of King Edward (Duarte) (1433–8) in particular being a disaster: the invading force was inadequate, the target city was very large and well defended (the governor being the same Salah ben Salah), there was no secrecy, the scaling ladders were too short, and the large army mobilised by the Moors ending up besieging the would-be besiegers on the shore and forced them to capitulate. They escaped by promising to return Ceuta, but reneged on the deal, leaving Henry's hostage younger brother to die in prison.[3] Civil war during the childhood of the successor, Afonso v (1438–81) was a further distraction, but it was in the 1450s that Portuguese expansion against the Moors resumed with the conquest of Alcácer Ceguir, partly in response to continued threats to Ceuta. The Portuguese fleet of more than two hundred ships

[1] Chronique du Roi João I, 69–77; Mattheus de Pisano, 45–56; J Nève (ed.). *Antoine de la Salle. Sa Vie et Ouvrages*, 1903, 141–46; Zurara, *The Chronicle of the Discovery and Conquest of Guinea*, trans. Beazley, Prestage, 1895, 15–17; Letter of João Mercader to Ferdinand I of Aragon, 13 September 1415, *Monumenta Henricina*, II, 222, reporting news received in Valencia via Granada. This report talks of the bombardment lasting 13 hours, which makes for a very long day, as the fighting on the beach, at the walls and gates and in the city is reported to have lasted four to five hours, but was over by "vespers" (de la Salle).

[2] Gomes Eannes de Zurara, *Chronica do conde Don Pedro de Menezes*, ed. Correa da Serra, 1792, 460. See also Zurara, *Chronicle of the Discovery and Conquest of Guinea*,17, reporting the relief of the city by Henry in 1418 and subsequent repairs to the defences.

[3] See good summary in Russell, *Prince Henry*, 161–82.

sailed in October 1458, and the troops had to fight their way ashore on arrival at Alcácer Ceguir, the defenders being driven back into their city. The invaders then landed their artillery and began to attack the defences with shot and incendiaries, and Prince Henry led the assault, breaking through the outer wall and reaching the city gate against which he deployed his "machines". As night fell, the attack was maintained, Henry setting up his big bombard to fire over the walls, more it was said to frighten than to harm, but the fierceness of the onslaught seems to have worked because the defenders now signalled their willingness to capitulate.[1] The rapid surrender of the strongly-walled town came as a shock. By November, the army of Fes was deployed outside, provided with large bombards, and the Portuguese had to organise the despatch of supplies and munitions to maintain their control of the place as the siege was sustained through the winter, spring and summer of 1459. There were large scale assaults and attempts at escalade that were driven off, while the governor, Duarte de Menezes, arranged repairs to the damage caused to the walls by the bombardment. At long last, a relief army from the Portuguese king forced the raising of the siege. It was noted that during 56 days, the Moors had fired 2,490 stones, but had killed only 25 of the defenders. Whether the figures are fact or fiction, they suggest either a well-prepared set of defences, or an inept artillery attack, rather than the less material protection implied by the Portuguese description![2] The success of the long defence by Menezes contrasts with the rapidity of the original capture, and can stand as further evidence of the technical and organisational superiority now held by the Iberian Christians over the Muslim rulers in north Africa. However, it should also be noted that the place, whose Arabic name Ksar-es-Seghir translates as "small castle", was not equivalent in size or importance to Ceuta, even less to Tangier.

In 1471, Afonso v achieved a long-held ambition with the capture of Tangier (Tanger). The most significant military action took place, in fact, at the nearby town of Arzila (Asilah), seven leagues away. A very large expeditionary force conveyed in more than 300 ships landed in August, with 24,000 fighting men. The army arrived outside Arzilla on 20 August, and at the outset had only two bombards with them when the assault was ordered. After four days' attack, the townsfolk displayed a flag of peace. The Portuguese then had to storm the castle, in which they captured 500 Moorish soldiers, and immense plunder. A few days later, Tangier was yielded to a Portuguese detachment.[3]

[1] Duarte Nunes do Leão, *Cronicas e vidas des Reys de Portugal D. Duarte undecimo, e D. Affonso o v, duodecimo*, 1780, II, 226–32, a seventeenth century account; J de Esaguy, *Tanger sous la domination portugaise*, 1937, 8–14; Russell, *Prince Henry*, 321–4.

[2] Nunes de Leão, 234–45.

[3] Nunes de Leão, 291–302.

Spain and the conquest of Granada

The realm of Granada had survived chiefly by admitting its tributary status. During a prolonged period of relative peace during the late fourteenth and early fifteenth centuries successive kings of the Nasrid dynasty constructed the magnificent palace of the Alhambra beside the older massive citadel (*alcazaba*) on the ruler's hill above the heavily populated city of Granada itself. But this ruling family also maintained a dense network of fortifications in their territory which had a frontier facing a potential enemy of no less than 650 kilometres in length. Inside the borders stood several significant cities, each provided with its own set of powerful city walls and a large citadel (and sometimes more than one): Ronda, Malaga, Almería, Guadix and Loja.[1] The purposes of these citadels included serving as refuges, and being the seat of local government and the garrison. But although renewed, their defences mainly dated back to the glory days of al Andalus. Warfare in the second part of the fifteenth century would assign not much of a contribution to a castle or castle-like enclosure in a full-scale siege, no matter how well constructed or highly located. Instead, it was the city wall and its outworks that would decide how readily a garrison could resist a modern attack. Along the lengthy border territory stood a line of castles, sited where feasible on the crests of hills or mountains and located around 12 kms apart, linked together by watch towers to create a continuous connected series. These frontier fortifications of the kingdom were more useful as places from which to observe, or as temporary refuges for local people. Generally, in design, the Nasrid dynasty work emulated the design of their Castilian neighbours, employing donjons and round, then later polygonal towers, although as the construction began during the thirteenth century and continued (through repair and modernisation) right to the end of the fifteenth, dating can be difficult.

Siege warfare in the frontier zone 1450s–60s

It would be more accurate to regard the long frontier more as a zone or region than as a border, with the Christian nobles and officials also being responsible either for defending or for attacking across the frontier zone as conditions, and the existence or otherwise of truces, determined. Fortified cities such as Jaen, Ubeda and Baeza also served as the bases for the Castilians. In 1458, Miguel Lucas de Iranzo was appointed Constable of Castile, with his base at Jaen, and it is his chronicle that provides much of the detail of the kind of warfare prevalent at this time. The year 1462 witnessed some frontier action, fairly typical of its kind. The constable had observed that La Guardia was very suitable for the construction of

[1] A Bazzana, 'Forteresses du royaume nasride de Grenade (XIIIe–XVe siècles): la défense des frontières', CG, XI, 1983, 29–37, for the evidence for much of this paragraph. See also Terrasse, Les forteresses de l'Espagne musulmane, *Boletin de la Real Academia de la Historia*, CXXXIV, 1954, 477–81; and L Torres Balbas, *Ciudades Hispano-Musulmanas*, 2nd ed., 1985, especially 495–500.

exactly the type of frontier zone fortification that we have noticed on the other side, *una torre y atalaya* (watch tower). The area was under continuous threat from troops based in the castle of Arenas, so he brought a force of 2600 men to attack it, leading 30 of them by night to begin a surprise escalade. The Muslims were alert, so he brought up the rest of his troops to begin a general assault, but was strongly resisted so, because the source notes he had not ordered up any artillery with him, he had to abandon the attempt.[1] In that same year a much more significant conflict was resolved in favour of the Castilians, when the fortress of Gibraltar, despite its heavy refortification since the mid-fourteenth century sieges, was yielded without a struggle by the Muslim commander to the duke of Medina Sidonia, to the consternation of the people of Granada.[2]

In 1463, another attempt at escalade was made, against the Granadan castle of Moclin, but this also failed, the approach of the attackers being discovered when it was heard by a dog. The constable meanwhile was being more successful in organising the protection of his side of the frontier, with cavalry patrols, and better pay to ensure sufficient numbers of officers and soldiers to guard the walled towns and forts and castles of the frontier under his command.[3]

It would be profound personal conflicts within the ruling family of Granada, skilfully exploited by the Spanish king, that would contribute to the speedy collapse of the kingdom. However, in 1464, that kingdom sat back and watched while a section of the Castilian nobility revolted against King Henry IV (1454–74), and caused massive disruption throughout Andalucia. The Constable remained loyal to the crown, but had to resist a powerful opposition led by the Master of the Order of Calatrava. To begin with, Don Miguel supplied the loyal town of Ubeda with extra provisions, gunpowder and ironware (cannon?) to resists the Master's siege. Then in 1465, the rebel forces closed in on Jaén itself. The chronicle goes into great detail to describe how the constable ensured that the city could maintain its food and water supplies and at the same time deny in particular use of the local mills to an attacker. The siege was characterised by frequent sorties and skirmishes, treasonable plots, and efforts to maintain contact with the loyal town of Andujar. If the account is correct, the plan worked, and by August the Master was forced to break up the siege because it was his army that was running short of provisions. During a subsequent truce, the constable ordered the strengthening of the city by enclosing the gardens lying outside between two of the gates, adding considerably to the place's defensive potential as well as protecting a means of providing sustenance.[4] The civil war resumed in 1466. The constable set out from Jaén to recapture the town of Baeza with 3800 men on 6 June, and camped overnight at a tower a league away. They tried to force their way into the city but the defenders withdrew inside, shutting the gates and manning the walls and towers. The constable attacked two of the gates and ordered escalade at four separate points,

[1] J de Mata Carriazo, ed., *Hechos de Condestable don Miguel Lucas de Iranzo*, 1940, 76–8.

[2] *Iranzo*, 84; Gayangos, *The History of the Mohammedan Dynasties in Spain (Al Makkari)*, 1843, II, 354.

[3] *Iranzo*, 145–6, 204 ff.

[4] *Iranzo*, 256, 273–84, 304.

and the garrison abandoned the walls and retreated to the castle (*alcazar*), which the royalists attacked with some kind of gunpowder weapons. But the constable had not brought siege artillery with him, and when the citadel resisted, he had to give up his attempt.[1]

The following year, an attack on the constable by rival nobles gave rise to another example of siege warfare. A party of opponents secretly entered Jaén through the Old Castle, one of three (no longer existing) that stood on the circuit of walls, and occupied it and the castle called Abehui. The constable, from his base in the New Castle decided to force the issue and organised an attack on the castle of Abehui. Two large bombards were mounted to fire on the castle, and quickly blew a breach in the wall. As the assault was being organised, the defenders of both sites capitulated. The constable set about repairing the damage, but also improved the defences by building a new postern gate and clearing away the mass of buildings around these old castles that had permitted the enemy entry in the first place.[2]

The Nasrids now made their own fate inevitable with a dynastic civil war of their own, with the new ruler Boabdil's (Abū 'abd Muḥammed XII's) uncle, the previous king, becoming an instrument of the Christian kingdom, alongside a suicidal act of breaking the truce. The well-recorded war that followed was nonetheless a testament to the defensive strength of Granada and the fierce courage of its inhabitants, as it took nine years for the new joint monarchy of Castile and Aragon (in the persons of Isabella of Castile (1474–1504), successor to Henry IV, who married Ferdinand V of Aragon (1479–1516)) to reduce the last Muslim state in Iberia, despite the absence of any overseas help. The conquest was overwhelmingly a progression of sieges by which one by one, region by region, Granada was conquered until only the capital itself remained. A variety of techniques were employed.

1482: capture of Alhama, defence of Loja

In 1482 the town and castle of Alhama were taken by surprise in a bold attack by the marquis of Cadiz who led his army of 4300 deep into hostile territory. Spies had alerted him that the rock-top castle, because of its seeming impregnability, was lightly guarded, and after a three day march a silent escalade two hours before dawn on 28 February killed the sentries, opened the gates to the rest of the army, and took the castle. The town was awoken by the noise and its inhabitants manned their walls and hastily built barricades to face the now enemy castle, armed with crossbows and arquebuses. The Castilian forces still outside now attacked the town walls with ladders, simultaneously with a sortie by the men in the castle, who seem to have been forced to open up a gap in their walls, since the castle gates were under sustained fire from the townspeople. The remarkable pictorial evidence of the siege preserved in the relief in the cathedral of Toledo shows gunners firing a large bombard, standing at about 45 degrees and pointed at

[1] *Iranzo*, 315–16.
[2] *Iranzo*, 338–50.

the city's gate, while another artillerist prepares a shot.[1] After a day's fighting the whole place fell. News of the catastrophe having reached Granada, troops raced from that city to try to recover the loss, but their attack was disorganised, and the Castilians were able to repel it with immense loss. An attempt to sap the walls was similarly a costly failure. Then, with a little more thought, the Granadans tried to divert the river that gave Alhama much of its water, and by posting a force of crossbowmen on the far bank were able to inflict heavy losses on the garrison as it tried to fight its way to the near bank. The marquis was also forced to increase the height of the defences on this side of the town by creating an extra barricade made of the town's wooden doors on top of the battlements. Shortly, the duke of Medina Sidonia arrived with relief.[2]

Next, Loja was the target of the king in person. Assembled supplies included provisions as well as large guns and gunpowder, and a fleet to intercept any relief from North Africa. Despite professional advice, however, it seems that Ferdinand was overconfident of his superiority, and he marched long before the whole army was ready (July). The city was strongly sited and the countryside around was scored with ravines and gorges. In addition, Ferdinand was careless in the siting of his batteries, making most of them useless. He was advised by his siege warfare expert, the duke of Villehermosa, to relocate his camp, but decided too late. A successful attack by Granadan troops put his army to flight. Some say that he left behind much of his artillery, others that he managed to rescue most of it, but in any case the first siege of Loja was a fiasco from the Christian viewpoint. Ferdinand scarcely compensated by recovering (in October) the castle of Zahara by secretly organised escalade, the capture of which by the Muslims had been the act that broke the truce originally. The Marquis of Cadiz, who commanded the attack, in fact had to lead a three-day assault using siege equipment to get his men to the foot of the walls, and sustain a bombardment with firearms and crossbows before the castle was yielded.[3]

1484/86: Alora, Ronda, Cambil, Illora, Moclin

The major objective in 1485 was the seaside fortress of Malaga, a significant centre of the kingdom the capture of which would also serve the strategic purpose of making communication between Granada and the sea very difficult. There was plenty of territory inbetween, well provided with castles and walled towns. Deploying their artillery effectively, the Spanish army was successful in reducing each

[1] The panels of the relief are reproduced in many modern histories. See e.g. J Vigon, *Historia de la artilleria Española*, 1947, I, 66.

[2] Gayangos, II, 371–2; Washington Irving, *A chronicle of the conquest of Granada, translated from the MSS of fray Antonio Agapida*,1829 republished 1986, I, 28–55. This account is despite its title a compilation of several contemporary chronicles and its stylish and embellished literary panache compensates for its easily recognised rhetorical flourishes; Anon, *Historia de los hechos de Don Rodrigo Ponce de Leon, Marques de Cádiz*, 1893, 204–8.

[3] Gayangos, II, 371–3; Irving, I, 81–93, 225–31. Makkari's text used the Morrocan Arab version of the word "midfa", definite confirmation that this was gunpowder artillery (Gayangos, II, notes). Zahara: Anon, *Hechos de Don Rodrigo*, 228–9.

of the smaller sites in their path, generally blowing breaches in the walls prior to assault, at which point, normally, the outnumbered garrison, if possible obtained terms allowing them to leave for Granada. One such attack was made on the fortress of Alora, standing on the way to larger objectives, and the marquis of Cadiz was sent ahead with a force to blockade it in June 1484, while the king followed with a massive train of carts carrying the artillery. The artillery being set up, the garrison was summoned to surrender but refused, replying with crossbows and firearms (hand weapons). The Spanish guns having knocked down two towers and great lengths of wall, and their ribaudequins and other lighter firearms having prevented the defenders from constructing a new defence of tapia behind the breach, and the attackers visibly preparing the mantlets and wooden towers to launch an assault, the Granadan troops inside required their *alcaide* to negotiate surrender.[1] The capture had taken only ten days.

The victorious Spanish army then tackled two closely linked cliff-top castles at Cambil and Albahar, which were joined by a bridge. A large cavalry force was sent ahead to set up a loose blockade to prevent the defenders re-provisioning. But to launch an attack on them called for the mobilisation of an immense force of labourers (6000 are mentioned, paid for by the bishop of Jaen) to clear the way through for the army and its artillery train, constructing a new road to get anywhere near the tops of the hills. The engineer Francisco Ramirez de Madrid organised the placement of the guns, the heavy weapons demolishing the walls and towers while lighter guns were used to discourage the defenders from trying to repair the breaches. Prudent surrender followed.[2]

The attack on Ronda was skilfully organised so as to be a surprise, the defenders being few, as Loja was believed to be the target. Its defences were formidable: an isolated rock-top citadel with triple walls and ravines on three sides, a town with two walled suburbs. The walls had been erected under the Nasrids, and included very solid walls and towers, and massive gateways (two of which survive), using bent entrances and the *torre albarrana*. The royal historian interestingly noted that the local people were trained from childhood in use of the crossbow. But the absence of much of the garrison was critical. Ferdinand arrived before it, deployed his army around the town and the castle in two distinct siege camps, and set up his batteries of heavy guns in three locations which, in four days, had demolished three towers and much of the outer wall of the first suburb (specifically, the wall and tower tops, and the intervening curtain wall, which made it impossible to defend), and this was then stormed. The artillery (and specifically the *lombardas* and *quartadgos*, on which see below) was now moved forward and set up against the wall of the city proper, in positions recommended by the master gunners. Returning forces of Ronda's army tried to fight their way in but were held off by the besiegers' well-protected siege camp. As the fighting progressed, the Spanish apparently discovered the source of the underground water supply that enabled the inhabitants of the rock to draw up water, and blocked it. Meanwhile, the guns

[1] F del Pulgar, *Cronica de los reyes Catolicos*, ed. J de Mata Carriazo, 1943, II, 120–22; Anon, *Marques de Cadiz*, 240–1.

[2] Pulgar, II, 126–8; Gayangos, II, 376; Irving, I, 302–11.

were now firing both iron and stone shot, including new types of ammunition being manufactured in the camp, and incendiary balls of tow, set in pitch, oil and powder, which could not be extinguished on landing. One of these was thrown by an engine with a terrifying noise over the wall, and the citizens, disheartened, decided to surrender.[1]

A similar fate was visited on Loja the following year, this time the king approaching the city with greater prudence. The siege camp was properly organised, with wooden bridges being constructed to allow communication between its various quarters in the rough terrain, and care being taken to defend against possible sorties. The guns were again decisive in wreaking destruction on the defences while incendiaries rained down on the houses inside. Loja surrendered after two days and nights of bombardment which had wreaked destruction and caused many casualties and much slaughter.[2]

The Spanish army marched next (June 1486) on the castle and town of Illora on its hill top, just four leagues distant from Granada. Illora had been often attacked before, most recently and in vain by Henry IV in 1456, and was an ancient structure frequently rebuilt. The walls were of concrete or tapia, with large rectangular towers at the corners of two circuits of walls, one of them larger and clearly the donjon. Attacking from two sides, the Spaniards managed to occupy the town, then set up three batteries each of eight bombards against the castle, the walls of which collapsed under the impact leading to surrender, not being capable of resisting guns the way they had resisted all previous attacks.[3]

The final victory of the campaign came against the castle of Moclin, on a high hill, and this time reinforced with additional defences in the form of trenches and bulwarks. Photographs show an irregular circuit of walls following the terrain of a sloping ridge, with a high-walled citadel with rectangular towers at the top of a steep incline.[4] The defenders also had guns, although their ribaudequins and *buzanos* (Pulgar) could not match the size of the Spanish battery once again comprising three separate sets of heavy weapons. These were effective in demolishing the tops of the walls and towers while the lighter artillery once again was used to interdict attempts by the garrison to repair the effects of the bombardment. Nonetheless, the fall of the castle came about when an incendiary flying threw the window of a tower ignited the castle's gunpowder magazine, causing such an immense explosion that none of the garrison had ever seen before, and inducing surrender. In these operations, the approach of the Spanish had been to send an advance guard to ring the threatened site and prevent any movement in or out, while the main royal army, accompanied by the heavy siege artillery, made its rather slower progress to the target.[5]

[1] Gayangos, II, 375–6; Irving, I, 260–66, 269–76; Anon, 243–44; Pulgar, II, 164–72. For modern descriptions of Ronda, G Goodwin, *Islamic Spain*, 1990, 97–100; R Collins, *Spain: an Oxford archaeological guide*, 1998, 226–7.

[2] Anon, 253–4; Irving, I, 349–63.

[3] Dotor, *Cien castillos famosos de España*, 196–7; Anon, 255.

[4] Echague, 14–15, 18 (plate 23); Anon, 258.

[5] Gayangos, II, 376–79; Irving, I, 364–81; Pulgar, II, 233–5.

1487: Velez Malaga and Malaga

The way was now clear for Ferdinand and Isabella to attempt the decisive blow against Malaga. First, in April 1487, they laid siege to Velez Malaga. The Muslim source ascribes the loss of the town to the civil war that erupted again in Granada, preventing any organised attempt at relief. The Spanish chronicles described an immense expeditionary army with a very large artillery train, which had to be conveyed to the target across inaccessible mountain roads further hindered by the damage caused to roads and bridges by the seasonal rains. The Spanish rulers hired 4000 pioneers to create a way through the mountains. Meanwhile, as they worked, the vanguard of the army was in place around Velez Malaga, and a fleet stood off the coast. The suburbs were stormed, although only after a six hour fight, and then turned into the base for further siege operations against the main part of the city. After ten days' hard labour and very slow progress, the effort to transport the heaviest guns was postponed and work was concentrated on moving the medium and lighter weapons. They finally arrived the day after a failed effort by a Granadan relief force to break through the siege camp, and the sight of the "lombardas, ribodequines, catapultas" was enough itself to bring surrender (27 April).[1] We will consider below the likelihood of the reference to catapults.

Nothing now stood in the way of a siege of the large, wealthy and well-fortified city of Malaga. Some accounts suggested that leading merchant citizens would have been willing to surrender the city peacefully, but were prevented from doing so by the ferocity of the commander of the garrison of the separate ridge-top fort of Gibralfaro, high above the city and which is connected to the *alcazaba* proper (itself with massive double-walls, towers and layers of gates) by a covered zig-zagging walled passage. A second *alcazar* stood on the city walls.[2] Malaga was well provided with provisions, arms and munitions, and fighting men.

King Ferdinand arrived in May, and posted his army in three encampments, the marquis of Cadiz being given the honour of attacking Gibralfaro. By sea, the fleet closed the harbour and threatened attack, while by land the troops set up siege lines of bulwarks and trenches while the workers could be heard day and night preparing the engines and the stone shot. The heavy guns were landed from the ships, and five of them were placed against Gibralfaro. A further battery of seven was called the "seven sisters of Ximenes". The defenders were diligent too, and had their own weapons of all kinds with which they maintained heavy fire on the besiegers, forcing them to work at night. These included *lombardas*, other fire-arms, and master gunners. Two ships arrived from Flanders in which the son of the Holy Roman Emperor was travelling, bringing with him two more heavy guns and all their necessary equipment and operatives (Pulgar). The bombardment was intense. Soon, one of the towers of the suburb was wrecked by artillery, and stormed. But the resistance was fierce, and the Spanish had to return to the attack

[1] Gayangos, II, 380; Irving, II, 3–40; Anon, 275; Pulgar, II, 261–78. The accounts do not tally exactly but all tell the same broad story.

[2] Torres Balbas, 499–500, summarises the defences of Malaga as being even more substantial than those of the larger Alhambra at Granada.

with ladders before finally capturing it. Granadan miners then counter attacked, demolishing the tower, but this created an even larger breach through which the besiegers stormed, and after a fierce two day battle the suburb was secured. Meanwhile, the marquis of Cadiz thought to take advantage of the breaches created by his artillery in his sector, and moved his lines closer to the wall of Gibralfaro, but was overrun in a devastating sally and forced to pull back to the original position. This time, the defenders were able to block the breach with palisades and ramparts of tapia.

The attack was sustained over several months, the large besieging army consuming great quantities of munitions and provisions and calling for a major logistics operation. Supplies of gunpowder were shipped in from Valencia, Barcelona, Sicily and Portugal to feed the batteries, in addition to which more gunpowder was manufactured in the camp, and carefully stored underground to protect it from the defenders' attentions. More old fashioned forms of weaponry were also used. Wooden siege towers, each capable of holding one hundred men, were constructed, along with many tortoises, or mobile mantlets. Mines were driven in towards the ramparts. The commander El Zegri led a vigorous and determined defence. It was noted that even if the weight of modern artillery attack had once been a surprise to the Granadans, they were by now accustomed to it and had worked out how to deal with its effects. Earthworks were thrown up behind the breaches, regular sorties were launched, and a defensive fire was kept up. The mines were located, and a countermine was dug which destroyed one of them.

But they could do nothing about the developing famine, as the city's supplies dwindled away. Meanwhile the attack was relentless. The chronicles tell of one unusual method deployed. When the chief of artillery, Francesco Ramiro de Madrid, was given command of the attack on one of the city's fortified bridges, he had a mine dug under the tower on the outer side, then moved in a gun (a *quartadgo*), had it arranged pointing upwards, and then fired it via a trail of gunpowder, the resulting shot breaking the tower's foundation, leading the defenders to flee to the inner end of the bridge, from whence they kept up a fire with their own heavy guns. But eventually, with the citizens hungry and mutinous, and having failed in a final sally to disrupt the siege, El Zegri withdrew to Gibralfaro while the city capitulated unconditionally, the final citadel giving up two days later.[1]

1489–91: Baza, Almería, Granada

In the early summer of 1489 the Spanish army descended on Baza. This siege was different again. The town, which stood in a plain, was surrounded by dense orchards and gardens, and in this case they served the occupants very well as an outwork, forcing the besiegers to pull back beyond them after bloody skirmishing and a failed attempt to construct a siege fort in which to base the artillery. This left them too far out to do any damage to the town, and also facing a very long circuit if they were to close off all avenues. But they did not give up. Instead, hav-

[1] Irving, II, 56–134; Gayangos, II, 380–1; Anon, 279–86; Pulgar, II, 283–325.

ing ensured a regular supply of provisions, they set about first cutting down the orchards, a long and wearying task, while they also set about constructing several well-protected siege camps on opposite sides, that would eventually be linked together with temporary forts and ditches. Finally, as the year progress, they built what had all the appearance of a permanent siege camp to enable them to remain there over the winter. With their own supplies now low, including the stock of arms and gunpowder, the citizens had no choice but to capitulate (4 December).[1] Soon after, Guadix and Almería were surrendered to the king without resistance (4 and 24 December). Many of the defenders of the latter had been sent to Baza, but the city was powerfully fortified.

Of the once proud kingdom only the city from which it took its name remained in the hands of its rulers. In 1490, with the Spanish occupied elsewhere, the king of Granada regained a little of his honour with the successful attack on the castle and town of Hamadan, which he took after bombardment with artillery demolished four towers including the principal, and the garrison of 180 surrendered.[2] But the coup de grâce was administered in 1491 with the final invasion of Granada and the siege of the city itself. The blockade lasted seven months, Ferdinand and Isabella perhaps concluding that there was no need to risk the heavy casualties and severe damage of an assault. Instead, with a large population consuming their diminishing resources, the Granadans would eventually have no choice but to negotiate, and as the Spanish repeated their action of building a permanent siege town (piously named Santa Fe) as winter approached, that is what happened. The end of Muslim al Andalus after seven centuries was remarkably tame, though the process of conquest had been immensely demanding of resources.

Siegecraft in the conquest of Granada

The war in Granada had served as a training ground in some respects for a new era. But in other respects, this war was also a last show for some quite ancient methods and weaponry. The fighting had concentrated overwhelmingly on sieges of major cities. Their defences had sometimes been reconstructed by the ruling dynasty, but in most cases the work dated from before the builders had become familiar with artillery. Thick walls could not resist sustained and accurate bombardment, and one feature of these campaigns is the arrangement of batteries of several guns together to concentrate on single stretches of wall or tower. Even where the walls were strong enough not to be breached, on several occasions the eye witness accounts told of the demolition of the battlements first, the effect of which was to make the wall useless as a defensive position. But it is also clear that the sustained and heavy bombardments were capable of doing more, and of creating breaches in the city and castle walls. The way to block up the resultant holes with temporary works was known, but the deployment of lighter guns to keep up a permanent interdiction of the breach to prevent the Granadans carrying out

[1] Irving, II, 170–240; Gayangos, II, 382–3; Pulgar, II, 413–23.

[2] Gayangos, II, 385.

such works is something we have seen repeatedly. It is apparent that a high level of skill in modern siegecraft was present.

During the struggle, the size of the Spaniards' artillery train grew relentlessly. Guns were in use in Spain right back to the mid-fourteenth century, but they were not consistently present until a century later, and even then, they were not regularly deployed. Whereas the treasury of Castile paid four gunners in 1479, the total of royal gunners was 14–17 in 1480, and 65 in 1482. During the same period, the manufacture of guns was begun at Córdoba, Medina del Campo and Ecija, and French and Breton engineers were hired to help. In an inventory of 1495, 179 firearms were listed in the arsenals of the conquered province, most of them dating back to the war of conquest. As elsewhere, the king seems to have taken a personal interest. The title of Captain of Artillery first appeared in 1465, but the establishment of an actual artillery corps is dated to 1479, and a sum is assigned to its salaries. The records show various individuals being paid for being "in charge" of the artillery from the turn of the century (such as Diego Rodriguez de Zapata (1406), Fernan Rodriguez de Monroy (1429), and numbers of *maestros bombarderos* starting in 1463. The significant figure in the conquest of Granada was Francisco Ramirez de Madrid, who had served King Henry IV and Queen Isabella and had been appointed in charge of works to the royal castle of Seville (1478) before becoming captain of artillery (1483). His skill and courage in the sieges described here led to his being knighted in 1487.[1]

Over time, the quality and accuracy of the guns improved. The types of weapon described are similar to those known from other parts, and the international character of the trade of master gunner was as common in Iberia as anywhere else. Although the records suggest the trend towards lighter and more portable weapons also featured here, and the scale of the task involved in transporting the *lombardas* was a common complaint, the particular and decisive role ascribed to the heavy weapons throughout this war was also clear. Among the unusual weapons that frequent the contemporary accounts should be mentioned the *quartadgos* (also *cortaos, cortagos*), which also appear in the Toledo cathedral reliefs. These seem to have been, in effect, mortars, weapons that lobbed shots on a curving trajectory over a wall, and played a major part in most of the sieges. They were however, as Pulgar took the trouble to note, remarkably inaccurate, sometimes falling inside, sometimes outside the targeted town (although see also the description of the possibly identical *courtau* in the glossary).[2]

If the kings of Granada had heavy artillery, it featured not a lot in the final conflict: the references have been to hand guns and crossbows. For a long time, the kingdom had been engaged mainly in raiding rather than wars of conquest, which might explain their apparent lack of concern with modernising their armoury, but every student of the struggle has agreed that the artillery of Ferdinand and Isabella (and the organisation that accompanied it) was a critical factor in determining the outcome of the particular sieges that characterised the war. The presence

[1] Vigon, 105–9.

[2] Vigon, *Artilleria*, I, 26–36, 41 (cortaos); M A Ladero Quesada, *Castilla y la conquista del reino de Granada*, 1967, 117–18; Pulgar, II, 235.

of heavy guns in the defences of Moclin and Malaga was sufficiently unusual as to merit specific mention by the Spanish accounts.[1]

Older weaponry was called upon, and the examples already given cite various examples from the chronicles. The construction of wooden shelters to cover approaches to the walls, the use of tortoises and siege towers, as at Malaga, the devising of new versions of old weapons, such as the *royal ladders* consisting of platforms that could be moved and then raised by poles against a wall, with shields covering men with crossbows and handguns on top, all featured. The continued role of the crossbow alongside the handgun (arquebus) is notable, confirmed both by the written records and by the Toledo cathedral reliefs, which show the two weapons in use side by side. Perhaps most remarkable of all are the references to the continued use of old siege artillery, in the form of mangonels, trebuchets and their cousins. It would seem either that these references are anachronistic inventions of the chroniclers or else simple mistranslations of original texts. Around 1450, the *lombarda* came into use as a new type of weapon and was called in some places a *lombarda trabuquera,* presumably in an attempt to describe the effect of its action.[2] But its regular presence from now on in the Spanish army does not prove that the trebuchet was still in use, an extremely unlikely hypothesis. And while it is not entirely impossible that mechanical slingshot weapons could still have been useful to hurl incendiaries, or items not intended to cause damage such as messages, the availability in large numbers of gunpowder weapons able to perform the same functions but with greater range is a telling argument against the idea. Nor do any of the illustrations of the sieges of the time indicate anything other than gunpowder artillery.

Developments in fortification

It is now time to attempt a review of the changes in techniques of fortification during the middle and final thirds of the fifteenth century. This is well-trodden ground, but, as the still very valuable study by J R Hale states,[3] definitive conclusions have still to be established on a number of issues. Partly this arises from the large number of sites still standing, but in many cases heavily altered. On the other hand, because of their continued occupation and use, many other fifteenth century works have been destroyed. Comparing works across Europe and the middle east has been done, but is vitiated by the competition among many earlier students leading them to "stretch" evidence in order to establish that their particular nation was definitely the originator of the new style. The period is certainly one of transition, but the builders and designers were not tied to any

[1] Quesada, 117–28. This study also collates evidence from archives and chronicles for the use and effect of the weaponry, its development and the required logistics.

[2] Vigon, 36, 41. Toledo cathedrals reliefs showing the *lombarda* and a smaller, wheeled long-barrelled gun at tablets 26 and 28, illustrating an unknown castle, and Malaga.

[3] J R Hale, The early development of the bastion: an Italian chronology *c.* 1450–1530, reprinted in *Renaissance War Studies,*1983, 2–10.

single state or ruler, and their ideas spread rapidly across both the Christian and Muslim worlds, so the search for the "first" may continue to be fruitless.

We will not fail to use the word "bastion" because of its later, more specific meaning. Bastion is a perfectly good general medieval term suggesting sometimes an outwork, either just another word for a tower projecting from a curtain wall, or a separate fort. Throughout the fifteenth century, bastions were one of a number of types of construction devised by engineers to cope with the growing power of gunpowder artillery. Time and again, we have met with defenders ordering the erection of temporary works in front of their medieval walls and gates. These often took the form of earthworks and palisades, provided with emplacements for artillery, and serving a number of purposes including that of keeping an enemy from being able to blast their way directly through the older masonry behind, while allowing (in many cases) for gunners and crossbowmen to provide a second tier of fire from the top of the wall behind. Naturally, these works did not last. On a number of occasions, however, as with the work done at Metz after the failed siege of 1444, the temporary works were replaced with permanent stone works afterwards, such as the *Porte des Alemans*. Such work amounts to improving existing defences, and was constrained by the presence of often very massive older work. While this may now have been out of date, it would quite often have been of fairly recent build, and the prospect, and cost, of replacing it was probably not an option for most towns and cities, and certainly not for most castle-owners. To find what the engineer of this period would do when given a blank sheet, therefore, it is necessary to find examples of work that was done from scratch. The examples there are confirm for certain that this was an age of transition.

Theory

Despite being a high period of the renaissance, there is actually not much in the way of written evidence of how designers and architects were thinking about fortress design. What there is, too, tends to look as much backwards as forwards: as "renaissance" has been understood to mean a desire to reach back beyond medieval obscurantism to the giants of the classical age, a return to the ideas of Vegetius and Vitruvius was only to be expected. However, given that the pressing need for new fortifications was to be able to resist a weapon entirely unknown to the ancients, how were the two things to be contained within a single approach?

The existing role of artists and engineers (a distinction incidentally not recognised by the age) in designing new fortifications has already been signalled with the role of Brunelleschi in the works done at Pisa. Other renaissance figures like Giotto have also been assigned the responsibility for new works such as that at Lucca by their historian Vasari. From the other side of the wall, Taccola, whose work on machines was published in 1449, has also already been identified for his clear understanding of the functioning (and weaknesses) of artillery. It is however from the 1440s onwards that artillery became an even greater challenge because of its greater accuracy and range, its greater destructive power, and critically, chronologically starting with the French but quickly becoming a feature of everyone from Burgundy to the Ottomans to Iberia, the development of permanent professionally organised state-run artillery trains. How would a fortress newly built from 1450 respond? Two Italian experts who wrote during this time should give some clues.

Alberti wrote his *Ten books on architecture* around this date, for Pope Nicholas.[1] He himself proclaimed how he was developing the ideas of Vitruvius, and was the first to do so. He was chiefly renowned as a mathematician. What is absolutely clear is that most of his arguments take as their authority the ideas of the ancients. Angled corners were vulnerable to siege engines, as Vitruvius had long since identified, and the best pattern for a curtain wall was one of "undulating bays", enabling effective flanking and presenting no corners: but even the authority cited for this recommendation was Tacitus.[2] In indicating that the shape of the fortress might be round, triangular or square, he wrote that the most important thing was to make sure that advantage was taken of the available site, and that the advice of Vegetius was followed, that the wall walk needed to be wide enough for two people, and the wall high enough to prevent escalade. The best defence against mining was a deep ditch, either wet or on hard rock underneath. The importance of flanking was well understood and projecting towers were needed every 50 cubits, open at the back, and provided with their own projections that were in effect machicolations. Inside the city, a separately walled quarter for the patrician class reflected perhaps the interests of his patrons, while the chief purpose of the citadel seems to have been as a separate strongpoint for the safety of the ruler.[3]

At a slightly later date, Avelino Filarete composed a treatise on architecture for Italian patrons too. His work pays more attention to contemporary realities than Alberti's. He provides a detailed blueprint for the building of a town, starting of necessity with the defences. These needed to comprise both an inner and outer wall with a ditch. Whereas the outer defence should be just six *bracchie* up to the merlons, and four thick (1 *bracchia* = 58 or 60 cms), the inner wall must be taller, and at least 20 *bracchie* distant, in order to reduce the effectiveness of enemy bombards against the masonry. Projecting towers were either round or rectangular, provided with distinctly old-fashioned loops designed for bows, and actually offering very little flanking. Again, the concern in the design of the citadel is solely with the security of the ruler, starting with the proposition that the operation is based on a labyrinth.[4]

There is nothing in these textbooks to suggest that around 1450–60, architects offering plans for the ideal city were paying any serious attention to the weaknesses of defensive plans that had been common for a very long time against modern artillery. Even where Filarete acknowledged the importance of an outer

[1] L B Alberti, *On the Art of Building in Ten Books*,trans. Rykwert, Leach, Tavernor, 1991, a recent edition but one of many.

[2] Alberti, 100 (book IV, chap. 3).

[3] Alberti, 102–5, 117–19 (book V, chap. 1), 122–5 (book V, chap. 5).

[4] W von Oettingen, *Antonio Avelino Filarete's Tractat über die Baukunst*, 1890, 148, 150–4, 160–1, 166–7, 178–80.

wall to protect against the wall-breaking power of the bombard, his own prescriptions make no allowance for the use of artillery in defence. This separation of the architect from the real world was a tradition that even the greatest would fall prey to, as with the drawings of fantasy fortifications by Leonardo da Vinci in the period 1480–1500.[1] The answer to the question of how the ancient and the modern were to be reconciled was not by following the instructions of Alberti and Filarete.

It is instructive to compare these proposals of the ideal with the proposals of someone who not only had knowledge of principles and the skills of an architect, but who also had attacked and defended towns and castles with the weapons of the age, and was able to express his learning (from practice) in writing. Such was someone we have encountered before, François Surienne (died 1462), known as L'Aragonnais, servant of several masters but also master of his trade. The Burgundian town of Dijon preserves a manuscript in which Surienne gave the burghers his recommendations for modernising their fortifications, on 25 June 1461. The town had been walled in the twelfth century, and rebuilt after 1358, at which time some fairly weak *boulevards* has been erected in front of the gates. Now, the council was asking for advice on how to bring their walls up to date from an acknowledged expert in artillery warfare. The chief proposals were these. The town had an outer ditch, created in the thirteenth century as an obstacle to mining, but now offering an attacker a ready-made approach trench: it should be filled in. In its place, a new outer defence should be constructed comprising a glacis raised from the outside of the inner ditch and sloping downwards. This would be covered by fire from two levels, lighter guns on the wall walk, and heavier guns placed in newly constructed emplacements (*canonnières*) at ground level, firing through embrasures in the town wall, but at sufficient height to cover the glacis and boulevards. Wall towers should be constructed, standing in the ditch and carefully provided with gun loops providing both forward fire and flanking fire. They should be straight-sided but with a rounded front, and were to hold serpentines and culverins, that is, smaller calibre but longer range artillery than the big bombards. The *boulevards* were to be provided with *canonnières* too, and arranged so as to provide flanking fire to cover the possible approaches to the wall.[2]

The advice seems not to have been taken. But the continuing movement in ideas is apparent. It may well be that Surienne did not think beyond suggesting transforming the existing old walls, or it may be that he had instructions as to what was feasible. But what he proposed was very much in the same tradition as the previous half century. What was changing was the conscious development of new methods of using modern artillery for effective defence, within old types of wall and tower. But the short memorandum is in marked contrast to the works of the two Italians, whose proposals scarcely acknowledged the significance of modern siege methods.

Francesco di Giorgio Martini was born at Siena in 1439 (and died in 1502). His text, which dealt with both civil and military engineering, may have been drafted

[1] Hale, 22–3.

[2] J Richard, Quelques idées de François de Surienne sur la défense des villes à propos de la fortification de Dijon (1461), *Annales de Bourgogne*,16, 1944, 36–43.

during the period of the struggles in which he himself was engaged, involving the major powers in Italy in 1479–82. He worked as an architect, engineer, sculptor and painter, and much of his writing deals with aqueducts and hydraulics generally, so he was broadly experienced.[1]

Giorgio's writings share his predecessors' commitment to a firm base in the classical authors, but also address modern needs directly. He wrote of the pressing need for a new fortress to deal with the devastating effects of attack by artillery, making the obvious point that this was something unknown to the ancients, whose experience of siege weaponry he indicated quite accurately. The drawings surrounding the manuscript include not only a range of different types of gun being deployed in a siege in the kind of emplacement that had been familiar for a long time, but also illustrate how to use a mine charged with barrels of gunpowder to devastate the foundations of a tower. The text explains that the defence against modern weapons required a double wall and large ditch, and places a lot of emphasis on the outworks, in particular ravelins *(rivellini)*, taking the form of lower, angled or semi-circular masonry defences in front of gates. Also prominent are caponiers stretching across the ditches, pierced with gun loops and positioned to sweep the whole of the ditch. Another drawing shows how to construct an internal defence behind a masonry wall shattered by artillery: in a curve between two strongpoints where the earth and timber structure joins the original stonework, embrasures have been provided for guns to sweep the space between, revealing a clear understanding of how to create a killing ground in the area where an enemy would expect to launch an assault through their successfully created breach.[2]

Giorgio's drawings of the ideal fortress show what he believed the modern castle should look like. They are a combination of modern and traditional. The traditional include machicolations and tall great towers. The modern include tiers of powerful defences based on semi-circular or pentagonal bastions, sometimes with scarped or sloping walls, something that represented a very effective defence against artillery, but rendered useless the machicolation simultaneously decorating the rampart above. Although the role of small guns in close-up defence featured in many of the designs, be it in the caponiers or in the flanks of the towers, the drawings offer no evidence that Giorgio had thought about the potential of his towers and bastions to become platforms for heavier guns. Although sometimes these are drawn at the same height as the wall walk, giving the design a similar appearance to the bastion trace common to the following century, this absence (confirmed by a suggestion that the towers in some of the designs are in fact hollow, while others fill the roof space with close-range defensive structures), alongside the presence of tall machicolated interior towers, suggests that even by the 1480s, military engineers had not yet worked out the most effective way to compete with

[1] Francesco di Giorgio Martini, *Trattati di architettura ingegneria e arte militare*, ed. C Maltese, 1967. See also the comments of Hale, 20–1, who dates the work to 1495.

[2] Giorgio, 3–7, and illustrations at ff. 5, tav. 6, 57 tav. 105 (bombard emplacements), 55 tav. 102 (mine), 64 tav. 119 (defence of a breach).

modern artillery.[1] But Giorgio's work show him grappling with some of the most pressing questions facing a fortress designer.

Practice in France and Burgundy

Fortunately, those who were actually building the walls seem to have had an appreciation closer to that of L'Aragonnais. New fortifications from the 1440s onwards in France and Burgundy offer plentiful examples.

The reality was varied. The city of Lille was Burgundian, and in an area that was from time to time in conflict. The rulers of Lille spent large sums of money over the years in maintaining their defences, and around 1400 had moved to purchase guns and projectiles (of a standard calibre) from Bruges and Antwerp. But the only real extra building work done here was to construct *boulevards* . The city already had a very long circuit of walls (4,500 metres) such that it would take a very large force to enclose it, and had very strong natural water defences. Although the fact that it was not besieged is not relevant to a discussion of the defences, it does suggest that its fine location (and perhaps well-maintained walls and effectively organised defenders) served as an effective deterrent.[2] This city, therefore, although in an area of potential or actual warfare on the frontiers between France and Burgundy much of the time, did not feel the need to upgrade its defences. Nearby, the city of Valenciennes also undertook some work to upgrade its defences during the fifteenth century, and a small survivor of this building remains from the massive sixteenth century reconstruction of the enceinte. The restored tower "de la Dodenne" was designed to achieve two functions. One was to stand over a sluice built to control the flow of one of the branches of the river Rhonelle which was prone to flooding. This natural threat was turned to the advantage of the defence by being diverted into the ditch, so that a substantial stretch could be turned into a wet moat. The other new work was the insertion of gun loops in the tower, but these were small embrasures designed to take *veuglaires*. Otherwise, this tall round tower represented nothing new in defensive technique. Later in the century, the city's citadel was provided with quadrilateral *canonnières* as a belated reflection of modern design.[3]

Quite in contrast to the relative conservatism of the work done at these important cities was the castle of Montcornet in the Ardennes, which came into the hands of the powerful Croy family from 1446, a kindred that served both the dukes of Burgundy, and later Louis XI of France. The original castle, of twelfth century origin, was completely rebuilt. It occupied a rocky spur 170 metres long and 50 wide, divided into a central mass and a lower court. The outer face of the central block was immensely thick, and provided with round towers, linked by an

[1] Giorgio, ff. 4v, tav. 4, and 5v, tav. 6, alongside numerous other drawings in which similar designs for castles and towns feature.

[2] D Clauzel, Lille et ses remparts à la fin du moyen âge (1320–1480), in Contamine, Guyotjeannin, (eds.), *La guerre ,la violence et les gens au moyen âge*, 1996, 273–93.

[3] A Salamagne, *À la découverte des anciennes fortifications de Valenciennes*,1999, 47, 68, with fascinating old photographs of the now demolished defences.

internal passage, and equipped with substantial gun emplacements. It is dated to the period between 1450 and 1475. Below it, the court features another massive tower, three storeys above the basement, protected internally with its own ditch and drawbridge. Its upper floors are liberally provided with loops for small guns. On the other side of the central block, a boulevard projects, itself of three levels and with ramparts varying between two and 5.3 metres thick, provided with 18 embrasures. This part is dated to around 1477–80. Montcornet's historian has compared it with the similar (disappeared) fortress built at Ham by the count of St Pol (1418–75), in which the predominant feature is the immensely powerful round tower constructed for artillery. Also noticeable as a feature is the large scale provision for smaller guns, suggesting a greater reliance on more rapid fire with smaller ammunition than hitherto.[1]

Nearby, in the province of Hainault, a similar trend in the development of towers supports the case for there being a general movement towards the substantial thickening of walls during the second half of this century, combined with a change in defensive techniques in which defence of the foot of the wall by means of hoarding or machicolation became less of a concern. The castle of Havre's walls are 3 metres thick, those of Solre-le-Château 3.2 metres, and those of Jeumont 3.9 metres, by contrast with the average one to 1.8 metres of previous times.[2] In the bishopric of Liège, the old castle of Logne in Stavelot, which stands on a summit and had been in regular defensive use since the tenth century, was rebuilt for artillery around 1480, with thickened walls, and new round towers, and altogether 56 *canonnières*.[3]

These Low Country examples are interesting in demonstrating that the period of transition towards the new style of fortification that would become established everywhere in the next century occupied several decades, and was not restricted to any single territory. If our study were to continue over a few more years, in fact, it would show that some of these sites (such as Franchimont near Liège) would be given not bastions, but low-lying and solidly roofed projecting casemates after 1500, and gun towers would also continue to feature. For small fortifications such as these castles, therefore, it was certainly not apparent that such methods were necessarily inferior.

The examples from within the French kingdom have been well studied. It is striking how plentiful they are. A particular selection of powerful new works can be found in the area where Brittany borders France, plagued by warfare in the 1480s. The temporary works of Surienne at Fougères had by now been replaced by the duke of Brittany with permanent defences: two gigantic gun towers were

1 A Salamagne, 'Le château de Montcornet dans les Ardennes et sa place dans l'archictecture militaire de la seconde moitié du XVe siècle', *Revue Historique Ardennais*, XVII, 1992, 137–90.

2 Salamagne, Les fonctions militaries des châteaux en Hainaut au début du XVIe siècle, *Châteaux et Chevaliers en Hainaut*, 1995, 49–50.

3 Hoffsummer, Hoffsummer-Brown, Wery, 'Naissance, transformation et abandon de trois places fortes des environs de Liège: Chèvremont, Franchemont et Logne', CG, XIII, 1987, 69–71.

added to provide flanking to the outer enceinte, with two floors for guns, but still topped not with a platform for guns, but with machicolations. Nearby, at Lassay, to the striking fortress rebuilt by its seigneur Jean de Vendôme from 1458, was added in the 1490s a vast prow-shaped barbican provided with ten *canonnières*, below a wall-walk also wide enough for guns, and provided below ground with a third level of loops connected by a gallery and covering the ditch. To the promontory-based old castle at Clisson was added during this time a new court protected by two circular gun towers, of great strength but again with their loops limited to flanking the enceinte rather than facing the field.[1]

Works ordered or inspired by the French crown were rising up on all sides of the kingdom. A very interesting example is Le Castillet, a citadel-gatehouse ordered by Louis XI after he had taken Perpignan in the south in 1475. Strong semicircular towers form a gatehouse made entirely of brick, each with a triangular projecting spur, catalan-style machicolations and plentiful supply of *canonnières*. The seigneurial residence at Bonaguil was transformed after 1483 with a vast provision for guns, but the probably emblematic role of this work, described in the last chapter, reminds us sometimes to look behind the appearances. At Pontivy, the reworking of the castle by its noble lord at this time involved defences largely based on passive strength, the walls of the circular towers being six metres thick and of the curtain walls four metres, limited numbers of *canonnières* and machicolations.

Louis XI had seized the duchy of Burgundy on the death of Duke Charles the Bold in 1477, and hastened to secure the loyalty of his new frontier territories with new modern fortifications known as the "three sisters" at Auxonne, Beaune and Dijon. Of the new citadels, only the first remains. The pentagonal citadel faced both the town and the exterior, and was given two new gun towers (called Notre-dame and des Moulins), while ravelins (one triangular and one horseshoe shaped) were added in front of the gates. The towers had a diameter of 19 metres and a wall thickness of five, provided with casemates for artillery.[2] What once stood at Dijon has been studied intensively from old illustrations (it was only destroyed by the citizenry at the end of the nineteenth century) and it followed the same approach, save that the purpose was blatantly to impose rule over the local inhabitants. The original work comprised two massive round towers designed for large gun embrasures, and linked by curtains, built inside the city wall facing the city. The next phase of building brought a mirror image of this outside the wall, which was then demolished inside the new castle. Gates to the city, and to the country, were each provided with powerful boulevards (completed around 1510). At Beaune, the town wall (of a length of 2.5 kms) had been substantially strengthened by the Bur-

[1] Mesqui, 129–30 (Clisson), 171–3 (Fougères), 211–12 (Lassay), for brief descriptions and bibliographies; and personal viewing.

[2] Mesqui, 37–40 (Auxonne), 66–8 (Bonaguil), 287–8 (Perpignan), 301–2 (Pontivy). Other examples from this period described in Mesqui's book include Berze-le-Chastel, Bressuire, the *Tour d'Argentan* at Chinon, the barbican at Montreuil Bellay, the boulevard of Sille-le-Guillaume, and Tremazan.

gundians after around 1450, and by 1475 there were some twenty new towers or gate towers. New works under Louis XI included two massive horseshoe-shaped gun towers, of an external diameter of eight to ten metres (internally, four to five), with sophisticated gun embrasures at two levels (and the roof) covering both the ditch (which was a formidable 30–40 metres wide), and facing the field. Strong boulevards stood in advance of the town gates.[1] Other people close to the king adopted the same building models. For example, the bishop of Cahors in his work at the castle of Rochechinard, dated to around 1480, oversaw the addition of one circular and one polygonal tower to the older castle. The latter had eleven gun ports in walls three metres thick, the former had eight ports on five floors offering in both cases wide flanking as well as the possibility of plunging fire.[2] Sufficient examples perhaps to illustrate the main features of the latest thinking in France during this period: these were not progenitors of the new fortresses of the following century, but rather they were immensely strong cases of a style that had been developed in the country over a period of years, concentrating on the circular gun tower with walls of enormous thickness, and outworks featuring boulevards and ravelins in the place of the older barbicans. The *Tour Boucle* of Mont St Michel, discussed in the last chapter, and now dated to the same building period as the fortresses just discussed, is undoubtedly a formidable work, but is perhaps best considered as a pentagonal gun tower that has an accidental resemblance to the angle bastions of the future, probably for reasons based on its actual location.

Practice in Italy

Similar defensive designs were being tried out in Italy. Initially, the same major features were to be put in place. There are many examples of old castles rebuilt, town walls replaced, or new defences erected during the troublesome second half of the century, followed by the devastating invasion of the French king in 1494. To isolate the original work of the beginning of this lengthy time of war from continuous redevelopment is a challenge not yet fully met. The popes' fortress in Rome, Castel Sant'Angelo, for example, underwent so many reconstructions that each previous phase is hard to distinguish.

The Romagna, unusually, played a precocious role for a brief time. The records of the small city of Foligno have been used to show planning decisions that represent an early commitment to the pentagonal bastion. The archives record a number of inventories over the 1440s and 1450s listing a varied collection of crossbows with small quantities of ammunition, and small supplies of powder for

[1] P Dangles, N Faucherre, 'Fortifications urbains ducales en Bourgogne. L'exemple de Beaune', *Congrès Archéologique de France*, 152, 1994, 317–34. For Dijon, the best source is N Faucherre, *Muraille de Dijon: le château de Dijon*, a now hard to locate exhibition catalogue of 1989. The main findings are reproduced in the catalogue for another exhibition, featuring the château, in the city's archaeological museum in 2007: E Jeangrand, *Le Château de Dijon: De la forteresse royale au château des gendarmes*, 1–30.

[2] J Derbier, 'Une rocca transfigurée: le château de Rochechinard (Drôme)', CG, 22, 2004, 103–8.

bombards, of which there were in 1452 nine of various sizes.[1] There is nothing unusual in the commune's arsenal of weapons. What has attracted attention was its decisions with regard to improving the town's defences. A contract of March 1440 specified that the town wanted both round and five-sided towers, a scarp outside the tower, three levels to enable bombards to be placed above and below, and a *revelino* to be built to protect the gate on the side where the abbey stood, and the same in front of the other gates, to be of angled shape. Inbetween, the walls were to be raised and a scarped *terragio* constructed.[2] Someone had proposed to the commune that the most modern fortification required new towers of round or pentagonal shape serving as gun towers, outworks of ravelins and scarping outside. The resulting building would have been more advanced than the designs of the Italian theorists who had not, at that time, actually published their work, so the concentration on use of artillery in structures designed specifically for them, combined with effective outworks for defensive purposes, was well ahead of the theory.

Were they actually built? Subsequent records suggest that something was done. A contract was made with master mason Salucio and his colleagues in 1447 for the repair of the town's towers. In 1450, 300 florins were to be spent for the further repair of the town walls, and for the "adaptation" (*reaptacionem*) of the five-sided tower. A further contract with the same master in 1456 agreed to pay him for the "making" of the five-sided tower with walls a modest four feet (1.3 metres) thick and for the construction of the vault of the tower. There are no remains from which to judge.[3] It is also possible that the particular skills needed for building this work were in short supply. Nonetheless, Foligno suggests that the ideas were now present.

At the other end of this critical period, defences were definitely built at Ostia, dated by inscriptions to the mid-1480s. The fortress was constructed was located to control access to the Tiber and therefore Rome itself. The shape is an irregular triangle, with machicolated round towers at two corners and a massive pentagonal bastion at the apex enclosing a powerful, tall donjon. The modern features include the thickness of the wall and its relatively low height (with outside protection from a river-filled brick-faced moat); the inter-reliance of the three projecting tower-bastions (the three sides being short, 23, 20 and 18 metres only) as means of defence; and inside, a continuous corridor in the thickness of the wall connecting the bastions. There were a score of gun positions facing in all directions, mainly for smaller weapons, and gun platforms on the tops. The machicolations for close defence, and the presence of the donjon, confirm that this was still in many ways a medieval castle. It was also significant that it fell without resistance in both 1494 and 1497. The evidence for its date seems strong, and its builder seems to have

[1] A Angelucci (ed.), 'Spigolatore militare dell'archivio communale di Foligno' *Archivio Storico per le Marche e per l'Umbria*, III, 1886, 473–4, 476, 480–1, 487, 492–4.

[2] Angelucci, 477–9.

[3] Angelucci, 483, 485, 489.

been Baccio Pontelli from Florence.[1] Ostia, described by its historian Verdier as a "step on the way to Vauban", was certainly transitional, and was particularly important in establishing a close relationship between strongly projecting bastions based on them being gun platforms. It was not new, however, in recognising the need to concentrate defence around a gun platform, as we have seen in the gun towers becoming commonplace elsewhere in Europe at an earlier date.

A similar type of design was noted at the hilltop castle of Brolio in Tuscany when it was rebuilt around 1484 with inter-related projecting square bastions at each of five corners, and other examples became more common during the 1490s, although in many cases there continued to be variety of shapes.[2] There is another very interesting example, again of triangular shape, at the hill-top fort of Sarzanello, north east of the town of Sarzana (itself provided with a strongly defended and modernised *rocca*). The Florentines began building the fort after 1493, instructing the constructor to erect three towers with a high wall and deep ditch in the place of the ancient fortification, "because of changes to the art of war", as clear a statement as there can be of the thinking behind the planning. The fort was not completed before coming into the possession of the Genoese after 1497, and to them is attributed the unusual and massive triangular ravelin as an outwork on the most vulnerable side, accessible only by a bridge from the fort itself, along with the demolition of the donjon of the previous castle.[3] Of another style altogether is the unfinished castle standing over the town of Matera in the deep south of the country, which seems to have been planned as a pentagonal structure with a rampart linking strongly projecting circular and machicolated gun towers, the whole work pierced with a large number of loops at several levels for (relatively) small artillery with interlocking fields of fire, and with the ditch additionally protected by means of gun turrets projecting from the base of the towers.[4]

By 1500, it is apparent that fortress builders in Italy (covering a wide range of individual rulers and city governments) had learnt from their experience and from one another that the keys to resisting a modern siege train were: effective outworks shielding a main rampart that was better able than an old-fashioned tall wall to withstand bombardment, and allowed by its design (but not yet having established the relative benefits of circular and pentagonal forms) to provide both effective flanking, in which smaller guns would predominate, and a platform for heavier guns that could be used to take the battle to the enemy. The French invasion would demonstrate that it took more than a good design and construction

[1] P Verdier, 'La rocca d'Ostie dans l'architecture militaire du quattrocento', *École Française de Rome. Mélanges d'Archéologie et d'histoire*, LVI, 1939, 282–316. See also Hale, 16.

[2] For a clear summary of these works and a list of references to detailed descriptions, see Hale.

[3] A Neri, 'Il forte di Sarzanello', *Archivio Storico Italiano*, 4th ser., XV, 1885, 345–52, with extracts from the relevant contemporary documents; Hale, 19 and plates 21, 22.

[4] As noted from personal observation. The work was begun in the 1490s, according to a local chronicler consciously modelled on the royal Castelnuovo of Naples. It was left incomplete when its oppressive owner was killed by his rebelling people in 1515: L Santoro, *Castelli angioini nel regno di Napoli*, 1982, 235–7, quoting Verricelli's Cronica.

techniques, however, to defend a fortress against the most effective siege train in western Europe. The close connections between the fortress designers and builders also meant that lessons from one construction might quickly influence plans for another.

Practice in the Czech lands and Central Europe

In the last chapter the claim of the Hussites in Bohemia to have pioneered the artillery bastion decades before anyone else was discussed. The addition of strong gun towers connected with outworks designed for the artillery age to existing castles was explored, along with the impact of these changes on neighbouring states. The novelty of this approach lay not so much with the strongly-built towers, but in their placement, such as to transfer the defence to an outer line as far as possible from the original, older buildings behind. How far had these ideas influenced other fortress building in central Europe in the period after the end of the Hussite wars? Although precise dating for many of the works carried out in the middle and later years of the fifteenth century is often lacking, archaeologists and historians have by patient study of many castles and town walls created sufficient evidence to produce an overall picture.

The impact on foreign contemporaries of the legacy of the Hussites is hard to define, although if the reaction of the future Pope Pius VI to seeing Tábor was the common response, one might have expected it to have been far-reaching. In a letter condemning the heresy shielded by these walls, Eneas Piccolomini drew attention to the strength ("twenty feet thick") of the barbican tower (*propugnaculum*) protecting the main entrance, and also noted that it had been designed for artillery on two levels.[1] However, material circumstances and artistic taste, as ever, determined what lords and rulers in Bohemia and surrounding lands would build, and it seems that after the 1430s, there was little attention paid to defensive measures again until the wars during which George Poděbrady (ruled 1458–71, following the death of Ladislas Posthumus in 1457) defeated his domestic opponents. The power of modern artillery once again revealed the indefensibility of traditional castles and led to a further round of construction. The castle of Český Šternberk was strengthened (around 1470) with powerful and forward-flung artillery defences in the form of a polygonal gun tower with a rounded end, containing casemates for guns on its first floor.[2] After 1480, the castle of Helfštejn was reinforced by its lord with a shield wall seven to nine metres thick, 12 metres high and 90 metres long provided with two projecting gun towers, the whole standing in front of

[1] R Wolkan, 'Der Briefwechsel des Eneas Silvius Piccolomini', *Fontes Rerum Austriacarum*, LXVII, 1912–18, part III, vol. I, 25–6 (Letter of 21 August 1451 to Cardinal Carvajal).

[2] Durdík, 'Abriss der Entwicklung der böhmischen Artillerieburgfortifikationen des 15 und Anfangs des 16 Jahrhunderts', *Castella Maris Baltici*, II, 1996, 38; the same, Batterietürme der böhmischen Burgen, in G László, *Nyolovariötödik Szöletésnapjára, Tanulmányok. Múvészettörténet-Múemvléddem*, VI, 1994, 235–43; Líbal, *Castles*, 145. This design may also have inspired the work (not closely dated) at Zelená Hora: T Karel, 'Předsunuté opevnění hradu Zelena Horá', *Castellologica Bohemica*, 10, 2006 (German summary, 179).

a ditch itself protecting a castle with gun-loop-provided U-shaped towers.[1] Major works were carried out at Svíhov (Schwihau), another private castle, and here the architect is known: Benedikt Ried. From 1480, the old castle was surrounded with a new wall with large, circular towers. One of these was 38 metres high and had six floors. A wide ditch was provided. It was in 1505 that Ried made the changes that transformed the defences of this castle, when, outside the only recently built walls, he had built an even stronger outer wall with horseshoe-shaped open bastions.[2] This phase encapsulated the change from a traditional to a modern defensive plan. Ried was also responsible for work done on the royal castle of Prague after King Ladislas II (1471–1516) had suppressed an urban revolt in 1485. His work comprised two new gun towers(those called Daliborka and Mihulka; Plate 25), with provision for heavy and lighter guns and handguns arranged to cover the whole ground outside, and, on the south, a whole new curtain wall with three horseshoe-shaped bastions. On the east, where the entrance stands, a new barbican was constructed beyond the ditch as an outwork to the Black Tower.[3] Elsewhere in the region, the military lessons of the Hussite period were not adopted by builders at this time, although the absence of firm dating evidence calls for caution. The town of Levoča in Hungary (now in Slovakia), a German settlement, erected a circuit of walls over a long period: the fifteenth century constructions along the most exposed side have the form of a double wall, with strong towers projecting from the taller inner circuit, and straddling the low outer wall, behind (originally) a substantial ditch. Overall, the circuit is similar to the contemporary work at Nürnberg: in this case, the best that could be achieved by a small population on a site constrained by its geography, but not especially forward-looking. More striking is the brand new work carried out at the magnificently sited hill top castle of Spiš. During the fighting between Ladislas of Poland and supporters of Ladislas Posthumus to secure the crown of Hungary, Spiš withstood a siege by the Czech nobleman Ján Jiskra in 1443, but was yielded by agreement. Jiskra then constructed what is now the outer ward, an enormous area designed to serve as a base for his troops outside and below the old castle. This long circuit of wall was provided with very few towers, and these with hardly any projection, and little provision for artillery apart from hand guns (Plate 18). Finally, in around 1490, at the time of the works in Prague, the Hungarian royal castle of Zvolen — itself a quadrangular palace rather than a fortress — was given defences in the form of an enclosing wall provided only with small circular gun towers at each corner, of modest projection.[4]

In the areas of Germany most exposed to the Hussite raids, large scale building had taken place, often led by the major potentates of the region such as the

[1] Líbal, 145; D Menclová, 'Vliv husitskych válekna pozdně gotickou fortifikečni architekturu', *Umění*, IX, 1961, 457 (illustration).

[2] Líbal, 146; G Fehr, *Benedikt Ried*, 1961, 46–7.

[3] Fehr, 16–19; Libal, 148; Durdík, *Abriss*, 40;

[4] I Chalupecky et al, *Spisský Hrad*, 23–30; Libal, *Castles*,152; *Zvolen Castle* (local guide leaflet); and personal observation of all three sites. Líbal gives a later date for the defences of Levoča than local information sources, but it appears he is right to suggest the fifteenth century for the towers.

bishops of Eichstatt, Würzburg and Bamberg. Mostly, the work was of modest scale, comprising the insertion of loops for guns or the building of additional towers to hold guns: dated examples have been located at the castles of Lichtenstein (both the Northern and Southern castles at this site), Raueneck and Altenstein (all in Franconia) and in the town of Zeil am Main, where the Bishop of Würzburg had erected a new castle, square with round corner towers, and on its vulnerable eastern side, a wide ditch and tall walls and towers provided with gun loops. Even after its repair after 1466, however, this castle continued to be characterised by thin walls, and provision only for hand guns. At Coburg, however, an outer ring wall with eight U-shaped towers provided with shuttered gun loops was added from 1482–9. At nearby castle Laubenbergstein, a powerful gun tower was erected around 1470, described as a donjon. Over the same period, the castle of Friedewald was reconstructed by the Master Builder of the Landgrave of Hesse, Hans Jakob von Ettlingen (died 1507). Ettlingen's design included on the quadrangular and low-lying site corner towers, one of which was built as a gun tower with walls 4.5 metres thick, arranged to mount guns inside. His skill in understanding how to provide for the weight of the guns, how to ensure that supplies could be brought up through the tower, and the shape of the loops required to enable the operators a good field of fire, were all demonstrated at this site.[1]

Benedikt Ried had undertaken a large-scale building project in Bavaria, at Burghausen, where the thirteenth century castle was modernised between 1480 and 1488, apparently for fear of the continuing advance of the Ottomans. The new development at this large ridge-top fortress was the wall (around three metres thick) provided with towers for artillery as the outermost defence. Other late fifteenth century changes in castles in the region also witnessed the adding of towers for artillery: Libal lists examples ranging from Haut Koenigsbourg in Alsace to Eltz in the Rhineland and the city of Heidelberg.[2] The barbican at the Spitaltor of Rothenburg ob der Tauber was consistent with contemporary trends. Generally, however, looking at the scale, dimensions, shapes and overall design of defences added to German fortresses during this period, it does not seem that many particularly advanced structures were put up. Such novelties as there were, were probably due to the impact of Italian designs.

Ried may also have been the designer of the famous barbican erected at the Polish city of Cracow from 1497. This was added to the city walls by the King following the failure of his military expedition into Moldavia in that year, apparently again for fear of the Ottomans. Instead, and alongside its rapid adoption as a ceremonial entrance for the city, its strength was tested — and found to be solid — against a much more powerful onslaught long after its construction. The barbican

[1] J Zeune, 'Hussitenzeitliche Schiessscharten an Fränkischen Burgen', *Castellologica Bohemica*, 10, 2006, 39–52, surveys a number of these sites; J Zeune, Die Baugeschichte des Veste Coburg, in M Henker, E Brockhoff (eds.), *Ein Herzogtum und viele Kronen. Coburg in Bayern und Europa*,1997, 11–13; R Gutbier, *Der Landgräfliche Hofbaumeister Hans Jakob von Ettlingen*, 1973, 9–19, 185.

[2] Líbal, 137–9; Fehr, 16; Krahe, *Burgen des Deutschen Mittelalters*,1996, for sketch plan and bibliography for all the sites in German-speaking or German-ruled areas.

was built as a 6/10 roundel, 30 metres in diameter, with walls 3.65 metres thick, linked by two straight walls to the city gate between six and ten metres behind. It stood beyond the city's double moat, and had its own moat, with sluices to control the flow of water. Built of brick on sandstone foundations, defence was provided by no fewer than 130 firing positions on four levels, accommodating in turn heavy guns, lighter guns, handguns and crossbows. The entrance route left the barbican in an angle of the structure. The seven watchtowers accessed from the top were possibly added for decorative effect around 1550. Not surprisingly, this erection provided a model for similar structures added to other Polish cities thereafter.[1]

That Benedikt Ried, a German working under the influence of Italian design, was selected to carry out the most prestigious building projects of the rulers of Bohemia and Poland may give a clue as to the limits of influence of the Hussites on fortress design. Gun towers were common across Europe by this time, and indeed their origins have been traced back as far as the late fourteenth century, although structures of that date were generally considerably weaker. What, then, happened to the bastions foreshadowed at Tábor? The nearest to a successor of that design has been identified at the small, privately-owned Bohemian town of Bechyně. It stands on a promontory, and the accessible side is defended by a wall containing twelve four-sided projecting bastions, open at the rear, standing behind a ditch and rampart, regularly spaced at the (close) distance of 25 metres apart. These remain, apart from those originally forming the gateway. According to one scholar, "it can be considered as the oldest known system of fortifications based on the bastion principle", and although its exact date is not known, several experts assign it to the 1470s. Remarkably, given such claims, there does not appear yet to be any detailed description or analysis of this frontage. The bastions are of modest dimensions. The wall of Bechyně does not appear to be the earliest scientific application of the idea of a bastion trace. It has no provision for guns, except from the now disappeared roof levels, and these can only have been very small weapons. If it was, it was not taken up by other builders, even in the same land, who continued to focus on round towers or horseshoe-shaped projections as the most modern exemplars to follow.[2] The name of the builder is not known.

[1] Described by T Torbus, 'Das Krakauer "Rondell" als Paradebeispiel der spätmittelalterlichen Barbarkane sowie seine polnischen Nachfolge bauten'. This article will appear in *Forschungen zu Burgen und Schlössern, Wartburg Gesellschaft von Burgen und Schlossern*,12, 2009, and I am most grateful to Dr Torbus for allowing me to read it before publication. See also, Hartmann Schedel, *Chronicle of the World. The complete and annotated Nuremberg Chronicle of* 1493, ed. S Füssel, 2001, for a description of Cracow and illustration of the walls before the barbican was built, fol. cclxiiii, plate CCLXV; S Mossakowski, 'The town walls of Old Cracow', *Europa Nostra Bulletin*,58, 2004, 141–44. Torbus is concerned to develop a rather tighter definition of what a barbican is than is common in western scholarship.

[2] Líbal, *Castles*, 144 (the quotation), and 'Burgen und Städte als tschechische Erbgut', *Castrum Bene*, 6, 1999, summary (211); Menclová, in *Umění*, IX, 1961, 471; plan, 463; Durdík, in *Europa Nostra Bulletin*, 53, 2000, 75. I am grateful for the comments of Tomáš Durdík on the uncertain date, and lack of detailed study (personal communication). Personal inspection confirms the small size and absence of provision for heavy weapons: see P Purton in *Castle Studies Group Journal* 22, 2008/9, 217–27.

Practice in the Spanish kingdoms

During the war against Granada, there was little opportunity, or need, to set about building modern fortifications of a permanent character, the demands of war leaving time only for temporary constructions. Whereas the crown of Castile and Aragon was not backward in reorganising its artillery, it seems to have paid less attention to modernising the defences, few examples of which show evidence of any work before the arrival of regular bastioned traces during the sixteenth century, with some exceptions mentioned below, and one genuinely transitional construction at Salses.

After the conquest, work was done at Almería to repair damage done by an earthquake. New triangular towers were constructed in the area on the western side of the outer enceinte with a view to making the immense circuit of fortifications, which were otherwise entirely unadapted to the modern age, more resistant to attack from guns in the area from which this seemed most likely.[1] The continued use of already built castles and town walls that had been maintained and used in all their various functions, including a defensive role increasingly less needed in the great majority of inland sites, led to few modifications of a limited character to allow the deployment of guns in defence.

One example with both a symbolic significance and a practical purpose was the castle of La Mota of Medina del Campo, one of the residences of Queen Isabella and also a royal prison. The outer wall of this originally thirteenth century creation, dominated by a very tall rectangular donjon, was ordered in 1475 to be provided with a powerful sloping escarpment inside which ran a gallery giving access to loops which looked into the very deep ditch, while the circular projecting corner towers were given flanking loops for small guns. The loops were organised at four levels, with a ventilation system, and half a kilometre of intramural galleries. The castle otherwise remained a typical model of the time of its first building. It had been put to the test during one of the violent civil wars of the time immediately before its adaptation.[2] The immense castle built at Coca between 1473–96 by the Fonseca family had a caponier covering the ditch, a protected system for raising water from underground springs, and similar barriers to hold artillery as at La Mota. A barrier for artillery also featured in the rebuilding of the old castle at Ponferrada, along with a half-buried "covered way" down to the river to protect a well. The castle is known to have faced artillery attack when it was held against the king in 1486 and had to be captured by a local army.[3] One of the few sites to be built anew during this time (by the Marquis de Moya) was the small but forward-looking site of Chinchón (Plate 27), located on an important internal route to the south of Madrid. The upper levels have been destroyed, but enough remains to

[1] L C Barrionuevo, *La Almería islamica y su alcazaba*, 1990, 248–52.

[2] F Cobos Guerra, J de Castro Fernandez, *Castilla y Léon: castillos y fortalezas*, 1998, 227–31. Clear illustrations in Ortíz Echagüe, 212–15.

[3] F Cobos Guerra, J de Castro Fernandez, 235–6 (Coca); same authors, *Castillo de Ponferrada*,2002, for a detailed history, description and analysis of this complicated multi-period fortress.

show that this was a regular quadrangle with projecting D-shaped corner towers and an outward-sloping lower stone scarp all round, provided at these levels with embrasures for artillery. The towers are small and do not have a great projection, and the loops are not numerous, indicating that the whole thinking behind the model was not perhaps understood but the similarity overall to the type of construction taking place in Italy by the 1490s is striking.

The rampart around Las Dressanes, the dockyards of Barcelona, would actually seem rather close to the prescriptions of the first two of the Italian architects just considered. The wall stands around ten metres high, and the towers are rectangular, standing another five metres higher still, projecting about two/three metres from the wall. A ditch stands in front. The sides and front of the towers contain gun loops, although not of a size to hold large weapons. Where the tower serves also as a gate, there was originally a drawbridge and a machicolation on the wall top directly above. The already conservative nature of this rampart, by contrast with the concept behind Chinchón, is also apparent.[1]

Quite distinct is the fortress of Salses (Plate 26), now in France, but erected after the territory of Rousillon was returned to Spanish control in 1497. The builder was Francisco Ramiro Lopez, head of artillery and an active participant in the conquest of Granada. Great resources were expended as marshes were drained (this was a low-lying site) and a remarkable new garrison fortress built within six years. Salses was a purely military building. Its garrison was listed as 1000 infantry and 350 cavalry. Its artillery comprised 17 heavy and 39 lighter guns along with 290 various handguns and arquebuses. In plan, it was a symmetrical rectangle with round corner towers, built with walls 4.7 metres thick. The towers had three levels of casemates for guns with a central ventilation shaft and water to absorb fumes, but the roof was originally crenellated. The ditch, 12–15 metres wide, could be filled with water. There were four detached outworks. These included D (or half-moon) shaped bastions, built out in the ditch, one providing a fairly traditional boulevard to protect the gate, another linked by a caponier to the curtain wall behind, and featuring the "ears" of future bastions, a design that enabled defenders to cover the ditch behind them without being exposed to attack from the field. In the middle of one face stood a rectangular donjon, 26 metres high, with seven floors, that was reached via drawbridges. Salses resisted a siege by the French in 1503, after which further changes were made to the design, including installing a caponier in the ditch, suppressing some of the outworks, adding a vast scarp to thicken the main walls, and reconstructing the tower tops as gun platforms. Salses was a truly mixed up fortress, combining elements of the most advanced design with backward-looking features such as the donjon, lack of internal circulation, gothic decorations, and restricted fields of fire. But it was an excellent example of design based on the real experience of its engineer, and its rapid reworking after it was first called upon to fulfil its purpose showed that its builders were willing to

[1] From personal observation. Chinchón is illustrated in Ortíz Echagüe, 111.

learn from experience. They did not conclude that a regular plan using the angle bastion was the best answer to their questions.[1]

Practice in England and Scotland

Very limited works were carried out in England and Scotland during these years. Reference has already been made to the striking new castle of Raglan, with its donjon designed to house artillery. The defences of Calais, a fortress the maintenance of which cost the crown a continuous heavy expenditure just to prevent it falling into the sea, underwent continuous alteration during the century, but without it always being clear what the work done actually led to. From the start of the English occupation, Calais had been supplied with artillery. Temporary works had been thrown up to resist the duke of Burgundy (1436) and in 1448–50, one of these bulwarks was rebuilt in stone in front of the Boulogne gate, although sadly nothing is known of this possibly precocious gun work.[2] The first works designed specifically (and only) to be an artillery defence may be the simple blockhouses erected around 1457 to house a chain to protect the harbour at Fowey. In 1481, the first major fortification specifically built as a gun fort was begun to protect the harbour of Dartmouth, on the initiative of the town, taking the form of linked rectangular and circular towers with batteries to north and south. These contained floor-level splayed embrasures for heavy guns to fire across the estuary, and the floor above was designed for hand guns. The gun ports are large, rectangular holes which must have been provided with shutters. The plan was determined by the shape of the existing site, adjacent to a small stone castle of fourteenth century date, and gave no thought to close defence or flanking. It remained in service for a long time because of the effective way it achieved its purpose (supported by another, simpler square tower at Kingswear on the opposite bank, begun in 1491), but it was not of the same character as the already more sophisticated fortresses being constructed over the Channel.[3] Also in the south-west, the castle at Berry Pomeroy was newly built towards the end of the century. The gatehouse was constructed with small gun loops with a limited field of fire, but in the other tower on the castle's front (St Margaret's), is found uniquely in England a loop designed to accommodate three guns at different heights, including the opportunity for considerable traverse. This castle is also interesting in showing (in the gatehouse) the need for a provision of an exceptionally strong timber lattice under the roof, which can only have been intended to support guns above.[4]

In Scotland, it used to be said that the new fortress at Ravenscraig (Fife) built

[1] L Bayrou, N Faucherre, R Quatrefages, *La forteresse de Salses* (guide book), 1998; A Sanchez-Gijón, J de Castro, *Luis Escrivá, Su apologia y la Fortificación Imperial* ,2000, 16.

[2] Colvin, *King's works*,I, 449.

[3] A Saunders, *Dartmouth Castle, Devon* (guide book), 3rd ed., 1991. On Fowey and its opposite number at Polruan, references in King, *Castellarium*,I, 73, 75.

[4] S Brown, 'Berry Pomeroy Castle, Devon', *Devon Archaeological Society Proceedings*,154, 1996, 27–9, appendix 14, G Parnell, The gunports, 325–8; S Brown, *Berry Pomeroy Castle* (guide book), 1997, 7–9, 26. I am indebted to Andrew Saunders for drawing this to my attention.

around 1460 by King James II was the first in that country to be designed for artillery, based on the provision of gun towers and gun loops in the remains of the promontory castle overlooking the Firth of Forth. It has been argued from dating evidence, however, that some of the provision for guns is later, and/or inserted. Certainly, there is little evidence of much being done to follow the example of Threave described above, with a few cases of loops being inserted into existing walls being the usual response during the second half of the century.[1]

Practice in the Eastern Mediterranean

The Ottoman conquest of the Byzantine empire, the fortifications of which were often based on now ancient defences, has been outlined already. What evidence is there for the adaptation of old walls to accommodate modern weapons during this time of invasions? Despite the occupation of much of the Morea by Venice and others with direct access to Italian know-how, not much. The ancient wall across the isthmus of Corinth, the Hexamilion, received some work to upgrade the defence, carried out initially in 1415 by Manuel Palaeologus, and again in very short time (suggesting that not much can have been accomplished) in the 1430s by the Byzantine despot Constantine following damage by the Turks in 1423. In 1446, it fell to escalade and mining, under cover of an artillery bombardment. The last use of the wall as a defence was when it fell to the Ottomans, however, in 1452. Excavations suggest that the irregularly spaced towers, which were both round and square, contained traditional slits for archery; but they probably had flat roofs suitable for mounting guns. It was no longer in use when the final conquest of the Morea followed in 1458, and a Venetian force counter-attacking in 1462 began to reconstruct it, but was allowed no time to complete the work. The original walls (between 2.3 and 2.9 metres thick) would have been helpless against the kind of bombardment we know the Ottomans were capable of mounting, although strengthening may have been carried out to some of the wall towers and to the citadel built half way along.[2]

On the island of Euboea (Negroponte), alongside the fortified town itself, the capture of which by the Ottomans has been described, the land was in the hands of local landowners. No fewer than 55 towers were built, mainly during the fourteenth century. It is likely that they were for individual lords, with a probable function of providing a defence for the local peasantry. After the capture of Constantinople, the Venetians purchased gunpowder, cannons and other weapons for Euboea, but this was all concentrated at Negroponte, and the towers played no part at all during the Ottoman conquest of the island. A similar conclusion should probably be applied to the Venetian possessions on the Morea. Here defence was concentrated in forts and the existing castles (in particular Coron (Koroni) and Modon (Methoni) and their outlying fortified sites) rather than the scattered towers and walled villages, designed to protect the population and its produce.[3] In the

[1] See the discussion by C Tabraham, *Scotland's Castles*,1997, 96–99.

[2] T Gregory, 'The hexamilion and the fortress', *Isthmia*, v, 1993, 20–24, 69–70.

[3] P Lock, The towers of Euboea: Lombard or Venetian, agrarian or strategic?, in Lock,

two first mentioned fortresses, much is obscured by subsequent rebuilding, and there is more to learn from the new work carried out at Nauplio (Nafplio), from the 1470s. On the original town site (the Acronauplia) further additions were made to the pre-existing Frankish and Byzantine defences, including the addition of revetments to the Frankish walls to strengthen them against modern weapons. New work comprised the Castello di Toro enclosing the headland beyond the Frankish walls, the chief surviving element of which is a powerful circular gun tower, with a splayed outward sloping wall. Similar design principles were also used in the construction of a new fort on the small island in the town's harbour, now known after its Turkish name of Bourtzi. This important work, begun in 1473, took the form of a hexagonal central tower between two polygonal lower batteries, with provision for guns at both ground and roof levels, as well as on the roof of the three-storey tower. Cisterns held water for the garrison, the floors of the tower were accessed only by movable wooden ladders, and there were two elaborate gateways with barbicans (again with defences designed for guns), one giving onto the little harbour that gives the only access. This truly formidable fort, remarkably similar to the work of the Hospitallers at Rhodes, can make a claim to be among the earliest of its type in the eastern Mediterranean world (Plate 21).[1]

At this time, the Turks were rather more concerned with knocking down walls than building them up. The extent of their understanding of the best design for fortresses was shown by their construction of the brand new work of Rumeli Hisar, built specifically to prepare the ground for the conquest of Constantinople itself. The traditional limitations of the landward-side design were considered above, to be balanced against the clear understanding of how to adapt the design to cover the water with heavy artillery.

In areas where further conflict could be anticipated, the Ottoman rulers generally responded with urgent measures determined by the strategic situation, whether this involved quickly erecting wooden forts, or a more extensive rebuilding of old stone walls to adapt them to modern needs. In 1482, for example, at a time of conflict with the princes north of the Danube, the new sultan Bayezid ordered the restoration of two wooden forts on that river previously destroyed, giving the task to his provincial commander.[2] At a more leisurely pace, but with a similar commitment of resources, Venetian chroniclers recorded that his father the Conqueror had deployed 10,000 workers to restore the defences of Constantinople, starting in 1472, and the main work was the excavation of the ditch in front of the triple walls.[3]

Sanders, *The archaeology of Medieval Greece*, 1996, 107–11; C Hodgetts, P Lock, Some village fortifications in the Venetian Peloponnese, same volume, 77–90.

[1] Andrews, *Castles of the Morea*, 90–105 (contains some small inaccuracies); E Spathari, *Nauplion-Palamidi* (guide book), 2004, throughout; G Tsekes, Castle of Akronafplia, and, Bourdzi, in A Triposkoufi, A Tsitouri, (eds.), *Venetians and Knights Hospitallers*, 2002, 90–4, 95; and personal inspection.

[2] Forres, ed., *Die Osmanische Chronik des Rustem Pascha*,1923, 16.

[3] Malipieri, 79.

Practice in Russia and the Black Sea

The impact of artillery on fortifications in Russia was a little tardy. This may simply have been because the works of earth and timber that had sufficed for so long were also relatively effective against gunpowder artillery, although questions of cost, access to suitable building materials, and, alongside these deterrents, the question of rebuilding cities and citadels in the most modern style for reasons of prestige, led to the move to stone anyway. In 1444, an attack by the Teutonic Knights on Novgorod with a full array of artillery was launched for five days, but the walls and their defenders resisted successfully. In 1450, in a war between the Grand Prince and Prince Dmitri, the city of Galich was attacked, and defended, with guns, but only fell to blockade. In 1451, the Tartars attacked Moscow, burning the suburbs (as they had done previously in 1439). But, noted the chronicler, the smoke from the fires concealed the places where the walls of the city were not of stone, and the citizens were well prepared to defend the city with weapons of all kinds including cannon, catapults and crossbows.[1]

At Moscow the critical period for new building was after the final overthrow of Tartar dominion (1481). The Grand Prince of Moscow-Vladimir, Ivan III (1462–1505) was responsible for the reconstruction of the Kremlin, laying the basis for the massive fortress-palace-church complex, which was large enough to contain a sizeable settlement as well, with which the world has become familiar. Standing on the hill at the confluence of two rivers, the walled circuit stretches one and a half miles (2.4 kms). The first stonework had been established by Dmitri Donskoi (from 1367). No doubt motivated by the sense in which he saw Moscow as the new Constantinople, Ivan actually turned to Italy for his designers. During the 1480s and 1490s, the new curtain wall and the immense towers soaring above them were gradually constructed, the names of several different architects being connected to each of them, and Pietro Solario from Milan and Ridolfo Aristoteli from Bologna featured. In front, a ditch of epic proportions was prepared, between ten and thirteen metres deep, and 31–36 metres wide, parts of which could be filled with water from the rivers. On the south side, the walls stand 20 metres above the river Moskva, and are four to five metres thick, constructed of well-made brick and strengthened on the interior by filled-in arches. The curtain wall was unpierced along its outer face, but more than wide enough, and strong enough, to mount guns along the wall walk, and its shape, with frequent changes of direction, as well as the frequent towers, suggest an awareness of the importance of flanking. The towers themselves were of different shapes, with three or four storeys and equivalent numbers of firing stations within. Very strong foundations were needed. During this time, they were topped with a final stage projecting outwards on corbels, the present upper stages being added much later. Many of the towers were capable of independent standing. There were five gate towers, provided with a barbican in front, of two or three stories themselves. The outer stone wall in front of the ditch was not completed until early in the sixteenth

[1] Nikonian Chronicle, V, 47, 70, 91–3.

century, but it was thought to have been preceded by a timber palisade.[1] Earlier accounts may have under-estimated the undoubted role of these works in stressing the power and prestige of the rulers who lived within, and their claim to have inherited the imperial splendour and Orthodox cloak of East Rome, but given the uncertain political and military position of that ruling family in the face both of the Tartars and of rival princes it would be reasonable to assume that there were also concerns about defensibility.

On the shores of the Baltic, Olavinlinna was started in 1475 by its noble owner, who moved gradually from timber to stone and then from a strong inner court to a fortress of several wards. The inner and outer were completed by the end of this initial phase of construction, in the early sixteenth century, and the inner castle was dominated by three very powerful gun towers presenting a rounded face to the outside.[2] The walls of the towers and curtain were extremely thick, and the defence was largely conducted from the top, demonstrating that whoever was the designer was familiar with the best European standards.

The principality of Moldavia achieved its period of greatest political influence during the rule of Prince Stephen the Great, whose long reign (1457–1504) saw the small country achieve solid steps towards establishing the kind of structures (physical and political) familiar to larger states, despite the pressure from the Ottomans to the south. The Moldavians were influenced by all their neighbours, Poland and Hungary in particular. In 1469, the Magyar monarch Matthias Corvinus had ceded two Transylvanian castles to Stephen, and victory over the Wallachians in 1474 led to the construction of new fortresses. In 1478, new work was done at the major merchant city of Cilia, the chronicle suggesting this was completing inside 24 days, clearly indicating that this was a repair job. The records continue however to tell of new fortifications founded by Stephen[3] – who retained sole legal right to build fortifications, and could call on the services of all his free subjects to assist in construction and maintenance – either before or after his defeat of large scale Ottoman invasions in 1475 and 1476. Most lay on the frontiers (in all directions – it was not only the Turks that Stephen had reason to fear) and on the coast, where the trading ports were of critical importance.

What do we know of the form of these works? Some Moldavian citadels were small. The citadel of the capital, Suceava, had an inner court 42.5 by 45 metres, and this was large by comparison with other similar defences. Most were rectangular, some polygonal and one was circular (Sovoca). Design to face firearms would not have been a consideration before the fifteenth century, where the first recorded use in what would become Romania was by Vlad Dracul in 1432, and in Moldavia itself in 1445. But the new or repaired fortifications seem to have had

[1] Among a very large number of descriptions, that of A Voyce, *The Moscow Kremlin, its history, architecture and art treasures*,1955, contains many plans and illustrations to accompany a written description of the defences, on which this paragraph is based: 4–7, 14, 20–26, and plates 17–27.

[2] Olavlinna, *Olofsborg*, guide, 1920, plan, photographs and summary in German.

[3] G Urechi, *Chronique de Moldavie*, trans. E Picot, 1878, 107, 135, 157, 163. See pp. 339–40 above for the Ottoman sieges of Cilia and Cetatea Alba in 1484.

this in mind. Wall thickness varied, but some were nearly five metres (Cetatea Alba), while Suceava itself was 2.5 metres. After 1476 (or perhaps in preparation for it), the wall of the outer closure was doubled in width, and semi-circular gun towers were added. At the inland fortress of Neamt, originally constructed in the late fourteenth century, as an irregular quadrangle with three metre thick walls, Stephen added a new enclosure to protect against artillery attack on the vulnerable north side, with an even thicker wall and four circular towers designed for guns. This, like Suceava, resisted Ottoman attack in 1476. Excavations at the castle of Roman, newly built by the Prince, have recovered a central rectangular fortress (55 by 45 metres) with eight circular towers, and walls four metres thick on deep, timber-framed foundations to take account of its low-lying, riverside location. Stephen then added (probably in the 1480s) an external enclosure with thick walls and three circular towers, and later still, a massive wet ditch 30 to 50 metres wide.

A common feature was the construction of tall, strong, projecting towers, although with little provision for artillery, leaving it perhaps to the roof tops to provide a base. As elsewhere, ditches and low outer walls provided a key element in protecting the main enceinte from attack. Suceava's ditch was 40 metres wide and five deep, a formidable obstacle. Not all Stephen's work was in stone – sometimes earth and timber fortresses were erected, as at the hill top fort of Birlad – and sometimes he also used monastic buildings to create refuges, such as the massive tower erected at the monastery of Putna. Visually, the similarity between places like the towers of the city wall of Sibiul and those of the Moscow Kremlin is striking, and as it is not clear who actually proposed the designs (although we know that thousands of Moldavians were mobilised to carry out the work), similar influences seem likely, perhaps through the continuous relationship with Genoese communities on the Black Sea coast.[1]

Ottoman offensives at the end of the fifteenth century

The threat to Christendom was Mehmed the Conqueror. There is further evidence for the development of the techniques, tactics and decisive factors of siege warfare in two dramatic and bloody sieges: the capture and recapture of Otranto, and the much better known defence of Rhodes by the Knights Hospitaller, both events starting in 1480. The three sieges are appropriate opportunities to summarise several developments at once, including the response of

[1] C C Giurescu, *Istoria Romanilor,* 1937, II part 2, 521–6, and illustrations on 420, 434, 446, 453, 457; G Anghel, 'Les forteresses moldaves de l'époque d'Étienne le Grand', CG, VII, 1974, 21–32, for summary archaeological reports and plans. Anghel also concluded that the most likely design influence was Italian. See also P Harrison, *Castles of God* ,2004, 129–30, for the monasteries, including photographs of the monasteries of Putna and Sucevita. Using well-built monastic establishments as sites for defended refuges for the local population was, of course, a common feature of those parts of the Balkan peninsula particularly exposed to attack or raid, as the rather better known fortified churches and monasteries of neighbouring Transylvania testify.

fortress builders to the actual experience of siege. At the same time, the Turkish sultan's greedy eyes were also focussed to the south, where the Mamluk sultanate in Cairo continued to vie with Istanbul to be the chief power in the Muslim world.

The defence of Rhodes

The island of Rhodes was a significant outpost of Christendom, and one requiring continuous defence. The Hospitallers, rulers since 1309, had maintained the island's defensive structures continuously. These included an extensive network of outer bulwarks, the most effective of which was the recently rebuilt fortress at Budrum on the mainland, but also forts and towers on all the surrounding islets and, on Rhodes itself, the ancient acropolis at Lindos was given a garrison, and no fewer than 30 castles dotted the terrain, while a further network of watchtowers ran around the coast connected by carrier pigeons and beacons. Some of these works had been put up in response to an assault, by an Egyptian fleet in 1444, known from a Catalan report and a Mamluk chronicle. This event, lasting for a month after the unopposed landing on 10 August, consisted of a series of probing attacks against different parts of the defences, in which both heavy guns (*bombardes grosses*) firing shot weighing up to six *quintals*, and older weapons (*trabuchs*) were deployed, alongside an attempt at mining of the outer works. Much of the assault was directed at the defences of the harbour, and one lesson learnt from the experience was the importance of keeping an attacker from being able to set up their batteries too close. The Mamluk account confirms that the invaders had brought a substantial siege artillery with them, although the significance of two different words for guns remains uncertain. The issue was decided on this occasion by a decisive victory for the Grand Master in the field.[1]

In the Ottoman attack in 1480, none of the outlying and smaller works played any part whatever in the proceedings: the castles provided a refuge for the country folk and their possessions against raiders, and the watchtowers watched. The arrival of a professional army with an artillery train meant that possession of the whole island rested solely on the fate of the city itself. One of the reasons for the Ottoman attack in the first place was information provided by a renegade Greek that the walls were old and crumbling. Certainly, in 1309, the Knights had inherited Byzantine walls, but not the resources to modernise them. That work really got underway with the beginning of the fifteenth century. The town walls were rebuilt over the period 1400–1440. In the next two decades, polygonal bulwarks were built in front of each of the five gates. A fort, called St Nicholas, was rebuilt on a mole that divided the commercial port from the galley port, funded by money from Burgundy. This fort took the form of a circular gun tower, with two storeys of gun platforms, 17.3 metres in diameter, surrounded by a new polygonal bulwark.

[1] L Nicolau d'Olwer, 'Un témoignage catalan du siège de Rhodes en 1444', *Estudi Universitaris Catalans*, XII, 1927, 377 f; references to artillery and trebuchets at lines 61, 100–2, 124, 195; Taghri Birdi, History of Egypt, trans. Popper, 19, 1960, 93–5. The original talks of *manjaniq* (catapults), *makahil, wal-madafi*,' interpreted by the translator as "guns and cannon".

The general design, although not the sophisticated sloping walls of the tower, bore a resemblance to the contemporary Venetian work at Nauplio, described above. On the other mole covering the commercial harbour, money provided by Louis XI from 1465 paid for the tower of the Mills, 12 metres in diameter with a gun platform in its lower storey. The new grand master appointed in 1476, the famous Pierre D'Aubusson, undertook even more work, the most important of which proved to be the completion of an outer defence all the way round in front of the city walls (themselves 2–4 metres thick and generally provided still with their old square towers). This *faussebraie* was high enough and wide enough to enable a strong outer defence and to shield the city wall from the effects of bombardment.[1] Defence was entrusted to different "nations" of knights and their soldiers, of whom there were perhaps 2000. Although the local population, increased by fleeing peasantry, would have been largely Greek, there were other groups and nations present as well, and most were to serve the Knights loyally. The latter dwelt in separate quarters in the city's citadel, which played little direct role in the siege.

The dangers of being persuaded by the honeyed promises of political exiles were never more strongly shown than by what now happened. Sultan Mehmed was unaware that D'Aubusson had spent three years non-stop modernising the defences and laying in supplies, in particular munitions, as well as raising a sufficient garrison to hold the walls and ships to prevent a complete blockade. Nonetheless, the army and fleet that was sent, commanded by the Greek now called Misac Pasha, was substantial, although the Christian accounts talking of numbers like 100,000 were pure fantasy. A more likely estimate was15,000 of all arms and occupations, including an artillery train and another foreigner serving the Turk, identified as a German (or perhaps Hungarian) master gunner called George.[2] Despite their under-estimation of the defences, however, the Ottomans came close to taking the city.

On 23 May 1480 the Ottoman fleet was seen from Rhodes and the defenders decided not to try to prevent a landing, allowing the invaders to come ashore close to the city and immediately begin its encirclement, and the construction of their camp. Almost at once, they set up a battery by the chapel of St Anthony, directly opposite the St Nicholas tower, of which the chief equipment was three immense brass bombards, firing shots (according to D'Aubusson) nine palms (an impossible seven feet, more than two metres) across.[3] The battery was itself protected by

[1] There is an excellent summary of what the defences probably looked like before the siege of 1480 in A Migros, 'Rhodes: the Knights' battleground', *Fort*, 18, 1990, 7–18; and A Stefanidou, The Medieval Town of Rhodes, in Triposkoufi and Tsitouri, 184–9; S Spiteri, *Fortresses of the Knights*, 2001, 28–33, 108–12.

[2] A letter from Naples dated 1 July conveyed news received in Venice from Rhodes that the Ottomans were at sea with 130 ships, 14 large bombards and 18,000 men: Foucard, ed., Fonti di storia Napoletana, *Archivio Storico per la province Napoletane*, VI, 1881, 80–1.

[3] Letter of D'Aubusson to Emperor Frederick III, 13 September 1480, in Porter, *A History of the Knights of Malta or the Order of St John of Jerusalem*, 1883, 706; H Fincham, ed., *Caoursin's account of the siege of Rhodes in 1480*, trans. J Kay, Poet Laureate to Edward IV, 1926, 10. These two accounts by participants provide crucial detailed information and largely confirm each other's versions.

an emplacement of palisades and earth, which was just as well as the Knights set up their own battery to fire back. The 300 monster shots hurled the short distance across the harbour were immensely destructive of the masonry once it had begun to crumble, but the defenders bravely worked to restore the curtain around the tower, using the labour of large numbers of citizens to help construct a new, temporary timber bulwark, and an initial assault launched from Ottoman ships after ten days was repelled (Plate 22). The great bombardment was resumed and the new defences were quickly demolished by their fire. The defenders dug a trench in the mole, and used the shattered stones as ready-made components for their new, temporary ramparts. Meanwhile, a more sophisticated means of attack was organised, in the form of a floating bridge which even D'Aubusson acknowledged was a work of "wonderful skill". A wooden siege tower on the far bank was erected to enable crossbowmen and arquebusiers to fire down on the men engaged in repairing the defences of the mole and tower. The commander was prepared, however, the new means of assault being impossible to conceal, even though it was boldly launched by night, and the massed fire of guns and crossbows based at the tower, along the mole and along the harbour walls caused heavy losses to the attackers as they neared, blasting the bridge to pieces and also sinking several of the supporting galleys (18/19 June).[1]

The Ottoman commander had not restricted his army's action to only one side, but was concerned to stretch the defenders by attacking from several directions. His batteries were in operation all the way round, and mortars were firing over the walls into the city in an attack designed to cause chaos and panic among the population rather than batter the walls. This example of terror tactics had been carefully prepared against by D'Aubusson, who had encouraged people to take shelter in cellars and caves and, as a result, only minimal casualties were suffered from this attack. At the same time, a big battery of eight large guns was set up on the opposite side of the city to attack the defences held by the Italian nation in front of the Jewish quarter. The bombardment was frightful, both accounts cited talking of several thousands shots fired, and apparently the noise could be heard a hundred miles away: a piece of embroidery, but the impression of an overwhelming artillery onslaught is well conveyed. The guns knocked down nine of the wall towers, and destroyed the Grand Master's palace. Here also were seen the trebuchets, part of the "many and diverse" weapons the Ottomans used. Caoursin wrote that they set up "certain instruments which he called slings" which hurled barrels full of stones over the wall onto the houses. D'Aubusson also wrote about the Turks shooting "fire balls" from "their balistae and catapults".[2] For five weeks this attack was sustained, and during it the attackers began to dig their approaches towards the outer defences with a view to draining and filling the ditch. The defenders were able to prepare counter-mines to frustrate the Turks, at one point taking away the stones the besiegers were piling into the ditch from underneath as the Turks placed them from above, leaving the attackers bewildered since their heap never seemed to rise.

[1] Letter of D'Aubusson, 707; Caoursin, 19–20; Malipiero, 124 ff.

[2] Letter of D'Aubusson, 700; Caoursin, 17; letter dated 30 June from Gabriele Postillo in Rhodes, in Foucard, 135, tells of "iii contalde che tirano a modo de trabucho."

In the course of the many-faceted attack, it was certainly the guns that were seen, by both sides, as the decisive element, a situation made even clearer since the Ottoman artillery on the side where the assault was expected were placed close to the walls, and the attackers had succeeded in occupying the outer works, enabling them to sustain their battery from across the ditch, a proximity that must have been terrifyingly dangerous to both assailant and defender. Yet despite the modern feel to the struggle, Caoursin reported that D'Aubusson was persuaded by a Greek master "wise and expert in sieges" to build his own trebuchet "like a sling" with which to hurl missiles against the Ottoman batteries and tents.[1]

D'Aubusson was well aware that an assault was imminent and dug ditches and erected palisades behind the wall of the Jewish quarter as it fell to pieces under the fire of the guns, creating an immense and inviting breach. The new defence behind the breached wall consisted of a crescent-shaped ditch in front of a rampart constructed of timber, earth and rubble bound together as best as possible, with a wide open space between it and the former defences in front, but of course joining up with the original rampart at each end. In this task, the labour of the citizens had been crucial. The work was however completed in time. The crisis came at the end of July. The Ottomans gave every notice of their intended attack, which duly followed on 28 July following a final blast from numerous guns, then a single mortar shot signalled the assault. Waves of bashi bazouks led the way across the ditch and the flattened old wall, following by disciplined columns of Janissaries, and a two-hour long combat of great viciousness ensued. The defences held, although only after a critical struggle at the point where the new rampart joined the old at the tower of Italy: if the Ottomans had captured this position, they would have been able to bypass the new defences into the city. It was held by D'Aubusson in person, with a handful of others, and the presence there of three standards of the Order seems to have played some part in causing a psychological panic among the attackers (most of whom, after all, were probably Greeks, Macedonians, Serbs or other former Christians from the Balkans) who turned and fled, the tight-packed formation of the Janissaries behind them now proving fatal as the whole force fell into chaos quickly turned into a massacre by pursuing defenders. Some 300 Janissaries were all who had actually fought their way into the Jewish quarter, where they were cut off and cut down.[2]

Misac Pasha pulled his army back from the walls preparatory to ordering their re-embarkation. The defenders' losses were heavy, but nothing like the several thousand dead lost by the pasha.

The sieges of Otranto

On the very day that the Ottoman assault on Rhodes came within a hair's breadth of success, another Ottoman force was sighted off the coast of southern Italy, landing without resistance and ravaging the countryside, the population fleeing to

[1] Caoursin, 24; Malipiero, 127 ("istrumenti da tirere piere").
[2] Caoursin, 28–31; Letter of D'Aubusson, 709–11; Malipiero, 124–8.

the walled towns that dotted the coast and the province of Puglia, chiefly Otranto, as on previous occasions. This time though the attack was not only a raid: the Ottoman force proceeded to march to Otranto and to capture it after 15 days, sending shock waves through the whole of Italy, and in particular to Naples, whose king Ferdinand I (1458–94, of the ruling house of Aragon) was ruler of the whole of the south, but who had failed dramatically to protect his land and people.

Otranto had once been enclosed by a powerful circuit of walls and towers, but that had been a millennium before this attack. Frederick II had ordered the erection of a castle, the result being the familiar rectangle with large angular towers at each corner. The city walls, however, had not benefited much from rebuilding, and a deep ditch represented the most significant obstacle by 1480.[1] There was apparently little or no modern artillery in the city, and probably not many trained fighters. The only local resource seems to have been oil, available to manufacture incendiary devices to support the defence, although there is no reference to any such being used. Several different accounts, including those sent by eye witnesses to the Sforza ruler of Milan, reported that having landed unopposed, the Ottoman force approached the city and cut down the surrounding gardens to clear the approaches and obtain from the olive trees the timber to create a rampart for themselves, then set up batteries of bombards, and also hand guns (springards): from which batteries 1,600 shots were fired. The seven large bombards had been conveyed across from Scutari. There is one reference to the local olive oil being used to cool the bombards between shots. First, however, the pasha had summoned the city to yield, a call rejected by the local commander Francesco Zurlo, apparently convinced of the likelihood of relief. However, within days the guns were demolishing the old walls and the castle too, leaving the defenders no leisure to effect repairs, and also filling the ditch. Otranto was stormed on 11 August, the commander and the archbishop being among the dead in the fighting, and with thousands dying or being seized as captives.[2]

Ferdinand's response was rapid, and neighbouring rulers including the pope, Florence and the duke of Milan offered support, including engineers whose role would turn out to be important. The Turkish army had been adventurous, pushing on from its conquest to try its luck at nearby cities, but found Taranto and Lecce defended and Neapolitan forces hurrying to the province to contain the threat. The report suggesting that at the outset there were also no armed men in any of the port-fortresses of Manfredonia, Trani and Barletta indicated a lack of preparedness by the royal government.[3] The king's son, Alfonso duke of Calabria, was given a powerful army of some 20,000, along with a substantial fleet of galleys

[1] G Bacile di Castiglione, 'Le mura ed il castello di Otranto', *Napoli nobilissima*, XIV, 1905, 1–2.

[2] Relazione della presa di Otranto, to Duke Ludovico Sforza, Bari, 13 October 1480, and Report to Costanzo Sforza, 25 September, in Foucard, 147–8, 162–66; Copia Idruntine expugnationis, in L G Rosa et al., *Gli umanisti e la guerra Otrantina*, 1982, 30–7; Giovanni Albino Lucano, De bello Hydruntini Alfonsi II Aragonei ducis Calabriae, same volume, 54–9; Malipiero, 130.

[3] Relazione de la presa, 168.

and the necessary munitions. However, the recapture was not brought about until 1481, in the meantime there being skirmishing, while the Ottoman commander hurried to repair the city's defences. These can only have been temporary works, but at the same time, given both the antiquated state of the town wall and the serious destruction it had sustained in the original siege, the extent of new work must have been considerable for the occupiers to have decided to try to hold it.

By 26 August, letters were reaching Naples explaining that the Turks were fortifying Otranto. By the end of September, the harbour had been strengthened by the sinking of ships in the port mouth, overlooked by a battery of guns emplaced to cover the entry, a likely route for a Neapolitan attack. On the land side, the work consisted mainly of a deepened ditch and patched up walls, and the occupation of neighbouring castles as outposts. A palisade was probably erected in front of the ditch to provide essential shielding for the walls, the weakness of which had been proven so recently, and, more importantly, the creation of earthwork and timber bulwarks in front of the gates to mount batteries of guns (and thereby also provide flanking), and the digging of an internal ditch behind the wall. The Ottomans had the time to bring in large quantities of victuals and munitions, and to ensure the availability of adequate water supplies in the cisterns. A live source of fresh water, a spring a short distance outside the walls, was heavily fortified with bombards and a numerous garrison.[1]

The second siege of Otranto began the following spring. The attack was set up by land and sea, with powerful batteries of artillery being deployed against the Ottoman defences. The bombardment included incendiaries, with engines being destroyed by fire started by nitre. The Neapolitan army, enjoying the support of other Italian and French troops, was nonetheless making little progress against resolute resistance. Trenches and approaches were under construction, involving large scale building and excavation work and the manufacture of vast quantities of fascines. In July, judging they were close enough, assaults were begun by the besiegers, and sustained by relays of fresh troops day and night. Inside the city, outbreaks of pestilence made things even harder for the defenders. Other attackers meanwhile began mining under the walls. Nonetheless, when a truce was sought, it was agreed, although (according to Lucano), it was used to quickly stabilise the foundations. All through the summer, resistance was maintained, until rumours arrived of the death of the sultan. The king shrewdly allowed the Ottomans to send a messenger to confirm the news, and on hearing it was true, the Turks abandoned Otranto (10 September 1481).[2] The significant element to note, like at Rhodes, was the success of this defence, in this case of a very old-fashioned fortification, against the most modern weapons, in a situation of complete isolation, for a prolonged period. Had it not been for the death of Mehmed, it is not impossible the garrison might have held out in a mirror image of Rhodes.

[1] Bacile de Castiglione, 2, with the limited evidence for the Ottoman work; Lucano, 66–7; Letter of Sadletto, in Foucard, 94, 101; Relazione, 169.

[2] Lucano, 75–87.

The new works at Rhodes and Otranto

The example of a real-world test for the defences of Otranto is especially interesting because of the role of Francesco di Giorgio Martini (see above) in the rebuilding of the defences. It is recorded that the duke of Calabria initially called on the services of the military engineer Ciro Ciri, loaned by his employer the duke of Urbino to help with the direction of the siege and the construction of siege weapons. The task now was to prevent a repeat of the events of 1480. This being a costly undertaking meant that it was carried out over a long period, and may not have been finally completed until the mid-sixteenth century, having been carried out perhaps in a series of fits and starts. Over these years, many changes will have been made in the design, in particular the creation of an enormous angular bastion at one end of the castle that was certainly not part of the works planned by Ciri or carried out by di Giorgio. Studies of the work helped by inscriptions indicate that part of the earliest new fortress were the two great towers (called Duchesa and Ippolita) on the town curtain, 12 metres in diameter and 12 metres high, scarped and projecting far into the ditch, and of equal height with the curtain, which was itself also further provided with projecting platforms and half-towers as needed. The castle, in its role as citadel, was also rebuilt, and in common Aragonese style with very large projecting circular towers at each corner, scarped, and standing behind a massive ditch. The principles demonstrated by this most modern of fortresses were, then, those of great size needed to cope with artillery and to mount it, defence from a wall top granting a continuous passage to defenders, and reliance on an immense ditch. The external works protecting the gates were essential to the plan. The simple quadrilateral plan of the castle reflected similar ideas. The straightforward way by which the fortress of 1490 was transformed into that of 1530 hints that fundamental principles were already established.[1]

The task facing the Hospitallers on Rhodes must have been even more daunting, as they contemplated the ruinous condition of their entire defence system. Being in no doubt of their continued vulnerability, the Knights set about rebuilding their dilapidated fortress. A key element in this was the low, solid outer rampart (eighty feet, 24.5 metres wide) ordered by D'Aubusson beyond the rock-cut ditch, as a means of keeping an enemy at a distance all round. Behind this, five of the curtain towers were rebuilt according to modern standards. In 1496, his successor built a huge pentagonal gun platform which provided two tiers of artillery in enclosed casemates, this being the construction leading to the idea that this was truly the first scientifically designed bastion. St Nicholas's tower as rebuilt was so powerful (24 feet, 7.4 metres thick, with an encircling outer wall of twice that strength), that it was not even attacked in the next siege (1522). The Naillac tower, at one end of the commercial port, was rebuilt 37 metres high on walls 3.7

[1] Bacile de Castiglione, 3–4, 20–23; Santoro, *Castelli Angioini e Aragonesi*, 1982, 206–10; M de Marco, *Torri e castelli del Salento. Itinerario Castrero*,1994, 59–60; M Cazzato, *Guida al Castelli Pugliesi*, 1, 1999, 11 (with illustrations and plans).

metres thick at the base, to control the harbour boom.[1] As fast as resources would allow, the city assumed the appearance of the most up to date fortification. As at Otranto, the key role was taken by the towers created to mount and to resist artillery. But in both fortresses, the understanding that the fortress represented a coherent whole in which each part of the defences had to be strong clearly underlay the respective designs: it was the reality, though, that through cost and resources, such works would inevitably take a number of building seasons to reach a state of defensibility. As it happens, other business kept the Ottomans away while Rhodes was rebuilt. There is a hint of irony in the fact that whereas the hastily-refurbished medieval defences of Rhodes, held by exceptional defenders, had succeeded in resisting the most modern attack in the world in 1480, forty two years later one of the most modern fortresses in the region fell to a more measured attack.

Ottomans and Mamluks

Ottoman ambitions reached in all directions, and in 1485 there began a war in the frontier region between the two empires in which the Mamluks came off best. Details are skimpy. Belying suggestions of the superiority of Ottoman arms over those of the much older dynasty to their south, Egyptian victories in several battles were the critical points in this six-year long conflict. But sieges also played a vital part. The state of Mamluk fortifications is not easily known. It had been noted as an exceptional step that sultan al-Ashraf Qa'itbay (1468–96) had carried out a four-month inspection tour of the fortresses and cities of Syria in 1478, and had ordered various repair works to the existing works, which may possibly have been as they had been left following the wars of the late thirteenth century.[2] In 1485, the Mamluk army crossed the mountains into Cilicia and recaptured the hill top castle of Kulak from the invading Ottomans, then (early 1486) besieged Adana. Fifty years earlier, this city had been described, along with its neighbour Tarsus, as being a fine, and well-protected town, with substantial and in places several walls. Constructing their artillery on the spot, the Egyptian army assailed Adana with guns and "mangonels". As this is affirmed in several sources, it may be assumed to confirm an exceptionally late use for old fashioned stone throwers. An Ottoman relief attempt failed to overcome the Mamluk siege works and led to a serious military defeat. The siege was pressed and the town surrendered. Tarsus then also yielded. The pattern was repeated in 1487, with a new Ottoman force repairing Adana, and the Mamluks recapturing it again. In 1490, Kuwara was taken by storm.[3] Peace was made in 1491. Just twenty five years later, a new Ottoman

[1] Among many descriptions see Migos, 14–18, and Spiteri (as per footnote 92 above).

[2] R L Devonshire (ed.), Rélation d'un voyage du sultan Qaîtbay en Palestine et en Syrie, *Bulletin de l'Institut Français d'Archéologie Orientale*, xx, 1922, 5–16; Ibn Iyas, *Histoire des Mamlouks Circassiens*, trans. G Wiet, 1955, II, 156.

[3] Ibn Iyas, 216, 244, 247, 286–7, 308; Forres, *Die Osmanische Chronik des Rustem Pascha*, 19; Kline, *The 'Voyage d'outremer' by Bertrandon de la Broquière*,59–60. For an analysis and account of the whole war, S Har-El, *Struggle for domination in the Middle East. The Ottoman–Mamluk war 1485–1491*, 1995.

sultan was to avenge this war and destroy the Mamluk empire altogether. But the events of 1485–91 had not predicted that outcome in any way.

General conclusions: attack and defence in Europe, Eurasia and the Middle East

The sieges of Rhodes and Otranto confirm several conclusions about the nature of siege warfare at the end of the fifteenth century. Both the actions, and even more the expensive building work undertaken after the fighting was done, illustrate that there was by now a commonly understood set of principles for modern defence, notably the gun tower and the inter-related role of outworks.

Fortress builders accepted the key role of the gun tower, which we may loosely define as a tower the primary purpose of which was the mounting of artillery. In its design the importance was recognised of being able both to have fields of fire to cover all approaches (including concealed positions for hand gunners to interdict the ditch), and to provide protected positions for one's own heavy guns. These demands posed problems that were now being addressed by flanking, and angled or cylindrical towers with carefully designed gun positions often on several levels. Much thicker walls were needed than in pre-gun days, and these in turn called for new designs for the gun embrasures to permit the required coverage of the ground, and also to meet the need for ventilation to keep the tower interiors healthy for the operating teams. To meet such a variety of demands required towers of unprecedented size, and this if anything was the single most striking feature of new defences of this time.

In itself, constructing larger towers was not a radical departure, but in the course of its evolution during the closing decades of the fifteenth century it would become merged with theoretical ideas obtaining new roots in the renaissance world, seeking to apply geometric (scientific) and classical principles to fortress design. The development of the lower-lying, but interlocking bastion trace with its similarly interwoven network of outworks designed through interlocking fields of fire to keep an enemy at a safe distance, however, was still not something that could be foreseen from the viewpoint of 1480 or 1490: the great fortresses of this period, even the most modern works such as Fougères or Rhodes, after all still bore a much greater resemblance to a high medieval castle than to a Vauban fortress. The works erected at this date were immensely strong, but in most cases they were limited to particular sectors of the circuit rather than applied to a complete new enceinte. Each work being so large meant that they were very costly in money and construction time. Even if late fifteenth century engineers had wanted to complete a regular geometrical trace of a wide, strong but lower-lying rampart connected all round with evenly-spaced gun towers, the task would have been beyond their resources.

The fortresses described here also featured a new style of outwork, even where the work was resticted to the most vulnerable positions such as in front of the gates. The traditional idea of concentric defences was being revisited for an age when the key means of defence was no longer the crossbow, and when the new chief weapon of attack (and defence) had to be resisted by means other than height. On top of walls and towers, the ubiquitous machicolations or hoarding for close defence now served no military purpose, instead actually representing an obstruction to the design of firing positions for heavy guns and not much use to hand gunners. In many cases, therefore, where such construction was to be found at this time, its purpose must have been as ornamentation. In a fortress with a military purpose, whether constructed of stone or from an earthwork and palisade, outworks like barbicans had to be able to mount artillery with a field of fire. A means had also to be provided for defending the artillery and its operators from being picked off, so having (in the more sophisticated constructions) levels from which hand gunners (and sometimes crossbowmen or archers too) could safely fire their weapons was an essential component. Where this could be doubled by overhead firing from higher inner walls behind, so much the more effective. The ability of an attacker to emplace their siege works close to the walls was critical in establishing whether the defence was going to be placed under immediate pressure. If the garrison could by means of effective firing from the walls and towers force the attacker to pitch their camp and to start their works out of range of the defenders' guns, the advantage gained was considerable. It would cost the attacker heavily in casualties and sheer effort to bring their own guns close enough, and to set them up securely. The defence could watch them doing so, and identify where their main points of attack would be, allowing time reinforcement or for second or third lines of defence thrown up behind. Although in some circumstances passive resistance might succeed, another effect of the power of fifteenth century artillery on siege tactics was to increase the value to a defence of an aggressive approach.

While the western world was undergoing dramatic changes in so many spheres and the Ottomans were in the process of creating their new empire, a more conservative era was under way in the east, that would ultimately lead to the subjection of the great civilisations of India and China to the expanding western nations.

The Far East, a brief overview

China and Korea

The Ming dynasty had initially presided over a period of Chinese imperial expansion, including the despatch of great fleets to survey the whole world. But the combination of continual military adventures, the construction of the Forbidden City in Beijing during the first two decades of the fifteenth century, the rebuilding and great extension of the Great Wall to shield the north, and simultaneous naval expeditions had brought about financial exhaustion. When this was compounded by the successful revolts of the Vietnamese seeking their independence (1407–28), a new Chinese ruler reversed the previous policy, and under his successors maintained an approach of closing China off to the rest of the world.

Therefore, where once the Chinese had been in the forefront of technical development in many spheres including siege warfare and gunpowder, the new and more peaceable policy brought this to an end. Development during the later fifteenth century was much more limited, although a trend towards larger weapons from smaller was notable. The traditional armoury of weapons otherwise was retained in use well into the sixteenth century: the centuries-old "thunderclap" bomb and its iron-cased successor (mainly with anti-personnel rather than siege uses), rockets, including multiple-firers mounted on carts (a sophisticated weapon observed by 1451), siege towers including versions with "flying bridges ("cloud ladders")" used to enable an assault on the top of a wall or tower, and gunpowder-filled mines, with, again, very advanced mechanisms to ignite them. Cannon were of both cast iron and cast bronze, with barrels constructed with reinforcing rings to preserve them.[1]

In Korea, where the Choson dynasty took power from 1392, the country took its lead from China in military matters.In 1413 there is the first reference to the use of a mortar being used to fire the Chinese explosive bombs, apparently for the very considerable ranges of some 500 metres (iron shot) or nearly 600 (stone).[2] The Koreans certainly had gunpowder weapons from the fourteenth century, but generally preferred to use powerful wooden arrows as the missiles, indicating that they were intended for use against ships. Nonetheless, by 1415, records show an arsenal of no fewer than 10,000 guns of all sizes (including hand weapons), deployed between the country's fleet and some one hundred castles, and from 1420 iron shot replaced stone as standard, and the guns were organised into four standard sizes.[3]

Japan

Although the Koreans were often engaged in fighting against pirates from Japan, the latter country otherwise remained in a backwater politically and militarily after the successful resistance against the Mongols, followed by a century of domestic violence, then in the fifteenth and sixteenth centuries a period of division into feuding warlordships, with samurai warriors dominating the country. The precise role and function of fortified buildings during this lengthy period no doubt changed with changing social conditions and cultural norms. There is a great deal of surviving source material, either in the form of scrolls with very detailed illustrations providing valuable visual evidence, or the texts of a genre of literature called "war tales", sometimes very long books which were often at least partly based in known historical fact, from which it is possible to derive evidence for what happened in actual conflicts: whether the detail given for each story is an

[1] G Menzies, *1421. The year China discovered the world*, 2002, 50–3, presents a clear picture of the changing policies of the Chinese emperors; S Turnbull, W Reynolds, *Siege weapons of the Far East (2), AD 960–1644*, 2002, 4, 9–19, 37–8.

[2] Turnbull and Reynolds, 14 (translating a Korean source).

[3] Turnbull and Reynolds, 20–1.

accurate version of the real events cannot always be verified, but can be assumed that even if fictional, the event *could* have happened.[1]

By the fifteenth century, a new period of castle building and siege warfare was getting underway in Japan, although firearms only became common in the sixteenth, after their introduction by the Portuguese. Featuring significantly were private castles. Warfare continued to be the domain of the warrior, so while peasants and city dwellers inevitably became involved, towns and cities rarely did so, and therefore remained, for the most part, unfortified. From the 1450s warrior lords began to construct their castles on inaccessible heights (a type of castle called *yamajiro*), but these sites made for uncomfortable residences, so it became normal for their lords to move to more pleasant dwellings, around which supporting services would spring up, at the foot of these hills, thereby creating the phenomenon of castle-towns (*jokamachi*). It was not long before the castles themselves began to be constructed on the more convenient plains, where their symbolic function became enhanced. The tall, richly decorated and majestic multi-storeyed donjons, were visible signs of authority across great distances. Although they were often given serious defences in the form of wet moats and extensive stone walls, and elaborate gateways and entrance passages, some of the greater examples of these *hirajiro* never faced an attack.[2] It is safe to conclude that the donjons at least served symbolic, ceremonial or sometimes residential functions, and very rarely defensive. This judgement might be applied beyond the donjon, too. The immense city built to house the imperial palace and the associated army of civil servants at Heian-kyo (modern Kyoto) had followed the Chinese model except for being simply demarcated with a light earth bank behind a moat. As part of the changes of the sixteenth century, this city was rebuilt behind an immense earthen rampart.[3]

Siege technology continued to be very basic, as if it were somehow dishonourable to use machines to attack a castle, as opposed to direct assaults which featured regularly. The war tales tell repeatedly of battles that revolved around the attack or defence of fortresses, as in the fourteenth century *Taiheiki*, discussed in Chapter 4. The form of the fighting could hardly have been more different from the carefully considered methods of attacking and defending fortresses familiar everywhere from the Atlantic to the Indian Ocean and the China Sea. The done thing for the attacking army was for its warriors to compete to be the first to charge straight at a fortress wall, seeking the honour of storming or scaling it first, or, just as likely, that of being the first to die from the arrows or stones of the defenders. That, however, was when the defenders themselves did not give up their walls and rush out to face the enemy in open battle. Such a culture made being a commander rather tricky (or in some ways, rather easy). But time and again, the stories tell of

[1] P Varley, *Warriors of Japan as portrayed in the War Tales*, 1994, for a thorough analysis of this theme, discussed in its application up to the fourteenth century in Chapter 4.

[2] K Hirai, *Feudal Architecture of Japan*,trans. Sato, Ciliotta, 1973, 15–17, for a summary of modern scholarship and many fine illustrations; Varley, 185–88.

[3] J Hall, Kyoto as historical background, in J Hall, J Mass, *Medieval Japan*, 1974, 6–12. See the description of the original plan from the *Taiheki* in Chapter 4 above.

the initial attack having failed, the assailant was left with no option but to starve out the defender. References to siege weapons being used are scant. There is evidence that the Japanese continued to use Chinese-style bombs, and these were launched from trebuchets until at least 1500. A Chinese "thunderbomb" noted in 1468 was made of paper and bamboo wrapping two pounds of gunpowder. When it exploded, it sent out a spray of iron filings to blind the defenders. Meanwhile, gunpowder artillery remained out of use. Chinese style siege towers also featured occasionally in fourteenth century sources, and there are also literary references to the use of other devices and tools.[1]

The prevailing culture of the fourteenth century did not change in the next, according to the fifteenth century romance *Yoshitsune,* which retells the myth around a real twelfth century hero, but in the course of a whole work devoted to fighting, never once involves a fortress, although there are references (as noted previously) to "secret" Chinese military treatises being used (this one being attributed to L'iu Tao, here kept by someone whose house was protected by a moat and watch towers, "guarded like a fortress").[2] Attack by bombardment, or by digging trenches, or mining, seem to have been rarities. Cultural considerations limited their employment. The skill and resources that went into the construction of castles suggest that it was not the shortage of knowledge or skill that could have been responsible.[3]

India

Meanwhile, in India, the latest military technology was late in arriving: or at least, if it was present, it failed to be noticed in most contemporary accounts. The mining skills available certainly seem to have been used regularly by Indian armies in some parts of the country, but early fifteenth century references to firearms are limited to 1412, 1423 and 1429, as already noted, although references to the use of gunpowder for ceremonial fireworks are older. By the middle of the century the descriptions are more definitely attributable to weapons. One of the accounts of the siege of Mandalgarh by Sultan Maḥmūd Khaljī in 1456 describes "the impact of the balls of the *r'ad* and the stones of the *manjaniq* and the rampart was demolished", while a description of a 1470 siege talks of "roaring *r'ad* having the effect of a thunderbolt". The first definite use of the word *r'ad* as a cannon can be dated to Chitor (Rajput) in 1442.[4] A Russian traveller described a Muslim siege of the

[1] A Hüllen, Burgen und Belagerungstechniken in Japan zur Zeit der Shôgune (14–19 Jh.), in Wagener and Laß, 400–405. This overall picture is also proposed in a recent study, S Turnbull, *Samurai. The world of the warrior,* 2003, 96–7, 115–16.

[2] *Yoshitsune, a fifteenth century Japanese chronicle,* ed./trans. H C McCullough, 1966, 97–8. Another title for the work is *Gikeiki.*

[3] Turnbull and Reynolds, 6, 24, 39.

[4] M Akram Makhdoomee, 'The art of war in medieval India', *Islamic Culture,* XI, 1937, 463, 475–78; I A Khan, 'Early use of cannon and musket in India 1442–1526', *Journal of the Economic and Social History of the Orient,* XXIV, 1984, 146–63, citing references from Nizamu'd Din and the *Firishta* (1607) on cannon, Ma'asir i Maḥmoud Shahi on Mandelgarh (writing 1467/8), and Maḥmud Gawan from 1470. See also J Sircar, 'Siegecraft in Medieval

strong hill top Hindu town of Bijanagar in 1473, following a month-long attack of bombardment by heavy guns, after which the walls were stormed successfully, but amid frightful carnage among the attackers which was repaid by a slaughter among the inhabitants.[1] A (much later) history of the Bahmani dynasty ruling the Muslim kingdom of the Deccan describes sultan Muḥammad III (1463–82), in pursuit of alleged wealth in diamonds, in 1474 successfully besieging hill top fortress Wairagadh by simple blockade. Then, facing a rebellion, his large army attacked Belgaum, a well-built stone fortress with deep wet ditches, with "cannon, guns and *manjanik*", bringing it to surrender, then repeated the process at Rajamundri with cannon, guns and arrows, with the same outcome.[2]

The earlier date is independently verified by the traveller Nicolo Conti's writings, no later than 1444, stating the inhabitants of central India used "balistae and those machines which we call bombardas".[3] This verification is helpful, because the scholars cited have sometimes arrived at different interpretations of the texts, and interpretation is not helped by the probability that words changed their meaning over time, and may have been transferred from describing stone throwers to describing gunpowder weapons. For example, Gode states that Bengal was actually the source of manufacture of guns in 1406, a claim not supported by the evidence of later years.[4]

All the sources describe the significant role played by tamed elephants, not just in the field (where they served as mobile forts carrying soldiers equipped with missile weapons including muskets), but also in sieges where from classical times onwards they sometimes served as mobile battering rams, trained to break down city gates and walls.[5] The continued attachment to catapult artillery alongside gunpowder weapons suggested by these sources well into the second half of the fifteenth century suggests that some value continued to be gained from the older weapons. It might also signify a slow process of providing sufficient numbers of the newer weapons to replace the older.

India', *Itihas*, IV, 1976, but much of this article seems to be based directly on Makhdoomee's work.

[1] Travels of Athanasius Nikitin of Twer, in R H Major, *India in the Fifteenth Century*, 1857, 29.

[2] J S King (ed.), *The History of the Bahmani Dynasty*, 1900, 106–9. The author, Ali bin Aziz-Ullah Tabataba, wrote in the later sixteenth century.

[3] Nicolo Conti, Travels . . . as related to Poggio Bracciolini in his work entitled Historia de Varietate Fortunae, in R H Major, 31.

[4] Gode, 'The use of guns and gunpowder in India from AD 1400 onwards', *Studies in Indian Culture and History*, V, 1960, 6.

[5] Makhdoomee, 469 (citing a reference in Amir Khusrau, Khaza'in al Futūh).

On the cusp of a new era: siege warfare in context

One feature of the long period we have studied has been the interchange of technological knowledge across seas and continents. Different nationalities, languages and religious belief systems have not stood in the way of places without particular knowledge being able rapidly to acquire the services of those able to introduce it. Means of attack and means of defence also quickly levelled themselves out over the same great distances. What then were the changes that made for the change of era at the end of the fifteenth century?

Apparently, it was not technology alone. Time and again the history of the fifteenth century, although demonstrating the improvements in manufacturing techniques and design that made the gun of 1500 a much different weapon to that of 1400, has shown that other factors still were the main determinants of the outcome of sieges. Certainly, the necessary changes to the size and design of fortifications to enable them to resist attack in the first place had led to improvements of such a scale that, by and large, only the powerful and wealthy could manage them. But such wealth and power was increasingly in the hands of rulers of more centralised states able to levy taxation, or possessed by the governments of cities where those who had achieved wealth through their economic activity had also secured political authority. Land and its exploitation remained the overwhelmingly largest source of wealth, so landowners continued to be a potent social and therefore political force. But the beginning of what historians have called the "early modern" age was underpinned by the consolidation of capital as the increasingly dominant economic power, as opposed to the medieval age where feudal relations between people was the predominant social and economic force. Whereas change (although change was continuous) over the millennium following the end of the western Roman empire was slow, in contrast the pace of development during the fifteenth century was dramatic. The century saw the beginning of western exploration, opening new trade routes and, soon afterwards, overseas empires. This process became consolidated during the following century and brought about the creation of economic, political and therefore military powers far in advance of anything seen in the previous thousand years, and beginning the development of a historically significant gap between capitalised Europe and the Asian world, while inbetween, the Muslim world was trying an alternative line of march.

Alongside, the explosion of scholarship, science and learning alongside art and writing, combined with the invention of printing, began a new period of spreading culture. Not unconnected with the latter, the new age would see a new generation of social and political radicalism, too, still largely dressed in religious terms. In Christianity, the Protestant reformation would be the first successful challenge to the universal church there had ever been, and the connections between this doctrinal revolution and social and political change made the early sixteenth century a different age from the late fifteenth.

The siege warfare carried out by Justinian and the first Caliphs would have been recognisable, in its essentials, to William the Conqueror, to Philip Augustus and Richard the Lionheart, to Frederick II Hohenstaufen and to Jaime of Aragon. Although Edward III and Henry V oversaw some changes in weaponry, changes in defence to match them would not have made the resulting fortresses look so substantially different to those of any preceding medieval ruler. The castles built by Mehmed the Conqueror to take, and control, Constantinople had new features unknown even a century before, but were visibly similar nonetheless. But not long after, the great new gun towers of Rhodes or Fougères and the growing outworks at fortresses newly constructed at the end of the fifteenth century, although still looking to height as a key factor, stood on the edge of a new period, one where some of the principles now being applied to these transitional defences would continue (while others would be further transformed) into the new principles behind the increasingly sophisticated geometric, bastion-traced garrison fortresses or fortified cities of the early modern age, held by garrisons of regular paid soldiers serving newly developed states. For all that, it has ever been the case that the courage, skill and determination, and the levels of organisation and provision of those involved in any siege has usually proved to have no less an influence on the outcome than the weaponry, bricks and stones.

8 New weapons and new defences

JUST AS A SUCCESSFUL ATTACK on a major fortification required a combination of different resources on a large scale, so also the defence needed to adopt a range of measures where possible. Whether there was a defence against mines depended in part on the site, and on how deep the ditch was. The defence against artillery relied both on the strength of the wall and towers, but also and increasingly on the design of outer layers of defence, ditches and wet moats, and second walls. In the ideal design, the combination of ditch and outer wall would protect the inner wall from bombardment, but to defend the outer defences against assault meant, if possible, providing a concentric design that enabled the defence to shoot from two levels simultaneously, which in turn limited the distance between the two walls. Gates were always important, and it is not to underestimate their role as a visual statement of the importance of the owner (or of the self-esteem of the city) to note that since time immemorial they had also been the subject of particular care when it came to defensive considerations. With the doubling up the curtain walls and the ditches, gateways also needed an outer layer if they were to fit the new configuration, requiring some kind of barbican.

Extreme measures might be taken, as with the building by the Moroccans of a complete new town with modern defences after the close-run siege of Algeciras in 1278–9, and if this massive undertaking was not only for these reasons, its timing, was nonetheless suggestive. Converting old defences or building new ones, even if using sites where some of the weaknesses could be compensated for by the topography was expensive. Fortresses built or rebuilt during the thirteenth century, such as the new castles and city walls in Capetian France, several of those put up or restored in the Latin east, or the work of Frederick II in Italy, or Edward I's in Wales, were of unprecedented physical strength. They were also prodigiously costly, so much so that the later generation of Edwardian castles was never completed.

Measures taken to resist the trebuchet and the more sophisticated combination of attacking options available to powerful rulers were not only static: defenders also had better means of attacking their attackers. Throughout, and particularly from the turn of the millennium, the important role of the bolt and arrow in defence (as well as attack) has been apparent. The medieval crossbow and the medieval ballista were weapons built on similar principles, the latter being a much larger weapon requiring a standing wooden frame and propelling large bolts, but both being operated by the force of tension rather than the original torsion of the Roman ballista. This is not accepted by some, who argue that the powerful ballistas noted in many accounts of sieges must have been torsion powered.[1] What is not in doubt, however, is that torsion power was definitely the propellant for a new weapon that became popular during the thirteenth century, the springald (French *esp(r)ingale*, German *Springu(o)lf* or *Selbschoss*, coming from the Italian *spingarda* via the Occitan *spingola*). This was a modern version of the torsion propelled anti-personnel catapult, hurling a weighty bolt for considerable distances using a bow string powered by a pair of twisted skeins held in a strong wooden frame. Studies of the weapon, of which few illustrations exist, suggest that it improved upon the Roman version by better quality metalwork and a screw to wind back the string, enabling tighter skeins and therefore greater power. The invention of the spinning wheel in Europe around 1250 also simplified the task of winding the skeins.[2] Tests carried out by Liebel with a model and computers to reproduce the ballistics, using horsehair for the skeins, led to the conclusion that the springald could propel a bolt weighing 1.4 kgs around 180 metres when shot from a tower top at an elevation of 15 degrees. Accounts tell of its frightening powers of penetration, even well-made plate armour offering little protection against a direct hit, while being hit when wearing less protection would almost always be fatal, making Thomas Gray's father's survival (see page 89), particularly remarkable.[3]

The evidence is also that crossbows of significant size were available. Of the normal crossbow shape and operated by the same principles, their greater scale meant that they required windlasses or something similar (such as the screw sometimes used with the springald) to wind them, and, as with the springald, required at least two people to operate them. Surviving examples listed by Liebel, and those where the dimensions were reported in the records, had spans of between 1.6 and almost 2 metres. By the fourteenth century crossbows using steel were to be found, producing even better results but making for a heavier machine. Shooting at maximum elevation, giant crossbows might reach a range of 250 metres, but at this range were extremely inaccurate: their effective range was significantly less than this, while their impact on the target was less than that of the springald.[4] Common in contemporary records were crossbows of "one" or "two"

[1] See J Liebel, *Springalds and Great Crossbows*, trans. J Vale, 1998, 2. Whatever the truth of Liebel's assertion, his further claim that the *petraria* was none other than the continuation of the Roman torsion-powered stone-throwing ballista is based on the evidence of these weapons needing large amounts of rope. But this should not be surprising if the weapon was in fact the manually-operated stone-throwing mangonel, which is certainly a more likely explanation. Furthermore, "ropes" in themselves would only provide torsion once twisted into the necessary skeins, a task that was complex, time-consuming and depended as well on what kind of material the rope was made from. There is little evidence (such as it available for the later weapon) to suggest that ropes of this kind were meant by the sources.

[2] Liebel, 8–9.

[3] Liebel, 16–17, 66–7. Note that the reported maximum range was not tested, but was calculated.

[4] Liebel, tables at 66–7. Payne-Gallwey reported an extreme range at 45 degrees of 370 yards (341 metres) for the fifteenth century steel weapon, but a maximum effective range of less than 200 yards (185 metres) for the earlier, still windlass-operated, machine, even one where the bow itself was of the most powerful then known composite construction: 1–7,

feet, a distinction capable of a number of possible interpretations, none of which can be definitively proved. Once commonly thought to signify how they were to be operated, it is no less likely that the terms refer to the size of the bolt or of the arms of the machine, as Liebel argues. Either way, they required machinery to span the weapon.

Such weapons became widely used in Christendom from the 1250s. Liebel noted a mention under Louis IX in 1249 and Gessler found that the arsenal at Reims had *espingales* by 1258. Henry III's army ordered horsehair for its weaponry at the siege of Kenilworth (1265–66). Other examples of its common use included presence in the arsenal at Berwick castle (1298) and use in the war in Flanders (1304). At Berwick, the inventory established that three springalds were deployed on the walls of the castle "with all their furniture".[1] This is a clue to their use. With their range and penetration, their chief value was to provide effective defence for fortress walls. They made it very dangerous for enemies to venture close to them, and close meant up to 180 metres, which was also the effective range for most siege artillery. Their continued popularity was confirmed by their continued place in arsenals. Dover castle's inventory of December 1344 included three large springalds with all their attachments, particularly cords, alongside eight smaller examples, some without cords. The same inventory included 135 crossbows, of which 34 were of "two feet", nine of "one foot", and three were large windlass-operated machines. In January 1361, the inventory included nine springalds with their compete equipment, all in good order, of which two were large, with three barrels full of quarrels for them. The quality of the other weaponry noted was much more varied, suggesting that the springald remained important.[2] The small proportion of springalds compared with crossbows suggests as well that they were either expensive to build, or required particular skills, Liebel calculating that machines built for the Avignon popes cost the equivalent of six months' wages of an unskilled labourer.[3]

Springalds still suffered from drawbacks. To absorb the shock of shooting required a heavy frame, while the component parts themselves were also weighty, given the strength they needed, making for restricted mobility. Skeins of horsehair continued to be vulnerable to wet weather, making it essential that they be protected. This meant that they had to be sheltered. The most effective means of doing so was to deploy them inside a defence, rather than on top, hence the preference for placing them inside towers. But the consequence was to limit their field

20–22. Some of the late medieval illustrations of these weapons are reproduced in Nicolle, Thompson, *Medieval Siege Weapons(1), Western Europe* AD *585–1385*, 2001, 20–1, 33. These are from Taccola's *De Ingeneis*, German fifteenth century MSS, and the fourteenth century *Romance of Alexander*. Liebel reproduces the illustration of a giant windlass crossbow from Milemete's *De Secreta Secretorum* (Liebel, 29), and Valturio's springald.

[1] *Documents illustrative of the History of Scotland from the death of King Alexander III to the accession of Robert Bruce*, ed. Stevenson, 1870, 323–25.

[2] A Way, 'Accounts of the Constable of the Castle of Dovor', *Archaeological Journal*, XI, 1854, 383–4. The stores also listed crossbow arms and spare cords, and hawsers for spanning the weapons.

[3] Liebel, 18.

to that allowed by the shooting slit, as well as the height above ground level: naturally, the nearer the ground, the less the field of vision and the shorter the range. Furthermore, they continued to suffer a problem that was serious for a weapon intended to protect from above against close range attack: that of not being able to be depressed beyond about 15 degrees without the bolt slipping out. Generally, though, the evidence overwhelmingly reports that springalds were used in towers to cover the approach to gates, be it in German towns and cities or English castles[1] and Gascon defences of all kinds. Their use must have been primarily to defend against attackers approaching too closely, to attempt to fill the ditch, begin sapping the wall or launch a general assault. They would have been of limited use against bombardment from trebuchets, provided the latter were emplaced at the limit of their effective range. Alternatively, they would also have been a valuable deterrent and protection against sudden attacks.

Either explanation fits the measures ordered by the French lord Hugues de Cardaillac for the defence of his castle at Brioule in 1346. While small cannon featured in his instructions, the very first weapon listed was an *espingala* to be placed on the first floor of the *meurtrière* in front of the gate (presumably some kind of barbican), with its crew of two men. Crossbows were to be placed in the upper floors of the towers, and in the towers of the inner defences and featured repeatedly at all points of the defence, the regulation specifying both large and smaller versions. The chief weapon at the other gate was another springald. Interestingly, and reflecting the fact that none of these weapons could be directed to fire downwards, the crew of the springalds were instructed to throw down big stones (*peyras grosses*) at enemies who had got too close to be shot at.[2] This was the arrangement for an old castle, being brought up to modern defensive standards in 1346. The important role of springalds and large crossbows at this date is also demonstrated by the explicit instructions issued for the defence of the new castle being built for the count of Savoy in the contested frontier region between Savoy and Dauphiné. The new fortress being erected at Remens (Ain), called Château Gaillard, in 1345–50 was a stone quadrangle with projecting corner towers, construction taking place behind temporary wooden palisades to prevent hostile interruption. During the first year of building, a local *balistarius* was paid for staying at the castle site for six weeks to string and arm a dozen *ballistae*, two of them large weapons with windlasses, for the defence of the castle. The large weapons were of "two feet", the others of one.[3] At the same time, the procurators for the king of Mallorca ordered that

[1] Or in addition to Welsh castles. The North ("engine") tower at Cricieth Castle, it is suggested, was designed for either a springald or other engine (R Avent, *Criccieth Castle* (guide book), 1989). It is a two-storey structure with wide steps to the roof and was surely too small (nine metres by five internally) for a trebuchet. Might it have had a timber roof to protect the weapon(s) that do cover the approach?

[2] E Forestie, 'Hugues de Cardaillac et la poudre à canon (XIVe siècle)', *Bulletin Archéologique et Historique de la Société Archéologique de Tarn-et-Garonne*, XXIX, 1901, 20, 204, where his regulation is reprinted.

[3] P Cattin, 'La construction d'un château au XIVe siècle: Château Gaillard (Ain)', *Cahiers Perce de Lucinge*, ser. 4, 26, 1990, 38, 40, 41.

the royal castle of Bellver and the city below be provided with ballistas and other items necessary for its defence.[1]

So in the middle of the fourteenth century crossbows were amongst the first requirements of a castle being built where attack could be expected. They continued to be valued weapons of defence well into the fifteenth century. Springalds, however, disappear from the record by around 1400, entirely replaced by gunpowder weapons. The arsenal of the duke of Orléans at Blois castle, inventoried in 1418, 1421 and 1434, listed large quantities of guns of various sizes, with their accompanying munitions, alongside substantial supplies of crossbows, including (in 1421) 29 newly-constructed and others listed as having been brought there from the siege of Tours, including it seems from the description some weapons capable of multiple discharges, but no springalds.[2]

It cannot be claimed from the evidence that defence of fortresses against attack by trebuchet was significantly strengthened by the deployment of the springald and the giant crossbow. There are precious few references in the many accounts of the full scale sieges of the thirteenth and fourteenth centuries in which the role of such defensive weapons is specifically indicated. Rather, the new bolt-shooting weapons must have been important in strengthening the defences against direct attacks in general, such as attempts at escalade, and possibly meant that a besieger had to take measures to neutralise them with stronger protection for the crews of his artillery, and for the soldiers and workmen deployed to approach towards the walls, in the form of cats, mantlets, sows and other such devices. The attack and defence of strong places continued to have the same form, and to operate on the same principles, as it had throughout. New weapons of attack such as the trebuchet, and of defence such as the springald, called for changes in the design of fortifications, chiefly in size and layering to keep an attacker at a greater distance. But these were modifications, not fundamental changes. The integration of the new with the existing is quite clear from the examples studied, stretching from the thirteenth through into the fourteenth century. Alongside deploying springalds and large crossbows, commanders were still equipping their defences with temporary defences such as the wooden hoarding attached to the wall tops to allow their people to drop missiles on attackers getting to the foot of the wall, or to cover a ditch below, while an essential component of the attackers' arsenal continued to be the simple ladder, a piece of equipment that had to be tall enough for its purpose, and strong enough for several armoured men to climb it at one time. Such was the specification of ladders commissioned from local carpenters in Savoyard accounts for 1312–14 and 1321–2, alongside other payments for military operations of that period that included timber for the construction of wooden shelters for the assailants (cats and sows), and, intriguingly, for oak for a "big engine" inside the castle of St. Germain, "in anticipation of a siege by the Dauphin."[3]

[1] Campaner y Fuertes, *Cronicon Mayoricense*, 1881, 109 (instruction dated November 1349).

[2] Dupré, 'Inventaires de l'artillerie du château de Blois en 1418, 1421 et 1434', *Revue des Sociétés Savantes des Départements*, ser. 4, v, 1867, 313–14.

[3] P Cattin, 'Le château de Saint-Rambert (Ain) au début du XIVe siècle. Les comptes de

The slow impact of cannon

These last examples occur in the same period as the first indisputable references to the use of cannon in medieval warfare and its rapid diffusion. However, its first serious impact on fortifications would not take place for many decades, and traditional forms of siege warfare including siege towers and battering rams continued to be employed right through the century. While defences were now being adapted to hold small cannon, through the addition of gun holes and methods of ventilation to towers, the walls and towers themselves were not immediately subject to a different design. After all, the steps needed to protect fortresses against attack by trebuchet and mine were also quite sufficient to resist early cannon, at least until the end of the fourteenth century. For a substantial period during the second half of the fourteenth century, gunpowder weapons were found operating in sieges, on both sides, alongside batteries of traditional stone throwing artillery. Several conclusions have been deduced from these events.

It was not until the last quarter of the fourteenth century that technical improvements made it possible to construct guns that could seriously damage properly-built fortifications. Before that, many rulers adopted the new weapons, but if we compare the use of cannon in the period from the sixth to the eighth decades of that century in Germany, Poland, Savoy, Muscovy, the Balkans and Mamluk Alexandria, it is clear that whether the forces involved possessed guns or not, the outcomes were not seriously influenced by these factors. It is evident from such dramatic conflicts as the siege of Zara in 1345 and the siege of Chioggia in 1379, both involving technically advanced Venice, that while the switch from the trebuchet to the gun as the predominant siege weaponry was clearly identified in the accounts, the sieges themselves were determined by factors other than the technology of the artillery. Even with the better quality guns available by 1400, analysis of the all-conquering campaigns of Timur suggested fairly clearly that these fearsome armies were mightily proficient in siege warfare, but did not seem to have used the gun. There is little in the history of their almost invariably successful attacks to indicate that this was a cause of weakness, either. Instead, with the advantage of massive numbers to provide labour, and access to large numbers of artisans and other experts conscripted from the conquered lands to devise and construct machines, Timur's armies won their sieges by use of all the traditional methods, including large batteries of stone throwing artillery, large scale deployment of mobile timber shelters to allow troops to fill ditches and reach the bottom of walls, and, especially, mining. The unstoppable Mongol was without doubt capable of doing anything he decided was needed, including reinventing the ancient Persian device of the siege mound, so there can be little doubt that had he wanted guns he would have found a way to procure them. But in the context of the forces and resources available to him, he found the methods he used, and refined to an

dépenses de la châtellenie de St Rambert 1299–1340', *Cahiers Perce de Lucinge*, 4th ser, 30, 1995, 58, 104–8. Other references in these accounts include the construction of a bricola for the defence of the castle (1312–14) and a number of crossbows, including payment for a cover for four ballistas (1314–15).

extreme pitch of horror, more than sufficient. One of the new powers of the age, albeit one whose rise was briefly halted by Timur, the Ottoman emirate, had learnt siege skills quickly as it faced sophisticated enemies to south and west, but also did not apparently take up the new weapon as a significant part of its arsenal until into the fifteenth century. Likewise, where the two types of weapon were both in use, as at the Florentine siege of Vicopisano (1405/6), or the Anglo-French fighting in Saintonge (1405), it was by no means guaranteed that the gunpowder artillery of this date was more useful in battering medieval walls than the great stones still hurled with great destructive impact by weapons of the trebuchet type.

So if there had not been through the fourteenth century a decisive change in the design of fortresses to deal with new weaponry, one change that has been noted was a revival of the tactic, albeit in a rather different form from before, of what was previously called a "siege fort". Often called a version of the name "bastion" in various languages (*bastia, bastita,* even a new meaning attached to *bastide*), these were not the much more narrowly defined bastions of the sixteenth century onwards, constructed as part of a scientifically designed, interconnected trace. Rather, they were often temporary forts built to blockade an enemy city, in which case they might well be linked together by trenches or palisades to provide a continuous encirclement, or sometimes free standing, relying on their garrisons to control intervening land, or else they had the same form but were erected by the defence to keep an enemy at a greater distance. The use of systems of outlying defences required an attacker to adopt a very different style of approach. Time and again, we have found that in the new world of the late fourteenth and fifteenth centuries, individual defences were only rarely held with success against attack from state forces, but in contrast, the siege of a resolutely held town provided with modern defences was usually a matter of long and expensive actions, frequently involving the attacker in the erection of a ring of temporary fortresses and bastions either to isolate and starve the besieged, or, at much closer range, to mount the weapons of attack in order to bombard the ramparts of the defenders' own bastions and walls, or to provide cover for mining operations. Around 1400, this was the reality of siege warfare everywhere from Valois France to Ming China. Such temporary works leave no traces, but there are sufficient accounts and occasional illustrations to confirm that they possessed varied characteristics, according to the materials available. Where large timbers were essential components, indeed, one would expect that the fort would be dismantled and the timber reused after its purpose was achieved. Whatever the outcome of the siege, neither a successful defender nor a successful attacker would want to leave intact an encircling ring of empty redoubts.

The advent of new, much larger guns was noted in the period from around 1380 onwards, capable of firing gigantic shot with absolutely devastating effect, once the logistical problem of bringing them to the target had been overcome. Examples of immense guns used around the early fifteenth century to demolish city and castle walls have been cited, as well as their limited availability: only wealthy cities (as in the Netherlands) or powerful princes were able to have such weapons. The role of heavy guns was recognised in the writing of Christine de Pisan around 1408.

Even this was not an entirely clear cut advantage for the attacker, as the analysis of Henry v's conquests in Normandy from 1415 shows. Possessing effective artillery was undoubtedly a critical factor in the English victories, but few were the sieges won by this element alone, and the demoralisation and absence of leadership among the French was probably more important even than the big batteries. The duration of several of the sieges (for example, Cherbourg and Rouen, or in different circumstances Melun in 1420) was evidence that well-built and bravely-defended ramparts could resist direct assault by any contemporary method, the two extremes of premature collapse and enduring resistance against similar odds yet again confirming the critical role of the intangible question of morale. The Anglo-French-Burgundian war that continued throughout the first half of the century, however, witnessed the continued development of ever more effective gunpowder artillery, until at the time of the final expulsion of the English from Normandy and Gascony in the 1450s the skilfully reorganised and even more effective batteries of the Bureau brothers signalled, finally, the end of older methods and weapons of siege warfare. No longer were the massive but impractical giant cannon the most important part of the ruler's arsenal, but instead batteries of rather more modest but still very powerful guns, importantly firing the same size of shot, and achieving a powerful impact were now the key to unlocking almost any defence. The form and effectiveness of what we may term modern artillery, in which the organisational element was no less important than the characteristics of the weaponry, was established first by the French kingdom, rapidly adopted by their Burgundian neighbours, and was also significant two or three decades later in the last stages of the conquest of Granada. To establish this arsenal involved a significant investment of state resources to pay for expert direction and administration, and required the consolidation of the means to design, manufacture, test, transport and store permanently the resulting weapons.

Only the powerful could manage this. It used to be thought that it was the French invasion of Italy in 1494 that witnessed the unquestionable arrival on the scene of the modern artillery train, but all the elements had been put in place, and their effectiveness demonstrated, as early as the 1450s. When the Ottomans took Constantinople using some really gigantic guns, these were already historical curiosities elsewhere, and the pummelling of the ancient city's magnificent defences was as much due to the centralised organisation of the Ottoman artillery train and its carefully-planned deployment as to the size of any individual weapons, and to the highly professional approach to the siege operation in its entirety, with its mines and earthworks of no less importance than the guns. The sources do not allow a reconstruction of the particular effects of individual weapons, but the fact that the monster guns could only shoot a couple of times a day itself justifies the conclusion that the rest of the battery was of no less importance in creating the breaches. One of the major differences between this city, which fell, and others of the age which did not, was that the defence seems not to have heavy guns of its own with which to batter the Ottoman siege works, that were in consequence permitted to be erected at what proved to be a fatally close distance from the outer wall. In this conflict, as with others of those studied above, the age of the walls was not the decisive issue. But this is not to deny that, other things being equal, to create fortifications that were designed to both mount, and face, powerful new siege artillery was a necessary innovation for fortress designers during the fifteenth century.

Moving towards "early modern" fortresses

The central European fighting between the Hussites and their enemies after 1420 saw a conflict that was marked by innovation and experimentation on the Bohemian side, where the Hussite leaders built fortifications to accommodate, and to resist, gunpowder weapons. If the creation of towers solely to hold guns was not new, having been predated by a generation in western Europe, the overall design of Hussite fortifications suggested much thought about how to deal with the challenge of powerful artillery, as well as how to use it. This was done in the heat of a desperate struggle, with the Hussite leaders facing overwhelming odds, and aware that whereas their imperial enemies could always gather new forces, for them a single decisive defeat must spell destruction. So their innovations on the battle field and in their defences must have been driven by urgent necessity. Not for them an abstract study of theoretical ideas, but rather the test of a violent reality in which they could not afford to make a single mistake. The main points of the new defences added to the old walls of Tábor in the 1420/30s by Žižka were examined above. If each of the elements was not in itself novel, the combination of doubled walls, and the provision of half-round gun towers on the inner wall, and of triangular projections for artillery set in the outer shield wall, with large polygonal gun towers every 250 metres added later after the initial works had been completed, indicated an understanding of the role of these connected defences. These defences, and their resolute defenders, proved sufficient to hold Tábor against an imperial army for three months until it gave up, chastened with heavy losses from the defenders' guns.

If the Hussite progress in fortress design did play a role in leading to the bastion *trace*, the actual first use remains unsolved and certainly dates to after the end of our period, although numerous transitional examples have been identified from across Europe, while the Ottomans were not backward either. The purpose of the new work was to provide a base for defensive artillery, such as the specially designed gun ports in new fortresses like Rumeli Hisar or the *canonnières* of French and Burgundian new building (1450s onwards). To provide protection (and a firm base for weapons on top), thinking of this time called for immense thickness of walls, making the modern fortress much stronger than ever before, and very expensive, and new forms of close-range protection, including caponiers from which a ditch could be defended with firearms, and ravelins as outworks (designed for guns) in front of gates, where the barbican had once stood. The space between the inner and outer walls now needed to be scarped, too. The further development towards the principles of the later design of having provision for a continuous wall-top defence including provision for heavy guns at this level, and a massive increase in the size and role of the ditch, was suggested in the new works done at the cities of Otranto and Rhodes after their sieges in 1480, the latter including in particular a two-storey pentagonal gun platform of enclosed artillery casemates (1496) which has often been described as the origin of the "true" bastion. It is not useful either to endorse or to refute that claim: there is some muddle involved in the question itself. If one asks, can we find examples of fifteenth century fortifications that appear to have the same form as sixteenth and seventeenth century bastions, the answer is yes, whether it be Tábor or Mont Saint Michel or Rhodes. But the question should really be: are there any forerunners of the scientifically designed *complete fortress* that uses a *system* of interlocking bastions? The answer is clearly no. Fortresses in this time and location, unquestionably military in purpose and having good reason to expect heavy attack, did not have the luxury of starting anew with an ideal design, but were compelled to maintain the defensibility of their circuit while work to modernise it continued. The existing topography and geography limited the prospects for undertaking a geometric reconstruction, as theoretical treatises of the fifteenth century were advocating, although even here, with the authors not impeded by existing realities on the ground, the recommendations did not demonstrate that they were based on an understanding of the principles underlying the later bastion trace (for example, Salses). These designers were (to varying degrees) grappling with the right questions: it was just that they had not reached the "right" answers. At the time, it was not apparent that the solution of the usually circular projecting gun tower was anything less than adequate to the needs of the time. That future fortress builders would adopt a different route as the *processes* already well underway from the middle of the fifteenth century evolved further (such as more powerful artillery) could not be known.

It would be a combination of a suddenly rapidly changing world around and after 1500 with the application of the experience learnt from using modern siege technology that had evolved throughout the fifteenth century that would lead to a form of fortress warfare in the sixteenth century, and onwards, that would have been unrecognisable to even their immediate ancestors. Many of the main elements in the design of the fortress of 1450 (whether in France, Asia Minor or China) would have seemed familiar to rulers and leaders nine centuries before (albeit with significant changes in scale): concentric defences, projecting towers, ditches, outworks to protect gates, battlemented walls and platforms for artillery atop the towers. Similarly the attacker would have recognised the usefulness of the mantlet, battering ram, siege tower and siege fort, along with the continuing popularity of the mine. Only the artillery would have been unfamiliar. But the operations taking place another century later were distinctly of a different epoch.

There are no ends or beginnings in history, but every study has to start and end somewhere. We have traced significant continuities between the classical and medieval worlds in Europe and the Middle East. We have shown where and how technologies and technical ideas from China were cross-fertilised with those of the Roman successor states via central Asia (carried either through peaceful trade or violent nomadic expansions) and the Muslim world, in turn acting on Africa and the Indian subcontinent, while Chinese influence was intermittently effective in Indochina and, of course, Korea and Japan. We have demonstrated that there were very few revolutionary changes throughout the period, but rather incremental changes and developments both in weaponry, and in the form and strength of fortifications, while making the point that it is necessary to put the latter in their historical context: it was the effective exploitation of the economic growth that created the wealth that enabled rulers to construct or modernise their walled and towered cities and fortresses during this whole era. Evolving technologies reached an apex in the twelfth and thirteenth centuries, without there having been a point at which any single dramatic alteration in siege warfare might be pinpointed.

The arrival of the counterweight-operated trebuchet around 1200 did have a significant impact on the effectiveness of siege operations, and in order to resist it defences built thereafter had to be of a different scale. Gunpowder weapons, on the other hand, underwent a very long period of improvement before they finally began to register a significant impact against well-built defences, but once they did so, starting during the fifteenth century, fortifications again had to undergo quite distinct changes in form, design and strength. Just as the counterweight trebuchet before it seems to have been reintroduced to China by the Mongols, so too gunpowder weaponry, long in use in that land, was reintroduced in devastating new forms from the west considerably after its first military use in Europe. The destructive force of large guns far exceeded that of any previous weapon, and called into being a time, perhaps half a century or even more in length, in which various fortress designs were tried. Finally, the geometric, interconnected and generally scientific fortress the main points of which were regular bastions of immense strength forming the keys of interlocking patterns of outworks, ramparts and ditches took over. Such fortresses fall outside of this study. Leaving aside their purely military intentions as opposed to the more varied roles intended for most of their predecessors, their construction and arming required the resources of centralised states, and their defence (and attack) required the attention of professional armies. There was clearly, nonetheless, a long period of continuity between the late medieval form of siege warfare and that now called early modern during the fifteenth century, at the end of which renaissance engineers, artists and rulers who based their innovations on an appeal to the rediscovered works of the ancient classics in fact produced the fortresses that fitted easily into a different, and almost modern world. The technology had certainly made its contribution to the developments we have charted, but technology was then and remains the servant of human action and wider social, economic and political change.

Time line

Notes:

HR = Holy Roman (Emperor)
"etc." means that other sieges in the same country or area also took place.
Not all sieges discussed in the chapters are included in this chronology, in some cases because the exact date is not known.

Year	*Significant reign or event*	*Western Europe, Iberia, Scandinavia*	*Central and Eastern Europe, Italy, the Baltic*	*Balkans, Middle East, North Africa*	*Central Asia, Russia, China, India, Japan*
1126	Southern Sung dynasty				K'ai –feng
1206	Genghis Khan (d. 1227)				Hsiang-yang, Te-an
1209					Chung hsing-fu
1211			Fellin	Ankara	
1212	Frederick II (HR Empire) (d. 1250)				
1213	James I (Aragon)	(d. 1276)	Albarracin (etc.)		
1218					Otrar, Suqnaq, Kargin
1219					Jand, Fanakat, Khojend, Zarnuq
1220		Entella			Bukhara, Samarkand
1221					Merv, Gurganj, Tirmuz, Balkh, Herat, Talaqan
1222			Oesel		
1223				Trebizond	
1225				Tabriz, Tiflis	

Year	*Significant reign or event*	*Western Europe, Iberia, Scandinavia*	*Central and Eastern Europe, Italy, the Baltic*	*Balkans, Middle East, North Africa*	*Central Asia, Russia, China, India, Japan*
1226		Avignon, Salvatierra		Kâhta, Cerniszkezak	
1228			Dorpat, Dünamunde		
1229	Ögodei Khan (d. 1241)	Mallorca	Gaeta		
1230		Rothesay; Jaén		St Hilarion, Kantara	
1231		Ceredigion region		Harran, Raqqa, Beirut	
1232				St Hilarion, Kyrenia	
1233		Brecon, Carmarthen, Usk; Burriana		Amida	Ts'ai Chou
1234					Vladimir, Torckok
1235		Ibiza			Gandzak, Shamkor, Isfahan
1236		Córdoba			
1237					Riazan, Koloma, Moscow
1238		Valencia	Brescia		
1239			Balga, Christburg	Jerusalem; Ani, Erzerum, Kayseri, Sivas, Konya	
1240		Carcassonne	Ravenna, Faenza	Beaufort	Kiev, Vladimir, Galich (etc.); Lahore (etc.)
1241			Vardin		
1242		Montreuil, Fontenay, Taillebourg (etc.)	Esztergom		

1243			Viterbo	Tyre	
1244				Damascus	
1247		Montségur		Ascalon	
1248		Seville			
1249	Sixth Crusade			Mansourah	
1252			Memel; Naples		
1255			Nakel		Dagon (Rangoon)
1256			Nakel (etc.); Foggia, Oria	Alamout	
1258				Irbil, Harran, Baghdad	
1259			Luck	Aleppo	
1260	Baybars I (Egypt) (d. 1277) Kubilai Khan (d. 1294)	Cefynllys	Königsberg (etc.)	Selymbria, Constantinople	
1262				Mosul	
1263		Rothesay; Dyserth, Deganwy		Kerak	Iuriev
1264		Northampton (etc.)	Sutri, Vico		
1265				Sinope; Bira, Arsuf, Caesarea	
1266		Kenilworth	San Germano	Saphet	
1267				Sis	
1268			Lucera	Beaufort	Hsiang-yang (–71), Fanch'eng (–73)
1270	Seventh Crusade			Tunis	

Year	*Significant reign or event*	*Western Europe, Iberia, Scandinavia*	*Central and Eastern Europe, Italy, the Baltic*	*Balkans, Middle East, North Africa*	*Central Asia, Russia, China, India, Japan*
1271				Safita, Crac, Bira (etc.)	
1272	Edward I (England) (d. 1307)		Gdańsk		
1273	Rudolf I (HR Empire) (d. 1291)				
1274				Sidjilmassa	
1275		Xerez		Neopatras, Tinmelal	Iki, Hakata
1276		Pamplona	Vienna		Jessulmeer
1277		Dolforwyn, Montesa			
1278		Malaga, Algeciras	Drossendorf		
1281		Aberystwyth, Hawarden		Berat	Nga-caung Khyam
1282	'Sicilian Vespers'		Messina		
1283					Mou Tcheng
1284		Albarracin			
1285		Gerona		Margat	
1286			Herbotenstein		
1287		Dryslwyn	Gaeta		
1288			Worringen		
1289				Tripoli (Syria)	
1290			Cracow	Mahdia	
1291			Messina	Acre	

1292				Qalaat ar-Rum	
1294		Caernarfon			
1295		Bayonne, Rions			
1296		Berwick			Delhi
1297			Riga		
1298				Nadjimah	
1299				Tarifa, Tlemcen; Damascus	Ranthambhor
1300		Caerlaverock	Syracuse		Myin-Saing
1301		Bothwell	Bingen		
1302		Waele			
1303		Tournai	Lublin	Arwad	Chitor
1304		Stirling; Zierikzee		Constantine (–05)	
1306		Kildrummy, Carlisle		Rhodes; Algiers	
1308	Henry VII (HR Empire) (d. 1313)			Ephesus, Madytus	Siwana
1309					Arunepal
1310			Swięcie		
1311			Brescia, Rostock, Wismar		
1312		Perth, Linlithgow	Florence		
1314		Roxburgh, Edinburgh, Stirling	Trapani		
1315		Carlisle	Christmemel		

Year	*Significant reign or event*	*Western Europe, Iberia, Scandinavia*	*Central and Eastern Europe, Italy, the Baltic*	*Balkans, Middle East, North Africa*	*Central Asia, Russia, China, India, Japan*
1316	Gediminas (Lithuania) (d. 1341)	Bristol			
1317		Dublin			
1318		Berwick; Nyköping			
1319		Berwick			
1320		Bouvignes			
1321		St Briavels, Newport, Leeds			
1322			Viborg (–24)		
1323		Iglesias			
1324	Orhan (Ottoman) (d. 1362)	Cagliari; La Réole, Metz		Ta'izz	
1325					Pskov
1327	Edward III (England) (d. 1377)	Pavia			
1328		Faucquemont		İzmir	
1329			Lucca		
1330			Pilten		
1331		Gibraltar			
1333		Berwick	Pavia, Lublin		
1334		Dalkeith, Edinburgh		Gafsa	
1335				Tlemcen	

1336		Stirling			
1337	'100 Years War'	Dunbar	Padua	Tlemcen	
1338		C Cornet (etc.)		Dhemer	
1339		Cambrai			
1340		Tournai, Mortagne, C Cornet; Tarifa	Terni		
1341		Roxburgh; Brest, Auray, Nantes (etc.); Alcalá la Real	Lucca		
1342	Louis the Great (Hungary) (d. 1382)	Rennes, Hennebont, Auray,			Izborsk
1343		Algeciras; Argilers (etc.)			
1344				İzmir (–48)	
1345		Utrecht; Bergerac, Monségur, La Réole	Wroclaw	Zara	
1346	Charles IV (HR Empire) (d. 1378)	Caen, Calais, Aigullon; Liddel Strength	Clermont		
1347	The 'Black Death' arrives	Argenteau	Aversa		
1348				Tunis	
1349			Eltville		
1350	Peter the Cruel (Castille) (d. 1369)	St Jean d'Angely	Pistoia (etc.)		
1351	Firuz Shah (Delhi) (d. 1388)		Montclair		

Year	Significant reign or event	Western Europe, Iberia, Scandinavia	Central and Eastern Europe, Italy, the Baltic	Balkans, Middle East, North Africa	Central Asia, Russia, China, India, Japan
1352		Guines; Aguilar	Zug, Zürich		Angkor Thom
1354		Alghero		Gallipoli	
1355		Berwick			
1356		Breteuil, Romorantin, Rennes (etc.)	Treviso, Giadra		
1358		Cormicy			
1359	Dmitri Donskoi (Russia) (d. 1389)				Nagakait; Shao-hsing
1360		Barcelona	Helsingborg		
1362		Calatayud	Kauve, Helsingborg		
1363		Mantes, Valognes (etc.)			
1364		Murviedro; La Charité (etc.)			
1365				Alexandria	
1366				Mesembria, Gallipoli	Samarkand
1367				Corycos	
1368	Ming dynasty (China)	León, Toledo, Algeciras			Izborsk
1369			Godeswerder, Kovno		
1370	Timur (d. 1405)	Ciudad Rodrigo; Limoges			Vladimir
1371		Carmona; Moncontour			
1372		Moncontour (etc.)			

1373			Suscinio	Derval, Sis	
1374				Kyrenia	
1375		St Sauveur/Vicomte		Sis; Kairouan	Tver
1376			Feltre; Speyer		
1377	Jagiello (Poland-Lithuania) (d. 1434)	La Réole, Bergerac (etc.)			
1378		Berwick; St Malo			
1379		Oudenarde	Chioggia (–82)		
1381		Louvain			
1382		Oudenarde			Moscow
1383		Gravelines, Ypres (etc.)	Vilnius; Syon	Thessaloniki	Sistan
1384		Lochmaben	Marienwerder		
1385		Damme			
1386		Brest		Tiflis	
1387		Brest; Montfort		Van	
1388		Grave (etc.)			Yunnan-fou
1389	Bayezid I (Ottoman) (d. 1402)				
1390			Vilnius	Mahdia	Grodno
1393		Gorinchem		Takrit	
1394				Mardin, Amida (etc.); Constantinople (etc.)	Angkor Thom
1396				Widdin, Nicopolis (etc.)	
1397		Loevestein			

Year	*Significant reign or event*	*Western Europe, Iberia, Scandinavia*	*Central and Eastern Europe, Italy, the Baltic*	*Balkans, Middle East, North Africa*	*Central Asia, Russia, China, India, Japan*
1398		Roermond, Wisby			Multan, Batnur
1399				Tiflis (etc.)	
1400				Sivas, Aleppo, Hama (etc.)	
1401		Groningen		Damascus, Baghdad (etc.)	Smolensk
1402		Caernarfon	Godeswerder	İzmir	
1403		Gorinchem; Carreg Cennen	Dünaburg		
1404		Coety	Wisby		Firuzkah; Smolensk
1405	Shahrukh (d. 1447) (Khurasan)	Marck, Sluis, Mortagne; Plymouth; Bourg, Blaye, Calais; Usk	Poma di Suixio, Pisa		
1406		Bouillon	Trixio		
1407		Arkel, Maastricht; Aberystwyth, Harlech			
1409					Moscow
1410	Sigismund (HR Empire) (d. 1437)		Marienburg		
1411		Coucy, Étampes (etc.)			
1412		Dun-le-Roi			

1413	Henry v (England) (d. 1422)	Autel, Brandebourg		
1414		Compiègne, Soissons, Arras		
1415		Harfleur		Ceuta
1417		Caen, Falaise, Cherbourg (etc.)		
1418		Rouen, Senlis (etc.)	Cividale, Udine (etc.)	
1419	Philip the Good (D Burgundy) (d. 1467)	Gisors, Château Gaillard (etc.)		
1420		Montereau/Yonne, Melun (etc.)	Feltre, Udine; Prague, Benesov (etc.)	Trogir
1421	Murad ii (Ottoman) (d. 1451)	Meaux	Pilsen, Chomutov (etc.)	
1422	Charles vii (France) (d. 1461)		Karlstein (etc.); Lautenburg, Gollub (etc.)	Van, Constantinople
1423		St Valéry, Mt St Michel		
1424		Le Mans, St Suzanne (etc.)		
1425				Limassol
1427		St James de Beuvron	Usti, Stříbro (etc.); Golubac	
1428		Le Mans, Pontorson (etc.)	Nissa	
1429		Orléans, Paris		Sultaniyeh

Year	*Significant reign or event*	*Western Europe, Iberia, Scandinavia*	*Central and Eastern Europe, Italy, the Baltic*	*Balkans, Middle East, North Africa*	*Central Asia, Russia, China, India, Japan*
1430		Compiègne, Clermont (etc.)		Thessaloniki	
1431		St Suzanne (etc.)	Tachov		
1433		Passy, Avalon			
1434		Belleville			
1435		Roxburgh			
1436		St Denis, Vincennes, Paris, Fécamp, Calais, Gisors			
1437		Pontoise, Nemours (etc.)		Tangier	
1438		Montargis	Brescia; Tábor		
1439		Meaux	Verona	Smederevo	
1440	Frederick III (HR Empire) (d. 1493)	Harfleur	Peschiera	Belgrade	
1441		Creil, Pontoise			
1442		Dax, La Réole, St Sever		Tartas	Chitor
1443		Dieppe, Luxembourg, Thionville (etc.)			
1444		Metz		Rhodes	Novgorod
1445			Soest, Rheinfelden		
1447			Soest, Lippe		
1448		Le Mans			

1449		Fougères, Pont de l'Arche, Conches, Verneuil, Rouen (etc.)			
1450		Harfleur, Honfleur, Caen (etc.)		Svetigrad	
1451	Mehmed II (Ottoman) (d. 1481)	Bourg, Blaye, Bordeaux (etc.)		Kroja	Moscow
1452		Oudenarde (etc.)	Wiener Neustadt (etc.)		
1453		Castillon, La Réole (etc.); Ghent (etc.)		Constantinople	
1454				Trepča, Novoberda	
1455		Abercorn, Threave (etc.)			
1456		Utrecht, Deventer		Belgrade	Mandelgarh
1457		Sandwich			
1458				Acrocorinth; Alcócer Ceguir	
1459				Smederevo	
1460		Roxburgh		Amasra	
1461	Louis XI (France) (d. 1483)	Thorpe Waterville			Trebizond
1462	Ivan III (Russia) (d. 1505)	Gibraltar	Monheim, Augsburg	Mitylene	
1463			Moclin		
1464		Bamburgh			
1465		Dinant; Jaén		Cetatea Alba, Cilia	

Year	*Significant reign or event*	*Western Europe, Iberia, Scandinavia*	*Central and Eastern Europe, Italy, the Baltic*	*Balkans, Middle East, North Africa*	*Central Asia, Russia, China, India, Japan*
1466		Baeza			
1467	Charles the Bold (D Burgundy) (d. 1477)	Jaen			
1468	Al Ashraf Qa'itbay (Egypt) (d. 1496)	Harlech			
1469		Caister			
1470				Chalkis	
1471		Arzilla		Tangier	Vijaya
1472		Beauvais			
1473		St Michael's Mt			Bijanagar
1474	Ferdinand (Aragon, m. Isabel of Castile) (d. 1516)	Venlo, Nijmegen	Neuss (–75)		Wairagadh
1475		Blâmont, Nancy			
1476		Nancy	Grandson		
1477				Kroja, Shköder	
1480			Otranto	Rhodes	
1481			Otranto		
1482		Alhama, Loja, Zahara			
1484		Álora, Cambil, Ronda (etc.)		Cilia, Cetatea Alba	
1485		Loja			
1486		Illora, Moclin		Adana, Tarsus	

1487	Velez Malaga, Malaga
1489	Baza, Guadix, Almería
1491	Granada

Glossary

Note: some of the words defined here remained in use throughout the coverage of the book, others were coined anew, or transferred from an older meaning. The lack of certainty attached to interpreting many of these terms may be frustrating, but unless the context makes it clear, there is a risk that the word in use conveyed a different meaning for the writer than it did for others at the same, or different, times. Sometimes, a word has a precise significance for one user, but is a generic term for another.

agger	(Latin). Literally, rampart of which earth is the main component.
alcazar, alcazaba	(Arabic). Citadel, usually in the form of the ruler's defended structure attached or within a town or city.
algarrada	(Catalan). Catalan version of *'arrada.*
'almajaneque	(Al Andalus), from Arabic *majdanik,* stone thrower.
arbale(s)tre à tour	(French). See also *arbalista.* The version *à tour* usually means a large scale crossbow-type bolt-shooting weapon mounted on a wooden frame.
arbalista (etc.)	(Latin), and numerous variant spellings. French *arblaste,* again numerous variant spellings. A crossbow, sometimes with implication of being a large frame-mounted device, more often a generic term. Derivative forms used for crossbowman, and crossbow slits.
arbotane	(Italian). Lit. "pea shooters", hand guns, fifteenth century (chapter 5).
arie (te-)s	(Latin) Battering ram(s). Sometimes in the generic sense of "siege weapon."
arquebus	Appearing late in the fifteenth century, handheld firearm, common infantry weapon in the sixteenth century and forerunner of the musket (German *Hackbut*).
'arrada	(Arabic, from Syriac). Arabic catapult, general term but may have meant a large crossbow-type bolt-firing weapon cf. *ballista.*
arrow slit	A slit cut in a wall to enable an archer to shoot out from behind, introduced in Byzantine fortresses of the seventh century and universally used, with numerous variants. French version *archère.*
arx	(Latin). Specifically, the citadel, but also a generic term for castle or fortress, implying a significant building.
assommoir	(French), murder hole, for dropping stones etc on attacker from above gate; see also machicolation.
atalaya	(Al Andalus) Watch tower.

Balearic sling	Trans. of *balearis funda,* artillery term in al-Andalus met in the twelfth and thirteenth centuries. No evidence to confirm what type of weapon is meant although the name suggests some kind of trebuchet. There is no necessary connection with the Balearic islands as it may derive from Greek *ballo,* throw.
ballesto de torno	(Spanish). *Ballista,* large crossbow, equivalent to the *arbale(s)tre à tour.*
ballista	(Latin). Originally, a stone-throwing torsion engine invented by the Greeks and used by the Romans. In medieval times, a large crossbow shooting bolts (etc).
ban	(Indian). Gunpowder-propelled rocket (chapter 4).
barbican	An outer defence erected in front, to protect a gate.
barud	(Arabic). Originally saltpetre, became general term for gunpowder.
bashura	(Arabic). Outer defensive wall, often signifying the outer enclosure.
bastide	(French). A deliberately created (and usually geometrically arranged) new settlement, common in SW France (and English Gascony) from thirteenth century. Despite earlier claims, it is now understood that only a small percentage were fortified from the outset. By fifteenth century also sometimes used to mean bastion or outwork or temporary siege work.
bastion	Any form of tower or projecting defensive work, occasionally, by transference, the whole work. From early modern times, was used much more precisely to signify an artillery platform projecting from the *enceinte.*
bawn	(Scotland, Ireland). Enclosure wall for the courtyard attached to a *tower house,* sometimes with fortifications.
belfry	A siege tower, designed to overtop the defensive wall to enable attackers to shoot down, and sometimes provided with a bridge to enable attackers to storm the defences.
bent entrance	Originally Muslim design of gate, requiring the entrant to turn at right angles one or more times before reaching the interior. Became widespread.
berefridum	(Latin). See *belfry.* Numerous variants of the name.
Bergfried	(German). Tall, often narrow tower in many German and central European castles, with a very restricted interior. Possible use as symbolic power statement and/or a watchtower.
berm	(From French). The space between a ditch and the rampart.
biffa	Type of trebuchet.
Blide, Blyde	(German). Trebuchet.
blifyeu	(Welsh). Catapult (generic term).
bombard	Type of gun, commonly applied to large or the very largest, chiefly in the first half of the fifteenth century.
bretêche, bretache	(French). See *hoarding* and *machicolation.*
burdj	(Arabic). Generic term for a tower (also spread into Urdu).
Bus(s), Buchse	(German). Gun. Thus *Büchsenmeister,* gun-master.

calabre, calibus — (Occitan). Trebuchet.

caltrops — Sharpened metal spikes spread in ditches to wound attackers.

canon perrier — Gun type of uncertain character (see *cannon*). (Chapter 5).

canonnière — (French). Opening in wall for use by gun.

cannon, canon — Prior to becoming the generic term for all non-manual gunpowder firearms, had a variety of meanings related to a particular type of gun, generally fairly small; sometimes attached to other names of gun types, usually making it impossible to be certain what weapon was meant.

caponier — Enclosed gallery across a ditch between the inner and outer defences, usually provided with shooting positions. Late fifteenth century onwards.

castellation (sometimes in-) — Lit., creating a *castle,* for example by fortifying an existing building, although applied generally to any kind of building or settlement.

castellum — (Latin) Literally, a diminutive of *castrum,* common word for *castle,* sometimes a purely military construction such as a siege castle/fort.

castle — Dwelling of a feudal lord (including ecclesiastical), or the monarch, serving one or many of multiple functions. In medieval usage, almost invariably implied fortified. Word also used by contemporaries for both permanent and temporary fortifications.

castrum — (Latin). Lit. a fortified place, sometimes used to describe a walled town, sometimes a castle. In medieval use, might apply also to a (sometimes only slightly) fortified village; see also (Greek) *kastron.*

cat — Wooden shelter for besieger (from Latin *gattus).*

contravallation — Outward facing rampart protecting a besieging army from attack from outside.

courtau, cortago — (and other variant spellings). Name for a type of gun, becomes common in last quarter of fifteenth century but difficult to determine its type or size. Iberian references imply something more like a mortar (see chapter 6).

crab — Mobile roofed shelter for a ram.

crapardeau — Type of gun. Long, small calibre when in common use (mid fifteenth century).

colovrine — Gun type, appears in early fifteenth century when seems to have been a handgun. The later (sixteenth century onwards) *culverin* was quite different, being a standard large artillery piece.

culverin — Gun type, of various sizes up the large (chapter 5) before eventually becoming one of the standard types in early modern armies.

crenellation — Parapet on top of wall, with openings for use by defending archers (etc.). The intervening stonework known as *merlons.*

donjon — Literally, the "lordship tower" or lord's residential or ceremonial tower (or hall), the most important part of the castle;

	also had a variety of uses. Rarely used as a citadel (or defence of last resort, as was once commonly held), but sometimes played that part in practice.
drawbridge	Bridge across a ditch or moat that could be raised and lowered by the occupants of the town or castle. Various designs.
enceinte	(French). The enclosing wall or rampart around town, castle, village, church (etc.).
fascine	Bundle of any suitable natural material, used for filling in ditches or to construct barricade.
foneval	(Catalan). Stone thrower.
fortalice	(numerous variants, from Latin). Usually small fortified place.
gorod	(Polish and Russian). Fortress
grad	(Russian). Fortress
Greek *fire*	Although a term used widely to describe incendiaries, exact meaning applied to Byzantine invention designed to burn even in water.
gun tower	A tower constructed specifically to mount gunpowder weapons either inside, and/or on the roof.
hackbut	Hand gun, forerunner of the arquebus.
Hauptburg	(German). Lit. main castle, applying to the inner section where the lord resided, equivalent to inner ward or bailey.
hisn, (pl.) husun	(Arabic). Fortified rural refuge(s).
hoarding	Timber gallery built out from the top of a wall to enable defenders to shoot or drop stones (etc.) onto attackers at the foot of the wall. French *bretêche*.
hui hui p'ao	(Chinese). Literally, Muslim catapult, signifying *trebuchet* (chapter 2).
Hus	(German). Lit. house, signifying a defensible place.
jokamachi	(Japanese). Castle-town (chapter 6).
kastal	(Swedish). Free-standing stone tower (chapter 3).
kastron	(Greek). See *castrum*. Definition varied, sometimes meaning a fortified refuge, and later more precisely a castle.
keep	Later word in modern English usage, describing a *donjon*.
kremlin	(Russian). Citadel (see chapter 4). Also *krom*.
licence to crenellate	Formally, the right to turn a building into a castle by providing it with battlements. In reality, usually signified recognition of noble status for the owner, whether or not the building had any defensive function.
lupus	(Latin), lit. "wolf", another term for a siege weapon.
machicolation	Murder hole. A permanent form of *hoarding*, comprising stone projection from front of tower or wall, with holes for dropping stones (etc.) onto attackers at the foot of the wall. In French, *meurtrière*. Might be continuous along a wall, or a single structure e.g. over a door, in which case it is a *box machicolation*.
madjanik	(Arabic). Arabic version of the *mangonel*, usually a generic term for stone-throwing artillery.

Mamluk	(Arabic), Lit., "owned", in the sense of slave soldiers, who went on to establish the long-lasting military dynasty ruling Egypt and Syria.
mangana	(Latin). See *mangonel.*
mangganikon	(Greek). See *mangonel.*
mangola	(Latin). See *mangonel.* One of many variant spellings.
mangonel	Both a generic term for medieval stone-throwing artillery, and to describe specifically manually (traction) powered weapons. sometime wrongly used as a translation for the late Roman *onager.*
mantlet	Roofed timber protection for a besieger, implying a smaller protection for one or two, but also generic term for siege implements; also a wooden shutter between merlons, protecting defenders of a wall.
midfa	(Arabic). Gun (see chapter 3).
miles (-ites)	(Latin). Roman soldier, medieval "knight".
mining	The excavation of the ground below a defence, with a view, usually, to cause the collapse of masonry above when the timber frames holding up the mine were burnt. One defence against this was the *counter mine,* involving digging out from inside the fortification to intercept the mine before it reached the wall.
ministeriales	(Latin). A specifically German institution of socially unfree knights, bound to service of their lord, but sometimes wealthy and powerful.
moat	Water-filled ditch.
mortar	Type of gun particularly used in siege warfare, for shooting projectiles with a steep trajectory over defences. References from mid-fifteenth century (chapter 6). Short in length with large calibre.
munitio	(Latin). Fortress.
murage, muragium	Grant of a tax to be levied by a town for the purpose of constructing a town wall.
mus (musculus)	(Latin). Lit. mouse, another term for a timber shelter for the besieger.
naft	(Arabic). Petroleum, with particular reference to naphtha incendiaries, later transferred to mean gunpowder.
p'ao, phao	(Chinese). A manually-powered stone thrower; many different sizes and specifications covered by this general term.
petraria, petrobolos	(Latin, Greek). French *perrière,* also *petrary* and many variants. General term for stone-throwing artillery, sometimes given a more specific meaning as a smaller, traction-powered weapon by comparison with the counterweight trebuchet.
porok	(Russian). Either battering ram, or stone thrower (see chapter 9)
portcullis	(from French). Commonly, a wooden or metal grille dropped from above to close an entrance. Universally used.
propugnacula	(Latin). General term for outer defences of a fortress, variously an outer wall, or bastions.

proteichisma	(Greek). Lower, outer defensive wall, sometimes with a more general significance of an outer defence.
r'ad	(Indian). Cannon (see chapter 6).
revelino	(Italian). Ravelin, projecting outwork, progenitor of the later smaller outworks set according to a geometric plan in front of bastions and the enceinte (chapter 7).
Ribald	German term for Lithuanian siege weapon, possibly ballista (see chapter 2).
ribaudequin	Wooden platform, wheeled, with several small guns mounted (the term probably means the mounting rather than the guns). See chapter 7.
robinet	Small stone thrower ca. 1300 (chapter 2).
rocca	(Italian). Castle.
sapping	See also *mining*. A form of attack on a defence involving trying to remove stones at the base of a wall to bring about the collapse of the masonry above.
schiopo, sclopo	(Italian). Gun, later specifically the arquebus (chapter 3).
serpentine	Type of gun. Of medium size, a little larger (perhaps) than the *crapaudeau*. Appeared later in the fifteenth century and common in the sixteenth.
siege castle	Temporary fortification erected to enable its garrison to attack, or cut off, a besieged place. Often had the same form as contemporary castles (esp. when these were predominantly earth and timber) but did not share other castle functions. By the fourteenth century, more often called a fort.
springald	(derivation uncertain). Torsion powered bolt-thrower constructed in large wooden frame, in use from early thirteenth century. See chapters 1, 8.
steen	(Flemish). Tower or more broadly, castle.
talus	Sloping outer face of wall or tower, keeping attacker at a distance, providing protection against sapping, also allowing defender to bounce stones off it.
tapial, tabiya	(Arabic, in Al Andalus). Compacted baked mud around wooden frames to make a very strong concrete for walls and towers.
testudo (-ines)	(Latin) Lit. "tortoise(s)", usually, by medieval times, a moveable roofed timber shelter to protect troops attacking the base of a wall; sometimes a generic term for a "siege weapon". In Roman times it had meant a military formation of soldiers using their shields to protect their heads and front.
tormentum	(Latin). Originally, an instrument for twisting, etc. By extension, applied to mean torsion-powered artillery. In time, became generic term for artillery.
torre albarrana	(Al Andalus, Spanish/Arabic). External tower linked at the level of the wall walk with the adjacent main wall.
tower house	Modern term coined to describe a residential tower, sometimes with serious defences, built usually by lesser nobility. Sometimes had outer buildings and a *bawn*, although traces of these have often not survived.

trebuchet	(uncertain derivation). The counterweight powered stone thrower from the end of the twelfth century. Many variant names.
tripontum	Type of trebuchet (chapter 1), of uncertain character.
trueño	(Spanish). Gun (see chapter 3).
Tumeler	(German). Throwing machine (also *Dummeler*) (chapter 3).
Tummeraer	(German) Stone thrower (chapter 2).
urbs (-es)	(Latin). City, but sometimes conveyed meaning of "fortified".
var	(Hungarian). Castle.
veuglaire	Type of gun (lit. "fowler"), of varying sizes, but generally medium-size. Common in the middle of the fifteenth century though in use from the late fourteenth.
Vorburg	(German). Outer castle area.
vinea	(Latin). Another word for a timber shelter at a siege.
yamajiro	(Japanese). Castle built on high ground (chapter 7).
Zwinger	(German). The space between the inner and outer walls.

Sources

There are many reference books containing information on the words used in this study, but no single glossary. Particularly useful have been:

Alonso, M, *Enciclopedia del idioma*, Madrid 1958.
Bradbury, J, *The Routledge Companion to Medieval Warfare*, London 2004.
Encyclopedia of Islam, (revised edition), Bearman P J (ed.), Leiden, ongoing.
Latham, R, (ed.), *Revised Medieval Latin Word-List from British and Irish Sources*, Oxford, 1994.
Lexer, M (ed.), *Mittelhochdeutsches Handwörterbuch*, Leipzig 1872–.
More-Figueroa, L de, *Glosario de Arquitectura Defensiva Medieval*, Cadiz 1994.
Schiller, K, Lübhen, A (eds.), *Mittelniederdeutsches Wörterbuch*, Bremen 1878.

For late medieval gunpowder artillery, particularly useful is

Smith, R D, DeVries, K, *The Artillery of the Dukes of Burgundy*, Woodbridge 2005.

Bibliography

Primary Sources

Abu'l Fazl al Baihati, Tarikhu's Subuktegin, *History of India*, Papers of H M Elliot II, ed. J Dawson, London 1867.

Aegidius: Aegidi Romani Bituriensis archiepiscopi, Libellus de re militari, S Hahn, *Collectio Monumentorum*, Brunswick 1724.

Agapida, Fray Antonio: *see* Irving, W

Ahimaaz, *The chronicle of*, trans. M Salzman, New York 1924.

'Aini, Al-: La Monte, J L, Ziada, M M, Bedr ed Din Al'Aini's acount of the conquest of Cyprus 1424–1426, *Annuaire de l'Institut de Philologie et d'Histoire Orientale et Slave*, VII, 1939–44.

Alberti, L B, *On the Art of Building in ten books*, trans. J Rykwert, N Leach, R Taverner, Cambridge (Mass)/London 1991.

Amadi, F, Chronique, ed. M René, M L de Mas Latrie, Paris 1891, in *Chroniques d'Amadi et Strombaldi*.

Anagnostae, Joannis, Narratio de extremo Thessalonicensi Excidio, Migne, *PG*, 156, Brussels 1866.

Anderson, A O, ed., *Early sources of Scottish history*, Edinburgh 1922.

Andreae Ungari, Descriptio victoriae a Karolo Provinciae comite reportate, *MGH SS* XXVI, G Waitz ed., Hannover 1882.

Anelier, G, *Histoire de la guerre de Navarre en 1276 et 1277*, trans. F Michel, Paris 1856.

Angelucci, A, Spigolatore Militari dell'archivio communale di Foligno, *Archivio Storico per le Marche e per l'Umbria*, III, 1886.

Annala Connacht, A Martin Freeman, ed., Dublin 1944.

Annales Capitoli Posnaniensis, *MGH SS* XXIX, ed. M Perlbach, Hannover 1892.

Annales Gandenses, ed. I M Lappenberg, *MGH SS* XVI, Hannover 1859; ed./trans. H Johnstone, London 1951.

Annales Moguntini, *MGH SS* XVII, G H Pertz ed., Hannover 1861.

Annales Monasterii de Burton 1004–1263, *Annales Monastici*, I, ed. Luard, 1864, translated in M Barber, K Bate, *The Templars*, Manchester/New York 2002.

Annales monasterii de Theokesberia, Annales Monastici, I, ed. H R Luard, RS, London 1864.

Annales Placentini Guelfi, *MGH SS* XVIII, ed. G H Pertz, Hannover 1863.

Annales Prioratus de Dunstaplia, in *Annales Monastici* III, ed. H R Luard, *RS*, London 1866.

Annales Ricardi secundi et Heinrici quarti, H T Riley ed., *RS*, London 1866.

Annales Ripenses, E Kruman, ed., *Danmarks Middelalderlige Annales*, Copenhagen 1980.

Annales Sandovogii, *MGH SS* XXIX, ed. M Perlbach, Hannover 1892.

Annales Scanici, E Kruman, ed., *Danmarks Middelalderlige Annales*, Copenhagen 1980.

Annales Toledanos I, II, *España Sagrada*, 2nd ed. Ed. H Florez, Madrid 1799.

Annals of Jessulmer, in J Tod, ed., *Annals and Antiquities of Rajasthan*, London 1829/32 repr. 1972.

Annals of Loch Ce, ed/trans. W M Hennessey, *RS*, London 1871.

Annals of Mewar, in J Tod, ed., *Annals and Antiquities of Rajasthan*, London 1829/32 repr. 1972.

Annals of Ulster, S Mac Art, G Mac Niocaill, eds., Dublin 1983.

ANON, A brief Latin chronicle, J Gairdner, ed., Three fifteenth century chronicles, *Camden Society NS* 28, 1880.

——, A short English chronicle, J Gairdner, ed., Three fifteenth century chronicles, *Camden Society NS* 28, 1880.

——, Chronik aus Kaiser Sigismund's Zeit, in *Die Chroniken der Frankischen Städte: Nürnberg* I, Leipzig 1862.

——, Chronique du siège d'Orléans et de l'établissement de la fête du 8 mars 1429, Salmon M A ed., *Bibliothèque de l'école de Chartres*, 2nd ser. Vol. 3.

——, Der Feldzug gegen die Hussiten im Herbst 1421, in *Die Chroniken der Frankischen Städte: Nürnberg* II, Leipzig 1864.

——, *Die Altosmanlischen anonymen Chroniken*, F Giese ed., Leipzig 1925.

——, La chronique de Lorraine,L Marchal, ed., Nancy 1859.

——, Obsidione Iadrensis, *Scriptorum Rerum Hungaricarum*, J G Schwandtner, ed., III, 1746.

Anonimalle Chronicle, V H Galbraith ed., Manchester 1927.

Anonimo Fiorentino: Molho A, Szmura F (eds.), *Alle Boche della Piazza. Diario di Anonimo Fiorentino (1382–1401)*, Florence, 1986.

Anonimo Pisano Cronichetta, in Odoardo G O, *L'Assedio di Pisa 1405–6*, Florence 1885.

Anonymi auctoris Chronicon ad annum Christi 1234 pertinens, ed. J C Chabot, *CSCO, Script. Syr. Series 3*, vol. 14 1937, vol. 15 1916.

ANONYMI, Gesta Hungarorum, *Scriptores Rerum Hungaricarum*, ed. E Szentpetery, Budapest 1938.

APPENWILER, Erhards von, Die Chronik 1439–71, *Basler Chroniken*, 4, Bernoulli A., ed., Leipzig 1890.

ASIK PASHA: Dervish Ahmed called Asik Pasha Sari, R Kreutel ed/trans. as *Vom Hintenzelt zum Hohen Pforte*, Graz 1959.

Assault of Mansura, the, *Excerpta Historica*, London 1831.

AZURARA, Gomes Eannes de, *Chronique du roi dom João I*, Virginia de Castro e Almeida ed., Paris 1933.

——, *The chronicle of the Discovery and Conquest of Guinea*, trans. C R Beazley, E Prestage, London 1895, 1899.

BAIN J, ed., *Calendar of documents relating to Scotland in the PRO*, 4 vols, Edinburgh 1884–7.

BAKER, Galfridi, de Swynbroke, *Chronicon*, E M Thompson, ed., Oxford 1899.

BALĀDHURĪ, abu 'Abbas Ahmad ibn-Jabir al, Kitab Futūh Al-Baldun, trans. P K Hitti, *Origins of the Islamic State*, New York 1916; and in *History of India*, Papers of H M Elliot I, ed. J Dawson, London 1867.

——, Dal Futūh 'al Buldan by Ahmed ibn Yahya 'al Balādhurī, *Biblioteca Arabo-Sicula*, ed. M Amari, Turin/Rome 1880 .

BARBARO, Nicolo, *Diary of the siege of Constantinople*, trans. J R Jones, New York 1969.

BARBOUR, J, *The Bruce*, ed./trans A A M Duncan, Edinburgh 1999.

BARLETIUS: Marini Barletii, *De obsidione Scodrensi ad Serenissimum Leonardum Lauretanun Aristocratie Venetiae principem*, 1504.

BARTOŠ, Kronika Bartoska z Drahonic, *Fontes Rerum Bohemicarum*, V, J Goll, ed., Prague 1893.

BASIN, Thomas, *Histoire de Louis XI*, L Halphen, ed., vol. 2, Paris 1966.

——, *Histoire de Charles VII*, ed./trans C Samaran, 2 vols, Paris 1933.

BAVIÈRE, Jean de, Chronique de la règne, *Chroniques Liègoises* I, ed. S Balau, Brussels 1913.

Bayeux Tapestry, ed. D M Wilson, London 1985.

BEL, Jean le, *Chronique*, J Viard, E Deprez, eds., Paris 1904.

BELLONO, A, Vitae patriarcharum Aquilejensium, *RIS* XVI, ed. Muratori, Milan 1730.

BENET, John, Chronicle for the years 1400 to 1462, Hariss G L & M A, *Camden Miscellany* XXIV, *Camden 4th series* 9, London 1972.

BERCHEN, W van, De Gelderse Kroniek, A J De Mooy, ed., *Gelre*, 24, Arnhem 1950.

BERRY, Hérault du Roy, Le récouvrement de Normandie, ed. J Stevenson, RS London 1863.

BLONDEL, R, De Reductione Normanniae, ed. J Stevenson, *RS*, London 1963.

BOUCIQUAUT, *Le livre de faits du bon messire Jehan le Maingre, dit Bouciquaut, maréchal de France et gouverneur de Jennes*, D Lafanda, ed., Paris/Geneva 1985.

BOUTEILLER, E de, ed., *La Guerre de Metz en 1324. Poème du XIVe siècle*, Paris 1875.

BOWER, Walter, *Scotichronicon*, ed. D E R Watt, volume 7 eds. A B Scott, D E R Watt, U Moret, N F Shead, Aberdeen 1996.

BREMEN: Die Bremer Chronik von Rinesberch, Schine und Hemeling. *Die Chroniken der niedersachsischen Städte: Bremen.* H Meinert, ed., Bremen 1968.

BŘEZOVÁ, Laurentius (Vaurina z Brezova), Kronike Husitska, *Fontes Rerum Bohemicarum*, V, Prague 1893.

BROQUIÈRE, Bertrandon de la, *The "voyage d'outremer"*, G R Kline, ed., New York, 1988.

BROSSET, M, Histoire de la Géorgie, St Petersburg 1849.

BRUGES: *Inventaire des archives*, L Gilliodts van Severen, ed., 1st ser., vol. III, Bruges 1875.

Brut y Tywysogyon (Peniarth), ed/trans. T Jones, Cardiff 1952.

Brut y Tywysogyon (Red Book of Hergest), ed/trans T Jones, Cardiff 1955.

BRUT: *The Brut or the chronicles of England*, Brie F W D, ed., London 1906–8.

BUEIL, J de, Le Jouvencel, intro/ed. C Favre, L Lecestre, 2 vols, Paris 1887.

BURCHARD OF MT SION, *Description of the Holy Land*, trans. A Stewart, 1896.

CABARET: Pedro, C del (ed.), La conquête savoyarde du Pays de Vaud d'après "les chroniques de Savoie de Cabaret", in Bagliani A P, Poudret J-F, eds., *La Maison de Savoie et le Pays de Vaud*, Lausanne 1989.

CAGNY, Perceval de, *Chroniques*, H Moranville, ed., Paris 1902.

Caithrein Cellachein Caisil, A Bugge, ed/trans., Oslo 1905.

Calendar of Close Rolls, Public Record Office, London, various dates.

Calendar of Liberate Rolls, Public Record Office, London, various dates.

Calendar of Patent Rolls, Public Record Office, London, various dates.

CAMPANER Y FUERTES, A, *Cronicon Mayoricense*, Palma de Majorca 1881.

CANARIUS, J, De bello Constantinopolitano anni Christi 1422, *Migne PG* 156, 1866.

CANON OF PRAGUE, Continuation of Cosmas of Prague, Annales Otatariani, *MGH SS* IX, G Waitz, ed., Hannover 1851.

CAOURSIN: *Caoursin's account of the siege of Rhodes in 1480, translated into English by John Kay, poet Laureate to king Edward IV*, H W Fincham, ed., London 1926.

CAPO STEFANO, Marchione di, Cronaca Fiorentines, *RIS* xxx(1) new ed., N Rodolico, ed., 1903.

CAPPONI, Gino de Neri, Commentari dell'Acquito, ovvere Presa di Pisa seguita l'anno 1406, *RIS*, XVIII, ed. Muratori, Milan 1731.

CARPINI, G di Plano, *The story of the Mongols whom we call Tartars*, trans. E Hildinger, Boston 1996.

Cartulaire Normande de Philippe Auguste, Louis VIII, St. Louis et Philippe le Hardi, ed. L Delisle, *Mémoires de la Société des Antiquaires de la Normandie*, ser. 2, volume 6, 1852.

CATTIN, P, Le château de Saint-Rambert (Ain) au début du XIVe siècle. Les comptes de dépenses de la châtellanie de St Rambert 1299–1340, *Cahiers Perce de Lucinge*, 4th series, 30, Paris 1995.

Chachnama, in *History of India*, Papers of H M Elliot I, ed. J Dawson, London 1867.

CHAN, Hok-lam, *The fall of Jurchen Chin. Wang E's memoir on Ts'ai Chou under the Mongol siege (1233/4)*, Stuttgart 1993.

CHAPLAIS, P, ed., The War of Saint Sardos. *Camden 3rd Series* LXXXVII, London 1954.

CHARLES IV, Emperor: E Hillenbrand, *Vita Caroli Quarti, Die Autobiographie Karl IV*, Stuttgart 1979.

CHARTIER, J, *Chronique de Charles VII, roi de France*, ed. V de Viriville, Paris 1858.

CHASTELLAIN, Georges, *Chronique*, Kervyn de Lettenhove ed., Brussels 1863–66.

CHENG-WOU TS'UN TCHENG LOU, *Histoire des campagnes de Gengis Khan*, trans. Pelliot, P, Hamlois, L , Leiden 1951.

CHINAZZO, Daniele di, *Cronica de la guerra de Veneziani a Zenovesi*, V Lazzario, ed., Venice 1958.

CHIOS, Leonard of, Letter to Pope Nicholas V, in J R Melville-Jones, *The Siege of Constantinople*.

CHRISTINE DE PISAN, *The Book of Fayttes of Armes and of Chivalry*, trans. W Caxton, ed. A T P Byles, London 1932.

Chronica de gestis principum a tempore Rudolfi regis usque ad tempora Ludovici imperatoris, *MGH SrG Chronicae Bavariae*, G Leidinger, ed., Hannover/Leipzig 1918.

Chronica de Walliae and other documents from Exeter Cathedral Library MS 3514, ed. T Jones, *Bulletin of the Board of Celtic Studies*, XII, 1946.

Chronica Jutenses, E Kruman, ed., *Danmarks Middelalderlige Annales*, Copenhagen 1980.

Chronica Latina Sabaudiae. *Monumenta Historiae Patriae, Scriptorum* II, D Promis, ed., Turin 1839.

Chronica Marchionum Brandenburgensium, ed. G Sells, *Forschungen zur Brandenburgischen und Preussischen Geschichte*, I, Leipzig 1888.

Chronica Regum Manniae et insularum, ed/trans. P A Munch, Dr Goss, Douglas 1874.

Chronica Romanorum PP et Imperatorun ac de rebus in Apulia gestis, auctore ignoto monacho cisterciensi, ed. A Gaudenzi, Naples 1888.

Chronica Sialandiae continuatio, E Kruman , ed., *Danmarks Middelalderlige Annales*, Copenhagen 1980.

Chronicle of Morea, ed./trans H E Lurier, New York/London 1964.

Chronicle of Novgorod 1016–1471, trans. R Michels, N Forbes, intro. C R Beazley, A A Shakhmatov, *Camden 3rd Series*, XXV, 1914.

Chronicon Bergomense Guelpho-Ghibellinum, *RIS* XVI (II), C Capasso, ed., Bologna 1940.

Chronicon Caesaraugustanorum reliquiae, ed. T Mommsen, *MGH Chronica Minora*, XI(2), Berlin 1894.

Chronicon Colmarense, *MGH SS* xvii, G Pertz, ed., Hannover 1861.
Chronicon de Lanercost, J Stevenson, ed., Edinburgh 1839.
Chronicon Episcopum Hildesheimensium, G Pertz, ed., *MGH SS*, vii, Hannover 1846.
Chronicon monasterii Campensis ordinis Cisterciensis, C Eckertz, ed., *Fontes adhuc inediti rerum Rhenanarum*, ii, Köln, 1870.
Chronicon Rottenburgense, transcripsit ex codice bibliotechae illus. DD Francisci L B de Hegemuller, in *Raimundi Duellie Miscellanorum*, ii, 1724.
Chronicon Vazatense, *Archives historiques du département de la Gironde*, xv, Bordeaux 1874.
Chronicon Ventense, ed. M E Piganeau, *Archives historiques du département de la Gironde xv* 1874.
Chronicon veteris universitatis Pragensis, *Fontes Rerum Austriacarum, scriptores* ii(i), K Haefler, ed. Vienna 1856.
Chronijk der Landen van Overmaas en der aangrenzende Gewesten, door eenen inwoner van Beek bij Maastricht, Habets J ed., *Publications de la Société d'Histoire et d'Archéologie dans le duché de Limbourg*, vii, 1870.
Chronique d'Arthur de Richemont, Connetable de France, duc de Bretagne 1393–1458, ed. G Gruel, Paris 1890.
Chronique d'Enguerran de Monstrelet 1400–1444, ed. L Douet-D'Arcq, Paris 1867.
Chronique d'un religieux de Saint Denys 1380–1422, trans. M L Bellaguet, 6 vols Paris 1839.
Chronique de Gislebert de Mons, ed. L Vanderkindere, Brussels 1902.
Chronique de Jean Le Fevre, Seigneur de St Rémy, ed. F Morand, 2 vols, Paris 1876.
Chronique du Mont St Michel, 1348–1468, Luce S ed. 2 vols, Paris 1879.
Chroniques de Savoye, *Monumenta Historia Patriae, Scriptores* ii, Turin 1839.
Cingis Khan, *Die Geschichte seines Lebens nach den chinesischen Reichsannalen*, trans. F Krause, Heidelburg 1922.
Clavijo, G de, *Embassy to Tamerlane 1403–6*, trans/ed. G Le Strange, London 1928.
Clyn, Friar John, *Annals of Ireland*, ed. Butler P, Dublin 1849.
Codex diplomaticus Brandenburgensis continuatus, Raumer, G W von, ed. *Sammlung ungedrükter Urkunden zur Brandenburgischen Geschichte*, Berlin 1831–3, repr. 1976.
Comines, Philippe de, *Mémoires*, new ed. by Godefroy Lenglet du Fresnoy, 4 vols, London/Paris 1747, containing the *Discours du siège de Beauvais par Charles Duc de Bourgogne en l'an 1472* (Beauvais 1622).
Compagni, Dino, *Chronicle of Florence*, trans. D E Bornstein, Philadelphia 1986.
Constructione castri Saphet, De, trans. in H Kennedy, H, *Crusader Castles*, Cambridge 1994, 190–98.
Conti, N, Travels in the early part of the 15th century, as related by Poggio Bacciolini in his work entitled Historia de Varietate Fortunae, lib. iv, in Major R H, ed. *India in the fifteenth century*, London 1857
Copia Idruntine expugnationis, in L G Rosa, I Nuovo, D Defilippis, *Gli umanisti e la guerra Otranteria. Testi dei secoli xv e xvi*, Bari 1982.
Corazzini, G O, *L'assedio di Pisa 1405–6*, Florence 1885.
Cousinot de Montreuil, *Chronique dite de la Pucelle*, ed. Vallet de Viriville, Paris 1892.
Cronica Alfonso x, ed. M Gonzalez Jimenez, Murcia 2001.
Cronica de San Juan de la Peña, Historia de la corona de Aragon, ed. T Ximenez de Embun, Zaragoza 1876.
Cronica del rey D Enrique segundo, Madrid 1877.

Cronica del rey D. Alfonso XI, Cayetano Rosell ed. Madrid 1875.
Cronica latina de los reyes de Castilla, ed. L C Brea, Cadiz 1984/Madrid 1999.
Cronica presulum et Archiepiscorum Coloniensis ecclesie. Fontes adhuc inediti rerum Rhenanorum, ed. G Eckertz, Köln, 1864.
Cronica S Petri Erfordensis Moderna, *MGH SS* XXX, O Holder-Egger, ed., Hannover 1896.
CUSIN, F, Documenti per la storia del confine orientale d'Italia nei secoli XIV e XV, *Archeografo Triestino*, ser. XXI, Trieste 1936.
CUVELIER, *La vie vaillant Bertran du Guesclin*, ed. Charriere E, Paris 1839.
CZARNKOW, Joannis de, Chronicon Polonorum, *Monumenta Poloniae Historica*, II, ed. A Bielowski, Lvov 1872.
DARDEL, Jean, Chroniques d'Arménie, *RHCA*, II, E Dulaurier et al eds, Paris 1906, repr. 1967.
DECEMBRIO, Petro Candido, Vita Philippi Mariae Vicecomitis, Mediolanensium Ducis, *RIS*, XX(1), A P Candido ed., Bologna 1925–58.
DEDEYAN, G, *La chronique attribuée au connetable Smbat*, Paris 1980.
DENNIS G T, ed., Three reports from Crete on the situation in Romania 1401–2, *Studi Veneziani* XII, 1970.
DESCLOT, B, *Chronicle of the Reign of King Pedro III of Aragon*, trans. F L Critchlaw, 2 vols, Princeton 1928–34.
DETMAR VON LÜBECK, Chronik, *Scriptores rerum Prussicarum*, III, T Hirsch et al eds, Leipzig 1866.
DEVONSHIRE, R L, trans., 'Relation d'un voyage du sultan Qaîtbay en Palestine et en Syrie', *Bulletin de l'Institut Français d'Archéologie Orientale*, XX, 1922.
DEX, Jacques de, Die Metzer Chronik, *Quellen zur Lotharingischen Geschichte*, IV, G Wolfram ed., Metz 1906.
DHAHABI: Somogyi J, Adh Dhahabi's record of the destruction of Damascus by the Mongols in 699–700/1299–1301, in S Lowinger, J Somogyi (eds.), *Ignace Goldziher Memorial Volume*, part 1, Budapest 1948.
Diplomatari de l'Orient Catala (1301–1409), A Rubro I Lluch, ed., Barcelona 1947.
DŁUGOSS, I, *Annales seu Chronicae inditi regni Poloniae*. 6 vols, ed. I Dabrowksi et al, Warsaw 1964–.
Documents illustrative of the History of Scotland from the death of King Alexander III to the accession of Robert Bruce, ed. J Stevenson, Edinburgh 1870.
DOUKAS, *The decline and fall of Byzantium to the Ottoman Turks (Historia Turco-Byzantina)*, trans. H J Magoulias, Detroit 1975.
DUCAS, M, Historia Byzantina, Migne, *PG*, 157, 1860.
DUDA, H W, *Die Seltschükengeschichtes des Ibn Bibi*, Copenhagen 1969.
DULAURIER, E, Les Mongols d'après les historiens Arméniens. Fragments traduits. *Journal Asiatique* 5th ser., XI, 1858.
DUPRÉ, M, Inventaires de l'Artillerie du château de Blois en 1418, 1421 et 1434, *Revue des Sociétés Savants des Départements*, 4th series, V, 1867.
DUSBURG, Petri de, Chronica Terre Prussie, ed/trans. K Scholz, D Wojtecki, R Buchner, *AQ* XXV, Darmstadt 1984.
DYNTER, E de, *Chronica nobilisimorum ducum Lotharingiae et Brabantiae*, ed. F X de Rain, Brussels 1854.
EBENDORFER, Thomas, Chronica Austriae, A Lhotsky ed., *MGH SrG NS* XIII, Berlin/Zurich 1967.
ELLIS, Henry, *Original Letters illustrative of English History*, 2nd series volume I, London 1827.

Elmham, Thomae de, *Vita & gesta Henrici Quinti, anglorum regis*, T Hearne, ed., Oxford 1727.

Erfurt: Bayer C, ed., *Urkundenbuch der Stadt Erfurt*, Halle, 1889.

Escouchy, Mathieu d', *Chronique*, G Du Fresne de Beaucourt, ed., 3 vols, Paris 1863, and Buchon J A C, *Choix de chroniques et mémoires sur l'histoire de France*, Paris 1838.

Essenwein, A, ed., Germanischen Museum. *Quellen zur Geschichte der Feuerwaffen*, Leipzig 1977.

Exchequer Rolls of Scotland, vol. iv, ed. G Burnett, Edinburgh 1880.

Fada'il, Al Mufaddal ibn abi, Histoire des Sultans Mamlouks, ed./trans. Blochet E, *Patrologia Orientalis* 12, vols. 14 (1920) and 20 (1929).

Fallmerayer, J P, Original-Fragmente, Chroniken, Inschriften und andere Materiale zur Geschichte des Kaiserthums Trapezont, *Abhändlungen der Historische Classe der königlich Bayerischen Akademie der Wissenschaften*, iv, 1846.

Fauquembergue: *Journal de Clement de Fauquembourgue, greffier du Parlement de Paris 1417–35*, A Tuetey, H Lacaille, eds., 2 vols Paris 1903.

Fesano, Abu-l Hasan 'Ali Ben Abd Allah ibn Abi Zer' *Annales regum Mauritaniae*, trans. C J Tornberg, Uppsala, 1845.

Fidā, Abu'l, *Annales Moslemici*, ed. J J Reiske, 1789.

Filangieri, R et al, eds, *I registri della cancellaria Angiona*, Naples 1950–.

Filarete: Oettingen, W van, *Antonio Avellino Filarete's Tractat über die Baukunst*, Vienna 1890.

Firuz Shah, Futuhat-I-Firuz Shah, Roy N B, ed., *Islamic Culture*, xv, 1941.

Fischer-Wieruszowki, Kriegerisches Einfall der Mongolen in Japan. Eine Japanische Bildrolle. *Ostasiatische Zeitschrift* NS 89, xi, 1935.

Flament, A J A, *Chroniek van Maastricht*, Maastricht 1915.

Fordun, J de, *Scotichronicon (continuation by Bower)*, ed. Hearne, and ed. W F Skene, Edinburgh 1872, repr. 1993.

Forestie, E, Hugues de Cardaillac et la poudre à canon (xive siècle), *Bulletin Archéologique et Historique de la Société Archaeologique de Tarn-et-Garonne*, xxix, 1901.

Foucard, C, Fonti di storia Napoletana nell'archivio di stato in Modena. *Archivio storico per le Province Napoletane*, vi, 1881.

Four Masters, the, *Annals of Ireland*, J O Donovan, ed., Dublin 1848–50.

Fowler, D C, Briggs C F, Remley, P G, *The governance of kings and princes, John of Trevisci's middle English translation of the De Regimine Principum of Aegidius Romanus*, New York 1997.

Franciscani Thorunensis, Annales Prussici 941–1410, *Scriptores rerum Prussicarum*, iii, T Hirsch et al, eds., Leipzig 1866.

Franke, H, (ed.), Die Belagerung von Shao-Hsing im Jahre 1359: ein Beitrag zur Kriegsgeschichte Chinas im Mittelalter, in Franke, H, *Krieg und Krieger im Chinesischen Mittelalter*, Stuttgart 2003.

——, Die Belagerung von Hsiang-Yang, in G L Ulmen, ed., *Society and History: essays in honour of K A Wittvogel*, The Hague 1978.

Froissart, J, *Chroniques*. Livre i ed. G T Diller, Geneva 1991/2; Livre ii, ed. G Reynard, Paris 1894–97; Livre iii, ed. L & A Mirot, Paris, 1931, 1957, 1966.

Gabrieli, F, ed., *Arab Historians of the Crusades*, London, 1969.

Galician-Volhynian Chronicle, ed. G A Perfecky, Munich 1973.

Gallus Anonymous, Cronica et Gesta Ducum sive Principum Polonorum, ed. C Maleczynski, *Monumenta Polonica Historica* NS ii, Krakow 1952.

Gatari, G & B, Cronaca Carreresi, *RIS* xvii(i), new ed., A Medin, G Tolomei, eds., 1909.

GAUTIER, P, ed., Un récit inédit du siège de Constantinople par les Turcs 1394–1402, *Revue des Études Byzantines* 23, 1965.

GAYANGOS, P de: *see* Ahmed ibn Mohammed al-Makkari

Gesta Henrici Quinti, trans/ed. F Taylor, J S Roskell, Oxford 1975.

GHAZI, Zahiru'd Muhammad Bulsur Padinah, *The Baburnama in English*, trans. A S Beveridge, London 1922.

GILBERT, J T, ed., Chartularies and Annals of St Mary's Abbey, Dublin, 2 vols, *RS*, London 1884.

——, Historical and municipal documents of Ireland AD 1172–1320, from the Archives of the City of Dublin, *RS*, London 1870.

GINGINS LA SARRA, F, *Depêches des Ambassadeurs Milanais sur les campagnes de Charles le Hardi (1474–77)*, Paris/Geneva 1858.

GIORGIO MARTINI, Francesco di, *Trattati di architettura ingegneria e arte militare*,C Maltese ed., 2 vols, Milan 1967.

Glyn Dŵr, Annals of, in J E Lloyd, *Owen Glendower*, Oxford 1931 and 1966, 150–54.

GOCKINGEN H, Sweeney J R, *Der Mongolsturm. Berichte von Augezeugen und Zeitgenossen 1235–50*, Graz/Wien/Köln 1985.

GRAY, T, *Scalachronica*, H Maxwell, trans., Glasgow 1907.

GREGORAS, Nicephoros, Byzantinae Historae, Migne, *PG*, 148–9, Brussels 1865.

GREGORY, William, Chronicle of London, J Gairdner, ed., The historical collection of a citizen of London, *Camden Society NS* 17, 1876.

GRIGOR OF AKANC', History of the nation of the archers, ed. R P Blake, R N Frye, *Harvard Journal of Asiatic Studies* 12, 1949.

GRUEL, Guillaume, Chronique d'Artus III, Comte de Richemont, Duc de Bretagne et Connestable de France, in J A C Buchon, *Choix de Chroniques et Mémoires sur l'histoire de France*, Paris 1835.

GUILLAUME DE NANGIS, *Chronique latine de 1113–1300*, ed. H Gérard, Paris 1843.

GUTIERREZ DIAZ DE GAMEZ, *The unconquered Knight. A chronicle of the deeds of Don Pero Nino, count of Buelna*, Evans J , ed/trans., 1928, reprinted Woodbridge 2004.

GYSEBURNE: Cronica Walteri de Gyseburne, Rothwell H ed., *Camden Series* LXXXIX, London 1957.

Hacon's Saga, in *Icelandic Sagas* IV, trans. G W Dasent, London 1894.

HAENISCH, E, Die letzten Feldzüge Cinggis Han und sein Tod, *Asia Major*, IX, 1933.

HAENISCH, E, Olbricht, P, Zum Untergang zweier Reiche. Berichte von Augenzeugen aus den Jahren 1232/3 und 1368/70, *Abhändlungen für die Kunde des Morgenlandes*, XXXVIII(4), 1969.

HANA, C, *Bericht über die Verteidigung der Stadt Te-an während der Periode K'ai-His 1205–8*, Wiesbaden 1970.

HARDYING, John, *The Chronicle from the first begynnyng of England*. Amsterdam 1976.

HARFF: Letts M (ed.), *The pilgrimage of Arnold von Harff, Knight*, London 1946.

HASSENSTEIN, W, ed., *Das Feuerwerkbuch von 1420*, Munich 1941.

HENRY IV, Royal and Historical Letters, F C Hingerton, ed., *RS*, London 1860.

HENRY OF LANCASTER, *Le Livre de Seyntz Medecines. Unpublished devotional treatise*, ed. E J Arnould, Oxford 1940.

HERMANNI DE WARTBERGE, Chronicon Livoniae, ed. E Strehlke, *Scriptores rerum Prussicarum*, Leipzig 1863.

HIMYARI, al, *La péninsule ibérique au moyen âge d'après le Kitab Ar-Rawd al Mi'tar fi Habar al-Akhtar*, trans. E. Lévi Provençal, Leiden 1938.

Histoire des patriarches d'Alexandrie rélatifs au siège de Damiette, *Revue de l'Orient Latin*, II, 1905/8.

HOCSEM, Jean de, *Chronique*, ed. G Kurth, Brussels 1927.

HOLLADAY, J A, *Illuminating the Epic. The Kassel Wilhelalm codex and the Landgraves of Hesse in the early fourteenth century*, Seattle/London 1996.

HRBEK, I, Ein Arabischer Bericht über Ungarn (Abu Hamid al Andalusi al Garnati), *Acta Orientalis Academiae Scientorum Hungaricae*, v, 1955.

HSIANG-YANG SHOU-CH'ENG LU, ti Chao Wan-nien. Trans. H Franke, *Studien und Texte zur Kriegsgeschichte der Südlichen Sungzeit*, Wiesbaden 1987.

HSIAO, Ch'i-ching, *The military establishment of the Yuan dynasty*, Cambridge (Mass.), London, 1978.

HSU MENG-HSIN, Son-ch'ao Pei ming Hui-pien. *Die Belagerung von K'ai-Feng im Winter 1126/7*, trans. S Werner, Stuttgart 1992.

HUGUENIN, J F, *Les chroniques de la ville de Metz*, Metz 1838.

HUILLARD-BREHOLLES, J L A, Luynes, H de Albertis de, *Historia Diplomatica Friderici Secundi*, 12 vols, Paris, 1852–61.

Hypatian Chronicle, selection, in M Laran, J Saussay, *La Russie ancienne*, Paris 1975.

IBN 'ABD AZ-ZAHIR, Muhi al-din, Sirat al Malik az Zahir (Life of Baybars), in Sadeque, S F, *Baybars I of Egypt*, Dacca 1956.

IBN AL 'AMĪD, al Makīn, *Chronique des Ayyoubides (602–658 / 1205/6–1259/60)*, ed./trans. A M Edde, F Micheau, Paris, 1994.

IBN AL ATHĪR, *The annals of the Saljuq Turks. Selections from the al-Kamil fi 'l Ta'rikh of 'Izz al-Din Ibn al-Athir*, trans. D S Richards, London 2002.

——, Annales du Maghreb et de l'Espagne, trans. E Fagnan, Algiers 1898.

——, Dal Kamil 'at tawarila, *Biblioteca Arabo-Sicula*, ed. M Amari, Turin/Rome 1880.

IBN AL FURĀT, Selections from the Tārikh ad-Duwal in U and M C Lyons, *Ayyubids, Mamluks and Crusaders*, Cambridge 1971, II.

IBN ARABSHAH, Ahmed, *Tamerlane or Timur the Great Amir*, trans. J H Sanders, London 1936.

IBN ASI'R, Kamilu 't Tawarikh, *History of India*, Papers of H M Elliot II, ed. J Dawson, London 1867.

IBN BATTUTA: *The travels of Ibn Battuta*, trans H A R Gibb, 4 vols, Cambridge 1938–94.

IBN IYAS, *Histoire des Mamlouks Circassiens*, trans. G Wiet, Cairo 1955.

IBN JATIB, Parangon de Malaga y Sale, trans. E Garcia Gomez, *Andalucia contra Berberia*, Barcelona 1976.

IBN JUBAYR, *Travels*, trans. R J C Broadhurst, London 1952.

IBN WASIL, Jamāl ad Din, Mufarrij al-kurūb fi akhbār Banu Ayyūb. Excerpts trans. in Gabrieli, F, *Arab historians of the crusades*, London 1964.

Inventaire de la Bastille de l'an 1428, *Revue Archéologique, 12, 1855*.

IRVING, W, *A chronicle of the conquest of Granada (from the MSS of Fray Antonio Agapida)*, repr. London 1986.

Jacobi de Verona: Rohricht, R, ed., 'Liber peregrationis Fratris Jacobi de Verona (1335)', *Revue de l'Orient Latin*, III, 1895.

JAMES I, King of Aragon, *Chronicle*, ed. J Forster, 1883.

JAMSILLA, N de, De rebus gestis Frederici II imperatoris ejusque filiorum Conradi et Manfredi, in G de Re, *Cronisti e Scrittori sincroni Napolitani*, Naples 1868, II.

JARRY L, *Le compte de l'armée Anglaise au siège d'Orléans 1428–9*, Orleans 1892.

JAZARI, Al, ibn al-Razzazz, *The book of knowledge of ingenious mechanical devices*, D R Hill trans., Dordrecht/Boston 1974.

JEAN D'OUTREMER, Chronique abregé, *Chroniques Liègoises*, ed. S Balau, E Fairen, II, Brussels 1931.

JOINVILLE, L de, *Life of St Louis*, ed/trans. M R B Shaw, Harmondsworth 1963.

JÖRG VON NÜRNBERG, *Nachricht von den Turken*, Nürnberg 1500.

Journal d'un bourgeois de Paris 1405–49, ed. A Tutuey, Paris 1881.

JUAN MANUEL, Don, El Conde Lucanor, *Cronica Abreviada*. Ed. J M Blecua, Madrid 1983.

JULIANO, Fragmenti chronici Forojuliensis, *RIS*, XXIV, Muratori ed., Milan 1738.

JUVAINI, 'Ala ad-din 'Ata Malik, *The History of the World Conqueror*, trans. J A Boyle, 2 vols, Manchester 1958.

KALHANA: *Kalhana's Rajatararigini. A chronicle of the Kings of Kas'mir*, ed/trans. M A Stein, London 1900.

KERKHOERDE, Johann von, Chronik, 1405–65. *Die Chroniken der westfälischen und niederrheinischen Städte*, 20, Leipzig, 1887.

KHALDŪN, ibn, *Histoire des Berbères et des dynasties musulmanes de l'Afrique septentrionale*, trans. Baron de Slane, 3 vols, Paris 1925–34.

KHANDAMIR, *Habeeb-us-Siyar (life of Tamerlane), a literal translation, parts V, VI*, Bombay 1900.

KHAZREJIYY, Aliyyu'bow'l Hasan, *The Pearl strings, A history of the Resuliyy dynasty of Yemen*, trans. J W Redhouse, ed. E G Browne, London 1906.

KHUSRAU, Amir, The Khaza'in al Futūh, The Campaigns of 'Alauddin Khaljī, Habib M (trans.), in Nizami, K A (ed).), *Politics and Society during the early Medieval period: collected works of Professor Mohammed Habib*, New Delhi, 1981, vol. 2 and Khusrau, Amir, Tarikh'l Alai, *History of India*, Papers of H M Elliot III, ed. J Dawson , London 1871.

KIKULLEW, J, Chronica Hungarorum, *SRH*, ed. J G Schwandtner, 1746.

KING, J S, ed., *The History of the Bahmani Dynasty, founded on the Bulhān-I Ma,āṣir*, London 1900.

King's Mirror, The — Speculum Regale, Konnings Skuggsja, trans. Larson L M, New York 1917.

KNAUZ, F, *Monumenta Ecclesiae Strigensis*, 3 vols, Esztergom 1874–1924.

KNEBEL, Johannis, capellani ecclesiae Basiliensis, Diarium, *Basler Chroniken*, 3 vols,W Vischer W, H Boos H, eds., Leipzig 1880.

Knighton's chronicle, ed./trans. G H Martin, Oxford 1995.

Koelhoffsche Chronik 1499, Hegel C ed., *Die Chroniken der niederrheinischen Städte: Coeln*, III, Leipzig 1877.

KONSTANTIN DEM PHILOSOPHEN, *Lebensbeschreibung des Despoten Stefan Lazarevic*, M Braun ed/trans., Wiesbaden 1956.

KORNER, N, Chronik, in *Die Chroniken der niedersachsischen Städte: Lüneburg*, Stuttgart 1931.

KRITOVOULOS, *History of Mehmed the Conqueror*, ed./trans. C T Ross, Princeton 1954.

KYESER, Konrad, *Bellifortis, Bayerische Staatsbibliothek Clm. 30150*, U Lindgren, C Waldburg Wolfegg , Graf zu, E Berninger, eds., Stuttgart 2000.

LALAING: Le livre des faits du bon chevalier messire Jacques de Lalaing, in Kervyn de Lettenhove, ed., *Oeuvres de Georges Chastellain*, vol. VIII, Brussels 1866.

LARKING, L B, Inventory of the effects of Roger de Mortimer at Wigmore Castle and Abbey, Herefordshire, dated 15 Edward II, AD 1322, *Archaeological Journal*, XV, 1858.

LAZZARINI, V, La presa di Chioggia, *Archivo Veneto*, LXXXI, 1952.

LEHMAN, C, *Chronica der freyen Reichs Statt Speyr*, Frankfurt am Main 1612.

LEONARDI CHIENSIS, De Lesbo a Turcis capta, in C Hopf, *Chroniques gréco-romanes*, Berlin 1873.

LESCUR, G, *Histoire de Gaston IV, Comte de Foix, chronique inédite du XV siècle*. Ed. H Courteault, Paris 1893.

Letters and Papers illustrative of the wars of the English in France during the reign of Henry VI, ed. J Stevenson, 3 vols, *RS*, London 1861.

Liber Pluscardensis, ed. F J H Skene, Edinburgh 1877.

LIEW, Foan Ming, *The treatises on military affairs of the Ming dynasty*, 2 vols, Hamburg 1998.

LOMELLINO, Angelo Giovanni, Letters, in J R Melville Jones, *The siege of Constantinople.*

LONGNON, A, ed., *Documents rélatifs au comté de Champagne et de Brie 1172–1361*, 3 vols, Paris 1901.

LOPES, F, Chronicle of Dom Fernando, extracts in F Lopes, *The English in Portugal 1367–87*, ed./trans. D W Lomax, R J Oakley, Warminster 1988.

LOPEZ DE AYALA, P, *Cronica del rey D. Pedro*,G Zurita, E Llaguno y Amirola, eds., Madrid 1875.

LOPEZ, F, *Cronique du roi D. Pedro I*, G Macchi, ed., J Steunou, trans, Paris 1985.

LUCANO, Giovanni Albino, Ioannis Albini Lucani De Bello Hydruntino Alfonsi II Aragonesi ducis Calabriae, L G Rose, I Nuovo, D Defilippis eds., *Gli umanisti e la guerra Otranteria.* Bari 1982.

LUCAS DE IRAUZO: Mota Carriazo, J de, ed., *Hechos del Condestable Don Miguel Lucas de Irauzo*, Madrid 1940.

LUDOLPH VON SICHEN, *Description of the Holy Land*, trans. A Stewart, 1896.

Lübeck: Die Chroniken der neidersachsischen Städte: C Kopmann, F Bruns, eds), 1910.

MACHAUT, G, La prise d'Alexandrie, *Société de l'Orient Latin, ser. Hist.* I, M L de Mas Latrie, ed., Geneva 1877.

MAILLA, J A M de Moyriac de, *Histoire générale de la Chine ou annales de cet empire traduites du Tong-kien-Kang-Mou*, Paris 1779.

MAKHEIRAS, Leontios, *Recital concerning the sweet land of Cyprus entitled "Chronicle"*, ed./trans. Dawkins R M, Oxford 1932.

MAKKARI, Ahmed ibn Mohammed al, *History of the Mohammedan dynasties in Spain*, trans. Gayangos, P de, London 1840.

——: *see also* Gayangos, P de

MAKRIZI, al, *Chronique des Ayyubides*, trans. A-M Edde, F Micheau, Abbeville 1994.

——, Histoire de l'Egypte, trans. E Blochet, *Revue de l'Orient Latin*, 8–11, 1900–8; and, *Histoire des sultans mamlukes de l'Egypte*, trans. M Quatremère, Paris 1837.

Malfu'za 't 'l Timuri, *History of India*, Papers of H M Elliot III, London 1871.

MALIPIERO, Domenico, Annali Veneti dall'anno 1457 al 1500, F Longo, A Sapredo eds., *Archivio Storico Italiano* VII, 1843.

MANELMI, E, *Commentariolum de quibusdam gestis in bello gallico ill. v. Francisci Barbari, praefecti praesidii Brixiao, seu de obsidione Brixaie an.* CICCCCCXXXVIII, Brescia 1728.

MARCHE, Olivier de la, Les mémoires de Messire, in *Collection complete des Mémoires rélatifs à l'histoire de France*, M Petitot, ed., vols. 9, 10, Paris 1825–.

MARINO SANUDO TORSELLO, Istoria del regno di Romania, in Hopf, *Chroniques gréco-romanes*, Berlin 1873.

Martoni: Le Grand, L, ed., Rélation du pélérinage à Jerusalem de Nicolas de Martoni, notaire italien (1394–95), *Revue de l'Orient Latin*, III, 1895.

MARVAZI, Sharaf al Zaman, *On China, the Turks and India*, trans. V Minorsky, London 1942.

MATTHEW OF WESTMINSTER, *Flores Historiarum*, H R Luard, ed., 3 vols, *RS*, London 1890.

MATTHEW PARIS, Chronica Majora, ed. H R Luard, *RS*, London 1876–7.

Matthew Paris, the illustrated Chronicles of, ed/trans. R Vaughan, Cambridge 1993.

MEDZPEH, Thomas de, *Exposé de les guerres de Tamerlane et de Shah Rokh dans l'Asie occidentale, d'après la chronique arménniene inédite de Thomas de Medzpeh*, F Neve, ed., Brussels 1860; and *T'ouma Metsobets'i's History of Tamerlane and his successors*, R Bedrosian, trans, New York 1987.

MELIK DANISMEND: I Melikoff, ed/trans., *La Geste de Melik Danismend*, 2 vols, Paris 1960.

MELIS STOKE, Rijmkroniek, Brill W G, *Historisch Genootschap te Utrecht*, 40, 42, 1885.

MELVILLE JONES, J R, *The siege of Constantinople 1453: seven contemporary accounts*, Amsterdam 1972.

MICHELS, Abel de, *Les Annales Impériales de l'Annam traduite en entier pour la première fois du texte Chinois*, Paris 1889–94.

MIHAILOVIC, Konstantin, *Memoirs of a Janissary*, trans. B Stolz, ed. S Soucek, Ann Arbor 1975.

MINERBETTI, Piero di Giovanni, Cronica, *RIS*, II, Florence 1770.

MINHĀJ UD-DIN, Abu Umar I-Usmari, *Tabakat-I-Nasiri: a general history of the Mohammadan dynasties of Asia from* AH *194 to 658 and the irruption of the infidel Mongols into Islam*, trans. H G Raverty, London 1881.

MIRZA MUHAMMAD HAIDAR, *A History of the Moghuls of Central Asia (Tarikh-i-Rashidi)*, trans. N Elias, E D Ross, London 1895/1972.

MOLINET, Jean, *Chroniques*, G Doutrepont, O Jodogne, eds., 3 vols, Brussels 1935.

MONDEXAS, Marques de, *Memorias historicas de la vida y acciones del Rey D. Alonso el Noble*, Madrid 1783.

Monumenta Henriciana, ed. M Lopes de Almeida, I Ferreira da Costa Brochada, A J Dias Dinis, Coimbra, 1960–.

MOROSINI: *Chronique d'Antonio Morosini, extraits rélatifs à l'histoire de France*, ed./trans. G Lefevre-Pontalis, L Dorez, 4 vols, Paris 1898–1902.

MOT, G J, L'Arsenal et le parc de matériel à la cité de Carcassonne en 1298, *Annales du Midi* 68, 1956.

MULICH, Hector, Chronik, *Die Chroniken der schwäbischen Städte*, Leipzig 1892.

MUNTANER: *Chronicle of Muntaner*, trans. Lady Goodenough, 2 vols, London 1920.

MURIMUTH, Adae, Continuatio chronicorum regum Angliae, E M Thompson, ed., *RS*, London 1889.

NAN-TCHAO YE-CHE, *Histoire particulaire. Traduction d'une histoire de l'ancien Yun-Nan*, C C Sainson, ed., Paris 1904.

NAVARRE, Philippe de, Monreal, Gerard de, *Les gestes des Chiprois*, G Raynaud, ed., Geneva 1887.

NEOCASTRO, B de, Historia Sicula, Paladino G ed., *RIS*, XIII (3), Bologna 1922.

NESAWI, Mohammed en, *Histoire du sultan Djelal ed-Din Mankobirti, prince de Kwarezm*, trans. Ottodas, Paris 1895.

NESRI, Auszuge aus Nesri's Geschichte des Osmanischen Hauses, Noldecke T ed., *Zeitschrift der Deutschen morganländische Gesellschaft*, XV, 1861.

NÈVE, J de, *Antoine de la Salle. Sa vie et ouvrages*, Paris/Brussels 1903.

NICOLAU D'OLWER, L, Un témoignage catalan du siège de Rhodes en 1444, *Estudis Universitaris Catalans*, XII, 1927.

NIKEPHORAS GREGORIAS, Historia Romaike (Rhomaische Geschichte), J L van Dieten, ed., Stuttgart 1973; and Migne *PG* 148, 1865.

NIKITIN OF TVER, Athanasius, Travels, in R H Major, ed., *India in the fifteenth century*, London 1857.

Nikonian Chronicle, ed. S A Zenkovsky, Princeton 1984– vols. I, II, III, IV, V.

NISHAPURI: C E Bosworth, ed., *Ths history of the Seljuk Turks from the Jami al-Tawarikh, an Ilhkanid adaptation of the Saljuqnama of Zahor al-Din Nishapuri*, Richmond 2001.

Normanniae Nova Chronica, 473–1378, e tribus Chronica MSS, S. Laudi, S. Catharinae, et Majoris Ecclesiae Rotomagensium collecta, ed Cheruel, A, in *Mémoires de la société des antiquaires de Normandie*, ser. 2, vol. 8, 1851.

NOVARE, Philip de, *The wars of Frederick II against the Ibelins in Syria and Cyprus*, trans. J L la Monte, New York 1936.

NUWENBURG, Mathiae de, Chronica, *MGH SrG NS* 84(2), Berlin 1936.

OMANI, Ibn Fadl Allah al-, *Masalik el Absar fi Mamalik el Amsar*, trans. G Demombryne, Paris 1927.

ORUC, and ANONYMOUS HANIVALDANUS, Sultan Bayezid, R F Kreutel, ed/trans., as *Der Fromme Sultan Bayezid (1481–1512)*, Graz, Vienna, Cologne 1978.

Ottokars Oesterreichische Reimchronik, ed. F Lichtenstein, J Seemuller, *MGH*, Hannover 1890.

PACHYMERES, G, De Michaele et Andronico Palaeologis, Migne *PG* 143–3, 1865.

PAGE, John, The siege of Rouen. Historical collections of a citizen of London, J Gairdner, ed., *Camden Society* 1876.

PALMERII, Mattheei, De captivitate Pisarum liber, *RIS*, XIX (II), G Scaramella, ed., Citta da Castella 1904.

Paston Letters, N Davis, ed., Oxford 1958.

PERE III OF CATALONIA, *Chronicle*, trans. M and J N Hillgarth, Toronto 1980.

PIERO, Nofri di Ser, Cronaca, in Corazzine, G O, *L'assedio di Pisa 1405–6*, Florence 1885.

PINKERTON J, *The history of Scotland from the accession of the House of Stuart to that of Mary*, 2 vols, London 1797 (for original letter of James II).

PISANO, Matthew de, Livro da guerra da Ceuta. *Collecao de libros ineditos de historia Portugueza*, J Correa da Serra, ed., I, Lisbon 1790.

PIUS II: The commentaries of Pius II, trans. F A Gragg, *South College Studies in History*, XXII, 1936–7.

POLO, Marco: *The book of ser Marco Polo*, trans/ed. H Yule, London 1903.

PONCE DE LEON: anon, Historia de los hechos de don Rodrigo Ponce de Leon, Marques de Cadiz (1443–88), *Coleccion de documentos ineditos*, 106, ed. Marques de la Fuensanta del Valle, Madrid 1893.

PORCHER, J, *French miniatures from illuminated MSS*, London 1960.

PORTER, W, *A history of the Knights of Malta or the Order of St John of Jerusalem*, revised ed. London 1883 (original letters in appendix).

POSILGE, Johan von, Chronik des Landes Preussens 1360–1419, *Scriptores rerum Prussicarum*, III, ed. T Hirsch et al., Leipzig 1866.

PRAILLON, Chronique dite de, in Saulcy de, Huguenon, *Rélation de la siège de Metz en 1444 par Charles VII et René d'Anjou*, Metz 1835.

Primera Cronica general de España, ed. R Menendez Pidal, Madrid 1955.

Privy Council of England, proceedings and ordinances, II, 1410–22, ed. H Nicolas, London 1834.

PROMONTORIO DE CAMPIS, Iacopo: F Babinger ed., Die Aufzeichnungen des Genuesen Iacopo de Promontorio de Campis über den Osmanen staat um 1475. *Bayerische Akademie der Wissenschaften — Philosophisch-Historische Klasse*, 1956, heft 8, Munich 1957.

PTOLEMY OF LUCCA, *On the government of rulers. De regimine principum*, trans. J M Blythe, Philadelphia 1997.

PULGAR, F del, *Cronica de los Reyes Catolicos*, J de Mota Carriazo ed., 2 vols, Madrid 1943.

PUYOL, J, Las Cronicas Anonimas de Sahagun, *Boletin de la Real Academia de la Historia*, 76–7, 1920.

Quaternus de Excadenciis et Revocatis Capitanatae de mandato imperialis maiestatis Frederici Secundi, ed. anon., Monte Cassino 1903.

RAPHAYNI DE CARESINAS, *Chronica*, Pastorello E, *RIS* XII(II), new ed., E Pastorello, ed., 1922.

RASCHID ELDIN, *Histoire des Mongols de la Perse*, trans. M Quatremère, Paris 1836; and

RASHID AL-DIN, *The successors of Genghis Khan, translated from the Persian* by J A Boyle, New York/London 1971.

RATHGEN, B, Schafer K H eds, Feuer- und Fernwaffen beim papstlichen Heere im 14 Jahrhundert, *Zeitschrift für Historische Waffenkunde*, VII, 1915–17.

RAZZAK: *Narrative of the journey of Abd-er-razzak, ambassador from Shah Rokh in India 1442*, R H Major, ed. , London 1847.

REDUSIIS, Andrea de, Chronica Tarvisinum, *RIS* XIX, 1731.

RICCHERIO, Cristoforo, Constantinople 1453. *See* J R Melville Jones, *The siege of Constantinople: seven contemporary accounts.*

ROBERTUS DE AVEBURY, De Gestis Mirabilis Regis Edwardi Tertii, ed. E M Thompson, *RS*, London 1889.

Rogerii Miserabile carmen super destructione regni Hungariae per Tartaros facta, *MGH SS* XXIX, ed. M Perlbach, Hannover 1892. 18.38, and *Scriptores rerum Hungaricum*, ed. E Szentpetery, Budapest 1938, II.

ROTHENBURG: *Die Urkundenbuch der Reichstadt Rothenburg*, ed. L Schnurrer, 2 vols, Wurzburg 1999.

Rotuli Normanniae in turri Londinensi asservati, Johanne et Henrico Quinto, ed. T D Hardy, London 1835.

Russian Primary Chronicle, ed/trans. S H Cross, O Shervowitz-Wetzov, Cambridge (Mass) 1953.

RUSTEM PACHA, *Die Osmanische Chronik des Rusten Pascha*, L Forres, trans., Leipzig 1923.

RYCCARDI DE SANCTO GERMANO, Chronica, *MGH SS* XIX, ed. G H Pertz, Hannover 1866.

SABA MALASPINA, Rerum sicularum historia, ed. W Koller, *MGH SS* XXXV, Hannover 1999.

SALIMBENE DE ADAM, Cronica, 2 vols, ed. G Scalia, Bari 1966.

SAMPIRO, su cronica, ed. J Perez de Urbel, Madrid 1952.

SANUTO, Marino, Vitae Ducum Venetorum, ab origine urbis (421–1493), *RIS*, XXII, Muratori ed., Milan 1732.

SARTOR, M, *Les tapisseries, toiles peintes et broderies de Reims*, Reims 1912.

Saxon Mirror, trans. M Dobezy, Philadelphia 1999.

SCHEDEL, Hartman, *Chronicle of the World. The complete and annotated Nuremberg Chronicle of 1493*, ed. S Füssel, Köln and elsewhere, 2001.

SCHILLING, D, *Die Berner Chronik des Diebold Schilling 1468–1484*, ed. G Tolder, Bern, 1897–1907.

——, *Luzerner Bilderchronik*, R Durrer,P Hilber eds., Luzern 1932.

SCHILTBERGER, J, *Bondage and travels of, in Europe, Asia and Africa, 1396–1427*, J Buchan Telfer, P Brown, trans., London 1879.

SCHOLL, J, *Annales Bingenses*, ed. Sander, Mainz 1853.

SCOLARI: Poggio, Jacopo di, Messer, Vita di Messer Filippo Scolari, cittadino Fiorentino, *Archivo Storico Italiano*, IV, Firenze, 1843.

Secret History of the Mongols, trans. F W Cleaves, Harvard/London 1982.

Servion, J, *Gestez et croniques de la maison de Savoye*, F-E Bollati, ed., Turin 1879.

Seudo Aristoteles, *Poridat de las Poridades*, ed. L A Lasten, Madrid 1957.

Shams 'l Siraj, 'Afif, Tarikh 'l Firuz Shah, *History of India*, Papers of H M Elliot iii, ed. J Dawson, London 1871.

Sharaf al-Din: Chereffedin Ali, *The history of Timur Bec known by the name of Tamerlane the Great, translated into French by M Peter de la Croix*, trans. J Darby, London 1723.

Shomonki: Rabinovitch, J N, ed/trans., *Shomonki. The Story of Masa Kado's rebellion*, Tokyo 1986.

Siege of Caerlaverock, The, ed. N H Nicholas, 1828.

Siegl, K, Briefe und Urkunde zur Geschichte der Hussitenkriege aus dem Egerer Stadtarchiv, *Zeitschrift des Deutschen Vereines für die Geschichte Mährens und Schlesiens*, 22, 23, 1918–19.

Siete Partidas, las, del Rey Alfonso el Sabio, ed. G Lopez , Paris 1847.

Silvii, Aenae, episcopi Senensis, postea Pii Papae ii, *Historia rerum Friderici tertii imperatoris*, J Henrici Boederi, ed., Argentorati 1685.

Simon de Saint Quentin, *Histoire des Tartares*, J Richard, ed., Paris 1965.

Soldo, Cristoforo da, La Cronaca, *RIS* xxi (iii), G Brizzalara, ed., Bologna 1938.

Song of Igor's Campaign, trans. V Nabokov, London 1960.

Sourdet-Thomine, J, Les conseils du Sayh Al-Harawi à un prince Ayyubide, *Bulletin d' Études Orientales*, xvii, 1962.

Specklin, D, *Les Collectanées*, R Reuss, ed., Strasbourg 1890.

Sphrantzes, G, *The fall of the Byzantine Empire*,trans. M Philippides, Amherst 1980.

St Denis: *Le premier volume des grandes chroniques de France conservées en l'église de St Denis*, ed. P Paulun, Paris 1836.

Stavelot, Jean de, Chronique latine, *Chroniques Liégoises* ed. S Balou, i, Brussels, 1913.

Stellae, Johannis, Annales Genuenses, *RIS*, xvii, 1730.

Stouff, L, *La déscription de plusieurs forteresses et seigneuries de Charles le Téméraire en Alsace et dans la haute vallée du Rhin par Maitre Mongin Contault, Maitre des Comptes à Dijon (1473)*, Paris 1902.

Strecche: Chronicle of John Strecche for the reign of Henry v (1414–1422), ed. F Taylor, *Bulletin of the John Rylands Library* 16, 1932. Reprinted Manchester 1932.

Surienne: Richard J, Quelques idées de Francois de Surienne sur la défense des villes à propos de la fortification de Dijon (1461), *Annales de Bourgogne*, 16, 1944.

Sweertius, F, ed., *Rerum Belgicarum Annales, Chronici et Historici de bellis, urbibus et moribus gentis*, Frankfurt 1620 (containing the chronicles of Johannes Gerbrand, Reyner Scotus, Aegidius Roya and anon.).

Taccola, Mariano, *De Machinis. The engineering treatise of 1449*. G Scaglia, 2 vols, Wiesbaden 1971.

Taghri Birdi, Abu-l Mahassin ibn, History of Egypt 1382–1469 ad, ed./trans. W Popper, *University of California Publications on Semitic Philology*, 17 (1957) – 23 (1960).

Taiheki. A chronicle of medieval Japan, H C McCullough, ed./trans., New York 1959.

Tcheou Ta-Kouien, Mémoires sur les coutumes de Cambodge, Pelliott P ed., *Bulletin économique Francaise de l'Extrême Orient*, 2, Hanoi 1902.

Tedaldi, G, account of the capture of Constantinople in J R Melville Jones, *The Siege of Constantinople 1453: 7 contemporary accounts*, Amsterdam 1972.

Tholomei Lucensis, Annales, *MGH SrG* NS 8, B Schmiedler, ed., Berlin 1930.

Thomae Historia pontificum Salonitarum et Spalatinorum, *MGH SS* xxix, ed. L de Heinemann, Hannover 1892.

THOMAE TUSCI, Gesta imperatorem et pontificum, *MGH SS* XXII, E Ehrenfeuchter, ed., Hannover 1872.

THORPE, I, ed., Maistre Richard, a 13th century translator of the de Re Militari of Vegetius, *Scriptorium*, VI–VII, 1952–3.

THUROCZ, Johannes de, *Chronica Hungarorum*, ed. E Galantai, J Kristo, Budapest 1985; trans F Mantella, P Engel, *Chronicle of the Hungarians*, Bloomington 1991.

TIMUR, *Malfūzāt-i Tīmūrī, or Tūzak-i Tīmūrī: the autobiography of Tīmūr*. In *The history of India, as told by its own historians*, III: 389–477, ed. H M Elliot and J Dowson, London 1871.

TRIVET, N, *Annales*, ed. T Hog, London 1845.

TSUNUDA, Ryusaku, Carrington Goodrich, L, *Japan in the Chinese Dynastic Histories*, South Pasadena 1951.

TURSUN BEG, *The history of Mehmed the Conqueror*, ed./trans. Halil Inalcik, Rhoads Murphy, Minneapolis 1978.

TYRE, William of, *Crusader Spain the thirteenth century: the Rothelin continuation of the history of William of Tyre with part of the Eracles or Acre text*, trans. J Shirley, Aldershot 1999.

——, the old French continuation of, in Edbury, P W, *The Conquest of Jerusalem and the third crusade*, Ashgate 1998.

UMAR IBN IBRAHIM AL-AWSI AL-ANSARI, *A Muslim Manual of War*, trans. G T Scanlon, Cairo 1961.

Umur Pacha: Melikoff-Sayar, I, ed., *Le Destan d'Umur Pacha (Dusturnami-i Enveri)*, Paris 1954.

URECHI, G, *Chronique de Moldavie*, E Picot, trans., Paris 1878.

URSINS, Juvenal des, Jean, *Histoire de Charles VI ,R oy de France (1380–1422)*, Michaud, Poujoulat, eds., Paris 1836.

USK, Adam of, *Chronicle*, E Maunde Thomson, ed/trans., 1904, reprint. 1990.

VALENCIENNES, Henri de, *Histoire de l'Empereur Henri de Constantinople*, J Longman, ed., Paris 1948.

VENETTE, Jean de, *The chronicle of*, ed. R A Newhall, trans. J Birdsall, New York 1953.

VIGNEULLES, Philippe de, *La chronique*, C Bruneau, ed., 4 vols, Metz 1917–33.

VILLANI, G, *Croniche Fiorentino (Nuova Cronica)*, ed. G Porta, Parma 1990; trans as *Villani's Chronicle*, by R E Selfe, P H Wickstead, London 1906.

VILLANI, Matteo, *Cronica*, A Racheli, ed., Trieste 1858.

VILLARD DE HONNECOURT, *Sketchbook*, ed. T Bowie, Westport 1959.

Vita Edwardi II, ed/trans. N Denholm-Young, London 1957.

WALSINGHAM, T, *Historia Anglicana*, H T Riley,ed., 2 vols, *RS*, London 1864.

——, Ypogdima Neustriae. Chronica Monasterii S Albani, ed. H T Riley, *RS*, London 1876.

WARKWORTH, J, Chronicle of the first thirteen years of the reign of Edward the Fourth, ed J O Halliwell, *Camden Society* London 1839, reprinted 1990.

WASSA'F, Abdullah, Tazjiyatu-l Amsarova Tajriyatu'l Asar, *History of India* III, Papers of H M Elliot, ed. J Dawson, London 1867.

WAVRIN, John de, Collection of the chronicles and ancient histories of Great Britain, now called England, trans. W and E Hardy, *RS*, London 1887.

WAY, A, Accounts of the Constables of the castle of Dovor, *Archaeological Journal*, XI, 1854.

WIERSTRAAT, C, Die histori des beleegs van Nuis zusamen, ed. C Noerrenberg, *Die Chroniken der Deutschen Städte*, 20, Leipzig 1887.

WIGAND VON MARBURG, Chronica nova Prutenica, *Scriptores rerum Prussicarum*, II, T Hirsch et al, eds., Leipzig 1863.
WILBRANDI DE ALDENBORCH, *Peregrinatio*, J C M Laurent, ed., Hamburg 1859.
WILLIAM DE RISHANGER, *Chronicle*, ed. J O Halliwell, ed., London 1860.
WILLIAMS, G, ed., Official Correspondence of Thomas Bekynton, secretary to Henry VI, vol.II, *RS*, London 1872.
WINDECKE, E, *Denkwürdigkeiten zur Geschichte des Zeitalters Kaiser Sigismunds*, W Altmann, ed., Berlin 1893.
WINKELMANN, E, ed., Acta Imperii Inedita seculi XIII et XIV. *Urkunden und Briefe zur Geschichte des Kaiserreichs und des Königreichs Sicilien in den Jahren 1200–1400*, 2 vols, Innsbruck 1885.
WORCESTRE, William, *Itineraries*, ed./trans. J H Harvey, Oxford 1969.
Yoshitsune. A fifteenth century Japanese Chronicle, ed./trans. H C McCulllough, Stanford 1966.
YUNINI, Al, Li Guo, *Early Mamluk Syrian historigraphy. Al Yunini's Dhayl Mir'at al-zaman*, Brill, 1998.
ZARI, Aben abi, *El Cartas. Noticia de los reyes del Magreb e historia de la ciudad de Fez*, trans. A Huici Miranda, Valencia 1918.
ZARKESHI (Muhammad ibn Ibrahim al Zarkashi), *Chronique des Almohades et des Hafcides*, trans. E Fagnan, Constantine 1895.
ZIA'L EL DIN BAMI, Tarikh'l Firuz Shah, *History of India*, Papers of H M Elliot III, ed. J Dawson , London 1871.
ZINK, Burkard, Chronik, *Die Chroniken der schwäbischen Städte*, Leipzig 1866.

Secondary sources

ABULAFIA, D, *Frederick II: A medieval emperor*, London 1988.
——, The role of trade of Muslim–Christian contact, in *The Arab influence in Medieval Europe*, ed. D A Agius, R Hitchcock, Reading 1994.
AGOSTON, G, 'Ottoman artillery and European military technology in the 15th and 17th centuries', *Acta Orientalia Academia Scientarum Hungaricae*, XLVII, 1994.
AINSWORTH, W, *Travels and researches in Asia Minor, Mesopotamia, Chaldea and Armenia*, 2 vols, London 1842.
ALBAN, J R, English coastal defence: some 14th century modifications within the system, in *Patronage, the Crown and the Provinces in later medieval England*, ed. R A Griffiths, 1981.
ALLEN BROWN, R, COLVIN H M, TAYLOR, A J, *History of the King's Works*, I, London 1963.
ALLSEN, T T, The circulation of military technology in the Mongolian Empire, in N di Cosmo, ed., *Warfare in Inner Asian History*, Leiden/Boston/Köln, 2002.
ALTOA, K et al, 'Narva castle — an outpost of the occident', *Castella Maris Baltici*, II, 1996.
AMATUCCIO, G, *Mirabiliter pugnaverunt. L'esercito del Regno di Sicilia ai tempo di Federico II*, Naples, 2003.
ANAMALI, S, 'La forteresse de Skanderbeg à Mati', *Studia Albanica* 4th year, 1967.
ANDERLE, J, PROCHÁZKA Z, ŠVÁBEK V, 'Hrad Hus a jeho obležení. Die Burg Hus und ihre Belagerung', *Castellologica Bohemica* 3, 1993.

Andrews, F B, The medieval builder and his methods, reprint from *Birmingham Archaeological Society Transactions* xlviii, 1925.

Andrews, K, *Castles of the Morea*, Princeton 1953.

Anghel, G, Les forteresses moldaves de l'époque d'Étienne le Grand, CG vii, 1974.

——, Quelques considérations concernant le développement de l'architecture des fortifications médiévales de Roumanie du xiiie siècle au début du xvie siècle, CG x *1980*, Caen 1982.

——, Typologie des églises fortifiées de Roumanie. CG *ix 1978*, Caen 1982.

Antonow, A, 'Die Johanniterburg Biebelried bei Würzburg', *Burgen und Schlösser*, 17, 1976.

Apell, F von, *Geschichte der Befestigung von Strassburg in Elsass*, Strassburg 1902.

Arnold, B, *Medieval Germany 500–1300: A political interpretation*. London 1997.

——, *Princes and territories in Medieval Germany*, Cambridge 1991.

Asher, C, Delhi walled: changing boundaries, in J Tracy ed., *City Walls: the urban enceinte in global perspective*, Cambridge 2000.

Ashtor, E, *A social and economic history of the Near East in the Middle Ages*, London 1976.

Asingh, P, Engberg N, The Castles on Hjelm Island — a nest of outlaws, CG 20, 2000 (pub. 2002).

Aubin, J, 'Comment Tamerlane prenait les villes', *Studia Islamica* xix, 1963.

Avent, R, *Castyll Tywysogion Gwynedd, Castles of the Princes of Gwynedd*, Cardiff, 1983.

——, *Cricieth Castle*, Cardiff, 1998.

——, *Dolwydelan Castle, Dolbadarn Castle, Castell y Bere*, Cardiff 2000.

Ayalon, D, *Gunpowder and firearms in the Mamluk kingdom*. London 1956.

Ayers, B S, Smith R, Tillyard, M, 'The Cow Tower Norwich: a detailed survey and partial re-interpretation', *Medieval Archaeology* xxxiii, 1988.

Ayers, B, *English Heritage Book of Norwich*, London 1994.

Babinger, F, *Beitrage zur Frühgeschichte der Türkenherrschaft in Rumelien (14–15 Jahrhundert)*, Munich 1944.

Bacili di Castiglione, 'Le mura ed il castello di Otranto', *Napoli Nobilissima*, xiv, 1905.

Baczkp, L, *Geschichte Preussens*, 6 vols, Konigsberg, 1792–5.

Bain, R Nisbet, 'The siege of Belgrade by Muhammed ii, July 1–23 1456', *EHR*, 7, 1892.

Bakirtzis, N, Oreopolis P, 'An Essay on Byzantine Fortification: northern Greece 4th–15th centuries', *Ninth ephorate of Byzantine Antiquities*, Thessaloniki 2001.

Barbetta, G, *Le mure e le fortificazione di Verona*, Verona 1978.

Barrow, G W S, *Robert the Bruce and the Community of the Realm of Scotland*, 3rd ed, Edinburgh 1988.

Barry, T B, *The Archaeology of Medieval Ireland*, London/New York 1987.

Barth, H, 'Reise von Trapezunt nach Skutari', *Mitteilungen aus Justus Perthes' Geographischer Anstalt*, Gotha, 1860.

Barton, K, *The archaeology of Castle Cornet, St Peter Port, Guernsey*, ed. N Jee. St Peter Port 2003.

Bartusis, M C, *The late Byzantine Army: arms and society 1204–1453*. Philadelphia 1992.

Basset, H, Terrasse, H, *Sanctuaires et forteresses Almohades*, Paris 1932.

Bates, C J, *Bamburgh Castle. Its history and architecture*, 3rd ed., Newcastle on Tyne, 1894.

Baudry, M-P, La politique de fortification de Plantagenets en Poitou 1154–1242, *ANS* xxiv, ed. J Gillingham, 2002.

Bauer, T C, Batenburg castle (the Netherlands), CG xvi, Caen 1994.

BAZZANA, A, Forteresses du royaume nasride de Grenade, la défense des frontières, CG xi, Caen 1983.

BAZZANA, A, Guichard, P, 'La conquête de la region valencienne d'après la chronique de Jacques i et les donnes archéologiques', *Castrum* 3, 1988.

BECET, M, 'Les fortifications de Chablis au xve siècle', *Annales de Bourgogne*, xxi, 1949.

BELDICEANU, N, 'La campagne ottomane de 1484', *Revue des Études Romaines*, v/vi, Paris 1960.

——, 'La conquête des cités marchandes de Kilia et de Cetatea Alba par Bayezid ii', *Sud-Ost Forschungen*, xxiii, 1964.

——, 'Récherche sur la ville ottomane au xve siècle', *Bibliotheque archéologique et historique de l'Institut Français d'archéologie d'Istanbul*, xxv, 1973.

BELDICEANU-STEINHERR, I, 'La conquête d'Adrianople par les Turcs: la pénétration turque en Thrace et la valeur des chroniques ottomanes', *Travaux et Mémoires* i, 1965.

BENADDUCI, G, *Della signoria di Francesco Sforza nella Marca, e peculiarmente in Tolentino*, Tolentino 1892.

BENNINGHOVEN, F, 'Die Gotlandfeldzüge des Deutschen Ordens 1398–1408', *Zeitschrift für Ostforschung 13, 1964.*

——, 'Zur Technik spätmittelalterliche Feldzüge im Ostbaltikum', *Zeitschrift für Ostforschung*, 19. 1970.

BENTIVOGLIO, E, Valtieri S, *Guida a Viterbo*, Bari 1982.

BERTHELOT, M, 'Histoire des machines de guerre et des artes mécaniques au moyen âge', *Annales de Chimie et de Physique*, 7e ser, 19, 1900.

——, 'Pour l'histoire des Arts mécaniques et de l'artillerie vers la fin du moyen âge', *Annales de Chimie et de Physique*, 6e ser, 24, 1891.

BERTHIER, S, EL-AJJI, E, 'Études et travaux à la citadelle de Damas 2000–2001: un premier bilan', *Bulletin d'Études Orientales*, liii–liv, Supplement, Damascus 2002.

BERTHIER, S, La citadelle de Damas: les apports d'une étude archéologique, in H Kennedy, ed., *Muslim Military Architecture in Greater Syria*, Leiden/Boston, 2006.

BEYAERT, M, Quelques problèmes de développement des armes à feu non portatives au moyen âge. *Military Studies in Medieval Europe, Papers of the medieval Europe Bruges Conference* vol. ii, Zellik 1997.

BHAKARI, S K, *Indian Warfare*, New Delhi, 1981.

BISSON, T, *The medieval crown of Aragon*, Oxford 1986.

BLACKMORE, H L, 'The Boxted bombard', *Antiquaries Journal*, 67, 1987.

BLOCKMAN, W, PREVENIER W, *The promised lands: the Low Countries under Burgundian rule*, trans. E Fackelman, Philadelphia 1988/99.

BOARDMAN, S. *The Early Stewart kings: Robert ii and Robert iii 1371–1406*, East Linton 1996.

BOAS, A, *Jerusalem in the time of the Crusades*, London/New York 2001.

BOASE, T, ed., *The Cilician Kingdom of Armenia*, Edinburgh/London 1978.

BOECKLER, A, *Deutsche Buchmalerei: Der Gotik*, Taunus 1989.

——, *Deutsche Buchmalerei: Vorgotische Zeit*, Taunus 1989.

BOGDANOWSKI, J, *Architektura obronna w krajobrazie Polski od Biskopina do Westerplatte (English summary)*, Warsaw/Krakow 2002.

BÖHME, H W et al, *Burgen in Mitteleuropa: ein Handbuch*, Stuttgart 1999.

BON, A, 'Forteresses médiévales de la Grèce centrale', *Bulletin de Correspondence Hellenique* lxi, 1937.

——, *La Morée franque*, Paris 1969.

——, 'The Medieval Fortifications of the Acrocorinth', *Corinth*, iii, 1936.

Bonaparte, L N, *Études sur le passé et l'avenir de l'artillerie*, 1851.

Bonde, S, *The fortress churches of Languedoc*. Cambridge 1994.

Borchardt, K, Tittmann, E, 'Mauern — Tore — Türme. Zeugnisse zur Geschichte von Rothenburg ob der Tauber', *Jahrbuch des Vereins Alt-Rothenburg*, I, 2005.

Borderie, A de la, 'Le siège de Brest en 1387', *Revue de Bretagne, de Vendee et d'Anjou*, I, 1889.

Boskovic, D, *Monuments et stations archéologiques en Serbie II: Serbie centrale*, Belgrade 1956.

Bouard, M de, 'Le château de Caen', *Congrès Archéologique de France*, 132, 1974.

Bouchier, E S, *A short history of Antioch, 300 BC – 1268 AD*, Oxford 1921.

Boudet, M, 'Assauts, sièges et blocus de Saint Flour par les Anglais 1356–91', *Revue d'Auvergne* 1893.

Boulay, F R H du, *Germany in the Later Middle Ages*. London 1983.

Boursier, D, *Histoire de la ville et châtellanie de Creil (Oise)*, Paris/Creil 1883.

Boyle, J A, The capture of Isfahan by the Mongols, in O Lattimere, ed., *The Mongol World Empire*, London 1977.

Brackenbury, H, 'Ancient Cannon in Europe', *Proceedings of the Royal Artillery Institution*, IV–V, 1865–7.

Bradbury, J, *The medieval archer*, Woodbridge 1985.

——, *The Medieval Siege*, Woodbridge 1992.

Braun, J, Faucherre N, Spieser J, 'La Tour Blanche et la tour du Trigonion de Thessalonique', *Byzantinische Forschungen* XII, 1987.

Braune, M, 'Die mittelalterliche Befestigungen der Stadt Tortosa/Tartus', *Damaszener Mitteilungen* 2 1985.

——, 'Die Stadtmauer von Damaskus', *Damaszener Mitteilungen* II, 1999.

Braune, M, Hirnam al-Zaym, Stadttore und Stadtmauer von Damaskus, in *Damaskus-Aleppo. 5000 Jahre Stadtentwicklung in Syrien. Beiheft der Archäologischen Mitteilungen aus Nordwestdeutschland*, 28, Mainz 2000.

Bretschneider, E, *Récherches archéologiques et historiques sur Pekin et ses environs*, trans. V Colein de Plancy, Paris 1879.

Brown, C, *The Second Scottish War of Independence 1332–1363*, Stroud 2002.

Brown, M, *James I*. Edinburgh 1994.

Brown, P, *Indian architecture: Buddhist and Hindu periods*, 2nd ed Bombay 1976.

——, *Indian architecture: Islamic Period*, Bombay 1968.

Brown, S, 'Berry Pomeroy Castle', *Devon Archaeological Society Proceedings*, 54, 1996.

——, *Berry Pomeroy Castle*, London 1997.

Brusten, C, *L'armée Bourguignonne de 1465 à 1468*, Brussels 1953.

Buffet, H-F, 'L'Enceinte de Saint Malo', *Congrès Archéologique de France*, CVII, 1950.

Burns, R, *Monuments of Syria, an historical guide*, London/New York, revised edition 1999.

Butler, L A, *Denbigh Castle and Town Walls*, London 1990.

——, The siege of Dolforwyn castle in 1277, CG, XIX, 2000.

Butler, L, Knight J, *Dolforwyn Castle, Montgomery Castle*, Cardiff 2004.

Buttner, R, *Burgen und Schlösser an der Donau*, Vienna 1964.

Büttner, R, Madritsch, R, *Burgen und Schlösser in Niederösterreich: von Bisamberg bis Laa/Thoya*, Vienna, 1987.

Cahen, C, Les changements de techniques militaires dans le proche orient médiéval et leur importance historique, in *War, technology and society in the Middle East*, ed. Parry V J and Yapp M 1975.

——, *The formation of Turkey. The Seljukid sultanate of Rum: eleventh to fourteenth century*, trans./ed P Holt, Harlow 2001.

CAIN, B le, Les fortifications de Harfleur au début du XVe siècle, in M-P Baudry, ed., *Les fortifications dans les domaines Plantagenet, XII–XIV siècles*, Poitiers 2000.

CAIRNS, C, 'Guns and Castles in Tipperary', *Irish Sword* XVI, 1985.

CARRINGTON GOODRICH, M, FENG Chia-Sheng, 'The early development of firearms in China', *Isis* 36, 1946.

CATARINO, H, 'O Castelo de Paderne (Albufeira): resultados de la primeira intervenção arqueologica', *Arquelogica Medieval* 3, 199?.

CATTIN, P, La construction d'un château au XIVe siècle: Château Gaillard (Ain), *Cahiers Perce de Lucinge*, Series 4, 26, 1990.

CAUNE, A, 'Steinburgen des 12–16 Jahrhunderts im Dünamunde ungsgebiet', *Castella Maris Baltici*, II, 1996.

CAZZATO, M, *Guida ai Castelli Pugliesi, (1) La provincia di Lecce*. Lecca 1999.

CHAMBERLIN, C, "The King sent them very little relief": the Castilian siege of Algeciras, 1278–9, in D Kagay, L Villalon, eds., *Crusaders, condottiere and cannon*, Leiden/Boston 2003.

CHEVEDDEN, P E, Fortifications and the development of defensive planning during the Crusader period, in *The Circle of War in the Middle Ages*, ed. D J Kagay, L J Andrew Villalon, Woodbridge 1999.

——, The Artillery of King James I the Conqueror, in *Iberia and the Mediterranean world of the Middle Ages*, ed. P E Chevedden, D J Kagay, P G Pedella, 2 vols Brill 1996.

CHIAUDANO, M, 'La Finanza Sabauda nel sec. XIII', *Biblioteca della Societa Storica Sudalpina*, Turin 1933–34, 1937, vols. 131–133.

CHRISTIANSEN, E, *The northern crusades*. London 1990.

CLAUZEL, D, Lille et ses remparts à la fin du Moyen Âge (1320–1480), in P Contamine, O Guyotjeannin, eds., *La guerre, la violence et les gens au Moyen Âge*, I, Paris 1996.

CLEPHAN, R C, 'The ordnance of the 14th and 15th Centuries', *Archaeological Journal*, LXVIII, 1911.

COBOS GUERRA, F, CASTRO FERNANDEZ, J J de, *Castillo de Ponferrada*, Leon, 2002.

COEDÈS, G, Boisselier, J, eds., Asie du Sud-Est, I: Le Cambodge, *Manuel d'archéologie d'extrème orient*, Paris 1966.

——, Asie du Sud-Est: II, Le Viet-nam, *Manuel d'archéologie d'extrème orient*, Paris 1972.

COLDSTREAM, N, 'Architects, Advisers and Design at Edward I's Castles in Wales', *Journal of the Society of Architectural Historians of Great Britain*, 46, 2003.

COLLINS, T E, 'The medieval town defences of Cashel', *Tipperary Historical Journal* 1997.

CONRAD, H, *Geschichte der deutschen Wehrverfassung*, I, Munich 1939.

CONTI, F, *Castelli del Piemonte*, 3 vols, Milan 1975.

CONWAY DAVIES, J, 'The Despenser War in Glamorgan', *Transactions of the Royal Historical Society*, 3rd Ser, IX, 1915.

CORNAN, R, 'Le Château de Saint Malo', *Congrès Archéologique de France*, CVII, 1950.

——, 'Dinan', *Congrès Archéologique de France*, CVII, 1950. Madrid 1987.

COSTE, M, *History, description, interpretation of the castle of Bonaguil*, Bonaguil 2001.

COULSON, C L H, Battlements and the bourgeoisie: municipal status and the apparatus of urban defence in later medieval England, *Medieval Knighthood* IV, ed. S Church, R Harvey, Woodbridge 1995.

——, 'Cultural realities and reappraisal in English castle-study', *Journal of Medieval History*, 22 1996.

COUSINERY, M E M, *Voyage dans la Macédonie*, Paris 1831.

Creighton, O, "Castles of Communities": medieval town defence in England, Wales and Gascony, CG 22, 2004.

——, 'Early Leicestershire Castles: archaeology and landscape history', *Leicestershire Archaeological and Historical Society Transactions* 71, 1997.

Creighton, O, Higham R, *Medieval Town Walls. An archaeology and social history of urban defence*, Stroud 2005.

Cressier, P, 'La fortification islamique au Maroc: éléments de bibliographie', *Archéologie Islamique*, 5, 1995.

Cresswell, K A C, 'Fortification in Islam before 1250', *Proceedings of the British Academy* xxxviii, 1952.

——, *A brief chronology of the Muhammadan monuments of Egypt to 1517*, Cairo 1919.

——. *The Muslim architecture of Egypt*, ii, *Ayyubids and early Bahrid Mamluks*. Oxford 1959.

Crozet, R, 'Un episode de la guerre de cent ans: le siège de Romorantin par le prince des Galles (1356)', *RH*, cliii, 1926.

Custoza, G V, d'Arcano, M G, *Castella: centodue opere fortificate del Friuli Venezia Giulia*, Udine 1995.

Dangles, P, Faucherre, N, Collet B, Morelière, A, 'Fortifications urbains ducales en Bourgogne. L'exemple de Beaune', *Congrès Archéologique de France*, 152, 1994.

Dangles, P, Faucherre, N, Collet, B, Speranza, M, 'Les tours albarannes de l'enceinte d'Auxonne', *Congrès Archéologique de France*, 152, 1994.

Dannecker, P, WLassety, J von, 'Combat anno 1399. Das Fragment des zweiten Tannenburg-buchse'. *Deutsche Waffen-Journal* 10/1981.

Davidsohn, R, *Forschungen zur alteren Geschichte von Florenz*, Berlin 1896–1908.

——, *Storia di Firenze. Le origini*. Firenze 1909.

Davies, N, *God's Playground. A history of Poland*, i. Oxford 1981.

Davies, R R, *The Age of Conquest: Wales 1063–1415*, Oxford 1987.

Davies, S, *Welsh military institutions 633–1283*, Cardiff 2004.

Davis, P R, *Castles of the Welsh princes*, Swansea 1988.

Demont, E, Leger A, *Histoire de la ville d'Harfleur*, Le Havre 1868.

Demotz, B, 'L'état et le château au moyen âge: l'exemple savoyard', *Journal des Savants* 1987.

——, *Le Comté de Savoie du xie au xve siècles. Pouvoir, Château et État au Moyen Âge*, Geneva 2000.

Dennis, G T, 'The reign of Manuel ii Palaeologus in Thessalonica 1382–87', *Orientalia Christiana Analecta*, 159, 1960.

Derbier, J, Une rocca transfigurée: le château de Rochechinard (Drôme), CG 22, 2004.

Deroko, A, *Medieval castles on the Danube*, trans. K Radovanović, J Dekker, Belgrade 1964.

Deschamps, P, *Les Châteaux des Croisés en Terre Sainte*. Vol. 1, *Le Crac des Chevaliers*, Paris 1934; Vol. 3, *La Défense du Comté de Tripoli et de la Principauté d'Antioche*, Paris, 1975.

——, *Terre Sainte Romane, Abbaye Ste Marie de la Pierre-qui-vive*, 1964.

Desnoyes, M, 'Les armes du siège de 1428', *Mémoires de la Société Archéologique et Historique de l'Orléanais*, 18, 1884.

Dietschy, J V, *Rheinfelden*, Rheinfelden 1880.

Dixon, P, Marshall, P, Palmer Brown, C, Samuels J, *Newark Castle Studies*, Newark 1994.

DIXON, P, MARSHALL, P, 'The great tower in the 12th century: the case of Norham Castle', *Archeological Journal* 150, 1993.

DODD, A, ed., *Oxford before the University*, Oxford 2003.

DOLLEN, B von der, 'Die Entwicklung des Burgenbaus in Irland seit der anglo-normannischen Eroberung', *Burgen und Schlösser* 37, 1996.

DONNELLY, C J, 'Frowning ruins: the tower houses of Medieval Ireland', *History Ireland* 4, 1996.

DRACOP, J, 'Edward's Castles, defence or aggression?', *Archeology Today* 8 (6), 1987.

DRAKE, K, 'Burg und Stadt in Schweden in der Zeit von 1250 bis 1320', *Castella Maris Baltici*, II 1996.

——, Stadt-Burg-Stadt Hameenlinna, CG XVI, Caen 1994.

DROUYN, L, *La Guienne Militaire*, 2 vols, Bordeaux/Paris 1865.

DUBLED, H, 'L'artillerie royale française à l'époque de Charles VII et au début du règne de Louis XI (1437–69). Les frères Bureau', *Memorial de l'artillerie française* 50, 1976.

DUFTY, A R, 'review of B H St J Neil, Castles and Cannon', *Archeological Journal* 119, 1962.

DUNBABIN, J, *Charles I of Anjou*, London 1998.

DUNCAN, A A M, 'A siege of Lochmaben castle in 1343', *Transactions and Journal of Proceedings of the Dumfriesshire and Galloway Natural History and Antiquaries Society* ser. 3, 31, 1954.

DURDÍK, T, 'Batterietürme der böhmischen Burgen' in Gerő László, *Nysolevanötödik Szöletésnapjára, Tanulmányok Múvészettónténet-Múemlékvéddem*, VI, 1994, 235–43.

——, 'Abriss der Entwicklung der böhmischen Artillerieburgfortifikationen des 15 und Anfangs des 16 Jahrhunderts', *Castella Maris Baltici*, II, 1996.

——, Böhmische Burgen der Hussitenzeit. *Interdisziplinäre Beitrage zur Siedlungsarchäologie. Gedenkschrift für Walter Janssen*, eds., P Ettel, R Friederich, W Schier, Rahden/Westfalen, 2002.

——, 'Česká hradní architektura doby husitské (Böhmische Burgenarchitektur zur Zeit des Hussiten)', *Casopis společnosti přatel starožitností*, 99, 1991.

——, 'French influence on Bohemian castle architecture of the thirteenth century', *Fortress* 15, 1992.

——, Kastellburgen des 13 Jahrhunderts in Mitteleuropa. Prague 1994.

——, The city walls of Tábor – a breakthrough in the development of Central European fortification, in G Perbellini, *The Town Walls in the Middle Ages, Europa Nostra Bulletin*, 53, 2000.

——, 'Vorgeschobene Bastionen der böhmischen Burg in der Hochblute des Mittelalters', *Historicka Geografie*, 10, 1973.

——, Zum derzeitigen Stand der Studien zu Belagerungswerken vor Böhmischen Burgen, in Wagener, Laß (eds.), *Beihefte zur Mediavistik*, 7, 2006.

DURDÍK, T, PERTL, M, Tarasnize z hradu Křivoklatu, in *Urbes Medii Aevi*, ed. M Richter, Prague 1984 (German summary).

EANNES DE ZURARA, Gomes, *Chronica do Conde Don Pedro de Menezes*, J Correa da Serra, ed., Lisbon, 1792.

EBHARDT, B, *Der Wehrbau Europas im Mittelalter*. Berlin 1939.

EDWARDS, R W, 'Review of C Foss, Survey of medieval castles of Anatolia, I: Kutayha', *Speculum*, 62, 1987.

——, *The fortifications of Armenian Cilicia*, Washington 1987.

EDWARDS, T, Hawley's Fortalice: Dartmouth's first castle. *Dartmouth History Research Group* paper 24, Dartmouth 1998.

EKDAHL, S, 'Der Krieg zwischen Deutschen Ordens und Polen-Litauen 1422', *Zeitschrift für Ostforschung* 13, 1964.

EKROLL, Ø, Norwegian medieval castles 1170–1320: building on the edge of Europe, CG XVIII, Caen 1998.

ELTIS, D, 'Towns and Defences in later Medieval Germany', *Nottingham Medieval Studies* XXXIII, 1989.

EMERY, A, 'The development of Raglan Castle and keeps in late medieval England', *Archeological Journal* 132, 1975.

——, 'Malbork Castle', *Castle Studies Group Journal*, 21, 2007–8.

EMMINGHAUS, T G G, *Memorabilia Susatensia*, Jena 1749.

Encyclopedia of Islam, Leiden/London 1960–.

ENGEL, P, *The realm of St Stephen. A history of medieval Hungary 895–1526*. London/New York 2001.

ENLART, C, *Gothic art and the renaissance in Cyprus*, trans. D Hunt, London 1987.

ERBEN, W, 'Beitrage zur Geschichte des Geschützwesens in Mittelalter', *Zeitschrift für Historische Waffenkunde*, 7, 1915/17.

EROLI, G M, *Eramo Gattamelata da Narni. Suoi monumenti e sua famiglia*, Rome, 1876.

ESAGUY, J de, *Tanger sous la domination Portugaise (1471–1663)*, Tanger, 1937.

EVANS, J A S, 'The Walls of Thessalonika', *Byzantion* 47, 1977.

FALCO, G, Sulla costruzione del castello di Fossano (1324–32), *Fonti e Studi di Storia Fossanese*, Torino 1936.

FANG-CHIH CHEN, A, 'China's northern frontiers', *Yenching Journal of Social Studies*, IV/1, Peiping 1948.

FARQUHAR, D, *The Government of China under Mongolian Rule. A reference guide*. Stuttgart 1990.

FARRIS, W W, *Heavenly Warriors. The evolution of Japan's Military, 500–1300*, Cambridge (Mass.), London, 1992.

FAWTIER, R, *The Capetian Kings of France 987–1328*, trans. L Butler, R J Adam, London 1960.

FEDDEN, R, *Crusader Castles*, London 1950.

FEHR, G, *Benedikt Ried*, Munich, 1961.

FERJANČIĆ, B, Byzantine and Serbian Serres in XIV century (in Serbo-Croat), Belgrade 1944, *Serbian Academy of Science and Arts* Monograph DCXXIX (21). (English summary 135–41)

FINE, J V A, A tale of three fortresses: controversies surrounding the Turkish conquest of Smederevo, of an unnamed fortress at the junction of the Sava and Bosna, and of Bobovac, in T S Miller, J Nesbitt, eds., *Peace and War in Byzantium: essays in honour of George T Dennis, SJ*. Washington DC, 1995.

FINK, O, 'Burg Burgschwalbach und die Einführing der Feuerwaffen in Mittelrheingebiet', *Burgen und Schlösser*, 14, 1973.

FINO, J-F, *Forteresses de la France Médiévale*, 3rd ed, 1977.

——, 'L'artillerie en France à la fin du moyen âge', *Gladius* XII, 1974.

——, 'Machines de jet médiévales', *Gladius* X, 1972.

FISENNE, O von, 'Eine Waffentechnische Kuriosität aus der Eiffel. Gusseiserne Bombarde aus dem 14. Jahrhundert', *Burgen und Schlösser*, 14, 1973.

FLENSBURG: *700 Jahre Stadt Flensburg 1284 bis 1984*, Flensburg 1984.

FONS-MELLICOCQ, De la, *De l'artillerie de la ville de Lille au XIVe, XVe et XVIe siècles*, Lille 1854.

FORESTIE, E, 'Hugues de Cardaillac et la poudre à canon (XIVe siècle)', *Bulletin Archéologique et Historique de Tarn-et-Garonne*, XXIX, 1901.

Foss, C, *Byzantine and Turkish Sardis*, Cambridge (Mass.) 1976.
——, Cities, *Fortresses and Villages of Byzantine Asia Minor*. Aldershot 1996.
——, *Ephesus after Antiquity: a late antique, Byzantine and Turkish city*. Cambridge 1979.
——, 'Late Antique and Byzantine Ankara', *Dumbarton Oaks Papers* 31, 1977.
——, 'Late Byzantine fortifications in Lydia', *Jahrbuch der oesterreichischen Byzantinistik*, 28, 1979.
——, *Survey of Medieval Castles of Anatolia*, II: *Nicomedia*, Ankara 1996.
——, Survey of Medieval Castles of Anatolia: I, Kutahya. *BAR* International Series 261, 1985.
Foss, C, Winfield D, *Byzantine Fortifications: an introduction*. Pretoria 1986.
Fowler, K, *The King's Lieutenant. Henry of Grosmont, First Duke of Lancaster 1310–61*. London 1969.
France, J, Fortification east and west, in H Kennedy, ed., *Muslim Military Architecture in Greater Syria*, Leiden/Boston, 2006.
——, *Western Warfare in the age of the Crusades 1000–1300*, London 1999.
Franke, H, Siege and defence of towns in Medieval China, in F A Kierman, J K Fairbank, *Chinese ways in Warfare*, Cambridge (Mass.) 1974.
Freeman, A Z, 'A moat defensive: the coast defence scheme of 1295', *Speculum* XLII, 1967.
——, 'Wall breakers and river bridgers', *Journal of British Studies*, X, 1971.
Fröhlich, J, 'Středověká opevněni kolem hradu Zvikova', *Castellologica Bohemica* 2, 1991 (German summary).
Frolík, J, 'Nálezy z oblehání hradu Lichnice v Letech 1428–1429', *Castellologica Bohemica*, 8, 2002.
Frye, R N, *The golden age of Persia*, London 1975.
Fugedi, E, *Castle and Society in Medieval Hungary 1000–1437*, Budapest 1986.
——, Medieval Hungarian Castles in existence at the start of the Ottoman advance, in *From Hunyadi to Rackoczi: War and Society in late Medieval and early modern Hungary*, ed. J M Bak, B K Karoly, Brooklyn 1982.
Fulda, J, *Crusader manuscript illumination at Saint-Jean d'Acre 1275–1291*, Princeton 1976.
Gabriel, A, 'Châteaux Turcs du Bosphore', *Mémoires de l'Institut Français d'archéologie de Stamboul*, VI, 1943.
——, *Monuments Turcs d'Anatolie*, 2 vols, Paris 1931–4.
Gaier, C, 'Art et organisation militaires dans la principauté de Liège et dans le comté de Looz au moyen âge', *Mémoires, Academie Royale de Belgique*, Brussels 1968.
Galperin, P, 'Die "Bremen-Büchse"', *Deutsche Waffen-Journal*, 1/1987.
Ganiage, J, ed., *Histoire de Beauvais et du Beauvaisis*, Toulouse 1987.
Gardelles, J, *Les châteaux de moyen âge dans la France de sud-ouest. La Gascogne anglaise de 1216 à 1327*. Paris 1972.
Genicot, L F, *Les grands châteaux de Belgique*, vol I, Brussels 1975.
Gëro, L, *Castles in Hungary*, trans. G Gulyas, B Gaster, Györ, 1969.
Gerola, G, *Monumenti Veneti nell'isola di Creta*, Venice 1906.
Gessler, E A, 'Der Springulf, ein mittelalterliches Torsionsgeschütz, in Gebiete der nahmaligen Eidgenossenschaft', *Basler Zeitschrift für Geschichte und Altertumkunde*, 20, 1922.
Giese, W, Waffengeschichtliche und -terminologische Aufschlüsse aus Katalanischen literarischen Denkmalern des 14 und 15 Jahrhunderts, in *Homenatge a Antonion Rubio i Lluch*, I, Barcelona 1936.
Gille, B, *The renaissance engineers*, Paris/London 1966.
Giurescu, C C, *Istoria Romanilor*, Bucharest 1937.

Gode, P K, 'Use of guns and gunpowder in India from AD 1400 onwards', *Studies in Indian Culture and History*, v, Poona 1960.

Gohlke, W, 'Das Geschützwesens des Altertum und des Mittelalters', *Zeitschrift für Historische Waffenkunde*, v, 1909/11.

Goodman, A, *The Wars of the Roses*, London 1990.

Goubea, A Almagro, 'El sistema defensiva de Albarracin', *Arqueologia Medieval Espanola II Congreso*, II, Madrid 1987.

Grand, R, 'L'Architecture Militaire en Bretagne jusqu'à Vauban', *BM* 110, 1952.

Grandjean, M, Un jalon essentiel de l'architecture de brique piémontaise: l'oeuvre d'Humbert le Bâtard au château de Cheneau à Estavayer (1433–43), in Bagliani, Poudret (eds), *La Maison de Savoie et la Pays de Vaud*, Lausanne 1989.

Gravett, C, *Medieval Siege Warfare*. London 1990.

Greb L, Bronner C, *Die Kunstdenkmaler im Volkstaat Hessen: Kreis Bingen*, Darmstadt 1934.

Gregory, T E, The Hexamilion and the fortress, *Isthmia*, v, Princeton 1993.

Grekov, B, Iakobauski, A, *La Horde d'Or*, trans. R Thuret, Paris 1939.

Grenet, F, Isamaddinov, M, 'La prise de Samarcande par Gengis-Khan', *L'Histoire*, 149, 11/1991.

Griffiths, R A, 'Owain Glyndŵr and the siege of Coity Castle 1404–8', *Morgannwg* XLV, 2001.

——, 'Prince Henry, Wales and the royal exchequer', *Bulletin of the Board of Celtic Studies* 32, 1985.

Groeneveldt, W, 'Notes on the Malay Archipelago and Malacca from Chinese sources', *Verhandelingen van het Bataviaasch Genootschap van Kunsten en Wetenschappen*, XXXIX, 1880.

Grosdidier de Matons, M, *Metz*, Paris 1957.

Guerin, T de, 'The early history and first siege of Castle Cornet', *Reports and Transactions, Guernsey Society of Natural Science and Local Research*, 1904.

Guerquin, B, *Zamek w Malborku*, English summary M Lalynski, Warsaw 1960.

Gulick, F W van, *Nederlandse Kastelen en Landhuizen*, Den Haag 1960.

Gutbier, R, *Der landgräfliche Hofbaumeister Hans Jakob von Ettlingen*, Darmstadt/Marburg 1973.

Gwei-Djen, Lu, Needham, J, Chi-Hsong, Phan, 'The oldest representation of a bombard', *Technology and Culture*, 29(3), 1988.

Hale, J R, The early development of the bastion: an Italian chronology *c.* 1450–1530, reprinted in *Renaissance War Studies*, London 1983.

Hall, B S, *The technological illustrations of the so-called "anonymous" of the Hussite wars, codex latinus Monacensis 197*, part I, Wiesbaden 1979.

——, The coming of gunpowder and the development of firearms in the Renaissance, in B Buchanan, ed., *Gunpowder: the history of an international technology*, Bath 1996.

Hall, J, Kyoto as historical background, in J Hall, J Moss, *Medieval Japan, Essays in Institutional History*, New Haven/London 1974.

Hall, A R, Military Technology, in *A History of Technology*, ed. C Singer, E Holmyard, A Hall, T W Williams, volume II, Oxford 1956.

Hansen, P V, 'Casting and firing a replica of the Loshult gun', *Journal of the Royal Ordnance Society*, 14, 2002.

Har-El, Shai, *Struggle for domination in the Middle East. The Ottoman–Mamluk war 1485–1491*, Leiden/New York/Köln, 1995.

HARRISON, P, *Castles of God: Fortified Religious Buildings of the World*, Woodbridge, 2004.

——, Church fortification in La Thiérarche, *Fort* 22, 1994.

——, 'The fortified village of Ushguli in the Georgian Caucasus', *Fort*, 25, 1997.

——, 'The Tower Churches of Gower', *Gower* 46, 1995.

HARVEY, J, *English Medieval Architects*, revised edition, Gloucester 1987.

——, *Henry Yevele, life of an English architect*, London 1944.

HASLUCK, F, 'Monuments of the Gattelusi', *Annual of the British School at Athens*, XV, XVI, 1908–9, 1909–10.

——, 'The Latin monuments of Chios', *Annual of the British School at Athens*, XVI, London 1909/10.

HAZELOFF, A, *Die Bauten des Hohenstaufen in Unteritalien*, vol 1, Leipzig 1920.

HEATH, I, MCBRIDE, A, *Byzantine Armies 1118–1461*, London 1995.

HEBRON, M, *The medieval siege: theme and language in Middle English romance*, Oxford 1997.

HEFNER-ALTENECK, J H von, WOLF, J W, *Die Burg Tannenburg und ihre Ausgrabungen*, Frankfürt am Main, 1850.

HELD, J, The defence of Nandorfehervar (Belgrade) in 1456: a discussion of controversial issues, in S M and A H Vardy, *Society in Change: Studies in honour of Bela K Kiraly*, New York 1983.

HELLENKEMPER, H, *Burgen der Kreuzritterzeit in der Grafschaft Edessa und im Königreich Kleinarmenia*. Bonn 1976.

HELLER, K, *Rothenburg ob den Tauber in Wehr und Waffen*, Rothenburg ob den Tauber, 1920.

HENNE, A, WAUTERS A, *Histoire de la ville de Bruxelles*, 1845 new ed. M Martens, 4 vols Brussels 1968.

HERDE, P, From Adolf of Nassau to Lewis of Bavaria 1292–1347, in M Jones, ed., *New Cambridge Medieval History*, VI, Cambridge 2000.

HERMAN, G, *Monuments of Merv*, London 1999.

HERNANDEZ, F, The alcazaba of Merida, in K Cresswell, *Early Muslim Architecture*, part 2, Oxford 1940.

HERRMAN, G, et al, The International Merv project, Preliminary report of the 5th season, *Iran* XXXV, 1977.

HETHERINGTON, P, *Byzantine and Medieval Greece: Churches, Castles and Art*. London 1991.

HEYMANN, F G, The crusade against the Hussites, in K Setton, *A History of the Crusades*, III, ed. H W Hazard, Wisconsin 1975.

HEYWOOD, C, Notes on the production of fifteenth century Ottoman cannon. Proceedings of the International Symposium on Islam and Science 1401/1980, reprinted in the same, *Writing Ottoman History*, Aldershot 2002, 3–15.

HIEBERT, F T, 'The oasis and city of Merv (Turkmenistan)', *Archéologie Islamique*, 3, 1992.

HILLERBRAND, C, *The Crusades, Islamic Perspectives*, Edinburgh 1999.

HILTON, R H, 'Building accounts of Elmley Castle, Worcestershire 1345–6', *University of Birmingham History Journal*, X, 1965/6.

HINDLE, *Medieval roads and tracks*, 3rd ed., Princes Risborough 1982.

HIRAI, K, *Feudal architecture of Japan*, trans. H Sato, J Ciliotta, New York/Tokyo 1973.

HLAVÁČEK, I, The Luxemburgs and Rupert of the Palatinate 1347–1410, in M Jones, ed., *New Cambridge Medieval History* VI, Cambridge 2000.

Hodgson, J, *Southampton castle*, Southampton, nd.

Hoffsummer, P, *Le château de Franchimont*, 1997.

Hogg, O F G, *Clubs to Cannon. Warfare and weapons before the introduction of gunpowder*. London 1968.

Höller, A, Burgen und Belagerungstechniken in Japan der Zeit der Shogune (14–19 Jh.), in Wagener, Laß, (eds.), *Beihefte zur Mediavistik* 7, 2006.

Holst, N von, *Der deutsche Ritterorden und seine Burgen von Jerusalem bis Sevilla, von Thorn bis Narva*. Berlin 1981.

Huber, E, 'Études indochinoises v: la fin de la dynastie de Bayan', *Bulletin de l'école Française de l'extrême orient*, ix, 1909.

Hughes, M, The fourteenth century French raids on Hampshire and the Isle of Wight, in A Curry, M Hughes, eds., *Arms, Armies and Fortifications of the Hundred Years War*, Woodbridge 1994.

Humel, V, Research in Prague: an historical and archaeological view of the development of Prague from the 9th to the middle of the 14th century, in *From the Baltic to the Black Sea*, ed. D Austin, L Alcock, London/New York 1990.

Humphries, P A, *Engines of War — replica medieval siege engines at Caerphilly Castle*. 2nd ed., Cardiff 1996.

Hunger, V, 'Le siège et la prise de Vire par Charles vii en 1450', *Annales de Normandie* 21, 1971.

Inalcik, H, Osman Ghazi's siege of Nicaea and the battle of Bapheus, in E Zachariadou, ed., *The Ottoman Emirate*, Rethymnion 1993.

——, 'The conquest of Edirne (1361)', *Archivium Ottomanium*, 3, 1971.

Irwin, R, Gunpowder and firearms in the Mamluk sultanate reconsidered, in M Winter, A Levanoni, eds., *The Mamluks in Egyptian and Syrian Politics and Society*, Leiden/Boston 2004.

Jacq-Hergoualich, M, *L'armement et l'organisation de l'armée khmère aux xiie et xiiie siècles, d'après les bas-reliefs d'Angkor Vat, du Bayon et de Bonteay Chmar*, Paris 1979.

Jansen, P, Rocche. Châteaux et villes fortifiées: leur rôle dans les guerres des Condotte italiennes à la fin du Moyen Âge d'après l'exemple des Marches, in M Combet, R Herin ,eds., *Château et Guerre. Actes des Rencontres d'Archéologie et d'Histoire en Périgord, 25–7 Septembre 1998*, Bordeaux 2004.

Joachim, E, 'König Sigismund und der Deutsche Ritterorden in Ungarn 1429–32. Mitteilungen aus dem Staatsarchiv zu Konigsberg', *Mitteilungen des Instituts für Osterreichische Geschichtsforschung*, 33, 1912.

Johns, C N, *Caerphilly castle*, Cardiff 1978.

Jones, G.R., 'The defences of Gwynedd in the 13th Century', *Transactions of the Caernarvonshire Historical Society*, 30, 1969.

Jones, M, L'armée Bretonne 1449–1491: structures et carrières, in B Chevalier, P Contamine, eds., La France de la fin du xve siècle Renouveau et Apogée, Paris 1985; reprinted in Jones, M, *The creation of Brittany, a late medieval state*, London 1988.

——, 'The defence of medieval Brittany', *Archeological Journal* 138, 1981.

Jones, R. Les fortifications municipales de Lisieux dans les chroniques et dans les comptes, in P Contamine, O Guyotjeannin,eds., *La guerre, la violence et les gens au Moyen Âge*, 1, Paris 1996.

Jordan, A J, 'Date, chronology and evolution of the county Wexford tower house', *Journal of the Wexford Historical Society*, 13, 1990/1.

Kafka, K, *Wehrkirchen niederösterreichs*. Vienna 1969.

KAJZER L, Salm J, 'Burg und Stadt in mittelalterlichen Polen', *Castrum Bene* 6, ed. T Durdík, Prague 1999.

KAREL, T, 'Předsunuté opevnění hradu Zelená Hora (Vorgeschobene Befestigung der Burg Zelená Hora)', *Castellologica Bohemica*, 10, 2006.

KAUFMANN, J E and HW, JURGA, R M, *The medieval fortress*, London 2001.

KEIL, J, *Ephesos. Ein Führer durch die Ruinenstatte und ihre Geschichte*, Vienna 1964.

KENNEDY, H, *Crusader Castles*, Cambridge 1994.

KENYON, J R, 'Early artillery fortifications in England and Wales', *Archeological Journal* 138, 1981.

——, 'Early gunports: a Gazetteer', *Fort*, 4, 1977 (revised edition 1995).

——, *Raglan Castle*, Cardiff 2003.

——, Gunloops at Raglan Castle, Gwent, in *Castles in Wales and the Marches* ed. J R Kenyon, R Avent, Cardiff 1987.

KEULER, I L, Roermond: de Geschiedenis zijner vestingwerken, monumenten en voorname gebouwen, in *Gedenkboek ter gelegenheid van het zevenhonderd-jarig bestaan van Roermond als stad*, Roermond 1932.

KEYSER, E, ed., Rheinische Städtebuch, in *Deutsches Städtebuch*, 3, Stuttgart 1956.

KHAN, I A, 'Coming of gunpowder to the Islamic world and north India: spotlight on the role of the Mongols', *Journal of Asian History*, 30, 1996.

——, 'Early use of cannon and musket in India AD 1442–1526', *Journal of the Economic and Social history of the Orient*, XXIV, 1984.

——, 'Origin and Development of Gunpowder Technology in India AD 1250–1500', *Indian Historical Review*, IV, 1977.

——, The role of the Mongols in the introduction of gunpowder and firearms in South Asia, in B Buchanan ed, *Gunpowder: the history of an international technology*, Bath 1996.

KIANG, Heng Chye, *Cities of Aristocrats and Bureaucrats. The development of medieval Chinese cityscapes*, Honolulu 1999.

KIEL, M, 'A note on the exact date of construction of the White Tower of Thessaloniki', *Balkan Studies*, 14, 1973, 352–7, reprinted in the same, *Studies in the Ottoman architecture of the Balkans*, Aldershot, 1990.

——, *Ottoman architecture in Albania 1385–1912*, Istanbul, 1990.

KING, D J C, *Castellarium Anglicanum*, 2 vols, New York and London 1983.

——, The defence of Wales 1067–1283: the other side of the hill, *AC*, CXXVI, 1977.

——, 'The taking of Le Krak des Chevaliers in 1271', *Antiquity* 90, 1949.

——, The trebuchet and other siege engines, CG X, 1980.

KIRCHSCHLAGER, M, Stolle, T, Das teufelische Werkzeug-Entstehung von Geschichte der Weissensee Steinschleuder, in Wagener, O, Laß H, (eds.), ... Werfen hin steine/ groze und niht kleine ..., *Beihefte zur Mediavistik*, 7, Frankfurt/Berlin 2006.

KIRPICNIKOV, A, 'Steinfestungen Nordrusslands aus der Sicht der neueren bautechnisch-archäologischen Forschungen', *Burgen und Schlösser* 24, 1982.

——, 'Die Russischen Waffen der 9–13 Jahrhunderts und orientalische und Westeuropaische Einflüsse auf ihre entwicklung', *Gladius* 7, 1968.

KOLODZIEJSKI, S, 'Château et Ville en Pologne de midi', *Castrum Bene*, 6, ed. T Durdik, Prague 1999.

——, 'De Tyniec à Jaroslaw — des études au éléments défensifs des monastères en Pologne de midi', *Castrum Bene* 5, 1996.

KÖTZEL, T, Belagerungsburgen in Niedersachsen, in Wagener, Laß, eds., *Beihefte zur Mediavistik*, 7, 2006.

Krahe, F W, *Burgen des Deutschen Mittelalters*, Würzburg 1994.

Krautheimer, R, *Early Christian and Byzantine Architecture*, revised ed., Harmondsworth 1981.

Kuncevicius, A, 'Why did Vilnius become capital of Lithuania?', *Castella Maris Baltici*, ii, 1996.

Kunt, I Mertin, The rise of the Ottomans, in M Jones, ed., *New Cambridge Medieval History* vi, Cambridge 2000.

Kypta, J, Richterová, J, 'Opevněné postaveni oblehatelu hradu Grabštejno', *Castellologica Bohemica*, 9, 2004.

Lacroix, P, *The arts of the Middle Ages*, London 1875.

Lafuente, J, Zozaya, J, Algunas observaciones sobre el castillo de Trujillo, *Actas del xxiii Congreso Internacional de Historia del Arte*, ii, Granada 1977.

Landgraf, A, 'Die Wasserburgen des 13. und 14. Jahrhunderts im osten Niederösterreichs', *Burgen und Schlösser*, 14, 1973.

Larchey, L, *Origines de l'artillerie française. Planches autographiées d'après les monuments du xiv et xv siècles*, Paris 1863.

Lauzun, P, *Le château de Bonaguil*, Paris/Agen 1884.

Lavardin, J de, *The Historie of George Castriot, surnamed Scanderbeg, king of Albanie*, London 1596.

Lawrence, A W, 'A skeletal history of Byzantine fortification', *Annual of the British school at Athens*, 78, 1983.

Lazenby, D, *Cannons: that diabolic instrument of war*, 1999, website at http://middelalderscentret.dk/cannon2.htm (accessed 2006).

Leicht, P S, *Breve storia del Friuli*, 3rd ed. Udine 1952.

Lemerle, P, *L'emirat d'Aydin*. Paris 1957.

Leng, R, 'Ars belli. Deutsche taktische und kriegstechnische Bilderhandschriften und Traktake im 15. und 16. Jahrhundert', *Imagines Medii Aevi*, 12 (1,2), 2002, 21–261.

Leng, R (ed.), 'Anleitung Schießpulver zu bereiten, Büchsen zu laden und zu beschießen. Eine kriegstechnische Bilderhandschrift im cgm 600 der Bayerischen Staatsbibliothek München', *Imagines Medii Aevi*, 5, 2000, 23–141.

Líbal, D, 'Burgen und Stadte als tschechisches Erbgut', *Castrum Bene*, 6, ed. T Durdík, Prague 1999.

——, *Castles of Britain and Europe*, Prague, 1992.

Liddiard, R, *Castles in Context. Power, symbolism and landscape, 1066 to 1500*. Macclesfield 2005.

Liebel, J, Springalds and giant catapults, trans. J Vale, Leeds 1998.

Lindemann, G, *Illustratierte Führer durch Lüneburg und die Lünerburger Heide*, Lüneburg 1924.

Little, D P, The fall of 'Akka in 690/1291, in M Sharon, ed., *Studies in Islamic history and civilisation in honour of Professor David Ayalon*, Jerusalem/Leiden 1986.

Lloyd, T E, *A History of Wales from the earliest times to the Edwardian conquest*, 2 vols, London 1912.

Lock, P, Hodgetts, C, Some village fortifications in the Venetian Peloponnese, in P Lock, G D R Sanders, *The archaeology of medieval Greece*, Oxford 1996.

Lock, P, 'The Frankish Towers of central Greece', *Annual of the British School at Athens*. 81, 1986.

——, *The Franks in the Aegean 1204–1500*. London 1996.

——, The medieval towers of Greece: a problem in chronology and function, in B Arbel,

B Hamilton, D Jacoby, eds., *Latins and Greeks in the eastern Mediterranean after 1204*. 1989.

——, The towers of Euboea: Lombard or Venetian, agrarian or strategic, in P Lock, G D R Sanders, eds., *The archeology of medieval Greece*, Oxford 1996.

LONGO, A A, 'L'assedio e la destruzione di Gallipoli (1268–9)', *Archivio Storico Italiano*, CXLVI, 1988.

LOREN, C, 'The round towers on Sweden's eastern coast', *Castella Maris Baltici*, II, 1996.

LUTTRELL, A, FALKENHAUSEN, V von, 'Lindos and the defence of Rhodes 1306–1522', *Rivista di Studi Bizantini e Neoellenici*, NS 22/3, 1985/6.

LUTTRELL, A, The Hospitallers at Rhodes 1306–1522, in W Hazard, *History of the Crusades*, general editor K M Setton, III, Wisconsin 1975.

LUXEMBOURG: National Monuments Commission, *Wenzel Itinerary*, Luxembourg 1995.

MAKHDOOMEE, M Akram, 'The art of war in medieval India', *Islamic Culture*, XI, 1937.

——, 'Mechanical Artillery in medieval India', *Journal of Indian History*, XV, 1936.

MALLETT, M E, HALE, J R, *The Military Organisation of a Renaissance State. Venice c. 1400 to 1617*. Cambridge 1984.

MANZ, B F, *The rise and rule of Tamerlane*. Cambridge 1989.

MARCO, M de, *Torri e castelli del Salento. Itinerario castrero*. Lecce 1994.

MARSHALL, C, *Warfare in the Latin East 1192–1291*. 1992.

MASON SMITH, J, 'Demographic considerations in Mongol siege warfare', *Archivium Ottomanicum*, 13, 1993, 329–34.

MASPERO, G, *Le royaume de Champa*, Paris/Brussels 1928.

MASSEY R, Lancastrian Rouen. Military service and property holding 1419–49, in *England and Normandy in the Middle Ages*, ed. D Bates, A Curry London 1994.

MATAROSSO, F, *The English Castle*, London 1995.

MATHIEU, J R, 'A new interpretation of Caernarvon Castle', *Medieval Life* 9, 1998.

McCARTHY, M R, SUMMERSON, H R T, ANNIS R G, *Carlisle Castle: A survey and documentary history*. London 1990.

McDONALD, R Andrew, *The Kingdom of the Isles: Scotland's western seaboard c1100–c1336*. East Linton 1997.

McNAMEE, C, *The Wars of the Bruces. Scotland, England and Ireland 1306–1328*. East Linton 1997.

McNEIL, T, *Castles in Ireland: Feudal Power in a Gaelic World*, London/New York 1997.

McNIALL, T E, 'The origins of tower houses', *Archaeology Ireland* 19, 1992.

MEDUNA, P, 'Povichorý průzkum komplacu obléhacíchprací u Konopiště z let 1467–1468', *Castellologica Bohemica*, 4, 1994.

MEGAW, A H S, The arts in Cyprus: military architecture, in K M Setton, *A history of the Crusades*, IV, ed. H W Hazard, Wisconsin 1977.

MEGAW, B R S, 'The Scottish invasion of Man 1456', *Journal of the Manx Museum* VI, 1957.

MENCLOVÁ, D, 'Husitské Opevnění Tábora', *Zpróvy Památkové Péčé*, XIII, 1953.

——, 'Vliv husitskych válekna pozdně gotickou fortifikeční architekturu', *Umění*, IX, 1961.

MENZIES, G, *1421: the yea China discovered the world*. London 2002.

MERCIER, A, *Le château de Ham*, *BM* 78, 1914.

MESQUI, J, *Châteaux Forts et Fortifications en France*, Paris 1997.

MESQUI, J, FAUCHERRE, N, 'L'enceinte medievale de Césarée', *BM* 164(1), 2006.

MEYER, W, WIDNER, E, *Das grosse Burgenbuch der Schweiz*, Zurich 1978.

MICHAEL, N, EMBLETON G A, *Armies of medieval Burgundy 1364–1477*, Oxford 1983.

MICHAUD, M, *Bibliothèque des Croisades*, 4 vols, Paris 1829.

MIGOS, A, 'Rhodes: the Knights' battleground', *Fort*, 18, 1990.

MIJATOV, K, 'Das Schloss der Bulgarischen zaren in Tarnovo', *Acta Archeologica Academiae Scientarum Hungaricae*, XVII, 1965.

MILLINGEN, A van, *Byzantine Constantinople. The walls of the city and adjoining historical sites*, London 1899.

MIOTTI, T, *Castelli del Friuli, II: gastaldie e giurisdizioni del Friuli centrale*, Udine 1977–88.

MIQUELL, J, L'armament des châteaux dans la Gascogne anglaise de 1250 à 1325, in M-P Baudry, *Les fortifications dans les domaines Plantagenet, XII–XIV siècles*, Poitiers 2000.

MOLA, S, *Trani: il castello*, Bari, 2000.

——, trans. C Jenkner, *Castel del Monte*, Bari 1991.

——, *Itinerario federiciano in Puglia*, Bari, 1994.

MONETTI, F, RESSA, F, *La costruzione del castello di Torino, oggi Palazzo Madama*, Torino 1982.

MONREAL, L, DE RIQUIER, M, *Els Castells medievals de Catalunya*, vol. 2 Barcelona 1958.

Monumenten van Geschiedenis en Kunst in de Provincie Gelderland, De, I: Het Kwartier van Nijmegen, Den Haag, 1932.

Monumenten van Geschiedenis en Kunst in de Provincie Limburg, De, I: Maastricht, Utrecht, Den Haag 1926.

MORGAN, D O, *Medieval Persia 1040–1797*. London 1988.

——, 'The Mongol Armies in Persia', *Der Islam* 56, 1979.

MORRIS, J E, *The Welsh wars of Edward I*, Oxford 1901.

MOSSAKOWSKI, 'The town walls of the old Cracow', *Europa Nostra Bulletin* 58, 2004.

MOULE, A C, *Quinsai: with other notes on Marco Polo*, Cambridge 1957.

MÜLLER-WIENER, W, *Castles of the Crusaders*. London 1966.

——, 'Die Stadtsbesfestigungen von Izmir, Sigacik und Candarli', *Istanbuler Mitteilungen* 12, 1962.

——, 'Mittelalterliche Befestigungen in sudlichen Jonien', *Istanbuler Mitteilungen* 11, 1961.

MUGAREVICS, E, 'Zur Archäologie mittelalterliche Burgen in Estland', *Lübecker Schriften zur Archaologie und Kulturgeschichte*, 12, 1986.

MUK, J, 'Táborsky Hrad Hradiste ve Středověku', *Husitsky Tábor* 5, 1982.

MURDOCH, J, *A history of Japan*, London 1925.

MUS, P, 'Études indiennes et indochinoises, III, Les balistes du Bayon', *Bulletin de l'École Française de l'extrême orient*, XXIX, 1929.

NEEDHAM, J, abridged Ronan, C, *The Shorter Science and Civilisation in China*, 5, Cambridge 1995.

NEEDHAM, J, China's trebuchets, manned and counterweighted, in *On pre-modern technology and science: Studies in honour of Lynn White*, 1976.

NEEDHAM, J, et al, *Science and Civilisation in China*, vol v part 7: Military technology: the gunpowder epic, Cambridge 1986.

NEEDHAM, J, Yates R S, eds. *Science and Civilisation in China*, volume 5 part VI, Cambridge 1994.

NERI, A, Il forte de Sarzanello, *Archivio Storico Italiano*, 4th ser, 15, 1885.

NEWHALL, R A, *The English Conquest of Normandy 1416–24*, New York/London 1924.

NICHOLAS, D, *The Growth of the Medieval City*. London 1997.

NICHOLSON, H, *The Knights Hospitaller*, Woodbridge 2001.

NICHOLSON, R, 'The siege of Berwick 1333', *Scottish Historical Review* 40, 1961.

NICOLLE, D, *Medieval Warfare Source Book vol. 2: Christian Europe and its Neighbours*, London 1996.

NORMAN, A V B, 'Notes on some early representations of guns and on ribaudequins', *Journal of the Arms and Armour Society* VIII/3, 1975.

Nossov, K S, *Indian Castles 1201–1526*, Oxford 2006.

——, *Medieval Russian Fortresses* AD *862–1480*, Oxford 2007.

Nunes de Leao, Duarte, *Cronicas e vidas des reys de Portugal D Duarte, undecimo e D Affonso o v duodecimo*, II, Lisbon, 1780.

Nystazapoulou, 'La dernière reconquête de Sinope par les Grecs de Trebizonde (1254–65)', *Revue des Études Byzantines*, 22, 1964.

O'Keeffe, T, Ballyloughan, Ballymoon and Clonmore: three castles of *c.*1300 in county Carlow, *ANS* XXIII, Gillingham J, ed., 2000.

O'Keeffe, T, 'The fortifications of Western Ireland AD 1100–1300 and their interpretation', *Journal of the Galway Archaeological and Historical Society* 50, 1998.

O'Neil, B H St J, *Castles and Cannon. A study of early artillery fortifications in England*, Oxford 1960.

——, 'Rhodes and the origin of the bastion', *Antiquaries Journal*, XXXIV, 1954.

Olavinlinna, *Olofsborg*, Helsinki 1920.

Opolovnikov, A and Opolovnikova, Y, *The wooden architecture of Russia. Houses, Fortifications, Churches*, trans. J and R Whitby, London 1989.

Ortíz Echagüe, J, *España: castillos y alcázares*, Madrid, 1956; repr. 1964.

Ostrogorsky, G, 'La prise de Serres par les Turcs', *Byzantion* 35, 1965.

Otway-Ruthven, A J, *A History of Medieval Ireland*. 2nd ed, New York 1980.

Oxley, J, ed., *Excavations at Southampton castle*, Southampton 1986.

Pächt, O, Thoss D, Franzosische Schüle, in *Die Illuminierte Handschriften und Inkunabeln der Österreichischen Nationalbibliothek*, Vienna 1974.

Paciaroni, R, 'La ricostruzione di un castello sanseverinate alla fine del sec. XV: Truschia', *Studia Maceratesi*, 24, 1991.

Palliser, D.M., Town Defences in medieval England and Wales, in *The Medieval military revolution*, A Ayton, J L Price, London 1995.

Pan, Jixing, The origin of rockets in China, in B Buchanan, ed., *Gunpowder: history of an international technology*, Bath 1996.

Panaitescu, P P, 'Les chroniques Slaves de Moldavie au XVe siècle', *Romanoslavica* I, Bucharest 1958.

Paner, H, 'The Teutonic castle in Gdansk', *Castrum Bene* 5, 1996.

Pant, G, *Studies in Indian weapons and warfare*, New Delhi, 1970.

Parmentier, H, *Inventaire des monuments Cams de l'Annam*, 2 vols, Paris 1909–18.

——, 'Notes d'archéologie Indochinoise', *Bulletin École Française de l'Extrême Orient*, 23, Hanoi 1923.

——, Tables géographiques des Monuments du Champa et du Cambodge, in G Coedès, *Listes générales des inscriptions et des monuments*, Hanoi 1923.

Partington, J R, *A history of Greek Fire and Gunpowder*, intro B S Hall, reprint London 1999.

Paterson, R C, *For the Lion: History of the Scottish Wars of Independence 1296–1357*. Edinburgh 1996.

Payne-Gallway, R, *The Crossbow, medieval and modern*, London 1903.

Peers, C J, Perry, M, *Imperial Chinese armies 590–1260*, London 1996.

Peers, C J, Sque, D, *Medieval Chinese armies 1260–1520*, Oxford 1992, reprint 2000.

Pelc, O, *Im Schütz von Mauern und Toren. Die Befestigung der schleswig-holsteinischen Städte in Mittelalter und Neuzeit*, Heide 2003.

Pelliot, P, 'La date de l'apparition en Chine des canons Folang-ki (Farangi? Ou machines franques)', *T'oung Pao Archives*, 38, 1948.

Pentz, P, 'A medieval workshop for producing "Greek Fire" grenades', *Antiquity* 62, 1988.

PEPPER, S, Ottoman military architecture in the early gunpowder era: a reassessment, in J Tracy, ed., *City Walls: the urban enceinte in global perspective*, Cambridge 2000.

PETERS, C H, *De Nederlandsche Stedenbouw*, Leiden 1909.

PETRIE, J, 'Historians and the role of Castles in the Wars of the Roses', *Fort* 12, 1984.

PETROVIC, D, Firearms in the Balkans on the eve of and after the Ottoman conquests of the 14th and 15th centuries, in *War, Technology and Society in the Middle East*, ed. V J Parry, M E Yapp, London 1995.

PETROVICS, I, The role of towns in the defence system of medieval Hungary, in P Contamine, O Guyotjeannin, eds., *La guerre, la violence et les gens au Moyen Âge*, 1, Paris 1996.

PEYROU, J, HENRY-CLAUDE, M, *Montségur. Last rampart of the Catharist Church*, Gavaudan 1998.

PICQ, Ardant du, *Histoire d'une citadelle annamite — Bac Ninh*, Hanoi 1935.

PIPER, O, *Burgenkunde. Bauwesen und Geschichte der Burgen zunachst innerhalb der deutschen Sprachgebietes*, Munich 1912 (reprinted 1967).

PLANITZ, H, *Die deutsche Stadt in Mittelalter*, Graz-Koln 1954.

PLATT, C, *The English medieval town*, London 1976.

POERCK, G de, 'L'Artillerie à ressorts médiévale, notes lexicologiques et etymologiques', *Bulletin du Cange* XVIII, 1945.

PORTER, P, *Medieval warfare in Manuscripts*, London ND.

POST, P, 'Das gotische Faustrohr in der Leibrustkammer in Stockholm', *Zeitschift für Historische Waffen- und Kostumkunde*, ns 3, 1930.

POUNDS, N J G, *An economic history of medieval Europe*, 2nd ed., 1994.

——, *The medieval castle in England and Wales, a social and political history*, Cambridge 1990.

POWERS, J F, *A Society organised for war: The Iberian municipal militias in the central Middle Ages 1000–1289*. London 1988.

PRENTOUT, H, 'La prise de Caen par Edward III 1346', *Mémoires de l'Academie Nationale des Sciences, Arts et Belles-lettres de Caen*, 1904.

PRESTWICH, M, *Armies and warfare in the Middle Ages: the English experience*, London 1996.

——, 'English castles in the reign of Edward II', *Journal of Medieval History* 8, 1982.

PRICE, G, 'Castle Ownership: politics and power in high medieval England', *Medieval Life* 4, 1996.

PRINGLE, D, et al, 'Qal'at Jiddin: a castle of the crusader and Ottoman periods in Galilee', *Levant* 26, 1994.

PRINGLE, D, 'Reconstructing the castle of Safad', *Palestine Exploration Quarterly*, 1985.

——, *Rothesay Castle*, ed. C Tabraham, Edinburgh 1995.

——, Templar castles between Jaffa and Jerusalem, in H Nicholson, ed., *The Military Orders*, vol. 2 Aldershot 1998.

——, Town defences in the Crusader kingdom of Jerusalem, in I A Corfe, M Wolfe, eds., *The Medieval City under Siege*, Woodbridge 1995.

PUISEUX, L, *Siège et prise de Caen par les Anglais en 1413*, Caen 1888.

PURTON, P, 'The myth of the mangonel: torsion artillery in the Middle Ages', *Arms and Armour Journal*, 3, 2006.

——, Bohemian fortresses and the evolution of artillery defences in the fifteenth century, *Castle Studies Group Journal*, 22 (2008–9), 217–27.

QUARITCH WALES, H, *Ancient south-east Asian warfare*, London 1952.

QUESADA, M A Lodero, *Castilla y la conquista del reino de Granada*, Valladolid, 1967.

RABBAT, N O, *The citadel of Cairo: a new interpretation of royal Mamluk architecture*, Leiden 1995.

RAPPAPORT, P A, 'Russian medieval military architecture', *Gladius* VIII, 1969.
RATHGEN, B, *Das Geschütz im Mittelalter*, Berlin 1928.
RAUCH, C, *Die Kunstdenkmaler im Volkstaat Hessen: Kreis Bingen*, Darmstadt, 1934.
REES, S E, Caple C, *Dinefwr castle, Dryslwyn castle*, Cardiff 1999.
REID, N H, *Scotland in the reign of Alexander III 1249–86*. Edinburgh 1990.
REILLY, B F, *The medieval Spains*, Cambridge 1993.
RENET, M, *Beauvais et le Beauvaisis dans les temps modernes: époque de Louis XI et de Charles le Téméraire*, Beauvais 1898.
RENN, D F, Castle fortifications in England and adjoining countries from 1150 to 1250, in A Salamagne, R Le Jan, eds., *Château médiéval et la guerre dans l'Europe du Nord-ouest*, Lille 1998.
——, 'The Southampton Arcade', *Medieval archaeology* VIII, 1964.
——, 'The earliest gun port in Britain?', *Archeological Journal* CXXV, 1968.
RENOUARD, Y, *Bordeaux sous les Rois d'Angleterre*, Bordeaux 1965.
REY, G, *Étude sur les monuments de l'architecture militaire des Croises en Syrie et dans l'île de Chypre*, Paris 1871.
REY, R, *Les vieilles églises fortifiées du midi de la France*. Paris 1925.
RICE, D S, 'Studies in Medieval Harran', *Anatolian Studies* II, 1952.
RODDI, G, 'Note sulla costruzione del castello di Ivrea', *Studi Piemontesi*, XI, 1982.
ROGERS, C J, The age of the 100 Years War, in M Keen ed., *Medieval Warfare: A history*, Oxford 1999.
——, 'The military revolution of the Hundred Years War', *Journal of Military History* 57(2), 1993.
ROLL, I, Arubas, B, 'Le Château d'Arsur: forteresse côtière pentagonale du type concentrique du milieu du XIIIe siècle', *BM*, 164(1), 2006.
ROSA, L G, Nuovo, I, Defilippis, D, *Gli umanisti e la guerra Otranteria. Testi de secoli XV e XVI*.
ROSS, J, ERICHSEN, N, *The story of Pisa*. London 1909.
ROVIRA I VIRGILI, A, *Historia Nacional de Catalunya*, vol III, Barcelona 1924.
ROWELL, S C, *Lithuania ascending: a pagan empire within east-central Europe*, Cambridge 1994.
RUSSELL, P, *Prince Henry 'the Navigator'. A Life*, New Haven/London 2001.
SAADE, G, *Le château de Saladin*, Damascus nd.
SAILHAN, P, 'Typologie des archères et cannonières. Les archères des châteaux de Chauvigny', *Bulletin de la Société des Antiquaires de l'Ouest et des Musees de Poitiers*, 4th series, XIV, 1978.
SAINT-ORGAN, Lefebvre, *Compiègne*, Paris 1887.
SALAMAGNE, A, 'L'attaque des places fortes au XVe siècle à travers l'exemple des guerres anglo- et franco-bourguignonnes', *RH*, 289, 1993.
——, *À la découverte des anciennes fortifications de Valenciennes*, Cambrai 1999.
——, 'À propos de l'adaptation de la fortification à l'artillerie vers les années 1400: quelques remarques sur les problèmes de vocabulaire, de typologie et de méthode', *Revue du Nord*, LXXV, 1993.
——, 'Le château de Montcornet dans les Ardennes et sa place dans l'architecture militaire de la seconde moitié du XVe siècle', *Revue Historique Ardennaise*, XXVII, 1992.
——, 'Les années 1400: la genèse de l'architecture militaire Bourguignonne ou la definition d'un nouvel éspace urbain', *Revue Belge d'histoire militaire*, XXVI, 1986.
——, Les fonctions militaires des châteaux en Hainault du XIIIe au début du XVIe siècle, in *Châteaux et Chevaliers en Hainault au Moyen Âge*, Valenciennces 1995.
SALCH, C-L, *Atlas des villes et des villages fortifiées en France*, Strasbourg 1978.

SALCH, C-L, *Dictionnaire des châteaux et des fortifications du moyen âge en France*, Strasbourg 1979.

SAMUEL, M, 'A tentative chronology for tower houses in West Cork', *Journal of the Cork Historical and Archaeological Society*, 103, 1998.

SANCHEZ-GIJON, A, COBOS, F, CASTRO, J de, eds., *Luis Escriva, su apologia y la fortificacion imperial*, Valencia 2000.

SAND, G, *Clés et défenses d'une ville: Bruxelles et son histoire*, Brussels 1984.

SANTORO, L, *Castelli angoini e aragonesi nel regno di Napoli*, Milan 1982.

——, The fortified royal palace of the Aragonese and Angevin sovereign, in *The Castelnuovo in Naples, IBI* Bulletin 50, 1997.

SARKAR, J, *The Art of War in medieval India*, New Delhi 1984.

SAUL, N, Bodiam Castle, in *History Today*, 45.1, 1995.

SAUNDERS, A D, Attitudes towards defence and English monastic houses and granges, in *IBI* Bulletin 50, 1994/5.

——, *Dartmouth Castle*, 3rd edition, London 1991.

——, 'Southampton: the introduction of gunpowder artillery to the town's defences', *Europa Nostra Bulletin*, 53, 2000.

——, 'The Cow Tower, Norwich: an East Anglian bastille?', *Medieval archaeology* XXIX, 1985.

——, 'The Defences of Southampton in the later Middle Ages, in L A Burgess, ed., The Southampton Terrier of 1454', *HMC, Southampton Record Series* XV, London 1976.

SAUNDERS, J J, *A history of Medieval Islam*, London/New York 1965.

SAUVAGET, J, *Les monuments historiques de Damas*, Beirut 1932.

SAWYER, B and P, *Medieval Scandinavia*, Minneapolis 1993.

SAYLES, G O, 'The siege of Carrickfergus castle 1315–16', *Irish Historical Studies*, X, 1956–7.

SCHACHT, T, *Aus und über Ottokar von Horneck Reimkronik*, Mainz 1821.

SCHMIDT, E F, *Flights over the ancient cities of Iran*, 1940.

SCHMIDTCHEN, V, *Mittelalterliche Kriegsmaschinen*, Soest 1985.

——, 'Riesengeschütze des 15 Jahrhunderts, technische Hochsleistungen ihrer Zeit', *Techniksgeschichte* 44, 1977.

SCHNEIDER, J, *La ville de Metz aux XIIIe et XIVe siècle*, Nancy 1950.

SCHNEIDER, R, 'Die Geschütze des Mittelalters', *Zeitschrift für Historische Waffenkunde* 5, 1909/11.

SCHÜTT, H–F, 'Die als Steinbau errichtete Türmburg in Flensburg', *Offa*, 53, 1996.

SCRIATTOLI, A, *Viterbo nei suoi monumenti*, Viterbo 1988.

SEETZEN, U J, *Reisen durch Syrien, Palastina (etc)*, ed. F Kruse, Berlin 1854.

SERROYS, H, 'Towers in the northern frontier defences of the Ming', *Ming Studies* 14, 1982.

SETTIA, A A, Un castello in Torino, reprinted in *Proteggere e Dominare*, Rome 1999.

SEWARD, D, *The Hundred Years War*, London 1978.

SHAW, M. CHAPMAN, A, SODEN, I, 'Northampton', *Current Archaeology* 13, 1997.

SHENTON, C, 'The location of the siege camp at Berwick upon Tweed 1333', *Archaeologia Aeliana*, ser. 5, 29, 2001.

SHPUZA, G, 'La lutte pour la défense de Shkodër dans les années 1474 et 1478–9', *Studia Albanica*, 1968 (5th year, no.1), 181–9.

SHRADER, C R, 'A handlist of extant manuscripts containing the De Re Militari of Flavius Vegetius Renatus', *Scriptorium* 33 1979.

SINCLAIR, T A, *Eastern Turkey: an architectural and archaeological survey*, 4 vols, London 1987–90.

SIRCAR, J N, 'Siegecraft in medieval India', *Itihas* IV, 1976.

SIREN, O, *The walls and gates of Peking*, London 1924.

SKOLJAER, S A, 'L'Artillerie de jet à l'époque Sung', *Études Song*, Series 1, vol. 2, 1971.

SMITH, L Toulmin, ed., 'Expeditions to Prussia and the Holy Land made by Henry earl of Derby in the years 1390–1 and 1392–3', *Camden Society New Series* LII, London 1894.

SMITH, R D, 'All manner of peeces: artillery in the late medieval period', *Royal Armouries Yearbook*, 7, 2002.

——, Artillery and the Hundred Years War: myths and interpretations, in A Curry, M Hughes, *Arms, Armies and Fortifications in the Hundred Years War*, Woodbridge 1994.

——, 'Good and Bold. A late fifteenth century artillery train', *Royal Armouries Yearbook*, 6, 2001.

——, The Ho experiments, report no. 2 (August 2003), Medieval Gunpowder Research group, *Middelaldercentret* website (accessed 2007).

——, 'The reconstruction and firing trials of a replica of a 14th century cannon', *Royal Armouries Yearbook*, 4, 1999.

SMITH, R D, BROWN, R R, *Bombards. Mons Meg and her Sisters*, London 1989.

SMITH, R D, DE VRIES, K, *The artillery of the Dukes of Burgundy 1363–1477*, Woodbridge 2005.

SORTINOV, G, 'Le château fort de Chloumetsi et son atelier monétaire', *Journal International d'archéologie numismatique*, 19, 1918/19.

SPATHARI, E, *Nauplion-Palamidi*, Athens, 2004.

SPITERI, S C, *Fortresses of the Knights*, Malta 2001.

SPUFFORD, P, *Handbook of medieval exchange*, London 1986.

SPULER, B, *Die Goldene Horde. Die Mongolen in Russland 1223–1502*, Leipzig 1943.

STAŇKOVÁ J, ŠTURSA J, VODĚRA S, *Pražska architektura*, Prague 1991.

STEFANIDOU, A, The medieval town of Rhodes, in Triposkofi and Tsitouri (eds.), *Venetians and Knights Hospitallers*, Athens 2002.

STEIN, A, *Innermost Asia. Detailed report of explorations in Central Asia, Kan-Su and Eastern Iran*, 4 vols, Oxford 1928.

STELL, G, *Dunstaffnage and the castles of Argyll*, Edinburgh 1994.

STENMANS P, Lange J, Bommels N, eds., *Neuss in Wandel der Zeiten*, Neuss 1969.

STEWART, A, "Qal'at al-Rum/Hromgla/Rumkale and the Mamluk siege of 691 AH / 1292 CE", in *Muslim military architecture in greater Syria: from the coming of Islam to the Ottoman period* [Aleppo 2003], 269–80, ed. H Kennedy, History of warfare, 35, Leiden 2006.

STIESDAHL, H, Die Motten in Danemark, CG II, Köln 1967.

——, Types of public and private fortifications in Denmark, in *Danish Medieval History: New Currents*, ed. N Skyom-Nielson, N Lund, Copenhagen 1981.

STOLBERG, F, *Befestigungsanlagen im und am Harz von der Frühgeschichte bis zum Neuzeit*, Hildesheim 1968.

STRICKHAUSEN, G, Bemerkungen zu frühen Feuerwaffen im 14 Jahrhundert, in Wagener, Laß, eds., in *Beihefte zur Mediavistik*, 7, 2006.

SÜFFLAY, M, 'Städte und Burgen Albaniens, hauptsächlich während des Mittelalters', *Akademie der Wissenschaften in Wien: Philosophisch-historische Klasse*, 63 (1), Vienna 1924.

SVENNEWALL, T, 'Even the tallest tower. Early medieval fortification builders and buildings on Sweden's east coast', *Castella Maris Baltici* II, 1996.

SWEETMAN, D, *Irish Castles and Fortified Houses*, Dublin 1995.

——, 'The hall house in Ireland', *Archaeology Ireland* 12, 1998.

SWEETMAN, D, *The medieval castles of Ireland*, Woodbridge 2000.
ŚWIĘTOSŁAWSKI, W, *Arms and Armour of the Nomads of the Great Steppe in the Time sof the Mongol expansion (12–14th centuries)*, Lodz, 1999.
SZAKALY, F, The Hungarian-Croatian border defence system and its collapse, in *From Hunyadi to Rackoczi: War and Society in late medieval and early modern Hungary*, ed. J M Bak, B K Karoly, Brooklyn 1982.
TABRAHAM, C, *Bothwell castle*, Edinburgh 1995.
——, *Kildrummy castle, Glenbuchat castle*, Edinburgh 1995.
——, *Scotland's Castles*, London 1997.
TAFRALI, O, *Topographie de Thessalonique*, Paris 1913.
TALBOT, H, *The English Achilles. Life and campaigns of John Talbot 1st earl of Shrewsbury*. London 1981.
TATTON-BROWN, T, *The Westgate, Canterbury*, Canterbury Archaeological Trust 1985.
TAYLOR, A J, *Harlech Castle*, Cardiff, 1988.
——, Master Bertram, Ingeniator Regis, in *Studies in medieval history presented to R Allen Brown*, ed. C Harper Brill, C Holdsworth, J Nelson ,Woodbridge, 1989.
——, *Raglan Castle*, London 1950.
TEDESCHI, N, PERBELLINI, G, BARBIERI, G, *Castelli Scaligeri nel Veronese*, Verona 1986.
TEUPSER, W, *Rothenburg, Dinkelsbühl, Nördlingen*, Leipzig, 1928.
THOMAS, A, *The walled towns of Ireland*, 2 vols, Blackrock 1992.
THOMO, Pirro, 'La forteresse de Skanderbeg au Cap du Rodon', *Studia Albanica*, 4th year, 1967.
THOMPSON, M W, *Kenilworth Castle*, London 1996.
THORAU, P, *The Lion of Egypt. Sultan Baybars I and the Near East in the 13th Century*, trans. P M Holt, London 1987/92.
TIKHOMINOV, M, *The towns of ancient Rus*, Moscow, 1959.
TILLMAN, C, *Lexikon der Deutschen Burgen und Schlösser*.4 vols, Stuttgart 1957–61.
TITTMAN, W, 'Die Eltzer Büchsenpfeile von 1331/3', *Waffen- und Kostumkunde*, 36, 1994 and 37, 1995.
TORBUS, T, 'Das Krakauer "Rondell" als Paradebeispiel der spätmittelalterlichen Barbakane sowie seine polnischen Nachfolgebauten', *Forschungen zu Burgen und Schlössern*, Wartburg Gesellschaft zur Erforschung von Burgen und Schlösser, 12, 2009.
TORNČIČ, Z, 'Fortifications of Orders of Knights in medieval Croatia', *Castrum Bene* 5, 1996.
TORREMACHA SILVA, A, Las Fortificaciones medievales de Algeciras, *Congreso Internacional: El Estrecho de Gibraltar*, II, Madrid 1987.
TORRES BALBAS, L, *Cuidades Hispano-Musulmanes*, 2nd ed. Madrid 1985.
TOY, S, *The fortified cities of India*, London 1965.
——, *The strongholds of India*, Melbourne/London/Toronto 1957.
TREPTOW, K, ed., *Dracula: essays on the life and times of Vlad Tepes*, New York 1991.
TRIPOSKOFI, A, TSITOURI, A, (eds.), *Venetians and Knights Hospitallers*, Athens 2002.
TURNBULL, S, *Castles of the Teutonic Knights*, (1) *The red brick castles of Prussia 1230–1466*, (2) *The stone castles of Latvia and Estonia 1185–1560*, Wellingborough 2005 and 2004.
——, *Samurai. The World of the Warrior*, Oxford, 2003.
——, 'The blunted arrowhead: the defensive role of the great medieval fortresses of Albania', *Medieval History Magazine*, 2/5, January 2005.
TURNBULL, S, REYNOLDS, W, *Siege weapons of the Far East (2) AD 960–1644*, Oxford 2002.

TURNER, H D, *Town Defences of England and Wales: an architectural and documentary study AD 900–1500*, London 1971.

TURVEY, R, *The Welsh Princes 1063–1283*, Harlow/London 2002.

TUULSE, A, *Die Burgen in Estland und Lettland*, Dorpat 1942.

UNDERWOOD, R, *Anglo-Saxon Weapons and Warfare*, Stroud 1999.

URBAN, W, 'The organisation of defense of the Livonian frontier in the 13th century', *Speculum* XLVIII/3, 1973.

VALE, M G A, New techniques and old ideals: the impact of artillery on war and chivalry at the end of the hundred years' war, in: C T Allmand, ed., *War, Literature and Politics in the late Middle Ages*, Liverpool. 1976.

——, 'The last years of English Gascony', *Transactions of the Royal Historical Society* 5th ser, 19, 1969.

VARLEY, P, *Warriors of Japan as portrayed in the War Tales*, Honolulu 1994.

VASARI, G, *The lives of the painters, sculptors and architects*, trans. W Gaunt, London 1963.

VASILIATOS, N, *Coastal Castles* (in Greek), 1996.

VASILIEV, A, Jörg of Nurnburg, *Byzantion*, 10, 1935.

VASOLD, M, *Geschichte der Stadt Rothenburg ob den Tauber*, Stuttgart 1999.

VAUGHAN, R, *Charles the Bold, the last Valois Duke of Burgundy*, new ed., Woodbridge 2002, reprint 2004.

——, *John the Fearless. The growth of Burgundian power*, London/New York 1966, republished 1979.

VELASCO, A Gutierrez de, Las fortalezas aragonesas ante le gran ofensiva castellana en la guerra de los dos Pedros, in J Zurita, ed., *Cuadernos de la historia*, 12–13, Zaragona 1961.

VERDIER, P, 'La rocca d'Ostie deans l'architecture militaire de Quattrocento', *Mélanges d'archéologie et d'histoire, École Française de Rome*, LVI, 1939.

VIGON, J, *Historia de la artilleria Española*, I, Madrid 1947.

VOYCE, A, *The Moscow Kremlin. Its history, architecture and art treasures*, London 1955.

VRIES, K de, Gunpowder weapons at the siege of Constantinople 1453, in *War and Society in the Eastern Mediterranean*, ed. Y Lev, New York 1997.

——, Impact of gunpowder weaponry on siege warfare in the 100 Years' War, in *The Medieval City under Siege* I A Curtis, M Wolfe, Woodbridge 1995.

——, *Medieval military technology*, Oxford 1992.

WAGENER, S, De Monckler destructione – Die Belagerung der Burg Montclair 1351 und die Burgstellen in ihrer Umgebung, in Wagener, Laß, eds., *Beihefte zur Mediavistik* 7, 2006.

WALKER R, SPURGEON C, 'The custody of the De Clare castles in Glamorgan and Gwent, 1262–1263', *Studia Celtica* 37, 2003.

WANG LING, On the invention and use of gunpowder and firearms in China, *Isis*, 37, 1947.

WEIGEL, S, 'Die Stadtbefestigung Neutitscheins', *Zeitschrift des deutschen Verienes für die Geschichte Mährens und Schlesiens*, 23, 1919.

WERNER, E, *Chinese Weapons*, Shanghai 1932.

WESTPHALEN, Ernest Joachim de, Rerum inter Theodorum Moereum Archiepiscopum Colonienssem et Rempublicam Susatensum gestarum adversia, in *Monumenta inedita rerum Germanicarum praecipue Cimbricorum et Megapolensium*, VIII, Lipsius, 1743.

WETZSTEIN, J G, *Reisebericht über Hauran und die Trachonen*, Berlin 1860.

WHITE, L, jnr, 'Tibet, India and Malaya as sources of western medieval technology', *American Historical Review*, LXV, 1960.

WHITE, L, jnr, The crusades and the technological thrust of the west, in *War, Technology and Society in the Middle East* ed. V J Parry, M Yapp, 1975.

WIGGINS, K, 'Warfare underground', *British Archaeology*, 71, 2003.

WILLEY, P, *Eagle's Nest. Ismaili Castles in Iran and Syria*, London/New York 2005.

WILLIAMS, A, Ottoman military technology: the metallurgy of Turkish armour, in *War and Society in the Eastern Mediterranean*, ed. Y Lev, New York 1997.

WINKELMANN, E, *Kaiser Friedrichs II Kampf um Viterbo. Historische Aufsatze dem Andanken an Georg Waitz gewidmet*, Hannover 1886.

WINNIG, A, *Der Deutsche Ritterordern und seine Burgen*, Konigstein im Taunus, 1956.

WLASSETY, J von, 'Combat anno 1399. Welche Schlüssleistung besass die "Tannenburg-Büchse"?' *Deutsche Waffen-Journal* 6/1977.

WÜRDINGER, J, *Kriegsgeschichte von Bayern, Franken, Pfalz und Schwaben von 1347 bis 1506*, 2 vols, Munich 1868.

WYLIE, T, 'The first Mongol conquest of Tibet reinterpreted', *Harvard Journal of Asiatic Studies*, 37, 1977.

XHPUZA, G, 'La lutte pour la défense de Shkoder dans les années 1474 et 1478–79', *Studia Albanica*, 5th year, 1968.

YAMADA, N, *Ghenko. The Mongol invasion of Japan*, London 1916.

YANG HONG, ed., *Weapons in ancient China*, New York/Beijing 1992.

YOUNG, A, *Robert the Bruce's rivals: The Comyns 1212–1314*. East Linton 1997.

ZABIELA, G, 'The end of wooden fortifications in Lithuania', *Castella Maris Baltici*, II, 1996.

ZACHARIADOU, E A, 'The conquest of Adrianople by the Turks', *Studi Veneziani* XII, 1970.

ZAKY, A R, 'Gunpowder and Arab firearms in the Middle Ages', *Gladius*, 6, 1967.

——, 'Military literature of the Arabs', *Islamic Culture*, 30, 1956.

ZEUNE, J, Die Baugeschichte der Veste Coburg, in M Henker, E Brockhoff, eds., *Ein Herzogtum und viele Kronen. Coburg in Bayern und Europa*, Augsburg/Regensburg 1997.

——, 'Hussitenzeitliche Schiessscharten an Fränkischen Burgen', *Castellologica Bohemica*, 10, 2006.

Index

Printed and bound by CPI Group (UK) Ltd, Croydon, CR0 4YY
13/07/2026
02165170-0001